T0323928

ION CHANNELS IN HEALTH AND DISEASE

This Volume is an entry in the

PERSPECTIVES ON TRANSLATIONAL CELL BIOLOGY series

Edited by P. Michael Conn

A complete list of books in this series appears at the end of the volume

ION CHANNELS
IN HEALTH
AND DISEASE

Edited by

GEOFFREY S. PITT

Ida and Theo Rossi Distinguished Professor of Medicine
Director, Cardiovascular Research Institute
Weill Cornell Medicine
New York, NY, USA

AMSTERDAM • BOSTON • HEIDELBERG • LONDON
NEW YORK • OXFORD • PARIS • SAN DIEGO
SAN FRANCISCO • SINGAPORE • SYDNEY • TOKYO

Academic Press is an imprint of Elsevier

ELSEVIER

Academic Press is an imprint of Elsevier
125 London Wall, London EC2Y 5AS, United Kingdom
525 B Street, Suite 1800, San Diego, CA 92101-4495, United States
50 Hampshire Street, 5th Floor, Cambridge, MA 02139, United States
The Boulevard, Langford Lane, Kidlington, Oxford OX5 1GB, United Kingdom

Notices
Knowledge and best practice in this field are constantly changing. As new research
and experience broaden our understanding, changes in research methods, professional
practices, or medical treatment may become necessary.

Practitioners and researchers must always rely on their own experience and knowledge
in evaluating and using any information, methods, compounds, or experiments described
herein. In using such information or methods they should be mindful of their own safety
and the safety of others, including parties for whom they have a professional responsibility.

To the fullest extent of the law, neither the Publisher nor the authors, contributors, or
editors, assume any liability for any injury and/or damage to persons or property as a
matter of products liability, negligence or otherwise, or from any use or operation of any
methods, products, instructions, or ideas contained in the material herein.

British Library Cataloguing-in-Publication Data
A catalogue record for this book is available from the British Library

Library of Congress Cataloging-in-Publication Data
A catalog record for this book is available from the Library of Congress

ISBN: 978-0-12-802002-9

For information on all Academic Press publications
visit our website at https://www.elsevier.com/

  Working together
to grow libraries in
developing countries

www.elsevier.com • www.bookaid.org

Typeset by TNQ Books and Journals
www.tnq.co.in

Contents

5. Diagnosis, Treatment, and Mechanisms of Long QT Syndrome

E. WAN AND S.O. MARX

6. Ion Channels in Cancer

W.J. BRACKENBURY

7. TMEM16 Membrane Proteins in Health and Disease

H. YANG AND L.Y. JAN

8. K_{ATP} Channels in the Pancreas: Hyperinsulinism and Diabetes

M.S. REMEDI AND C.G. NICHOLS

9. The Role of Sperm Ion Channels in Reproduction

P.V. LISHKO, M.R. MILLER AND S.A. MANSELL

10. Mutations of Sodium Channel SCN8A ($Na_v1.6$) in Neurological Disease

J.L. WAGNON, R.K. BUNTON-STASYSHYN AND M.H. MEISLER

11. Alternative Splicing and RNA Editing of Voltage-Gated Ion Channels: Implications in Health and Disease

J. ZHAI, Q.-S. LIN, Z. HU, R. WONG AND T.W. SOONG

12. Computational Modeling of Cardiac K⁺ Channels and Channelopathies

L.R. PEREZ, S.Y. NOSKOV AND C.E. CLANCY

13. Connexins and Heritable Human Diseases

S.A. BERNSTEIN AND G.I. FISHMAN

14. Complexity of Molecular Genetics in the Inherited Cardiac Arrhythmias

N. LAHROUCHI AND A.A.M. WILDE

List of Contributors

G.W. Abbott University of California, Irvine, CA, United States

S.A. Bernstein New York University School of Medicine, New York, NY, United States

W.J. Brackenbury University of York, York, United Kingdom

R.K. Bunton-Stasyshyn University of Michigan, Ann Arbor, MI, United States

W.A. Catterall University of Washington, Seattle, WA, United States

C.E. Clancy University of California, Davis, CA, United States

G.I. Fishman New York University School of Medicine, New York, NY, United States

S.L. Hamilton Baylor College of Medicine, Houston, TX, United States

A.D. Hanna Baylor College of Medicine, Houston, TX, United States

T.T. Hong Cedars-Sinai Medical Center, Los Angeles, CA, United States; University of California Los Angeles, Los Angeles, CA, United States

Z. Hu National University of Singapore, Singapore

L.Y. Jan University of California, San Francisco, CA, United States

N. Lahrouchi Amsterdam Medical Center, University of Amsterdam, Amsterdam, The Netherlands

Q.-S. Lin National University of Singapore, Singapore

P.V. Lishko University of California, Berkeley, CA, United States

S.A. Mansell University of California, Berkeley, CA, United States

S.O. Marx Columbia University, New York, NY, United States

M.H. Meisler University of Michigan, Ann Arbor, MI, United States

M.R. Miller University of California, Berkeley, CA, United States

C.G. Nichols Washington University School of Medicine, St. Louis, MO, United States

S.Y. Noskov University of Calgary, Calgary, AB, Canada

L.R. Perez Polytechnic University of Valencia, Valencia, Spain

M.S. Remedi Washington University School of Medicine, St. Louis, MO, United States

L.J. Sharp Baylor College of Medicine, Houston, TX, United States

R.M. Shaw Cedars-Sinai Medical Center, Los Angeles, CA, United States; University of California Los Angeles, Los Angeles, CA, United States

T.W. Soong National University of Singapore, Singapore; NUS Graduate School for Integrative Sciences and Engineering, Singapore; National Neuroscience Institute, Singapore

J.L. Wagnon   University of Michigan, Ann Arbor, MI, United States

E. Wan   Columbia University, New York, NY, United States

A.A.M. Wilde   Amsterdam Medical Center, University of Amsterdam, Amsterdam, The Netherlands; Princess Al-Jawhara Al-Brahim Centre of Excellence in Research of Hereditary Disorders, Jeddah, Kingdom of Saudi Arabia

R. Wong   National University of Singapore, Singapore; NUS Graduate School for Integrative Sciences and Engineering, Singapore

H. Yang   University of California, San Francisco, CA, United States

J. Zhai   National University of Singapore, Singapore

Preface

This is the first edition of *Ion Channels in Health and Disease*. I am honored to serve as its Editor and grateful to Dr. Michael Conn, the series editor for *Prospectives in Translational Cell Biology*, for his invitation to work on this project. It is an invitation that I readily accepted, since this is the right time to focus on ion channels and disease. Channelopathies, diseases resulting from abnormal channel function, were some of the first and best characterized Mendelian and de novo genetic disorders. Now, with the explosion of new structural information about ion channels, made more accessible by advances in structural biology such as cryoelectron microscopy, and a flood of new genetic information flowing from cheaper and faster sequencing technologies, ion channels and channel dysfunction have entered the spotlight of multiple disease investigations. Simultaneously, as our understanding of ion channel function has matured, there has been an increased focus on the pathophysiology resulting from channel disorders and an increasing recognition that some of our long-accepted paradigms need to be revised. For instance, voltage-gated ion channels (particularly sodium and calcium voltage-gated ion channels) were long considered to function solely in electrically excitable tissues, such as muscle and brain. Recent evidence, however, has demonstrated unexpected roles of these and other channels in cancer and in development of nonexcitable tissues, suggesting even broader functions of these channels. Given these and other new advances, this edition focused on the role of ion channels in health and disease is timely and should be a valuable resource to many. Moreover, ion channels as a class are proven drug targets. Thus, a growing appreciation of the role of ion channels in health and disease provides an opportunity to develop new therapeutic strategies.

The individual chapters, authored by leaders in their fields, demonstrate the broad contributions of ion channels to health and disease. New takes on how ion channels contribute to cardiac arrhythmias or neurological disorders, areas in which channel roles are well established, are provided in chapters from Geoff Abbott; Bill Catterall; Elaine Wan and Steven Marx; Jacy Wagnon, Rosie Bunton-Stasyshyn, and Miriam Meisler; Lucia Romero Perez, Sergei Noskov, and Coleen Clancy; Scott Bernstein and Glenn Fishman; and Najim Lhrouchi and Arthur Wilde. Maria Remedi and Colin Nichols focuses on K_{ATP} channels, the targets of sulfonylureas used to treat diabetes. Roles of channels in cancer, a burgeoning field, are discussed by William Brackenbury. Also, Geoff Abbott's chapter focuses on how KCNE channel proteins regulate ion transport in the various secretory processes. Amy Hanna, Lydia Sharp, and Susan Hamilton addresses the roles of ryanodine receptor mutations in various muscle diseases, such as malignant hyperthermia, central core disease, and congenital myopathies. Huanghe Yang and Lily Jan provides a review on the on the rapidly expanding roles of a new class of channels, the TMEM16 family, that has now been implicated in multiple disease processes. Melissa Miller, Steven Mansell, and Polina Lishko discusses another novel family of channels, the Catsper channels, and how they control fertility. TingTing Hong and Robin Shaw; Jing Zhai, Qing-Shu Lin, Zhenyu Hu, Ruixiong Wong, and Tuck Wah Soong tackle more general processes, ion channel trafficking and alternative splicing, and how dysregulation contributes to an array of disease processes. Each of these chapters, providing the latest in structural, genetic, and biophysical information, emphasizes the central roles of ion channels in physiology and disease and offer insights into new therapeutic and diagnostic possibilities. I want to sincerely thank each of these authors, and their coauthors, for the thoughtful contributions. This collection will serve as a valuable resource for students, researchers, and physicians interested in this fascinating field.

Geoffrey S. Pitt
Weill Cornell Medicine, New York, NY, United States

The *KCNE* Family of Ion Channel Regulatory Subunits

G.W. Abbott

University of California, Irvine, CA, United States

INTRODUCTION

The *KCNE* gene family was first recognized in 1988, when Takumi and colleagues fractionated rat kidney mRNA, injected it into *Xenopus laevis* oocytes, and recorded electrical currents from the oocytes using the two-electrode voltage-clamp recording technique. When they clamped oocytes containing a certain fraction of mRNA to depolarized membrane potentials, Takumi et al. recorded very slowly activating, K$^+$-selective currents. From this fraction, they cloned the *KCNE1* cDNA, which expressed a 130 amino acid, single-pass transmembrane protein they termed IsK (slow K$^+$ current) and which was later referred to as minK, for minimal K$^+$ channel.[1,2] Today, IsK is most commonly referred to by its gene nomenclature (KCNE1) for both the gene and its protein product; we will use this system here, italicizing when referring to the gene rather than the protein.

KCNE1 was an enigmatic protein—it was small for a channel gene and possessed no homology with previously cloned channel genes. Mutagenesis studies confirmed that KCNE1 point mutants altered channel properties, alleviating concerns that the results were an artifact. Significantly, currents generated by *KCNE1* RNA injection into oocytes closely resembled the native cardiac slowly activating K$^+$ (I_{Ks}) current, which also fails to saturate even after several seconds of membrane depolarization at room temperature.[3] But how could this miniscule protein generate this K$^+$-selective current? The crucial breakthrough came in 1996. In that year, the Keating group reported positional cloning of a gene (KVLQT1) that is strongly expressed in the heart and "encodes a protein with structural features of a voltage-gated potassium channel." They linked KVLQT1 (now usually termed KCNQ1 or Kv7.1) to long QT syndrome (LQTS), a potentially lethal ventricular arrhythmia characterized by impaired ventricular repolarization and therefore lengthening of the electrocardiogram (ECG) QT interval[4] (Fig. 1.1A). Later that year, two teams independently reported that KCNE1 does not generate currents on its own, but modifies currents passed by KCNQ1 by coassembling with it to form a heteromeric, voltage-gated potassium (Kv) channel with dramatically altered properties[5,6] (Fig. 1.1B and C).

KCNE1 cRNA injection into *X. laevis* oocytes generates I_{Ks}-like currents because oocytes express endogenous KCNQ1, which is not readily distinguishable from other endogenous oocyte currents until it is upregulated and its activation slowed by exogenous KCNE1.[5–8] KCNE1 was thus

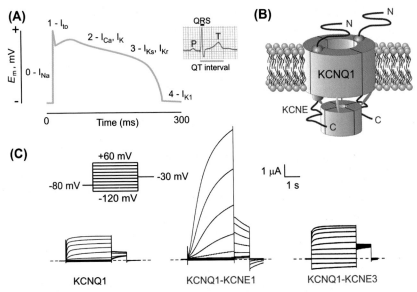

FIGURE 1.1 **KCNE regulation of KCNQ1 and cardiac electrophysiology.** (A) The upstroke (0) of the human ventricular myocyte action potential is facilitated by Na^+ influx. The initial repolarization notch is caused by the transient outward K^+ current (I_{to}), and the plateau phase (2) is a balance between residual Na^+ current, Ca^{2+} influx, and K^+ efflux (I_K). Phase 3 repolarization is primarily achieved by I_{Ks} (mediated by KCNQ1-KCNE1 and possibly other KCNQ1-KCNE combinations) and I_{Kr}. Inward rectifiers (I_{K1}) help to determine resting membrane potential between action potentials. *Upper right inset:* surface electrocardiogram showing P wave, QRS complex, T wave, and QT interval. (B) Schematic of a KCNQ1-KCNE Kv channel complex. (C) Functional effects of KCNE1 and KCNE3 on KCNQ1. KCNQ1, KCNQ1-KCNE1, and KCNQ1-KCNE3 currents in response to 3 s pulses to a range of voltages between −120 and +60 mV, with 10 s interpulse intervals, from a holding voltage of −80 mV, with a 1 s −30 mV tail pulse, as shown in voltage protocol (*upper left*). Channels were expressed by injection of corresponding cRNAs into *Xenopus laevis* oocytes and currents recorded 3 days postinjection by two-electrode voltage-clamp. Zero current levels indicated by dashed lines. *Adapted from Abbott GW. Biology of the KCNQ1 potassium channel. New J Sci 2014;2014.*

recognized as a K$^+$ channel regulatory or β subunit. Three years later, we and others reported cloning of four other KCNE genes,[9,10] and the KCNE family is now recognized as a five-member class (in the human genome) of ion channel β subunits that exhibit ubiquitous expression, promiscuous and versatile activity, and multiple disease correlations. Each of the KCNE subunits has an extracellular N-terminus, an intracellular C-terminus, and a single transmembrane domain[11] (Fig. 1.1B).

THE MECHANISTIC BASIS FOR FUNCTION OF KCNE PROTEINS

KCNE proteins cannot generate current themselves, but they alter the behavior and facilitate the native functions of certain ion channels. Their primary partners are Kv channels. Each KCNE isoform is capable of regulating multiple Kv α subunits at least in vitro, and for some, we and others have established their promiscuity in vivo.[12] Not all Kv α subunits are known to be regulated by KCNEs, but several can be regulated by multiple KCNE isoforms. Therefore, there are many possible permutations of α-KCNE complexes, and these likely contribute greatly to the diversity of Kv currents in vivo.[13]

KCNE Regulation of KCNQ1 Channel Gating, Conductance, Ion Selectivity, and Pharmacology

The most commonly recognized modes of KCNE regulation of ion channels involve formation of complexes with altered channel activity at the plasma membrane compared to channels formed by the pore-forming subunits alone. This typically involves changes in the voltage dependence and/or kinetics of gating. The best known examples, and most diverse known functional outcomes, involve KCNQ1—which is differentially regulated by each of the five KCNE isoforms.[14] KCNE1 eliminates KCNQ1 inactivation,[15] increases its unitary conductance,[16] right-shifts its voltage dependence of activation, and—most noticeably—slows its activation five-to tenfold.[16,17] KCNQ1-KCNE1 complexes also have slightly altered ion selectivity compared to KCNQ1 homomers. The molecular mechanisms by which KCNE1 regulates the gating of KCNQ1 have been the subject of intense research efforts for many years and are still not fully resolved. The stoichiometry of KCNQ1 complexes has also long been debated and even to this day there is not full agreement. Evidence from toxin binding, dominant-negative mutagenesis, and fluorescence spectroscopy strongly suggests that 4:2 (KCNQ1:KCNE1) complexes exist[8,18,19] (Fig. 1.1B). However, some groups contend that 4:4 complexes can also exist, depending on subunit expression ratios.[20]

In terms of structure, the KCNE1 transmembrane domain (residues 45–71) adopts an α helical conformation with a slight curve. This curvature may allow the C-terminal, intracellular part of KCNE1 to interact with the S4-S5 linker region of KCNQ1 to slow channel activation, while permitting its extracellular N-terminal portion to sit in the cleft between the voltage sensor and pore domains of KCNQ1 near the extracellular face.[21] S4-S5 connects the voltage sensor to the activation gate and extensive mutagenesis, binding, and modeling studies support its interaction with the cytosolic domain of KCNE1.[22–24] Positioning of KCNE1 in this groove could stabilize the closed state in favor of the open state, resulting in the signature retardation of opening and also the right-shift in voltage dependence observed after KCNQ1-KCNE1 coassembly, although it is still debated whether KCNE1 slows voltage sensor movement or pore opening itself.[25–27]

Another compelling possibility fielded recently is that KCNQ1 channel voltage sensors exhibit an additional stable intermediate (not fully "activated") state that can nevertheless also open the pore, unlike the canonical *Shaker* Kv channel which is thought to require full voltage sensor activation before pore opening. It is contended that KCNE1, by altering the manner in which the KCNQ1 voltage sensor and pore interact, prevents formation of the intermediate state and also changes the conformation of the fully activated state, to achieve the array of effects it exerts on KCNQ1—including altered gating, permeation, and even pharmacology.[28]

KCNE2 effects are dramatically different. KCNQ1-KCNE2 channels are constitutively active,[29] probably resulting from left-shifted voltage dependence of activation and/or greatly slowed deactivation at a given voltage. In physiological conditions, this means that KCNQ1-KCNE2 conditions are much less voltage-dependent than KCNQ1 or KCNQ1-KCNE1 channels. As discussed later, this property permits KCNQ1-KCNE2 channels to serve a variety of essential functions in nonexcitable, polarized epithelial cells. This ability is facilitated by other facets endowed by KCNE2. For example, KCNQ1 channels are inhibited by low extracellular pH; in contrast, KCNE2 coassembly results in KCNQ1 current augmentation by extracellular low pH, an essential property of KCNQ1-KCNE2 channels in parietal cells, where they support activity of the gastric H^+/K^+-ATPase by recycling K^+ ion back into the stomach lumen. KCNQ1-KCNE2 macroscopic currents are relatively small compared to even homomeric KCNQ1,[29] and analysis of unitary conductances from all KCNQ1-based channels is technically very challenging because of a low, flickery conductance. Thus, our functional understanding of these important channels at the unitary level lags behind that of other, easier to study channels.

As with KCNE2, KCNQ1-KCNE3 channels are constitutively active,[30] the mechanism appearing to be primarily a strong left-shift in the voltage dependence of activation.[31,32] Work from mutagenesis studies in which KCNE1 transmembrane residues were swapped with those in equivalent positions in KCNE3 strongly suggests that a portion of the KCNE3 transmembrane domain influences its ability to hold KCNQ1 open at rest.[33] In addition, we performed mutant cycle analysis and found that KCNE3 residues in the N-terminal domain that presumably lie near the extracellular side of the lipid bilayer interact with the KCNQ1 S4 helix of the voltage-sensing domain (VSD), helping to stabilize the activated conformation of the VSD.[34] The ability of KCNQ1-KCNE3 to open constitutively is probably essential to its function in the basolateral side of cells in the intestine, where it regulates cAMP-stimulated chloride secretion and also in the airway and mammary epithelia.[30]

KCNE4 and KCNE5 (also known as KCNE1L) are widely expressed but thus far relatively less studied than KCNE1, 2, and 3.[10,35–39] KCNE4 inhibits KCNQ1 function, possibly by right-shifting its voltage dependence of activation to the extent that KCNQ1 cannot be opened at voltages normally visited even under experimental conditions.[36,40] Likewise, KCNE5 is known to right-shift KCNQ1 voltage dependence of activation >140 mV more positive than that of homomeric KCNQ1.[38] Interestingly, KCNE4 and KCNE5 can each modulate KCNQ1-KCNE1 complexes, although the precise molecular mechanisms are uncertain and add further interest to the idea of stoichiometric flexibility and what we will refer to here as "subunit mosaicism" within KCNQ1-KCNEx (I_{Ks}) complexes.[37]

ROLES OF KCNE SUBUNITS IN CARDIAC ION CURRENTS AND ARRHYTHMOGENESIS

The most widely acknowledged role of the KCNE protein family is in the heart.[41] In human ventricular myocytes, two main Kv currents serve to repolarize the myocyte membranes to end each action potential.[42] hERG (for human ether-a-go-go related gene product) is the α subunit required for the primary repolarization current, I_{Kr} (rapidly activating K$^+$ current). KCNQ1 is the α subunit required to generate the slowly activating current, I_{Ks} (see previous discussion). I_{Ks}, like I_{Kr}, is also essential for human ventricular repolarization and appears to be particularly important when I_{Kr} is diminished, or during certain exercise activities including swimming.[43] Crucially, KCNE subunits modulate KCNQ1, hERG, and other important human cardiac Kv channel α subunits and are thought to be important for their correct function in the context of ventricular and/or atrial myocytes. We know this because inherited mutations in KCNE subunits cause similar ventricular or atrial arrhythmias to those observed in people with α subunit gene mutations (Table 1.1).

Cellular electrophysiology studies have shown how point mutations in KCNE genes can fundamentally alter the functional properties of heteromeric α-KCNE complexes, including the current density (for reasons

TABLE 1.1   Subset of *KCNE* Gene Variants Associated With Cardiac Arrhythmia for Which Cellular Electrophysiology Analyses Were Reported

Gene	Mutation	Disease	In Vitro Cellular Effects	References
KCNE1	A8V	LQT5, AF	I_{Ks} +9 mV positive shift in $V_{1/2}$ activation, no change in deactivation or activation time constants. I_{Kr} loss-of-function	44,45
KCNE1	G25V	Lone AF	I_{Ks} gain-of-function	46
KCNE1	R32H	LQT5	Gating not altered	45,47,48
KCNE1	38G versus 38S	AF, LQT5	I_{Ks} loss-of-function, reduced membrane K_V7.1 (KvLQT1) channel	49–53
KCNE1	V47F	LQT5	hERG gain-of-function	54
KCNE1	L51H	LQT5	Does not process properly, not found in membrane	54
KCNE1	G52R	LQT5	I_{Ks} reduced 50%	55
KCNE1	G60D	Lone AF	I_{Ks} gain-of-function, faster deactivation	46
KCNE1	S74L D76N	LQT5 (JLNS + RW)	D76N: I_{Ks}—smaller unitary currents and open probabilities Dominant negative I_{Kr}—loss-of-function S74L: I_{Ks} loss-of-function	56 57 16 54
KCNE1	Y81C	LQT5?	I_{Ks} loss-of-function, positive shift for voltage of activation	58
KCNE1	D85N	LQT5? DiLQTS DiTdP	No effect on I_{Ks}, $V_{1/2}$ of activation, or deactivation properties	59–67
KCNE1	W87R	LQT5	I_{Ks} loss-of-function, altered gating	54
KCNE1	R98W	LQT5?	Disrupts I_{Ks} trafficking, right-shifts voltage dependence of activation	68
KCNE1	V109I	LQT5	I_{Ks} loss-of-function (36%)	69
KCNE2	T8A Q9E	LQT6, diLQTS	Increases drug sensitivity of I_{Kr}	9,70
KCNE2	T10M	LQT6	I_{Kr} slow inactivation recovery and inactivation	71
KCNE2	M23L	AF	I_{Q1-E2} and I_{Ks} gain-of-function	72
KCNE2	R27C	AF	I_{Q1-E2} gain-of-function, $I_{Ca,L}$ suppression	73,74

TABLE 1.1   Subset of *KCNE* Gene Variants Associated With Cardiac Arrhythmia for Which Cellular Electrophysiology Analyses Were Reported —cont'd

Gene	Mutation	Disease	In Vitro Cellular Effects	References
KCNE2	M54T I57T A116V	LQT6, diLQTS, AF (I57T)	Pathological I_{Kr} loss-of-function $\pm$ drug. M54T and I57T also slow $K_V2.1$ activation; I57T increases $I_{Q1\text{-}E2}$ and I_{Ks} function	9,45,70,72
KCNE2	V65M	LQT6, Syncope	I_{Kr} loss-of-function, accelerated inactivation time course	75
KCNE3	T4A	LQTS? BrS	No effect on $I_{Q1\text{-}E3}$ I_{to} gain-of-function	76 77
KCNE3	R53H	AF	$I_{Q1\text{-}E3}$ gain-of-function	78
KCNE3	R99H	BrS LQTS?	$K_V4.3$, I_{to} gain-of-function $I_{Q1\text{-}E3}$ loss-of-function	79 76
KCNE4	E145D	AF	$I_{Q1\text{-}E4}$ gain-of-function	61
KCNE5	L65F	AF	I_{Ks} gain-of-function	80
KCNE5	Y81H D92E;E93X	BrS/ Idiopathic VF	$K_V4.3$ gain-of-function; No change in $I_{Q1\text{-}E5}$ density	81

AF, atrial fibrillation; *AV block*, atrioventricular block; *BrS*, Brugada syndrome; *diLQTS*, drug-induced long QT syndrome; *diTdP*, drug-induced torsade de pointes; *fs*, frame shift; *JLNS*, Jervell and Lange-Nielsen syndrome; *RWS*, Romano-Ward syndrome; *VF*, ventricular fibrillation. *Adapted from Crump SM, Abbott GW. Arrhythmogenic KCNE gene variants: current knowledge and future challenges.* Front Genet 2014;5:3; *other sources include cited papers,* http://www.genomed.org/lovd2/home.php, *and* http://www.fsm.it/cardmoc.

including trafficking defects and reduced unitary conductance) and the voltage dependence and kinetics of gating.[9,16,41] The perturbations disrupt electrical activity required for correct action potential duration and morphology leading to abnormal myocyte repolarization. What might appear a relatively innocuous event on a human body-surface ECG, eg, as lengthening of the QT interval, can serve as a substrate for reentrant arrhythmias that can rapidly degenerate into a condition termed *torsades de pointes*. This, in turn, can develop into ventricular fibrillation, a condition in which concerted, rhythmic contraction of the ventricles is replaced by chaotic smaller circuits that result in disorganized quivering of the heart rather than rhythmic pumping. This curtails the supply of essential oxygen via the bloodstream to important organs including the brain. This can result in loss of consciousness and sudden cardiac death; unless the individual is rapidly resuscitated using CPR and/or electrically defibrillated, there is little to no chance of recovery.[83,84] In contrast, atrial fibrillation (AF) is not generally acutely lethal but can increase the risk of life-threatening blood clots and stroke, and is the most common chronic arrhythmia, with an estimated 2–3 million sufferers in the United States alone. As people age, there is a much greater likelihood of developing AF, with the majority of cases stemming from structural heart disease.[85] In addition, postoperative AF, which typically lasts a month, is a common side effect of cardiothoracic surgery.[86] However, some cases, referred to as "lone AF," occur in the absence of these predisposing factors; a subset of cases of lone AF have been attributed to mutations in ion channel genes, including *KCNE* genes (Table 1.1).

KCNE1 in Human and Mouse Heart

KCNQ1 was the first known partner for KCNEs and the large majority of the KCNE literature still revolves around regulation of KCNQ1. Whether this is historical bias or proportional to the importance of KCNQ1 in KCNE biology remains to be seen. The first-discovered, best-understood, and perhaps most important such complex in human heart is KCNQ1-KCNE1, which generates I_{Ks}. The uniquely slow-activating nature of I_{Ks} positions it to deliver a robust repolarizing force to end phase 3 of the ventricular action potential (Fig. 1.1). KCNE1 mutations, like those in KCNQ1, are most commonly associated with LQTS (for KCNE1 this is classified as LQT5) and rarely with AF (Table 1.1). LQTS-linked KCNE1 variants cause loss-of-function of KCNQ1-KCNE1 channels, either from misfolding/misprocessing, impaired trafficking to the surface, reduced unitary conductance, right-shifted voltage dependence of activation, or altered gating kinetics (Table 1.1). It is important to mention that native I_{Ks} complexes also contain several other proteins in addition to KCNQ1 and KCNE1, including Yotaio, calmodulin, PKA, 4D3, PP1, and AC9[87–89] (Fig. 1.2).

Particularly severe loss-of-function of KCNQ1-KCNE1 channels, such as that caused by a pathogenic mutation in both alleles for either gene, can cause the cardioauditory Jervell Lange-Nielsen syndrome, which comprises LQTS and bilateral sensorineural deafness.[90,91] This revelation uncovered another essential role of KCNQ1-KCNE1, that of K+ secretion into the endolymph of the inner ear. While *KCNE1* mutations have taught us much about the obligate role of KCNE subunits in vivo, it is important to remember that only around 1% of sequenced LQTS cases arise from *KCNE1* variants (compare with >40% each for HERG and KCNQ1), and the other *KCNE* genes (especially *KCNE3-5*) are probably even rarer causes of LQTS. However, *KCNE* coding regions, including that of *KCNE1*, are comparatively small (~400 bases) meaning that the probability of mutations arising in them is proportionally lower than in much larger channel α subunit genes.[41] Even if one includes both introns and exons, *KCNE2*, for example, is per base of more pathophysiological significance that the *SCN5A* sodium channel gene that accounts for about 10% of LQTS cases.

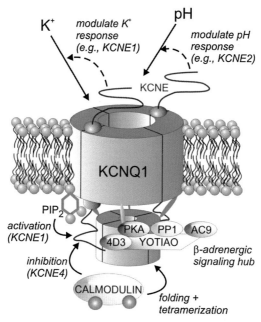

FIGURE 1.2 I_{Ks} **complexes comprise multiple subunit types.** Schematic of the various KCNE and non-KCNE subunits that regulate KCNQ1 and their crosstalk. *PDE4D3,* phosphodiesterase 4D3; *AC9,* adenylyl cyclase 9; *PIP$_2$,* phosphatidylinositol 4,5-bisphosphate; *PKA,* protein kinase A; *PP1,* protein phosphatase 1. *Adapted from Abbott GW. Biology of the KCNQ1 potassium channel.* New J Sci 2014;*2014.*

Human KCNE1 variants are associated with lone AF; most notably, the ratio of KCNE1-38G versus 38S polymorphisms an individual harbors appears to influence AF predisposition.[49,92] AF is thought to be caused by shortening of the atrial effective refractory period, and therefore gain-of-function mutations in KCNQ1, and the KCNE subunits that regulate it, are typically observed in the rare cases when AF can be linked to a specific gene variant. We previously discovered, using transcriptomics, that open chest surgeries associated with postoperative AF remodel a number of atrial genes in pigs, most strikingly *KCNE1*. This finding was subsequently mirrored in a human study, in which the *KCNE1* 112G>A polymorphism was linked to increased susceptibility to postoperative AF.[93]

Aside from regulation of KCNQ1, KCNE1 also modulates hERG, the other major repolarizing Kv channel in human ventricles. KCNE1 doubles hERG currents upon their heterologous coexpression in mammalian cell lines,[94] and *KCNE1* mutations impair native hERG currents in isolated myocytes.[54] The mechanism for KCNE1 augmentation of hERG is unknown, as is the relative contribution of KCNE1 to KCNQ1 versus hERG channels in native cardiac myocytes. It is interesting to note that KCNQ1 may form complexes with hERG channels in the heart, suggesting that KCNE1 regulation of either subunit, and effects of *KCNE1* mutations on channel function, may be even more complex than first suspected.[95,96]

KCNE2-5 in Human Heart

KCNE2 and KCNE3 each dramatically alter the voltage dependence of KCNQ1 gating such that KCNQ1-KCNE2 and KCNQ1-KCNE3 channels are constitutively active, forming K^+-selective leak channels at resting membrane potentials[29,30] (Fig. 1.1C). In the heart, the role of these complexes is unclear, with suggestions that KCNE2 and KCNE3 might contribute to regulation of KCNQ1-KCNE1[97–99]; such mixed-KCNE complexes likely do not have constitutive activation. KCNE2 and KCNE3 are important in human and mouse heart physiology, however, via different mechanisms.

Human *KCNE2* gene variants are associated with inherited and drug-induced ventricular arrhythmias. We discovered that KCNE2 regulates the hERG K^+ channel, modifying its unitary conductance, gating kinetics, and pharmacology.[9] KCNE2 gene variants, including both rare mutations and more common polymorphisms, are deleterious to hERG-KCNE2 channel function; those associated with LQTS reduce hERG-KCNE2 current density and/or increase sensitivity to drug block, by various mechanisms.[9] One of the more interesting sequence variant effects is that of KCNE2-T8A. This variant is absent in African Americans but present in >1% of Caucasian Americans. hERG-KCNE2-T8A channels function normally at baseline but are more susceptible to blockade by the sulfamethoxazole component of the antibiotic Bactrim.[70] This increased sensitivity was discovered to arise from loss of a glycosylation site in KCNE2 that, when intact, presumably partially shields hERG-KCNE2 channels from block by sulfamethoxazole.[100]

The pathogenic cardiac effects of KCNE2 mutations may be even more complex, however, because KCNE2 can regulate a number of other cardiac ion channels and arrhythmogenic KCNE2 mutations have been shown to alter the function of some of these complexes as well. Thus, KCNE2 modulates Kv1.5, Kv2.1, Kv4.2, and Kv4.3,[101–103] monovalent cation nonspecific HCN pacemaker channels,[104,105] and even reportedly the cardiac L-type calcium channel, Cav1.2.[73] This suggests a master regulatory role for KCNE2 in the heart and therefore the potential for a multifaceted arrhythmia syndrome when KCNE2 malfunctions (and see Ref. 106). KCNE2 mutations R27C, M23L, and I57T have each been associated with lone AF; these variants exert gain-of-function effects on KCNQ1 (and KCNQ1-KCNE1) (Table 1.1); in addition, R27C increases the suppressive action of KCNE2 on Cav1.2.[73] Either of these effects could contribute to atrial myocyte action potential shortening, a substrate for AF.

Three KCNE3 variants have been associated with cardiac arrhythmias (Table 1.1). KCNE3-T4A is associated with Brugada syndrome, a lethal ventricular arrhythmia, and causes gain-of-function of I_{to}, which is generated by Kv4.2 and/or Kv4.3 in human ventricles. Brugada syndrome is most commonly linked to loss-of-function of the cardiac voltage-gated sodium channel SCN5A, and I_{to} gain-of-function could mirror this phenomenon. KCNE3-R99H is also associated with Brugada and increases Kv4.3 current yet inhibits KCNQ1-KCNE3. KCNE3-R53H associates with AF and increases KCNQ1-KCNE3 current. In addition, we previously found that KCNE3 regulates Kv3.4, an important α subunit in skeletal muscle and that KCNE3-R83H associates with periodic paralysis.[107] This conclusion has been challenged, however, by groups that also detected KCNE3-R83H in asymptomatic individuals.[108] Whether this represents variable penetrance, necessity of a further modifier gene for pathology, or a misclassification, remains to be firmly established. KCNE4-E145D and KCNE5-L65F are both associated with AF and both cause gain-of-function of channels formed with KCNQ1. KCNE5 gene variants Y81H and D92E; E93X increase Kv4.3 activity without altering KCNQ1 function and have been linked to Brugada syndrome and idiopathic ventricular fibrillation (Table 1.1).

Consequences of Kcne Gene Knockout in Mice

Much of what we know about the many roles of *KCNE* genes in mammalian physiology has been gleaned from knockout mouse models, thereby simultaneously providing insights into the pathogenic ramifications of *KCNE* gene disruption. The *Kcne1* null mouse was generated and characterized before the rest of the gene family had been discovered and was found to exhibit a mild cardiac phenotype.[109] In adult mouse ventricles, I_{Kr} and I_{Ks} are much less important than in human heart. mERG and I_{Kr} exhibit little to no expression; I_{Ks} can be weakly detected (a current with similar pharmacological sensitivity to that of KCNQ1 is present, albeit small) in adult mouse ventricles but does not necessarily contain KCNE1.[110] KCNE1 is either absent or expressed at low levels in adult mouse ventricular myocytes, but it may be present in the conduction system and/or atrial myocytes, and both AF and enhanced QT-RR adaptability have been reported in *Kcne1* null mice.[111,112] It is possible that KCNE1 regulates subunits other than mERG and KCNQ1, such as Kv2.1, in mouse heart; we previously detected KCNE1-Kv2.1 complexes in rat heart.[101] *Kcne1*$^{-/-}$ mice are bilaterally deaf, have impaired endolymph production, and have thus provided an excellent model system to study effects upon the inner ear as seen in human *KCNE1*-linked Jervell Lange-Nielsen syndrome.[111] KCNE1 may also play a role in mouse pancreas and intestine although this is less well established, and the putative partner in these tissues, KCNQ1, is regulated by KCNE3 in the intestine and may be regulated by other KCNE isoforms in the pancreas.[30,113,114] *Kcne1* gene deletion also causes urinary and fecal salt wasting, hypokalemia and hyperaldosteronism, some of which may arise from disruption of a direct role for KCNE1 in the kidneys although the precise nature of this role is still being debated.[115–120]

KCNE2 augments Kv4.x and Kv1.5 current in mouse ventricles, and *Kcne2* deletion correspondingly reduces I_{to} and $I_{Kslow,1}$ density in mouse ventricular myocytes. In older adult mice, or in young adult mice under sevoflurane anesthesia, *Kcne2* deletion delays ventricular repolarization to the extent that it is readily observable on a body-surface ECG. In the case of Kv1.5, KCNE2 appears to be required for localization to the intercalated discs, and the deleterious effects of *Kcne2* knockout on $I_{Kslow,1}$ may be primarily from α subunit mistrafficking.[121]

Kcne2$^{-/-}$ mice also exhibit achlorhydria because KCNQ1-KCNE2 channels in the apical membrane of parietal cells are required for normal gastric acid secretion by the H^+/K^+-ATPase[122] (Fig. 1.3A). *Kcne2* deletion leads to KCNE3 upregulation in parietal cells, which in turn results in KCNQ1 being hijacked and taken to the basolateral membrane, where it cannot serve its normal function.[123] Mice with both *Kcne2* and *Kcne3* germline-deleted exhibit apically expressed parietal cell KCNQ1 (Fig. 1.3B), but their gastric phenotype is even more severe than *Kcne2*$^{-/-}$ mice. This is probably because KCNE2 is required to prevent KCNQ1 activity being inhibited by low extracellular pH and also loss of KCNE2 gating effects on KCNQ1.[31] *Kcne2* null mice develop gastritis cystica profunda, severe gastric hypertrophy and even gastric neoplasia; some of this may arise from bacterial overgrowth in the stomach although there also appears to be a gastric pH-independent function for KCNE2 in regulation of cell cycling and proliferation.[124,125]

KCNQ1-KCNE2 channels are also required for normal function of the thyroid and choroid plexus epithelium, and *Kcne2* gene deletion disrupts these roles. In thyroid cells, basolaterally expressed KCNQ1-KCNE2 is necessary for optimal activity of the sodium iodide symporter, which facilitates iodide uptake for thyroid hormone biosynthesis (Fig. 1.4). In periods of increased thyroid hormone requirement, such as during gestation and lactation, KCNE2-free thyroid cells cannot match this requirement, and *Kcne2*$^{-/-}$ dams and their pups suffer from hypothyroidism. This manifests as alopecia, cardiac hypertrophy (which are also observed in 12-month-old *Kcne2*$^{-/-}$ mice from heterozygous parents), and retarded growth of pups (which resolves in the fourth month).[126,127]

In the choroid plexus, which is the primary site of cerebrospinal fluid (CSF) production and secretion, apically localized KCNQ1-KCNE2 forms a newly recognized type of signaling complex with the sodium-coupled

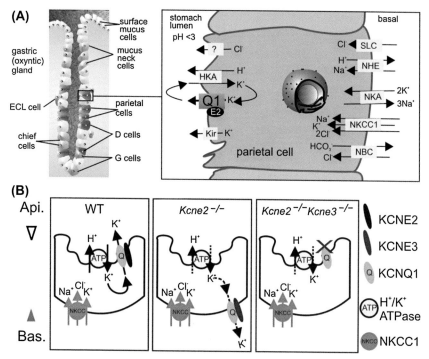

FIGURE 1.3 **KCNQ1-KCNE2 location and function in parietal cells and its genetic disruption.** (A) *Left*, schematic of a gastric gland; *right*, the gastric H^+/K^+-ATPase requires a luminal K^+ recycling pathway for proton secretion and gastric acidification, formed by complexes of KCNQ1 (Q1) and KCNE2 (E2). (B) Schematic *(adapted from Hu Z, Kant R, Anand M, King EC, Krogh-Madsen T, Christini DJ, et al. Kcne2 deletion creates a multisystem syndrome predisposing to sudden cardiac death. Circ Cardiovasc Genet 2014;7(1):33–42 and Roepke TK, King EC, Purtell K, Kanda VA, Lerner DJ, Abbott GW. Genetic dissection reveals unexpected influence of beta subunits on KCNQ1 K^+ channel polarized trafficking in vivo. FASEB J 2011;25(2):727–36)* showing the putative short circuit K^+ current in parietal cells of $Kcne2^{-/-}$ mice *(middle panel)* compared to $Kcne2^{+/+}$ mice *(left-hand panel)*. This circuit is thought to arise from the switch in KCNQ1 trafficking from apical ($Kcne2^{+/+}$, *left-hand panel*) to basolateral because of KCNE3 upregulation *(middle panel)*.[123] In double-knockout $Kcne2^{-/-} Kcne3^{-/-}$ mice, KCNQ1 is apical but presumed nonfunctional because it lacks KCNE2 and KCNE3, which confer constitutive activation and limit inactivation *(right-hand panel)*.[123]

myo-inositol transporter, SMIT1[128,129] (Fig. 1.5). *Kcne2* deletion results in increased seizure susceptibility and decreased immobility in the tail suspension test.[128] Global metabolite profiling by mass spectrometry revealed that $Kcne2^{-/-}$ CSF has lower levels than wild-type CSF of *myo*-inositol, an important osmolyte and precursor to cellular signaling molecules including phosphatidylinositol phosphates. *Myo*-inositol administration to $Kcne2^{-/-}$ mice alleviates differences in seizure susceptibility and behavior. Our present model is that loss of KCNE2 from apical KCNQ1-KCNE2-SMIT1 complexes causes an aberrant increase in activity that pulls too

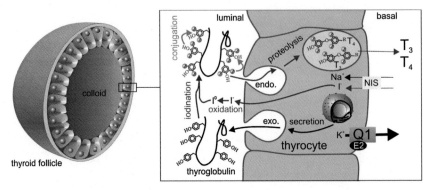

FIGURE 1.4 **Role of KCNQ1-KCNE2 in thyroid epithelial cells.** Left, schematic of a thyroid follicle with a perimeter consisting of thyroid epithelial cells (thyrocytes). Right, thyroid hormone (T_3 and T_4) biosynthesis requires that iodide ions pass across the thyroid cells from the blood into the colloid, where iodide is oxidized, and organified by incorporation into thyroglobulin (iodination and conjugation). KCNQ1-KCNE2 channels on the thyroid cell basolateral membrane are required for optimal function of the basolateral sodium iodide symporter (NIS). *Adapted from Abbott GW. KCNE2 and the K (+) channel: the tail wagging the dog.* Channels (Austin) 2012;*6(1).*

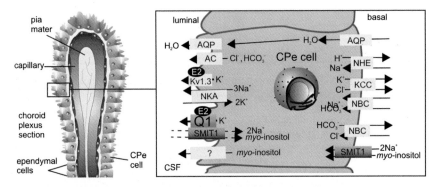

FIGURE 1.5 **Role of KCNQ1-KCNE2 in choroid plexus epithelial cells.** Left, representation of the choroid plexus epithelium (CPe). Right, in our current model, KCNQ1-KCNE2 channels form complexes with the sodium *myo*-inositol transporter in the apical membrane of the choroid plexus epithelium, which regulate CSF *myo*-inositol homeostasis.

much *myo*-inositol back out of the CSF, into which it is normally actively transported through the choroid plexus epithelium from the blood. This idea is reinforced by *Xenopus* oocyte coexpression studies showing that in bipartite complexes, KCNQ1 and SMIT1 augment each other's activity, whereas KCNQ1-KCNE2 inhibits SMIT1 activity.[128]

Intriguingly, *Kcne2* deletion causes a number of other pathologies for which the mechanistic underpinnings are not all yet entirely clear. *Kcne2−/−* mice exhibit hypercholesterolemia, hyperkalemia, decreased glucose tolerance, anemia, and elevated serum angiotensin II.[106] The hyperkalemia may arise from rewiring of parietal cells such that K+ leaks from the stomach lumen through to the bloodstream,[106,123] and this in turn could chronically elevate angiotensin II, although aldosterone is not elevated in adult $Kcne2^{-/-}$ mice (but they exhibit adrenal remodeling and lipid accumulation). Their anemia is caused by iron deficiency because of reduced iron absorption arising from achlorhydria, and this likely also causes their observed splenomegaly. Not unexpectedly, this multisystem syndrome impacts on cardiac function. $Kcne2^{-/-}$ mice are highly susceptible to ischemia/reperfusion injury-induced ventricular fibrillation, and unlike their $Kcne2^{+/+}$ counterparts, after fasting they suffer from hypoglycemia and are susceptible to atrioventricular block.[106] The litany of defects arising from $Kcne2$ deletion and their connectedness illustrates the potential complexity of human cardiovascular channelopathies, although it must be stressed that biallelic gene deletion is typically associated with a much more severe phenotype than the more common situation experienced in human channelopathies, ie, monoallelic gene mutations or polymorphisms. To counter this point, KCNQ1 SNPs are strongly linked to human diabetes[130–134] and associated with hypochlorhydria (evidenced by hypergastrinemia, a response to insufficient gastric acid secretion) as well as to LQTS and AF. The interplay between these extracardiac and cardiac perturbations has not been fully explored, but in one study, individuals with more severe KCNQ1-linked QT prolongation also had the highest levels of serum gastrin.[135]

$Kcne3$ deletion results in an indirect cardiovascular pathology. Although $Kcne3$ is expressed in neither the adrenal glands nor the heart of mice, $Kcne3$ deletion precipitates an autoimmune attack on the adrenal glands (but not other tissues tested) detectable as an activated lymphocyte invasion and upregulation of the B-cell chemoattractant chemokine CXCL13. $Kcne3^{-/-}$ mice also exhibit hyperaldosteronism, although the cause-and-effect relationship with the adrenal-targeted autoimmune response is unknown.[136] In turn, $Kcne3^{-/-}$ mice exhibit aldosterone-dependent QT prolongation and increased susceptibility, compared to wild-type littermates, to postischemic ventricular arrhythmias when challenged with ischemia/reperfusion injury. KCNE3 appears to be expressed in some blood cells, but the mechanistic basis for KCNE3-linked adrenal autoimmune targeting is a mystery. $Kcne3^{-/-}$ mice are normokalemic and exhibit normal glucose tolerance, but they have high serum LDL cholesterol, the mechanistic basis for which is also not known.[136] $Kcne3$ deletion also impairs intestinal activity of basolaterally located KCNQ1 and disrupts the cAMP-stimulated chloride secretion that KCNQ1-KCNE3 normally regulates.[137] The consequences of $Kcne4$ and $Kcne5$ gene deletion in mice have not yet been reported.

CONCLUSIONS AND FUTURE QUESTIONS

The scope of this chapter is to summarize the known and suggested physiological roles of KCNE subunits and what arises from KCNE gene sequence variation or gene deletion. The existing literature is somewhat biased toward cardiac physiology and disease because of the history of how the role of I_{Ks} was discovered and because of the excellent work of cardiovascular human geneticists in identifying the molecular correlates of cardiovascular ion channelopathies. More recently, we and others have examined epithelial roles for KCNE2 and KCNE3 in mice, but the potential scope and severity of human KCNE-linked extracardiac disorders, and their impact on the heart, has been little studied. Whether this reflects a greater tolerance of epithelial tissues for monoallelic SNPs in KCNE genes, greater redundancy of function of human epithelial KCNEs compared to the situation in mice, or just a relative lack of exploration in this arena, remains to be seen. KCNE2 SNPs have, however, been linked to increased incidence of early-onset myocardial infarction,[138] atherosclerosis,[139] and pulmonary dysfunction.[139,140] The precise mechanistic bases for these conditions, in the context of KCNE2 dysfunction, are unknown, but atherosclerosis and myocardial infarction certainly resonate with our recent findings of hyperlipidemia in $Kcne2^{-/-}$ mice.

The other notable bias in the KCNE field is that their regulation of KCNQ1 has been studied more than that of all other ion channels combined.[14] Is this a reflection of the relative biological importance of KCNQ1-KCNE complexes compared to other KCNE-based channels, or the dramatic effect of KCNE1 on KCNQ1, or again a historical bias because of the groundbreaking initial discovery of KCNQ1-KCNE1 complexes in 1996? Time will tell and will also eventually reveal how many of the myriad KCNE-α subunit complexes that can form in vitro are expression-system phenomena that do not form in vivo. Similarly, because of the bias toward KCNQ1-KCNE channels, extremely little is known about the mechanistic basis for functional effects of KCNEs on other α subunits that they regulate.

Other important questions persist. What is the primary purpose of KCNEs that drove their evolution in the first place? Was it regulation of channel gating kinetics or voltage dependence that was so crucial? Or was it their array of influences on channel trafficking—which include ensuring correct localization, endocytosis, and even dictating which α subunits mature to form functional complexes at the plasma membrane? Alternatively, was it a different role unrelated to ion channels, or a cellular signaling purpose that requires channel formation but is still yet to be discovered? The simplest life form in which they have been reported is Caenorhabditis elegans,[141–143] but do precursors exist in bacteria or other unicellular organisms? Finally, while structural advances using NMR have provided pictures of KCNEs without their partners, and these have been compared to mutagenesis data to produce models of their relative disposition in channel complexes, we wait with bated breath for the first high-resolution KCNE-α subunit structure. When this milestone is eventually reached, it will answer many questions, settle many arguments, and inevitably create many new ones.[1]

---

[1] Since this chapter was written, the author's lab has reported the cardiac effects of Kcne4 deletion in mice.[144]

References

1. Takumi T, Moriyoshi K, Aramori I, Ishii T, Oiki S, Okada Y, et al. Alteration of channel activities and gating by mutations of slow ISK potassium channel. *J Biol Chem* 1991;**266**(33):22192–8.
2. Takumi T, Ohkubo H, Nakanishi S. Cloning of a membrane protein that induces a slow voltage-gated potassium current. *Science* 1988;**242**(4881):1042–5.
3. Busch AE, Lang F. Effects of $[Ca^{2+}]_i$ and temperature on minK channels expressed in *Xenopus* oocytes. *FEBS Lett* 1993;**334**(2):221–4.
4. Wang Q, Curran ME, Splawski I, Burn TC, Millholland JM, VanRaay TJ, et al. Positional cloning of a novel potassium channel gene: KVLQT1 mutations cause cardiac arrhythmias. *Nat Genet* 1996;**12**(1):17–23.
5. Barhanin J, Lesage F, Guillemare E, Fink M, Lazdunski M, Romey G. K(V)LQT1 and lsK (minK) proteins associate to form the I(Ks) cardiac potassium current. *Nature* 1996;**384**(6604):78–80.
6. Sanguinetti MC, Curran ME, Zou A, Shen J, Spector PS, Atkinson DL, et al. Coassembly of K(V)LQT1 and minK (IsK) proteins to form cardiac I(Ks) potassium channel. *Nature* 1996;**384**(6604):80–3.
7. Blumenthal EM, Kaczmarek LK. The minK potassium channel exists in functional and nonfunctional forms when expressed in the plasma membrane of *Xenopus* oocytes. *J Neurosci* 1994;**14**(5 Pt 2):3097–105.
8. Wang KW, Goldstein SA. Subunit composition of minK potassium channels. *Neuron* 1995;**14**(6):1303–9.
9. Abbott GW, Sesti F, Splawski I, Buck ME, Lehmann MH, Timothy KW, et al. MiRP1 forms IKr potassium channels with HERG and is associated with cardiac arrhythmia. *Cell* 1999;**97**(2):175–87.
10. Piccini M, Vitelli F, Seri M, Galietta LJ, Moran O, Bulfone A, et al. KCNE1-like gene is deleted in AMME contiguous gene syndrome: identification and characterization of the human and mouse homologs. *Genomics* 1999;**60**(3):251–7.
11. McCrossan ZA, Abbott GW. The MinK-related peptides. *Neuropharmacology* 2004;**47**(6):787–821.
12. Abbott GW. KCNE2 and the K (+) channel: the tail wagging the dog. *Channels (Austin)* 2012;**6**(1).
13. Lewis A, McCrossan ZA, Abbott GW. MinK, MiRP1, and MiRP2 diversify Kv3.1 and Kv3.2 potassium channel gating. *J Biol Chem* 2004;**279**(9):7884–92.
14. Abbott GW. Biology of the KCNQ1 potassium channel. *New J Sci* 2014;**2014**.
15. Tristani-Firouzi M, Sanguinetti MC. Voltage-dependent inactivation of the human K+ channel KvLQT1 is eliminated by association with minimal K+ channel (minK) subunits. *J Physiology* 1998;**510**(Pt 1):37–45.
16. Sesti F, Goldstein SA. Single-channel characteristics of wild-type IKs channels and channels formed with two minK mutants that cause long QT syndrome. *J Gen Physiol* 1998;**112**(6):651–63.
17. Yang WP, Levesque PC, Little WA, Conder ML, Shalaby FY, Blanar MA. KvLQT1, a voltage-gated potassium channel responsible for human cardiac arrhythmias. *Proc Natl Acad Sci USA* 1997;**94**(8):4017–21.
18. Chen H, Kim LA, Rajan S, Xu S, Goldstein SA. Charybdotoxin binding in the I(Ks) pore demonstrates two MinK subunits in each channel complex. *Neuron* 2003;**40**(1):15–23.
19. Plant LD, Xiong D, Dai H, Goldstein SA. Individual IKs channels at the surface of mammalian cells contain two KCNE1 accessory subunits. *Proc Natl Acad Sci USA* 2014;**111**(14):E1438–46.
20. Yu H, Lin Z, Mattmann ME, Zou B, Terrenoire C, Zhang H, et al. Dynamic subunit stoichiometry confers a progressive continuum of pharmacological sensitivity by KCNQ potassium channels. *Proc Natl Acad Sci USA* 2013;**110**(21):8732–7.

21. Sahu ID, Kroncke BM, Zhang R, Dunagan MM, Smith HJ, Craig A, et al. Structural investigation of the transmembrane domain of KCNE1 in proteoliposomes. *Biochemistry* 2014;**53**(40):6392–401.
22. Choveau FS, Abderemane-Ali F, Coyan FC, Es-Salah-Lamoureux Z, Baro I, Loussouarn G. Opposite effects of the S4-S5 linker and PIP(2) on voltage-gated channel function: KCNQ1/KCNE1 and other channels. *Front Pharmacol* 2012;**3**:125.
23. Lvov A, Gage SD, Berrios VM, Kobertz WR. Identification of a protein-protein interaction between KCNE1 and the activation gate machinery of KCNQ1. *J Gen Physiol* 2010;**135**(6):607–18.
24. Kang C, Tian C, Sonnichsen FD, Smith JA, Meiler J, George Jr AL, et al. Structure of KCNE1 and implications for how it modulates the KCNQ1 potassium channel. *Biochemistry* 2008;**47**(31):7999–8006.
25. Nakajo K, Kubo Y. KCNE1 and KCNE3 stabilize and/or slow voltage sensing S4 segment of KCNQ1 channel. *J Gen Physiol* 2007;**130**(3):269–81.
26. Rocheleau JM, Kobertz WR. KCNE peptides differently affect voltage sensor equilibrium and equilibration rates in KCNQ1 K+ channels. *J Gen Physiol* 2008;**131**(1):59–68.
27. Ruscic KJ, Miceli F, Villalba-Galea CA, Dai H, Mishina Y, Bezanilla F, et al. IKs channels open slowly because KCNE1 accessory subunits slow the movement of S4 voltage sensors in KCNQ1 pore-forming subunits. *Proc Natl Acad Sci USA* 2013;**110**(7):E559–66.
28. Zaydman MA, Kasimova MA, McFarland K, Beller Z, Hou P, Kinser HE, et al. Domain-domain interactions determine the gating, permeation, pharmacology, and subunit modulation of the IKs ion channel. *eLife* 2014;**3**:e03606.
29. Tinel N, Diochot S, Borsotto M, Lazdunski M, Barhanin J. KCNE2 confers background current characteristics to the cardiac KCNQ1 potassium channel. *EMBO J* 2000;**19**(23):6326–30.
30. Schroeder BC, Waldegger S, Fehr S, Bleich M, Warth R, Greger R, et al. A constitutively open potassium channel formed by KCNQ1 and KCNE3. *Nature* 2000;**403**(6766):196–9.
31. Heitzmann D, Grahammer F, von Hahn T, Schmitt-Graff A, Romeo E, Nitschke R, et al. Heteromeric KCNE2/KCNQ1 potassium channels in the luminal membrane of gastric parietal cells. *J Physiol* 2004;**561**(Pt 2):547–57.
32. Heitzmann D, Koren V, Wagner M, Sterner C, Reichold M, Tegtmeier I, et al. KCNE beta subunits determine pH sensitivity of KCNQ1 potassium channels. *Cell Physiol Biochem* 2007;**19**(1–4):21–32.
33. Melman YF, Krumerman A, McDonald TV. A single transmembrane site in the KCNE-encoded proteins controls the specificity of KvLQT1 channel gating. *J Biol Chem* 2002;**277**(28):25187–94.
34. Choi E, Abbott GW. A shared mechanism for lipid- and beta-subunit-coordinated stabilization of the activated K+ channel voltage sensor. *FASEB J* 2010;**24**(5):1518–24.
35. Abbott GW, Goldstein SA. A superfamily of small potassium channel subunits: form and function of the MinK-related peptides (MiRPs). *Q Rev Biophys* 1998;**31**(4):357–98.
36. Grunnet M, Jespersen T, Rasmussen HB, Ljungstrom T, Jorgensen NK, Olesen SP, et al. KCNE4 is an inhibitory subunit to the KCNQ1 channel. *J Physiol* 2002;**542**(Pt 1):119–30.
37. Manderfield LJ, George Jr AL. KCNE4 can co-associate with the I(Ks) (KCNQ1-KCNE1) channel complex. *FEBS J* 2008;**275**(6):1336–49.
38. Angelo K, Jespersen T, Grunnet M, Nielsen MS, Klaerke DA, Olesen SP. KCNE5 induces time- and voltage-dependent modulation of the KCNQ1 current. *Biophysical J* 2002;**83**(4):1997–2006.
39. Bendahhou S, Marionneau C, Haurogne K, Larroque MM, Derand R, Szuts V, et al. In vitro molecular interactions and distribution of KCNE family with KCNQ1 in the human heart. *Cardiovasc Res* 2005;**67**(3):529–38.
40. Grunnet M, Olesen SP, Klaerke DA, Jespersen T. hKCNE4 inhibits the hKCNQ1 potassium current without affecting the activation kinetics. *Biochem Biophysical Res Commun* 2005;**328**(4):1146–53.

41. Abbott GW. KCNE genetics and pharmacogenomics in cardiac arrhythmias: much ado about nothing? *Expert Rev Clin Pharmacol* 2013;6(1):49–60.

42. Sanguinetti MC, Zou A. Molecular physiology of cardiac delayed rectifier K+ channels. *Heart Vessels* 1997;(Suppl. 12):170–2.

43. Ackerman MJ, Tester DJ, Porter CJ. Swimming, a gene-specific arrhythmogenic trigger for inherited long QT syndrome. *Mayo Clin Proc* 1999;74(11):1088–94.

44. Ohno S, Zankov DP, Yoshida H, Tsuji K, Makiyama T, Itoh H, et al. N- and C-terminal KCNE1 mutations cause distinct phenotypes of long QT syndrome. *Heart Rhythm* 2007;4(3):332–40.

45. Kapplinger JD, Tester DJ, Salisbury BA, Carr JL, Harris-Kerr C, Pollevick GD, et al. Spectrum and prevalence of mutations from the first 2,500 consecutive unrelated patients referred for the FAMILION long QT syndrome genetic test. *Heart Rhythm* 2009;6(9):1297–303.

46. Olesen MS, Bentzen BH, Nielsen JB, Steffensen AB, David JP, Jabbari J, et al. Mutations in the potassium channel subunit KCNE1 are associated with early-onset familial atrial fibrillation. *BMC Med Genet* 2012;13(1):24.

47. Splawski I, Shen J, Timothy KW, Lehmann MH, Priori S, Robinson JL, et al. Spectrum of mutations in long-QT syndrome genes. KVLQT1, HERG, SCN5A, KCNE1, and KCNE2. *Circulation* 2000;102(10):1178–85.

48. Westenskow P, Splawski I, Timothy KW, Keating MT, Sanguinetti MC. Compound mutations: a common cause of severe long-QT syndrome. *Circulation* 2004;109(15):1834–41.

49. Ehrlich JR, Zicha S, Coutu P, Hebert TE, Nattel S. Atrial fibrillation-associated minK38G/S polymorphism modulates delayed rectifier current and membrane localization. *Cardiovasc Res* 2005;67(3):520–8.

50. Fatini C, Sticchi E, Genuardi M, Sofi F, Gensini F, Gori AM, et al. Analysis of minK and eNOS genes as candidate loci for predisposition to non-valvular atrial fibrillation. *Eur Heart J* 2006;27(14):1712–8.

51. Xu LX, Yang WY, Zhang HQ, Tao ZH, Duan CC. Study on the correlation between CETP TaqIB, KCNE1 S38G and eNOS T-786C gene polymorphisms for predisposition and non-valvular atrial fibrillation. *Zhonghua liu xing bing xue za zhi* 2008;29(5):486–92.

52. Prystupa A, Dzida G, Myslinski W, Malaj G, Lorenc T. MinK gene polymorphism in the pathogenesis of lone atrial fibrillation. *Kardiol Pol* 2006;64(11):1205–11. discussion 12-3.

53. Husser D, Stridh M, Sornmo L, Roden DM, Darbar D, Bollmann A. A genotype-dependent intermediate ECG phenotype in patients with persistent lone atrial fibrillation genotype ECG-phenotype correlation in atrial fibrillation. *Circ Arrhyth Electrophysiol* 2009;2(1):24–8.

54. Bianchi L, Shen Z, Dennis AT, Priori SG, Napolitano C, Ronchetti E, et al. Cellular dysfunction of LQT5-minK mutants: abnormalities of IKs, IKr and trafficking in long QT syndrome. *Hum Mol Genet* 1999;8(8):1499–507.

55. Ma L, Lin C, Teng S, Chai Y, Bahring R, Vardanyan V, et al. Characterization of a novel Long QT syndrome mutation G52R-KCNE1 in a Chinese family. *Cardiovasc Res* 2003;59(3):612–9.

56. Splawski I, Tristani-Firouzi M, Lehmann MH, Sanguinetti MC, Keating MT. Mutations in the hminK gene cause long QT syndrome and suppress IKs function. *Nat Genet* 1997;17(3):338–40.

57. Duggal P, Vesely MR, Wattanasirichaigoon D, Villafane J, Kaushik V, Beggs AH. Mutation of the gene for IsK associated with both Jervell and Lange-Nielsen and Romano-Ward forms of long-QT syndrome. *Circulation* 1998;97(2):142–6.

58. Wu DM, Lai LP, Zhang M, Wang HL, Jiang M, Liu XS, et al. Characterization of an LQT5-related mutation in KCNE1, Y81C: implications for a role of KCNE1 cytoplasmic domain in IKs channel function. *Heart Rhythm* 2006;3(9):1031–40.

59. Nielsen NH, Winkel BG, Kanters JK, Schmitt N, Hofman-Bang J, Jensen HS, et al. Mutations in the Kv1.5 channel gene KCNA5 in cardiac arrest patients. *Biochem Biophys Res Commun* 2007;**354**(3):776–82.

60. Zeng ZY, Pu JL, Tan C, Teng SY, Chen JH, Su SY, et al. The association of single nucleotide polymorphism of slow delayed rectifier K+ channel genes with atrial fibrillation in Han nationality Chinese. *Zhonghua xin xue guan bing za zhi* 2005;**33**(11):987–91.

61. Zeng Z, Tan C, Teng S, Chen J, Su S, Zhou X, et al. The single nucleotide polymorphisms of I(Ks) potassium channel genes and their association with atrial fibrillation in a Chinese population. *Cardiology* 2007;**108**(2):97–103.

62. Porthan K, Marjamaa A, Viitasalo M, Vaananen H, Jula A, Toivonen L, et al. Relationship of common candidate gene variants to electrocardiographic T-wave peak to T-wave end interval and T-wave morphology parameters. *Heart Rhythm* 2010;**7**(7):898–903.

63. Paulussen AD, Gilissen RA, Armstrong M, Doevendans PA, Verhasselt P, Smeets HJ, et al. Genetic variations of KCNQ1, KCNH2, SCN5A, KCNE1, and KCNE2 in drug-induced long QT syndrome patients. *J Mol Med (Berl)* 2004;**82**(3):182–8.

64. Nishio Y, Makiyama T, Itoh H, Sakaguchi T, Ohno S, Gong YZ, et al. D85N, a KCNE1 polymorphism, is a disease-causing gene variant in long QT syndrome. *J Am Coll Cardiol* 2009;**54**(9):812–9.

65. Lin L, Horigome H, Nishigami N, Ohno S, Horie M, Sumazaki R. Drug-induced QT-interval prolongation and recurrent torsade de pointes in a child with heterotaxy syndrome and KCNE1 D85N polymorphism. *J Electrocardiol* 2012;**45**(6):770–3.

66. Kaab S, Crawford DC, Sinner MF, Behr ER, Kannankeril PJ, Wilde AA, et al. A large candidate gene survey identifies the KCNE1 D85N polymorphism as a possible modulator of drug-induced torsades de pointes. *Circ Cardiovasc Genet* 2012;**5**(1):91–9.

67. Gouas L, Nicaud V, Berthet M, Forhan A, Tiret L, Balkau B, et al. Association of KCNQ1, KCNE1, KCNH2 and SCN5A polymorphisms with QTc interval length in a healthy population. *Eur J Hum Genet* 2005;**13**(11):1213–22.

68. Harmer SC, Wilson AJ, Aldridge R, Tinker A. Mechanisms of disease pathogenesis in long QT syndrome type 5. *Am J Physiol Cell Physiol* 2010;**298**(2):C263–73.

69. Schulze-Bahr E, Schwarz M, Hauenschild S, Wedekind H, Funke H, Haverkamp W, et al. A novel long-QT 5 gene mutation in the C-terminus (V109I) is associated with a mild phenotype. *J Mol Med (Berl)* 2001;**79**(9):504–9.

70. Sesti F, Abbott GW, Wei J, Murray KT, Saksena S, Schwartz PJ, et al. A common polymorphism associated with antibiotic-induced cardiac arrhythmia. *Proc Natl Acad Sci USA* 2000;**97**(19):10613–8.

71. Gordon E, Panaghie G, Deng L, Bee KJ, Roepke TK, Krogh-Madsen T, et al. A KCNE2 mutation in a patient with cardiac arrhythmia induced by auditory stimuli and serum electrolyte imbalance. *Cardiovasc Res* 2008;**77**(1):98–106.

72. Nielsen JB, Bentzen BH, Olesen MS, David JP, Olesen SP, Haunso S, et al. Gain-of-function mutations in potassium channel subunit KCNE2 associated with early-onset lone atrial fibrillation. *Biomarkers Med* 2014;**8**(4):557–70.

73. Liu W, Deng J, Wang G, Zhang C, Luo X, Yan D, et al. KCNE2 modulates cardiac L-type Ca channel. *J Mol Cell Cardiol* 2014;**72**:208–18.

74. Yang Y, Xia M, Jin Q, Bendahhou S, Shi J, Chen Y, et al. Identification of a KCNE2 gain-of-function mutation in patients with familial atrial fibrillation. *Am J Hum Genet* 2004;**75**(5):899–905.

75. Isbrandt D, Friederich P, Solth A, Haverkamp W, Ebneth A, Borggrefe M, et al. Identification and functional characterization of a novel KCNE2 (MiRP1) mutation that alters HERG channel kinetics. *J Mol Med (Berl)* 2002;**80**(8):524–32.

76. Ohno S, Toyoda F, Zankov DP, Yoshida H, Makiyama T, Tsuji K, et al. Novel KCNE3 mutation reduces repolarizing potassium current and associated with long QT syndrome. *Hum Mutat* 2009;**30**(4):557–63.

77. Nakajima T, Wu J, Kaneko Y, Ashihara T, Ohno S, Irie T, et al. KCNE3 T4A as the genetic basis of Brugada-pattern electrocardiogram. *Circ J* 2012;**76**(12):2763–72.
78. Zhang DF, Liang B, Lin J, Liu B, Zhou QS, Yang YQ. KCNE3 R53H substitution in familial atrial fibrillation. *Chin Med J* 2005;**118**(20):1735–8.
79. Delpon E, Cordeiro JM, Nunez L, Thomsen PE, Guerchicoff A, Pollevick GD, et al. Functional effects of KCNE3 mutation and its role in the development of Brugada syndrome. *Circ Arrhyth Electrophysiol* 2008;**1**(3):209–18.
80. Ravn LS, Aizawa Y, Pollevick GD, Hofman-Bang J, Cordeiro JM, Dixen U, et al. Gain of function in IKs secondary to a mutation in KCNE5 associated with atrial fibrillation. *Heart Rhythm* 2008;**5**(3):427–35.
81. Ohno S, Zankov DP, Ding WG, Itoh H, Makiyama T, Doi T, et al. KCNE5 (KCNE1L) variants are novel modulators of Brugada syndrome and idiopathic ventricular fibrillation. *Circ Arrhythm Electrophysiol* 2011;**4**(3):352–61.
82. Crump SM, Abbott GW. Arrhythmogenic KCNE gene variants: current knowledge and future challenges. *Front Genet* 2014;**5**:3.
83. Jackman WM, Friday KJ, Anderson JL, Aliot EM, Clark M, Lazzara R. The long QT syndromes: a critical review, new clinical observations and a unifying hypothesis. *Prog Cardiovasc Dis* 1988;**31**(2):115–72.
84. Jackman WM, Clark M, Friday KJ, Aliot EM, Anderson J, Lazzara R. Ventricular tachyarrhythmias in the long QT syndromes. *Med Clin N Am* 1984;**68**(5):1079–109.
85. Riley AB, Manning WJ. Atrial fibrillation: an epidemic in the elderly. *Expert Rev Cardiovasc Ther* 2011;**9**(8):1081–90.
86. Roselli EE, Murthy SC, Rice TW, Houghtaling PL, Pierce CD, Karchmer DP, et al. Atrial fibrillation complicating lung cancer resection. *J Thorac Cardiovasc Surg* 2005;**130**(2):438–44.
87. Marx SO, Kurokawa J, Reiken S, Motoike H, D'Armiento J, Marks AR, et al. Requirement of a macromolecular signaling complex for beta adrenergic receptor modulation of the KCNQ1-KCNE1 potassium channel. *Science* 2002;**295**(5554):496–9.
88. Terrenoire C, Houslay MD, Baillie GS, Kass RS. The cardiac IKs potassium channel macromolecular complex includes the phosphodiesterase PDE4D3. *J Biol Chem* 2009;**284**(14):9140–6.
89. Li Y, Chen L, Kass RS, Dessauer CW. The A-kinase anchoring protein Yotiao facilitates complex formation between adenylyl cyclase type 9 and the IKs potassium channel in heart. *J Biol Chem* 2012;**287**(35):29815–24.
90. Schulze-Bahr E, Wang Q, Wedekind H, Haverkamp W, Chen Q, Sun Y, et al. KCNE1 mutations cause jervell and Lange-Nielsen syndrome. *Nat Genet* 1997;**17**(3):267–8.
91. Tyson J, Tranebjaerg L, Bellman S, Wren C, Taylor JF, Bathen J, et al. IsK and KvLQT1: mutation in either of the two subunits of the slow component of the delayed rectifier potassium channel can cause Jervell and Lange-Nielsen syndrome. *Hum Mol Genet* 1997;**6**(12):2179–85.
92. Lai LP, Su MJ, Yeh HM, Lin JL, Chiang FT, Hwang JJ, et al. Association of the human minK gene 38G allele with atrial fibrillation: evidence of possible genetic control on the pathogenesis of atrial fibrillation. *Am Heart J* 2002;**144**(3):485–90.
93. Voudris KV, Apostolakis S, Karyofillis P, Doukas K, Zaravinos A, Androutsopoulos VP, et al. Genetic diversity of the KCNE1 gene and susceptibility to postoperative atrial fibrillation. *Am Heart J* 2014;**167**(2):274–80. e1.
94. McDonald TV, Yu Z, Ming Z, Palma E, Meyers MB, Wang KW, et al. A minK-HERG complex regulates the cardiac potassium current I(Kr). *Nature* 1997;**388**(6639): 289–92.
95. Brunner M, Peng X, Liu GX, Ren XQ, Ziv O, Choi BR, et al. Mechanisms of cardiac arrhythmias and sudden death in transgenic rabbits with long QT syndrome. *J Clin Invest* 2008;**118**(6):2246–59.

96. Ren XQ, Liu GX, Organ-Darling LE, Zheng R, Roder K, Jindal HK, et al. Pore mutants of HERG and KvLQT1 downregulate the reciprocal currents in stable cell lines. *Am J Physiol Heart Circ Physiol* 2010;**299**(5):H1525–34.

97. Jiang M, Xu X, Wang Y, Toyoda F, Liu XS, Zhang M, et al. Dynamic partnership between KCNQ1 and KCNE1 and influence on cardiac IKs current amplitude by KCNE2. *J Biol Chem* 2009;**284**(24):16452–62.

98. Toyoda F, Ueyama H, Ding WG, Matsuura H. Modulation of functional properties of KCNQ1 channel by association of KCNE1 and KCNE2. *Biochem Biophys Res Commun* 2006;**344**(3):814–20.

99. Wu DM, Jiang M, Zhang M, Liu XS, Korolkova YV, Tseng GN. KCNE2 is colocalized with KCNQ1 and KCNE1 in cardiac myocytes and may function as a negative modulator of I(Ks) current amplitude in the heart. *Heart Rhythm* 2006;**3**(12):1469–80.

100. Park KH, Kwok SM, Sharon C, Baerga R, Sesti F. N-Glycosylation-dependent block is a novel mechanism for drug-induced cardiac arrhythmia. *FASEB J* 2003;**17**(15):2308–9.

101. McCrossan ZA, Roepke TK, Lewis A, Panaghie G, Abbott GW. Regulation of the Kv2.1 potassium channel by MinK and MiRP1. *J Membr Biol* 2009;**228**(1):1–14.

102. Roepke TK, Abbott GW. Pharmacogenetics and cardiac ion channels. *Vasc Pharmacol* 2006;**44**(2):90–106.

103. Zhang M, Jiang M, Tseng GN. Mink-related peptide 1 associates with Kv4.2 and modulates its gating function: potential role as beta subunit of cardiac transient outward channel? *Circulation Res* 2001;**88**(10):1012–9.

104. Qu J, Kryukova Y, Potapova IA, Doronin SV, Larsen M, Krishnamurthy G, et al. MiRP1 modulates HCN2 channel expression and gating in cardiac myocytes. *J Biol Chem* 2004;**279**(42):43497–502.

105. Yu H, Wu J, Potapova I, Wymore RT, Holmes B, Zuckerman J, et al. MinK-related peptide 1: a beta subunit for the HCN ion channel subunit family enhances expression and speeds activation. *Circulation Res* 2001;**88**(12):E84–7.

106. Hu Z, Kant R, Anand M, King EC, Krogh-Madsen T, Christini DJ, et al. Kcne2 deletion creates a multisystem syndrome predisposing to sudden cardiac death. *Circ Cardiovasc Genet* 2014;**7**(1):33–42.

107. Abbott GW, Butler MH, Bendahhou S, Dalakas MC, Ptacek LJ, Goldstein SA. MiRP2 forms potassium channels in skeletal muscle with Kv3.4 and is associated with periodic paralysis. *Cell* 2001;**104**(2):217–31.

108. Sternberg D, Tabti N, Fournier E, Hainque B, Fontaine B. Lack of association of the potassium channel-associated peptide MiRP2-R83H variant with periodic paralysis. *Neurology* 2003;**61**(6):857–9.

109. Charpentier F, Merot J, Riochet D, Le Marec H, Escande D. Adult KCNE1-knockout mice exhibit a mild cardiac cellular phenotype. *Biochem Biophys Res Commun* 1998;**251**(3):806–10.

110. Knollmann BC, Casimiro MC, Katchman AN, Sirenko SG, Schober T, Rong Q, et al. Isoproterenol exacerbates a long QT phenotype in *Kcnq1*-deficient neonatal mice: possible roles for human-like *Kcnq1* isoform 1 and slow delayed rectifier K+ current. *J Pharmacol Exp Ther* 2004;**310**(1):311–8.

111. Warth R, Barhanin J. The multifaceted phenotype of the knockout mouse for the KCNE1 potassium channel gene. *Am J Physiol Regul Integr Comp Physiol* 2002;**282**(3):R639–48.

112. Temple J, Frias P, Rottman J, Yang T, Wu Y, Verheijck EE, et al. Atrial fibrillation in KCNE1-null mice. *Circulation Res* 2005;**97**(1):62–9.

113. Demolombe S, Franco D, de Boer P, Kuperschmidt S, Roden D, Pereon Y, et al. Differential expression of KvLQT1 and its regulator IsK in mouse epithelia. *Am J Physiol Cell Physiol* 2001;**280**(2):C359–72.

114. Warth R, Garcia Alzamora M, Kim JK, Zdebik A, Nitschke R, Bleich M, et al. The role of KCNQ1/KCNE1 K(+) channels in intestine and pancreas: lessons from the KCNE1 knockout mouse. *Pflugers Arch* 2002;**443**(5–6):822–8.

115. Vallon V, Grahammer F, Richter K, Bleich M, Lang F, Barhanin J, et al. Role of KCNE1-dependent K$^+$ fluxes in mouse proximal tubule. *J Am Soc Nephrol* 2001;**12**(10): 2003–11.

116. Puchalski RB, Kelly E, Bachmanov AA, Brazier SP, Kuang J, Arrighi I, et al. NaCl consumption is attenuated in female KCNE1 null mutant mice. *Physiol Behav* 2001;**74**(3):267–76.

117. Barriere H, Rubera I, Belfodil R, Tauc M, Tonnerieux N, Poujeol C, et al. Swelling-activated chloride and potassium conductance in primary cultures of mouse proximal tubules. Implication of KCNE1 protein. *J Membr Biol* 2003;**193**(3):153–70.

118. Millar ID, Hartley JA, Haigh C, Grace AA, White SJ, Kibble JD, et al. Volume regulation is defective in renal proximal tubule cells isolated from KCNE1 knockout mice. *Exp Physiol* 2004;**89**(2):173–80.

119. Neal AM, Taylor HC, Millar ID, Kibble JD, White SJ, Robson L. Renal defects in KCNE1 knockout mice are mimicked by chromanol 293B in vivo: identification of a KCNE1-regulated K$^+$ conductance in the proximal tubule. *J Physiol* 2011;**589**(Pt 14):3595–609.

120. Vallon V, Grahammer F, Volkl H, Sandu CD, Richter K, Rexhepaj R, et al. KCNQ1-dependent transport in renal and gastrointestinal epithelia. *Proc Natl Acad Sci USA* 2005;**102**(49):17864–9.

121. Roepke TK, Kontogeorgis A, Ovanez C, Xu X, Young JB, Purtell K, et al. Targeted deletion of kcne2 impairs ventricular repolarization via disruption of I(K,slow1) and I(to,f). *FASEB J* 2008;**22**(10):3648–60.

122. Roepke TK, Anantharam A, Kirchhoff P, Busque SM, Young JB, Geibel JP, et al. The KCNE2 potassium channel ancillary subunit is essential for gastric acid secretion. *J Biol Chem* 2006;**281**(33):23740–7.

123. Roepke TK, King EC, Purtell K, Kanda VA, Lerner DJ, Abbott GW. Genetic dissection reveals unexpected influence of beta subunits on KCNQ1 K$^+$ channel polarized trafficking in vivo. *FASEB J* 2011;**25**(2):727–36.

124. Yanglin P, Lina Z, Zhiguo L, Na L, Haifeng J, Guoyun Z, et al. KCNE2, a down-regulated gene identified by in silico analysis, suppressed proliferation of gastric cancer cells. *Cancer Lett* 2007;**246**(1–2):129–38.

125. Roepke TK, Purtell K, King EC, La Perle KM, Lerner DJ, Abbott GW. Targeted deletion of Kcne2 causes gastritis cystica profunda and gastric neoplasia. *PLoS One* 2010;**5**(7):e11451.

126. Purtell K, Paroder-Belenitsky M, Reyna-Neyra A, Nicola JP, Koba W, Fine E, et al. The KCNQ1-KCNE2 K$^+$ channel is required for adequate thyroid I-uptake. *FASEB J* 2012;**26**(8):3252–9.

127. Roepke TK, King EC, Reyna-Neyra A, Paroder M, Purtell K, Koba W, et al. Kcne2 deletion uncovers its crucial role in thyroid hormone biosynthesis. *Nat Med* 2009;**15**(10):1186–94.

128. Abbott GW, Tai KK, Neverisky DL, Hansler A, Hu Z, Roepke TK, et al. KCNQ1, KCNE2, and Na$^+$-coupled solute transporters form reciprocally regulating complexes that affect neuronal excitability. *Sci Signal* 2014;**7**(315):ra22.

129. Roepke TK, Kanda VA, Purtell K, King EC, Lerner DJ, Abbott GW. KCNE2 forms potassium channels with KCNA3 and KCNQ1 in the choroid plexus epithelium. *FASEB J* 2011;**25**(12):4264–73.

130. Lee YH, Kang ES, Kim SH, Han SJ, Kim CH, Kim HJ, et al. Association between polymorphisms in SLC30A8, HHEX, CDKN2A/B, IGF2BP2, FTO, WFS1, CDKAL1, KCNQ1 and type 2 diabetes in the Korean population. *J Hum Genet* 2008;**53**(11–12): 991–8.

131. Liu Y, Zhou DZ, Zhang D, Chen Z, Zhao T, Zhang Z, et al. Variants in KCNQ1 are associated with susceptibility to type 2 diabetes in the population of mainland China. *Diabetologia* 2009;**52**(7):1315–21.

132. Unoki H, Takahashi A, Kawaguchi T, Hara K, Horikoshi M, Andersen G, et al. SNPs in KCNQ1 are associated with susceptibility to type 2 diabetes in East Asian and European populations. *Nat Genet* 2008;**40**(9):1098–102.

133. Yasuda K, Miyake K, Horikawa Y, Hara K, Osawa H, Furuta H, et al. Variants in KCNQ1 are associated with susceptibility to type 2 diabetes mellitus. *Nat Genet* 2008;**40**(9):1092–7.

134. Zhou JB, Yang JK, Zhao L, Xin Z. Variants in KCNQ1, AP3S1, MAN2A1, and ALDH7A1 and the risk of type 2 diabetes in the Chinese Northern Han population: a case-control study and meta-analysis. *Med Sci Monit* 2010;**16**(6):BR179–83.

135. Rice KS, Dickson G, Lane M, Crawford J, Chung SK, Rees MI, et al. Elevated serum gastrin levels in Jervell and Lange-Nielsen syndrome: a marker of severe KCNQ1 dysfunction? *Heart Rhythm* 2011;**8**(4):551–4.

136. Hu Z, Crump SM, Anand M, Kant R, Levi R, Abbott GW. Kcne3 deletion initiates extracardiac arrhythmogenesis in mice. *FASEB J* 2014;**28**(2):935–45.

137. Preston P, Wartosch L, Gunzel D, Fromm M, Kongsuphol P, Ousingsawat J, et al. Disruption of the K+ channel beta-subunit KCNE3 reveals an important role in intestinal and tracheal Cl-transport. *J Biol Chem* 2010;**285**(10):7165–75.

138. Kathiresan S, Voight BF, Purcell S, Musunuru K, Ardissino D, Mannucci PM, et al. Genome-wide association of early-onset myocardial infarction with single nucleotide polymorphisms and copy number variants. *Nat Genet* 2009;**41**(3):334–41.

139. Sabater-Lleal M, Malarstig A, Folkersen L, Soler Artigas M, Baldassarre D, Kavousi M, et al. Common genetic determinants of lung function, subclinical atherosclerosis and risk of coronary artery disease. *PLoS One* 2014;**9**(8):e104082.

140. Soler Artigas M, Loth DW, Wain LV, Gharib SA, Obeidat M, Tang W, et al. Genome-wide association and large-scale follow up identifies 16 new loci influencing lung function. *Nat Genet* 2011;**43**(11):1082–90.

141. Bianchi L, Kwok SM, Driscoll M, Sesti F. A potassium channel-MiRP complex controls neurosensory function in *Caenorhabditis elegans*. *J Biol Chem* 2003;**278**(14):12415–24.

142. Cai SQ, Hernandez L, Wang Y, Park KH, Sesti F. MPS-1 is a K+ channel beta-subunit and a serine/threonine kinase. *Nat Neurosci* 2005;**8**(11):1503–9.

143. Cai SQ, Park KH, Sesti F. An evolutionarily conserved family of accessory subunits of K+ channels. *Cell Biochem Biophys* 2006;**46**(1):91–9.

144. Crump SM, Hu Z, Kant R, Levy DI, Goldstein SA, Abbott GW. Kcne4 deletion sex- and age-specifically impairs cardiac repolarization in mice. *FASEB J*. 2016;**30**(1):360–9.

Ion Channel Trafficking

T.T. Hong[1,2], R.M. Shaw[1,2]

[1]Cedars-Sinai Medical Center, Los Angeles, CA, United States; [2]University
of California Los Angeles, Los Angeles, CA, United States

INTRODUCTION

During the average human lifetime, a heart beats more than 3 billion times. For each of those heartbeats to be effective, each second, several billion heart cells need to contract in synchrony. Well-synchronized contraction is achieved by electrical coordination: the rapid propagation of action potentials throughout each cardiomyocyte ensuring an efficient and near simultaneous activation. Once triggered across each cardiomyocyte membrane, action potentials lead to intracellular calcium release and cellular contraction. Both rapid propagation of action potentials and intracellular calcium release require sophisticated cascades of ion channel opening and closing which, to be successful, depend on the proper localization of particular ion channels at particular subdomains of the plasma membrane. The formation and correct delivery of ion channels to their membrane subdomain is referred to as protein trafficking and altered trafficking is a dangerous complication of diseased myocardium. Regulation of protein trafficking and its restoration in diseased myocardium presents opportunities for novel therapeutic development.

Each cardiomyocyte is a long-lived cell with a highly organized, robust, and dynamic intracellular environment. The nuclei of cardiomyocytes continuously transcribe genes to mRNA strands that are then translated into proteins. Particular proteins are then assembled into channels and shuttled throughout the cytoplasm to specific organelles and functional subdomains of surrounding cell membrane. The cardiomyocyte cell membrane has particular subdomains including regions designated for cell–cell communication (intercalated discs) and initiating contractile signaling (T-tubules). Proteins and channels that function at intercalated disc regions are different from proteins and channels that function at T-tubules. The question for cardiomyocytes is how, in normal physiologic conditions, particular regions of membrane are populated with the appropriate proteins. It is critical to understand the dynamic intracellular environment that contains rapid protein movements and very frequent protein turnover.

Cardiomyocytes in diseased and failing hearts are subjected to environments with less energy and more stress. In response the cells undergo not just pathologic remodeling of membrane structures and proteins, but also alterations of protein movement and turnover. As the physiologic movements of cardiac intracellular proteins are elucidated, and then disease related changes of these movements are understood, interventions can be designed to promote positive intracellular remodeling. Identification of intracellular pathways critical in stress response during heart failure progression will provide new targets for development of interventions to help restore normal myocyte function.

In this chapter, we briefly introduce the relevant principles of cardiomyocyte organization and then focus on channel trafficking. Ion channel trafficking in both healthy physiology and disease pathophysiology is discussed with special emphasis on connexin 43 (Cx43) gap junctions and L-type calcium channels (LTCCs). Cx43 channels are important for cell–cell electrical communication, and LTCCs are important for contraction. Cx43 channels need to localize to intercalated discs at the longitudinal ends of cardiomyocytes whereas LTCCs need to localize at T-tubules which are at the transverse borders of cardiomyocytes. An analysis of Cx43 and LTCCs therefore covers different regions and many fundamental concepts of cardiomyocyte organization.

CARDIOMYOCYTE ORGANIZATION

As illustrated in Fig. 2.1, adult ventricular cardiomyocytes are large, usually bi-nucleated contractile cells with a highly polarized phenotype. The basic contractile units are cytoplasmic myofibrils composed of specialized mechanical structures known as sarcomeres consisting thin filament actin, myosin, and accessory proteins. As a cell that contracts once every second, the cardiomyocyte has a high energy need which is supplied by large numbers of mitochondria in the cytoplasm. All myocytes contain a unique form of smooth endoplasmic reticulum (ER) called the sarcoplasmic reticulum (SR), which stores a large amount of calcium ions. Cardiomyocytes also have a rough ER and Golgi apparatus, near which microtubules are anchored and extend toward the peripheral of the cell. Nonsarcomeric actin cytoskeleton is spread out across the whole cytosol and enriched in submembrane cortical regions, providing structural support to sarcolemma membrane. The cardiomyocyte sarcolemma is a

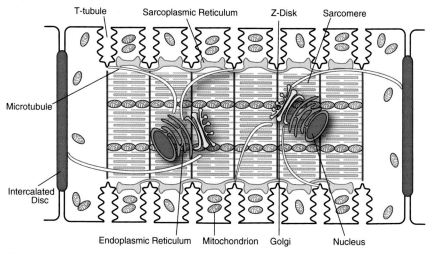

FIGURE 2.1 **Organization of an adult ventricular cardiomyocyte.** Intercalated discs are located at the longitudinal sides of each ventricular cardiomyocyte which mediate the cell-to-cell propagation of action potentials. T-tubules, which are rich in voltage-gated L-type calcium channels, are positioned closely to the sarcoplasmic reticulum, the internal calcium store. Sarcomeres form myofibrils, which are responsible for cardiomyocyte contraction upon intracellular calcium release. The Golgi apparatus and microtubules serve as the "loading dock" and "highways," respectively, to deliver ion channels to specific subdomains on the plasma membrane. Mitochondria provide the energy needed for the contraction of cardiomyocytes.

continuous lipid bilayer enclosing the whole myocyte with subdomains such as intercalated discs at longitudinal ends and T-tubule invaginations at the transverse borders. Small membrane microdomains also exist including flask-shaped caveolae structures as well as T-tubule minifolds formed by a scaffolding protein bridging integrator 1 (BIN1, or amphiphysin 2).

Cardiac muscle contracts on average more than 3 billion times. Mitochondria occupy ~30% of the volume of a cardiomyocyte and are positioned beneath the sarcolemma, between myofibrils, and near the nucleus[1] (Fig. 2.1). These different subgroups of mitochondria, identified by spatial location in the myocyte, appear to have different morphologies and properties.[1,2] In addition to providing energy, mitochondria play an important role in regulating cell death, propagating Ca^{2+} signaling, and mediating cardiac protection to ischemia/reperfusion and other stresses.[3,4]

Heart contraction is accomplished, at the subcellular level, by contraction of sarcomeres. A single sarcomere occupies the region between two Z-lines (Fig. 2.1) and consists of thick filaments of myosin and thin filaments of actin that are anchored to Z-discs. Myofibrils are connected to the sarcolemma along the longitudinal axis at the adherens junctions (fascia adherens) of the intercalated discs and across the lateral axis at the costameres connecting Z-discs to T-tubules.[5,6] The outer cell membrane is linked to cortical cytoskeletal desmosomes and actin microfilaments (F-actin), which provide mechanical support to maintain cell shape and specialized subdomains such as intercalated discs and T-tubules.[7,8] For a myofibril to contract, an elevation of cytosol calcium ions is required, which can be achieved by action potential–induced release of calcium from their storage site of SR lumens. Calcium ion release from and reuptake into SR regulates intracellular cytosolic calcium level, gating the contraction and relaxation of cardiomyocytes.

Cardiomyocytes also use common intracellular organelles to produce proteins and shuttle proteins to specific organelles and functional subdomains of surrounding cell membrane. After gene transcription in nuclei, proteins are translated and subjected to posttranslational modification in the ER and further modified in the Golgi apparatus. For membrane proteins, sorting and delivery to their subcellular destination begins in the Golgi apparatus (Fig. 2.1). The Golgi complex is usually found adjacent to the lateral side of each nucleus in mammalian ventricular cardiomyocytes. Colocalized with each Golgi is the centrosome at which microtubules are nucleated and extend throughout the cell.[9] Sorting of proteins mainly takes place at the *trans*-Golgi network (TGN).[10] Cargo proteins are sorted into post-Golgi carriers, which are docked onto molecular motors and delivered to the cell periphery along microtubules.[11] These extending microtubules form an intricate and dynamic outgoing network capable of shuttling ion channel containing vesicles to their destinations. In the context of trafficking, one can consider the Golgi to be the "loading dock" and microtubules the "highways" along which packets of channels are delivered to the plasma membrane. Along the microtubule highway, cytoplasmic and cortical actin cytoskeleton provides rest stops[12] to guide and redirect moving microtubules, facilitating and tailoring en route cargo delivery as well as protein insertion into sarcolemma surface destination.

Differential sarcolemma subregional organization plays a critical role in protein capture specificity as well as in protein activity at the membrane once captured. Sarcolemma subregions are usually composed of different lipid species, which then recruit membrane scaffolding proteins to shape and organize membrane subdomains. For instance, T-tubules are preferentially enriched in phospholipids[13,14] and contain tiny membrane folds sculpted by a membrane curvature formation protein BIN1.[6] Caveolae flasks found throughout sarcolemma surface, on the other hand, are cholesterol and sphingolipid rich bilayers[15] supported by a structural protein caveolin-3.[16] The intercalated discs are membrane structures supported by adherens junctions and desmosomes.

For each heartbeat, the normal sequence of events leading to contraction is (1) initiation of an action potential at the sinoatrial node, (2) myocyte-to-myocyte spread of action potentials through the atria, across the AV node, and through the His-Purkinje systems, (3) myocyte-to-myocyte spread of action potentials in ventricular myocardium in apex to base order, (4) individual cellular action potentials allowing calcium entry through calcium channels at T-tubules, triggering larger calcium release from the corresponding internal SR, and (5) calcium release causing myofibril contraction. The myocyte-to-myocyte spread of action potentials occur across intercalated disc regions. Calcium-induced calcium release is triggered at cardiac T-tubules. Thus, in general, the intercalated disc regions are responsible for myocyte-to-myocyte electrical communication and the T-tubule regions are responsible for triggering and regulating the strength of each contraction (Fig. 2.1). In this section, we will emphasize the organization of both intercalated discs and T-tubules, as well as membrane subdomains including caveolae and BIN1-sculpted T-tubule minifolds.

Intercalated Discs

Individual cardiomyocytes couple mechanically and electrically to each other through specialized structures termed intercalated discs which locate at the longitudinal ends of cardiomyocytes, perpendicular to myofibrils. Mechanical coupling of cardiomyocytes is achieved through desmosome and adherens complexes connecting the intercalated disc to structural components of the cytoskeleton such as intermediate filaments and F-actin.

Gap junctions are primarily responsible for electrical coupling of the cardiomyocytes as they allow direct communication between the cytoplasms of adjacent cells, permitting free and voltage-driven diffusion of ions.[7] The polarized localization of gap junctions at cardiomyocyte intercalated discs is essential for appropriately orchestrated propagation of action potentials throughout working myocardium. Altered localization and losses in cell–cell gap junction coupling occur during cardiac disease,[17] and contribute to arrhythmogenic substrates.[18] Gap junctions exist in plaques that are comprised of hundreds to thousands of connexon hemichannels, each of which is made up of six connexin molecules. Connexons on the surfaces of opposing cells must coalesce to form continuous conduits spanning both lipid bilayers. Over 20 isoforms of connexin have been identified in humans, of which connexin 43 (Cx43) is the primary connexin expressed in ventricular myocardium.[19]

Other ion channels, such as voltage-gated sodium channels (Na_v), are also enriched at the intercalated disc. The primary cardiac α subunit $Na_v1.5$ is preferentially localized to intercalated discs with auxiliary β2 and/or β4 subunits, mediating rapid action propagation from cell to cell across the myocardium. Their loss at intercalated discs may contribute to arrhythmogenesis in acquired and inherited diseases.[20]

T-Tubules

Another cardiac sarcolemma domain, which can be distinguished from the intercalated discs, is T-tubule invaginations. Continuously extended from surface sarcolemma, T-tubules are lipid bilayers embedded with transmembrane or lipid-associated proteins.[21] Cardiac T-tubules occur at regular intervals along the lateral sides of the cell, closely coincident with the sarcomeric Z-discs.

The physiological function of cardiac T-tubules depends on the proteins that are localized at and within the vicinity of the T-tubules, including transmembrane ion channels and ion-handling proteins. Specific membrane scaffolding proteins and cytoskeletal structural proteins are required to localize to T-tubules for the organization and regulation of T-tubule network and structure. By differentially compartmentalizing proteins involved in ion handling and signaling, T-tubules serve as a signaling hub-like organelle to regulate myocyte function. The expression of transmembrane ion channels, ion transporters, and pumps has been well characterized in cardiac T-tubules.[22]

The calcium-handling proteins that are important in cardiac excitation–contraction coupling, in particular the voltage-gated LTCCs, are mostly enriched in T-tubules. Enrichment of the LTCCs (with pore forming subunit Cav1.2) helps bring these channels in close proximity (~15 nm) to intracellular SR-based calcium sensing and releasing channel ryanodine receptors (RyR) for efficient calcium-induced calcium release (CICR) during each heartbeat. Upon membrane depolarization, initial calcium influx occurs through Cav1.2 channels and the close association between Cav1.2 and RyR permits efficient CICR and subsequent sarcomeric contraction.[23]

Membrane Microdomains

The sarcolemma is a lipid bilayer that is not uniform but rather separated into distinct membrane microdomains. Caveolae "little caves" exist in both T-tubule and lateral sarcolemma of ventricular cardiomyocytes. A caveolae is a flask-shaped structure enriched with cholesterol and sphingolipids formed by the cholesterol-binding scaffolding protein Caveolin-3 (Cav-3). Biochemical fractionation and electron microscopy studies have identified a subpopulation of many ion channels at caveolae, and loss of caveolae is associated with arrhythmogenesis.[24] The precise role of caveolae on ion channel regulation and its significance still awaits further investigation. It has been reported that a subset of Cav1.2 channels is localized within caveolae, regulating calcium signaling.[24] The mechanisms effecting ion channel enrichment at caveolae are unknown, but close interactions between caveolae and the cytoskeleton present an appealing possibility of targeted ion channel delivery to these sarcolemmal microdomains.[25]

We identified a new subdomain within cardiac T-tubules, the T-tubule minifolds sculpted by a membrane curvature formation protein BIN1.[6] BIN1-induced T-tubule folds regulate extracellular calcium diffusion, affecting ion channel activity at the T-tubule surface. In addition to folds formation, BIN1 protein also facilitates microtubule-dependent targeting of Cav1.2 channels to T-tubules.[26] Thus, by regulating Cav1.2 channels, BIN1 formed T-tubule folds may have an important role in excitation–contraction coupling.

TRAFFICKING IN HEALTHY AND DISEASED HEARTS

In the complex internal organization of cardiomyocytes, thousands of individual proteins contribute to a functional equilibrium. In inherited genetic diseases, mutations in a single protein disrupt the equilibrium that, over time, can manifest later in life as a generalized myopathy. Numerous inherited channelopathies are caused by mutations negatively affecting trafficking. In fact, trafficking mutations may be more common than mutations that affect channel or gating. For instance, Anderson et al. have found that of 28 clinically relevant mutations in Kv11.1, most reduce hERG current not by altering Kv11.1 expression or kinetics, but by diminishing Kv11.1 trafficking to the membrane.[27] Mohler et al. have identified that mutations in $Na_v1.5$ that limit binding of $Na_v1.5$ to a membrane anchor protein ankyrin-G cause aberrant $Na_v1.5$ trafficking and result in human Brugada syndrome.[28] Patel and colleagues have found that mutations in desmosomal desmoplakin affect Cx43 trafficking to intercalated discs, contributing to arrhythmogenesis in arrhythmogenic cardiomyopathy.[29,30] Much in the way trafficking mutations can take years to manifest; for acquired muscle disease such as ischemic heart disease, cellular remodeling begins to occur after an acute insult, but then can take weeks to months or even years to develop into a myopathy and heart failure syndrome. To understand the cellular basis of acquired myopathy, it is therefore useful to separate the initial insult from the longer-term changes that the insult induces. We believe much of the long-term changes occur in the cytoskeleton-based delivery apparatus, which then affects trafficking itself.

A remarkable aspect of cardiac ion channel biology is that individual ion channels have half-lives on the order of hours. Connexin 43 (Cx43) gap junction proteins, which form the gap junctions that occur at intercalated discs of ventricular cardiomyocytes and are responsible for cell–cell communication, have a half-life of 1–3 h.[31,32] For high-resolution visualization of Cx43 trafficking, interested readers are referred to the supplemental videos in references.[33–35] Similarly, potassium and calcium channels and sodium-calcium exchanger have half-lives that are reported in the 2–8 h range.[36–39] Channels such as the sodium channel may have a relatively more extended half-life of about 35 h[40;] however, these times likely vary based on cell type and membrane partners. It is not yet clear whether channel half-lives on the plasma membrane are different from overall protein half-life. It is already being discovered that different subpopulations of channels exist within the cardiomyocyte.[41] Half-life is likely not just a function of intrinsic protein susceptibility to degradation molecules but is heavily influenced by location of the channel in a particular membrane subdomain or region in the cell. Proteins that are spatially protected from being in contact with degradation processes will experience slower turnover. Despite all the variability between individual channels and channel locations, however, the overall turnover of ion channels is still on the scale of several to tens of hours. Thus, in the course of any day, a large proportion of ion channels in each cardiomyocyte is regenerated and has to be redelivered to their proper cellular location. To maintain such dynamic turnover, the trafficking of channels has to be a constantly ongoing process. The short life span of ion channels suggests there needs to be efficiency in their life cycle and movements which follow the order of: formation, delivery to the correct subdomain on plasma membrane, behavior once in membrane, and internalization back into the cell.

We have divided the sections on trafficking into three main types which are forward (antegrade) trafficking to the membrane, channel behavior once in the membrane, and reverse (retrograde) trafficking from the membrane. Each subsection is divided by healthy physiology and what is known of the diseased response.

Ion Channel Forward Trafficking

Newly translated cardiac ion channel proteins are folded and usually oligomerized in the ER and further modified in Golgi apparatus. Mature proteins are then sorted in TGN, inserted into membrane vesicles, and loaded onto microtubule highways for delivery to the plasma membrane.[34,42–46] A critical aspect of forward transport is localization to membrane subdomains. It is possible that channels may be delivered to random regions of the plasma membrane, and only then laterally diffuse to reach their functional subdomains.[43] However, the temporal and stochastic inefficiency of random channel insertion, together with subsequent lateral diffusion to subdomains, suggests that specificity of delivery from the Golgi to the surface domains also occurs. We have found that the Cx43 gap junctions are targeted from the cytoplasm to cell–cell border regions by a combination of the plus-end-tracking proteins EB1 and the dynactin subunit p150[Glued], the channel itself, and the adherens junction complex at the membrane. In this case, the intercalated disc subdomain membrane anchor adherens junction captures the EB1-tipped microtubules, allowing for cargo protein Cx43 hemichannels to be offloaded onto adherens junction–containing membrane[12,33,34] (Fig. 2.1). This targeted delivery paradigm of ion channel forward trafficking has been generalizable to other cardiac ion channels and explored in terms of other cytoskeletal elements and anchor proteins.[47] Discussed below includes the full pathway of ion channel forward trafficking starting from endoplasmic reticulum packaging to final membrane insertion at sarcolemma subdomains via directed cytoskeleton. The model of targeted delivery is then explored with particular emphasis on Cx43 and LTCC delivery as well as the key components in the trafficking machinery.

Forward Trafficking: Normal Physiology

Maturation and Exiting Endoplasmic Reticulum

After translation and posttranslational modification, newly synthesized plasma membrane proteins, including ion channels, must pass through the ER-Golgi apparatus and be packaged into membrane vesicles to be loaded onto transportation highways before their final destination. Ion channel messages (mRNA transcripts) are translated to proteins by ribosomes. Amino acid signal sequences within the nascent forming polypeptide localize the ribosome complex to the rough ER, where translocation of the newly forming ion channel to a lipid bilayer can occur. Most of the protein maturation processes including folding and posttranslational modifications such as phosphorylation and core-glycosylation take place in the ER apparatus. Once completed, ion channel proteins are transported to the Golgi apparatus, from where they are further modified and sorted for delivery.

Individual ion channel pore-forming (α) subunit proteins often must associate (oligomerize) with other (β) channel subunits to traffic and function correctly. Pore-forming K^+ channel α-subunits are typically comprised of four polypeptides while the voltage-gated Ca^{2+} and Na^+ channels are translated as a single large full-length protein. β-subunit association predominantly occurs within the ER and often aids the α-subunits to exit the ER.[48,49] A common mechanism of ER release is through masking of ER-retention motifs within the ion channel polypeptide following binding with another subunit protein, as is the case with Cav1,[49] or cellular chaperones such as 14-3-3 proteins, as is the case with some K^+ channels1.[50] Six connexin proteins oligomerize to form hemichannels in an event which, for different connexin subtypes, occurs at different points in the trafficking pathway.[19]

Regulation of channel formation can also be achieved through post-translational modifications. For example, several protein–protein interaction motifs (including 14-3-3 binding domains[50]) are phosphorylation dependent. Substantial evidence exists for regulation of anterograde vesicular traffic by protein kinase A (PKA). Indeed, increases in PKA activity have been associated with enhanced gap junction coupling.[19] Such effects may not be direct and most likely involve cooperation from other signaling cascades within the cardiomyocyte.

Importantly, the ER-Golgi checkpoint regulates both exit of ion channels from the ER and channel degradation, as subunits failing to traffic (eg, due to misfolding) may be redirected to an ER-associated degradation pathway.[19] As such, following transcription and translation, transport of de novo ion channel proteins from the ER to the Golgi apparatus represents a major rate-limiting step in their trafficking to the sarcolemma. Once exit from the ER is approved, ion channels are packaged into COPII-coated vesicles and transported along microtubules to the ER-Golgi intermediate compartment (ERGIC) by motor proteins.[51] Further sorting occurs within the ERGIC from where vesicular traffic continues on to the *cis*-face of the Golgi apparatus. The Golgi apparatus is comprised of several discs, or cisternae, of which the *cis*-face is most internal and close to the ERGIC and the *trans*-face is most external and represents the point of vesicular exit.

Sorting in the *Trans*-Golgi Network

As ion channels progress through the cisternae of the Golgi apparatus, further posttranslational modifications including complex-glycosylation, phosphorylation, and cleavage can occur to generate a completed functional ion channel ready for delivery.[52] The *trans*-face of the apparatus represents the complex vesicular cargo sorting compartment known as the *trans*-Golgi network (TGN). Here, clathrin-coated vesicles enriched with membrane proteins form and are loaded onto microtubules and trafficked to specific cellular destinations.[53] The mechanisms governing membrane protein sorting in the Golgi are better understood in epithelial cells (eg, apical versus basolateral membrane targets) than in cardiomyocytes.[54]

Different connexin isoforms can combine to form heteromeric connexon hemichannels and such isoform composition is understood to influence gap junction conductance properties. An added level of complexity is the possibility of heterotypic gap junctions, whereby isoform-distinct connexons on opposing cells may bind. Interestingly, it is at the TGN that the major ventricular gap junction forming protein Cx43 oligomerizes into hemichannels. This is relatively late for such an event to occur, as other connexins oligomerize in the ER, but it may represent a means of controlling heteromeric hemichannel formation with other connexin isoforms.[19] Upon exiting the TGN, vesicles containing de novo ion channels must navigate the complex cardiomyocyte intracellular environment, a feat they achieve by trafficking along dynamic microtubules.

Vesicular Traffic on the Cytoskeleton Highway

Microtubules act as "highways" within the cardiomyocyte. They emanate from the centrosome and provide directionality and specificity to particular compartments within the cell. Vesicles are transported on microtubules by energy-dependent motor proteins such as kinesin which travel toward microtubule plus-ends.[55] Microtubule plus-ends exist in a state of dynamic instability, whereby tubulin rapidly polymerizes and depolymerizes under the regulation of plus-end binding proteins, altering microtubule velocity[56] and binding partners.[34] Microtubules can therefore respond to stresses or stimuli and regulate delivery of their vesicular cargo to cellular compartments. Actin is also understood to regulate vesicular transport at early (ER-Golgi) and later (TGN-sarcolemma) stages and can transport vesicles in a directional manner via myosin motor proteins.[57,58] Interestingly, vesicles can switch to different microtubules at microtubule–microtubule intersections, or from microtubule to actin filament and vice versa.[58] The majority of cardiac ion channels have now been shown to traffic in a microtubule and/or actin-dependent manner. It is likely that a combination of both actin and microtubule-based vesicular trafficking are responsible for trafficking de novo ion channels to the sarcolemma by a process now increasingly understood to occur in a targeted fashion.

Insertion of Ion Channels Into the Sarcolemma

Vesicular fusion with sarcolemma surface is a final regulatory step in ion channel forward trafficking. Cortical actin is known to negatively regulate exocytic vesicle fusion, and its reorganization is likely to influence distal ion channel trafficking.[59] At the cell cortex, unloading of vesicles onto F-actin may occur which would introduce another level of potential targeting specificity. To achieve delivery of its ion channel cargo, a vesicle must fuse with the plasma membrane. The specifics of these mechanisms in cardiomyocytes, and how they may differ for individual ion channels, are poorly understood. In a recent study it was found that nonsarcomeric actin in cardiomyocytes *facilitates* forward delivery and insertion of Cx43 in the plasma membrane,[12] as discussed below.

Introduction to Targeted Delivery

The mechanisms by which microtubules exert their specificity in interacting with membrane subdomains are now being elucidated. Our original report in 2007[34] and subsequent studies[26,33,60] have led to the Targeted Delivery model of channel delivery. Targeted Delivery is the understanding that channels, once formed and exiting the Golgi, can be rapidly directed across the cytoplasm to their respective specific membrane subdomains. The highways for transport are microtubules whose negative ends originate at Golgi-oriented microtubule-organizing centers and whose positive ends are growing outward and can be captured at the plasma membrane by membrane anchor proteins and complexes. Specificity of delivery is a combination of the individual channel, the plus-end-tracking proteins at the positive ends of microtubules which guide microtubule growth

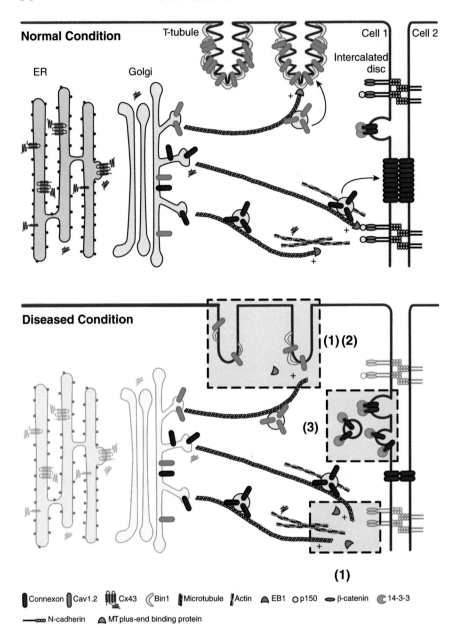

FIGURE 2.2 Channel trafficking in healthy and failing hearts. Ion channel proteins are synthesized by ribosomes. In the case of connexin 43 (Cx43), multiple isoforms are produced as a result of alternative translation. Nascent transmembrane proteins are translocated to the membrane of the rough endoplasmic reticulum, transported through the Golgi apparatus and to the *trans*-Golgi network. Channel proteins are sorted into vesicular carriers and docked onto microtubules at the TGN and subsequently delivered to their subcellular destinations in cooperation with actin "way stations" along the route. Microtubule plus-end

and capture and the membrane bound anchor complex which captures the microtubule, completing the highway for channel delivery (Fig. 2.2). Actin cytoskeleton serves as en route rest-stop stations to redirect microtubules,[12] providing additional sorting site. Targeted Delivery is explained below with two different types of channels: Cx43 gap junction hemichannels to intercalated discs and Cav1.2 channels trafficking to T-tubules, followed by an exploration of the critical components in the machinery.

Targeted Delivery of Cx43 and Cav1.2 Channels

The first example of targeted delivery of ion channels to cardiomyocyte surface subdomains is the delivery of Cx43 connexons to the intercalated discs. A major plus-end binding protein, EB1, is known to be necessary for targeted delivery of connexons to adherens junction complexes.[34] Through interaction with another plus-end protein, p150[Glued], the EB1-tipped microtubule complexes specifically with β-catenin molecules at the fascia adherens of intercalated discs. Vesicular cargo is unloaded and subsequently inserted into the plasma membrane at nearby gap junctions. Other reports propose a less specific paradigm of connexin delivery, whereby connexons are inserted indiscriminately into the lateral membrane of the cell and freely diffuse to gap junction structures.[61] Both models can exist in parallel. However, the inefficiency of lateral diffusion to a few specific subdomains, the short half-life of connexins, and the complex interactions between a single cell with multiple neighboring cardiomyocytes, all suggest directed targeting can be a more effective form of connexon localization to the intercalated disc.

Enrichment of Cav1.2 channels in the T-tubules is essential for the efficient contractile function of the myocardium. We found that trafficking of Cav1.2 vesicles from the TGN to T-tubules also occurs in a

◄ binding proteins interact with anchor proteins of specific membrane subdomains, allowing targeted delivery of cargo proteins. The interaction between EB1, a microtubule plus-end binding protein, and the adherens junction complex ensures the targeted delivery of Cx43 hemichannels to the intercalated discs, whereas the association of microtubules and bridging integrator 1 (BIN1), a membrane scaffolding protein, warrants the delivery of Cav1.2, a voltage-gated L-type calcium channel protein, to T-tubules. Channel proteins on the plasma membrane undergo internalization for degradation. In failing cardiomyocytes, expression of ion channels on the cell surface and the morphology of T-tubules are altered. As highlighted in light yellow, possible mechanisms underlying these changes include: (1) dissociation of microtubule plus-end binding proteins from microtubules; (2) reduced expression of membrane scaffolding proteins; and (3) increased internalization. Under oxidative stress, EB1 dissociates from the tip of microtubules, impairing the attachment of microtubules to the adherens junction and the delivery of Cx43 to the intercalated discs. During acute cardiac ischemia 14-3-3 mediated internalization of Cx43 is increased, diminishing the amount of Cx43 channels on the plasma membrane. In failing hearts the expression of BIN1 is significantly reduced, resulting in detachment of microtubules from sarcolemma and reduction in Cav1.2 delivery to the T-tubules. The dense membrane folds in T-tubules are also lost as a result of low expression of BIN1.

microtubule-dependent manner.[26] Moreover, just as with Cx43 connexons, dynamic microtubules preferentially interact with a specific membrane anchor BIN1 to effect targeted delivery of Cav1.2. BIN1 contains a membrane curvature BAR-domain (which confers the ability to form membrane curvature), a coiled-coil domain, and an SH3 protein–protein interaction domain. Perhaps most compelling for BIN1 utilizing the cytoskeleton is the finding that deletion of the coiled-coil and SH3 domains does not affect membrane invagination, but abrogates Cav1.2 colocalization with these structures. Therefore, it is through interaction specifically with the BIN1 protein, and not T-tubule structures, that targeting of Cav1.2 delivery is achieved.

The Targeted Delivery model has been supported by multiple subsequent reports.[26,62–65] At the same time, important details of this delivery paradigm continue to be elucidated. For instance, Patel et al. recently reported that the desmosome-associated linker protein desmoplakin binds to EB1 and is involved in targeted delivery of Cx43 to intercalated discs.[29] Also, nonsarcomeric actin appears to be involved in channel delivery. Actin is necessary for Cx43 forward delivery.[60,66] It remains to be determined how actin interacts with channels and the microtubule apparatus. At any given point in time, the majority of intracellular Cx43 channels are not moving rapidly on microtubules, but rather are stationary and associated with nonsarcomeric actin.[60]

Membrane Anchors in Targeted Delivery

In Targeted Delivery, specificity of delivery occurs near the membrane, in which a key aspect is the differential membrane bound anchor proteins at distinct membrane subregions. Membrane anchors are critical in capturing with specificity a subgroup of microtubules allowing channel delivery directly to regions of membrane that happen to contain the particular anchor. For Cx43 delivery to the intercalated disc, EB1-tipped microtubules bind to N-Cadherin associated β-catenin and also p150[Glued].[34] Desmoplakin may also be involved in capturing the EB1-tipped microtubule for Cx43 delivery,[29] although the transmembrane domain still appears to be N-Cadherin rather than desmosomal desmoglein.[30] For Cav1.2 channel delivery to T-tubules, a membrane anchor is the lipophilic membrane scaffolding protein BIN1[26] (Fig. 2.2). A subpopulation of Cav1.2 channels, on the other hand, can be delivered to caveolae through interaction between subunits of LTCC channel complex and the caveolae structural protein caveolin-3.[67]

Several other membrane scaffolding proteins have also been uncovered as critical regulators of targeting and maintenance of ion channels at cardiomyocyte sarcolemmal subdomains. For instance, ankyrin-G binds to and regulates $Na_v1.5$ localization[68,69] and ankyrin-B regulates the membrane targeting and subsequent regulation of the Na^+/Ca^{2+} exchanger

(NCX), Na^+/K^+ ATPase (NKA), and IP3 receptor (InsP3).[70] Ankyrin-G has also been shown to target $Na_v1.2$ and $Na_v1.6$ to specialized excitable membrane domains in Purkinje and granule cell neurons. In the heart, ankyrin-G localizes $Na_v1.5$ to the intercalated disc, a model supported by a human Brugada Syndrome *SCN5A* mutation abolishing the $Na_v1.5$ ankyrin-binding motif (for brevity, Mohler lab publications contained in[20]). An additional ankyrin isoform found in ventricular cardiomyocytes, ankyrin-B, is associated with targeting and maintenance of the NCX and NKA channels at T-tubules where they proximate with the InsP3 receptor of the SR and regulate Ca^{2+} export. Mutations in ankyrin-B ablating its interaction with NCX/NKA/InsP3 result in arrhythmogenic cardiac disorders in humans, including type-4 long-QT syndrome.[71] Ankyrin-B is also expressed in the pacemaking cells of the sinoatrial node. In addition to NCX/NKA/InsP3, Cav1.3 targeting is perturbed by loss-of-function mutations and thus extracellular calcium entry deregulated, possibly contributing to sick sinus syndrome.[20] In addition, fibroblast growth factor homologous factors are also potent regulators of $Na_v1.5$ and Cav1.2 localization to the sarcolemmal membrane[72,73] by interacting with C-terminal domains of ion channels.

Cytoskeleton in Targeted Delivery

Microtubules are the trafficking highways in Targeted Delivery for both Cx43 connexons and LTCCs. The important role of dynamic microtubules in delivery of $Na_v1.5$ to the sarcolemma was also recently highlighted in a study that found deleterious effects on $Na_v1.5$ function following microtubule stabilization with Taxol.[74] The specificity of Targeted Delivery is contributed by the +TIP proteins at the plus-ends of growing microtubules. For example, EB1 works in concert with p150[Glued] to target Cx43 channels to adherens junctions at intercalated discs. Further elucidation of this microtubule-based trafficking pathway could uncover mechanistic insight into differential targeting of ion channels to the intercalated discs. The other +TIP protein ClIP170 has been reported to interact with BIN1,[75] possibly facilitating BIN1 directed delivery of LTCCs to T-tubules. The +TIP proteins involved in delivery of sodium channels are still under investigation.

Other than microtubules, there is increasing appreciation of the involvement of nonsarcomeric actin cytoskeleton in Targeted Delivery. The fundamental question remains with regard to why actin is involved in channel trafficking. If vesicles containing channels can depart the Golgi and ride a microtubule highway straight to its proper subdomain, is there a need for actin filaments that appear to slow down vesicle transport? Actin can have at least two important roles in forward delivery. The first is to help contribute specificity to delivery. Vesicles transported along microtubules on kinesin motors move rapidly, at a rate of about $1\,\mu m/s$.[34] Thus delivery to most locations at a cell membrane can occur within a minute. Association with important accessory proteins and posttranslational modification of channels, both of which can affect delivery destination, probably also happen en route between the Golgi and membrane. Hopping off the microtubule highway on an actin "way station," which is analogous to a highway rest stop with convenience stores, could allow the channel and vesicle containing it to pick up accessory proteins and allow for needed posttranslational modification. Such rest stops would occur at Z-disc, subcortical locations, or other important cytoskeleton intersections in the cytoplasm. The rest stops could also serve as a channel reservoir. For instance the cardiac Z-disc and costameres are actin-rich structures. If LTCCs are needed acutely in T-tubule membrane, there could be a signal for rapid delivery to occur from the Z-disc and costamere reservoir to the nearby T-tubule membrane. Finally, an actin rest stop could allow the channel containing vesicles to use multiple microtubule highways in their delivery path. The Golgi exiting microtubule could be destined for an actin rest stop, allowing for a different membrane domain specific microtubule to finish the delivery. The second potential role for actin in microtubule-based forward delivery pertains to the microtubules themselves. In nonmyocyte systems, actin can help stabilize and guide microtubules.[76,77] Actin could be the blueprint along and across which microtubule highways are patterned. In this respect, actin involvement could be an upstream to microtubules in determining location of channel delivery.

Accessory Proteins in Targeted Delivery

With regard to accessory proteins, Cx43 hemichannels are notable for, despite extensive examination, not being associated with their own unique β-subunits that assist in their trafficking. Recently, it was identified that these hemichannels do have important subunits that are N-terminal truncated isoforms of the same full length protein.[78] The smaller proteins are created through the phenomenon of alternative translation.[78] To understand alternative translation, it should be recognized that traditional translation of mRNA begins with the first coding triplet, which is always an AUG (methionine). Most transcribed genes (mRNA strands) have other AUG sites downstream of the first one. The Cx43 gene *GJA1* has six in frame AUG triplets beyond the first one. Alternative translation occurs when ribosomal translation initiates not at the first triplet, but at a downstream triplet. By initiating translation at downstream sites, alternative translation creates N-terminal truncated proteins that lack the respective nontranslated upstream (N-terminal) portions of the proteins. In the case of Cx43, *GJA1* mRNA produces the expected full-length 43 kDa protein as well as proteins that are approximately 32, 29, 26, 20, 11, and 7 kDa in size.[78] Cx43 is the first mammalian ion channel shown to be subjected to alternative translation, which has already been confirmed by two separate reports.[79,80] It has also been found that the 20 kDa isoform (GJA1-20K) assists with Cx43 forward trafficking. We currently understand the smaller Cx43 isoforms to effectively be the "β-subunits" that promote and thus autoregulate full length Cx43 trafficking to the plasma membrane.[78]

In the case of LTCC, accessory β-subunits exist with four isoforms of β1–β4 and their expression at myocardium is different across species. In the mouse hearts, only β2 subunit (with five splice variants β2a-2e)[81] has been detected, whereas all the four isoforms have been detected in canine myocardium.[82] By masking the ER retention signal at the intracellular I-II loop of Cav1.2 protein, β-subunits are critical in facilitating the ER exiting of Cav1.2 channel. However, the role of β-subunit in targeted delivery of LTCCs remains unclear. We speculate that β-subunit may be the subunit that directly binds to membrane anchor proteins to facilitate delivery of LTCCs to membrane subdomains, and the specificity can be determined by binding of BIN1 or caveolin-3 like membrane anchor proteins with different β-subunit isoforms and splice variants.

Forward Trafficking: Pathophysiology

Recent studies have explored forward trafficking of cardiac ion channels in different disease states. A difficulty with such studies is that the short life spans of individual channels (hours) are dwarfed by the chronicity (months, years, and decades) of most failing hearts. A further complication is that dynamic channel behavior is best studied with in vitro cellular studies, well removed from intact animals and humans. A simple knockout mouse model establishes the importance of a channel or trafficking partner, but does not reveal the signaling interplay that occurs with other proteins during protein movements.

Trafficking related channel studies need to occur with cardiomyocytes intact, yet in environmental conditions that mimic those of failing heart. One important such condition is oxidative stress that occurs in ischemia-reperfusion injury as well as ischemic and nonischemic heart failure. When isolated cardiomyocytes are subjected to oxidative stress, Cx43 gap junction delivery to intercalated discs is impaired due to disruption of the forward trafficking machinery.[33] Specifically, oxidative stress causes the microtubule plus-end protein EB1 to disassociate from the tips of microtubules, impairing microtubule attachment to adherens junction structures and subsequent delivery of Cx43 hemichannels to plasma membrane[33] (Fig. 2.2). Manipulation of EB1 as well as the upstream regulators of EB1 localization at microtubules could potentially preserve or enhance gap junction coupling during stress. As many ion channels rely on microtubules for their transport, it is likely that such disruption of microtubule trafficking machinery inhibits delivery of many essential channels to the sarcolemma.

Such studies provide evidence that forward trafficking of Cx43 is impaired in acquired heart failure. At present we do not know how oxidative stress causes EB1 displacement and disassembly of the forward trafficking apparatus. We have preliminary investigations on the role of actin in maintaining EB1-based microtubule integrity and the response of these cytoskeletal fibers to stress conditions. This remains an active area of investigation.

In failing heart, forward trafficking of Cav1.2 channels to T-tubules is also impaired.[60] Biochemical assessment of Cav1.2 channel content in failing heart indicates no difference in total channel content compared to healthy muscle, yet channel localization to T-tubules is impaired.[60] A difference between impaired forward delivery of Cx43 channels and Cav1.2 channels in failing hearts exists with their respective anchor proteins. Even in diseased heart muscle, the adherens junction structures for Cx43 delivery to intercalated discs remain intact,[33] whereas transcription of BIN1 protein, needed to anchor microtubules for Cav1.2 delivery to T-tubules, is reduced by half.[60] In animal models, successful treatment of heart failure and recovery of function correlates with recovery of muscle BIN1 levels.[83,84]

In the preceding section, we introduced recent reports that the actin cytoskeleton could be an important regulator of forward trafficking that is potentially upstream of microtubule based trafficking.[60,66] In acute ischemic injury, Cx43 dissociates from actin and forward trafficking is impaired.[60] Despite these data, it remains to be determined whether actin filaments can serve as Cx43 sorting centers and microtubule organizers, or whether actin can actively bring Cx43 and other channels to the membrane surface. In another recent study, it was identified that the actin motor myosin 5B assists with forward trafficking of Kv1.5 and Cx43,[85] suggesting actin could do more than sorting proteins and organizing microtubules in the delivery of channels to the membrane surface.

Channels Regulation Once in Membrane Subdomains

Once channels are inserted into the plasma membrane, it is possible that they undergo lateral diffusion to other regions of membrane. However, multiplexing with scaffolding proteins and cytoskeleton elements will limit subsequent diffusion, and the overall extent and rate of lateral diffusion are unclear, with reports that vary significantly. In this section, we provide an overview of the current understanding of ion channel behavior once they are placed at the membrane subdomains including intercalated discs and T-tubules. Diseased alteration of ion channels at the membrane is also discussed.

Membrane Subdomains: Normal Physiology

Intercalated Discs

Intercalated discs are located at the longitudinal ends of cardiomyocyte to form cell–cell coupling critical in propagation of action potentials. Both Cx43 gap junctions and voltage-gated sodium channels are present at intercalated discs. For Cx43 hemichannels, it was previously understood that the channels are randomly placed in plasma membrane to subsequently diffuse laterally to plaque regions.[34] Even with Target Delivery for Cx43 insertion into intercalated discs, it is still likely that most channels move laterally within the plasma membrane, but in confined local zones rather than widely traversing regions of the cell surface. In 2011, Rhett et al. identified and named the "perinexus" as a region adjoining gap junction plaques in which hemichannels can collect and diffuse with movement into the plaque regulated by ZO-1.[86] Such local movements of Cx43 and other channels are highly likely and areas such as the perinexus could be transitional zones in which further posttranslational modification and clustering occur.

A recent development in cardiac membrane biology is the finding that T-tubule invaginations are not simply straight and planar, but instead contain complex folds, tight and narrow enough to limit the free flow of extracellular ions.[6] High-resolution imaging of intercalated disc regions reveal that intercalated disc associated membrane is also nonlinear, with finger-like and low-frequency undulations.[87,88] We speculate that membrane morphology such as location of critical curvature and inflection points may compartmentalize trafficking domains as well as support structures such as the perinexus, thereby affecting channel activity.

T-Tubules

T-tubule membrane is associated with a unique set of ion channels supported by membrane structural subdomains.[23] Immunocytochemistry and patch clamp data indicate that 80% of the sarcolemma LTCCs are concentrated at T-tubules. Localization of LTCCs to T-tubules is critical in bringing LTCCs to the vicinity of calcium sense and release channel RyR at SR membrane. Approximation of LTCCs and RyRs form a functional microdomain, the cardiac dyads, which are the individual calcium-releasing units in cardiomyocytes. As discussed earlier, Targeted Delivery also applies to Cav1.2 channel localization to T-tubule surface, which requires a membrane scaffolding protein BIN1.[26] Interestingly, BIN1 also forms minifolds within T-tubules, affecting extracellular ion diffusion thereby controlling the driving force of Cav1.2 channel activity.[6] BIN1-folded subdomains within T-tubules may also limit LTCC lateral diffusion once the channels are inserted into T-tubule membrane, to maintain functional LTCC-RyR dyads. Therefore, BIN1-like membrane scaffold proteins may help localize particular pools of ion channel proteins to membrane subdomains for compartmentalized regulation of ion channel activity and function. Similarly, ankyrin B can play a role similar to BIN1 in clustering NCX/NKA/InsP3 complex to microdomains distinct from the dyads.[71] Integrity of NCX/NKA/InsP3 complex is critical in removing calcium from the clefts between T-tubule surface and SR membrane, facilitating calcium decline during relaxation. Therefore, intracellular calcium equilibrium can be achieved by balancing calcium entry through LTCC clustered at BIN1-membrane and calcium removal by NCX anchored at ankyrin B-induced membrane.

Membrane Subdomains: Pathophysiology

In diseased hearts, gap junctions remodeling leads to decreased expression at intercalated discs and appearance at lateral borders. The mechanisms of pathologic gap junction redistribution to lateral sides of cardiomyocytes remain poorly understood. It is possible that intercalated disc membrane signals for targeted delivery of connexons may themselves be relocated to lateral membrane during disease, thus attracting hemichannel delivery.[65] The other possibility is that during disease, intercalated disc connexons can lose tethering to plaques and diffuse within the membrane to lateral regions.[89] The high rate of connexin 43 turnover, rapid rate of connexon delivery to plasma membrane, and lack of direct visualization of connexon movements once in the plasma membrane, together severely limit the ability for us to understand the mechanisms by which remodeled and poorly localized connexons arrive at their new destination. A real-time live-cell recording of surface connexon activity will be necessary to determine their fate once in the plasma membrane of health or stressed cardiomyocytes.

In failing hearts, LTCCs also have diminished forward trafficking resulting in intracellular accumulation of the channels.[60] There already exists significant evidence that gross T-tubule network remodeling occurs in failing heart.[90–92] It is an area of active research with regard to the mechanisms of T-tubule remodeling in failing hearts. Junctophilin-2 trafficked by microtubules has been implicated in impaired T-tubule maintenance during heart failure.[93] However, the role of junctophilin-2 in T-tubule remodeling during heart failure has been questioned due to a lack of decrease with heart failure as T-tubule structures are diminished[83,84] or return with recovery of T-tubule structures in treated heart failure.[83] In these same studies, BIN1 decreases with decrease in T-tubule density in heart failure,[83,84] and BIN1 recovers along with T-tubule density during functional recovery of myocardium.[83] During extended in vitro culture, isolated mature ventricular myocytes lose T-tubules in 3 days. Interestingly, actin stabilization by cytochalasin D can preserve T-tubules in cultured myocytes.[6,94,95] The cardiac isoform of BIN1 we described recently is found to be able to promote N-WASP dependent actin polymerization.[6] Exogenous BIN1 introduced by adenovirus not only rescues T-tubule membrane intensity[6] but also surface Cav1.2 channels[60] in isolated cardiomyocytes cultured in vitro. Taken together, these data support the concept that actin is important to T-tubule maintenance, and the mechanism can be mediated by cardiac BIN1.[6] It is worth noting that loss of BIN1 to levels that occur in end-stage heart failure[60] results in impaired actin associated T-tubule minifolds[6] and in BIN1-deificient mice T-tubule remodeling similar to that of failing hearts.[6]

Channel Internalization

Internalization: Normal Physiology

Since ion channel proteins are short lived and dynamically replaced, channel internalization from the plasma membrane also represents an important regulatory step in determining the level of gap junction coupling. Ion channel internalization can be induced by posttranslational modification including ubiquitination, acetylation, and phosphorylation. In the case of Cx43, phosphorylation is most well studied, and the importance of phosphorylation has been highlighted by recent findings that casein kinase-dependent phosphorylation alters gap junction remodeling and decreases arrhythmic susceptibility.[96] Many residues on the C-terminus of Cx43, specifically 22 serines, 5 tyrosines, and 4 threonines, are potentially subjected to phosphorylation. To make matters even more complex, Cx43 exists as a hexamer on the plasma membrane, and it is currently not known how phosphorylation differs between individual connexins of the same connexon. A cascade of phosphorylation occurs preceding ubiquitination of Cx43, which then leads to channel internalization.[97]

Endocytosis of Cx43 can occur either through internalization of uncoupled hemichannels or entire gap junctions, which requires engulfment of gap junctions from the opposing neighboring cell plasma membrane as well. The internalized double-membrane intracellular structures are known as nonfunctional annular gap junctions. Both the lysosome and the proteasome have been implicated in degradation of Cx43 and interestingly, autophagy is now known to be involved in degradation of annular gap junctions in failing hearts.[98] Studies have shown that recycling of gap junctions occur during cell cycle progression in cell lines,[99] but whether gap junctions are recycled in cardiomyocytes remains a controversial issue. It is exciting to consider the possibility that posttranslational modifications of Cx43 may be acting as checkpoints within the same connexin molecule, or connexon hemichannel, requiring a specific series of events to permit ubiquitination and internalization of a hemichannel, or annular gap junction.

General internalization of LTCCs is poorly understood with particular lack of studies in cardiomyocytes. In oocytes, the LTCC β-subunit can enhance dynamin-dependent internalization,[100] and in neurons Cav1.2 channels may undergo depolarization and calcium dependent internalization.[101] We found in cardiomyocytes that a dynamin GTPase inhibitor dynasore can increase surface LTCC expression, indicating dynamin-dependent endocytosis of cardiac Cav1.2 channels.[26]

Furthermore, a small GTPase Rab11 is implicated in endosomal transport of LTCCs, thereby limiting surface expression of LTCCs.[102]

Internalization: Pathophysiology

Our experience with Cx43 protein is that posttranslational modification preferentially affects ion channel internalization. Pathological gap junction remodeling is strongly associated with altered phosphorylation of Cx43.[19,103,104] Rather than individual independent phosphorylation events of singular residues at the C-terminus, it is likely that internalization results from a sophisticated cascade of posttranslational modifications. The Cx43 C-terminus contains a phosphorylation-dependent 14-3-3 binding motif at Serine 373 (within 10 amino acids of the end of the protein). 14-3-3 proteins are known to regulate protein transport and have been implicated in facilitating de novo Cx43 transport from ER to Golgi apparatus.[105,106] Phosphorylation of Ser373 and subsequent 14-3-3 binding provide a gateway to a signaling cascade of downstream phosphorylation of Ser368, leading to gap junction ubiquitination, internalization, and degradation during acute cardiac ischemia.[32]

The C-terminus of Cx43 is the main protein–protein interaction domain responsible for Cx43 binding to its partners within the cell.[107] In close proximity to the Cx43 14-3-3 binding motif is a PDZ domain at the distal end of the C-terminus. It is through this PDZ domain that Cx43 interacts with ZO-1,[108] and this interaction has been demonstrated to regulate Cx43 gap junction plaque size and assembly.[86,109] Disruption of Cx43/ZO-1 complexing has been reported to increase gap junction plaque size in cultured cells.[110,111] Phosphorylation of Cx43 Serine 373 can disrupt interaction with ZO-1,[112] and indeed it would be sterically unlikely for both 14-3-3 and ZO-1 to bind the same Cx43 protomer simultaneously. However, increased Cx43/ZO-1 interaction has also been associated with gap junction remodeling, highlighting the complex nature of these dynamic posttranslational and protein complexing events.[89,113]

CONCLUSIONS

The individual cardiomyocyte is a highly complex and dynamic system with internal organization designed to maintain efficient cell–cell communication and excitation–contraction coupling. To maintain intracellular homeostasis as well as overall synchrony across the myocardium, cardiomyocytes regulate ion channel intracellular movement and localization through highly sophisticated and highly efficient protein trafficking machineries. In diseased hearts, cardiomyocyte structures and organization are negatively affected by environmental conditions of stress, impacting channel trafficking and function.

Existing therapies focus on blocking these external signals. New therapies for failing heart will be focused on the specific organelles and pathways that regulate cardiomyocyte protein trafficking.

Acknowledgments

This work was supported by NIH/NHLBI R00 HL109075 (TTH), NIH/NHLBI R01 HL094414, and AHA EIA 13EIA4480016 (RMS).

References

1. Hwang SJ, Kim W. Mitochondrial dynamics in the heart as a novel therapeutic target for cardioprotection. *Chonnam Med J* 2013;**49**(3):101–7.
2. Beikoghli Kalkhoran S, et al. 4 Characterisation of mitochondrial morphology in the adult rodent heart. *Heart* 2014;**100**(Suppl. 1):A2–3.
3. Gustafsson AB, Gottlieb RA. Heart mitochondria: gates of life and death. *Cardiovasc Res* 2008;**77**(2):334–43.
4. Eisner V, Csordas G, Hajnoczky G. Interactions between sarco-endoplasmic reticulum and mitochondria in cardiac and skeletal muscle – pivotal roles in Ca(2)(+) and reactive oxygen species signaling. *J Cell Sci* 2013;**126**(Pt 14):2965–78.
5. Boateng SY, Goldspink PH. Assembly and maintenance of the sarcomere night and day. *Cardiovasc Res* 2008;**77**(4):667–75.
6. Hong T, et al. Cardiac BIN1 folds T-tubule membrane, controlling ion flux and limiting arrhythmia. *Nat Med* 2014;**20**(6):624–32.
7. Noorman M, et al. Cardiac cell-cell junctions in health and disease: electrical versus mechanical coupling. *J Mol Cell Cardiol* 2009;**47**(1):23–31.
8. Itoh T, et al. Dynamin and the actin cytoskeleton cooperatively regulate plasma membrane invagination by BAR and F-BAR proteins. *Dev Cell* 2005;**9**(6):791–804.
9. de Forges H, Bouissou A, Perez F. Interplay between microtubule dynamics and intracellular organization. *Int J Biochem Cell Biol* 2012;**44**(2):266–74.
10. Gu F, Crump CM, Thomas G. Trans-Golgi network sorting. *Cell Mol Life Sci* 2001;**58**(8):1067–84.
11. Luini A, et al. Morphogenesis of post-Golgi transport carriers. *Histochem Cell Biol* 2008;**129**(2):153–61.
12. Smyth JW, et al. Actin cytoskeleton rest stops regulate anterograde traffic of connexin 43 vesicles to the plasma membrane. *Circ Res* 2012;**110**(7):978–89.
13. Lau YH, et al. Lipid analysis and freeze-fracture studies on isolated transverse tubules and sarcoplasmic reticulum subfractions of skeletal muscle. *J Biol Chem* 1979;**254**(2):540–6.
14. Rosemblatt M, et al. Immunological and biochemical properties of transverse tubule membranes isolated from rabbit skeletal muscle. *J Biol Chem* 1981;**256**(15):8140–8.
15. Anderson RG. The caveolae membrane system. *Annu Rev Biochem* 1998;**67**:199–225.
16. Song KS, et al. Expression of caveolin-3 in skeletal, cardiac, and smooth muscle cells. Caveolin-3 is a component of the sarcolemma and co-fractionates with dystrophin and dystrophin-associated glycoproteins. *J Biol Chem* 1996;**271**(25):15160–5.
17. Saffitz JE, Hames KY, Kanno S. Remodeling of gap junctions in ischemic and nonischemic forms of heart disease. *J Membr Biol* 2007;**218**(1–3):65–71.
18. Saffitz JE. Arrhythmogenic cardiomyopathy and abnormalities of cell-to-cell coupling. *Heart Rhythm* 2009;**6**(8 Suppl.):S62–5.
19. Hesketh GG, Van Eyk JE, Tomaselli GF. Mechanisms of gap junction traffic in health and disease. *J Cardiovasc Pharmacol* 2009;**54**(4):263–72.

20. Hashemi SM, Hund TJ, Mohler PJ. Cardiac ankyrins in health and disease. *J Mol Cell Cardiol* 2009;**47**(2):203–9.
21. Lindner E. [Submicroscopic morphology of the cardiac muscle]. *Z Zellforsch Mikrosk Anat* 1957;**45**(6):702–46.
22. Brette F, Orchard C. Resurgence of cardiac t-tubule research. *Physiology (Bethesda)* 2007;**22**:167–73.
23. Orchard C, Brette F. t-Tubules and sarcoplasmic reticulum function in cardiac ventricular myocytes. *Cardiovasc Res* 2008;**77**(2):237–44.
24. Balijepalli RC, Kamp TJ. Caveolae, ion channels and cardiac arrhythmias. *Prog Biophys Mol Biol* 2008;**98**(2–3):149–60.
25. Head BP, et al. Microtubules and actin microfilaments regulate lipid raft/caveolae localization of adenylyl cyclase signaling components. *J Biol Chem* 2006;**281**(36):26391–9.
26. Hong TT, et al. BIN1 localizes the L-type calcium channel to cardiac T-tubules. *PLoS Biol* 2010;**8**(2):e1000312.
27. Anderson CL, et al. Most LQT2 mutations reduce Kv11.1 (hERG) current by a class 2 (trafficking-deficient) mechanism. *Circulation* 2006;**113**(3):365–73.
28. Mohler PJ, et al. $Na_v1.5$ E1053K mutation causing Brugada syndrome blocks binding to ankyrin-G and expression of $Na_v1.5$ on the surface of cardiomyocytes. *Proc Natl Acad Sci USA* 2004;**101**(50):17533–8.
29. Patel DM, et al. Disease mutations in desmoplakin inhibit Cx43 membrane targeting mediated by desmoplakin-EB1 interactions. *J Cell Biol* 2014;**206**(6):779–97.
30. Shaw RM. Desmosomal hotspots, microtubule delivery, and cardiac arrhythmogenesis. *Dev Cell* 2014;**31**(2):139–40.
31. Beardslee MA, et al. Rapid turnover of connexin43 in the adult rat heart. *Circ Res* 1998;**83**(6):629–35.
32. Smyth JW, et al. A 14-3-3 mode-1 binding motif initiates gap junction internalization during acute cardiac ischemia. *Traffic* 2014;**15**(6):684–99.
33. Smyth JW, et al. Limited forward trafficking of connexin 43 reduces cell-cell coupling in stressed human and mouse myocardium. *J Clin Investig* 2010;**120**(1):266–79.
34. Shaw RM, et al. Microtubule plus-end-tracking proteins target gap junctions directly from the cell interior to adherens junctions. *Cell* 2007;**128**(3):547–60.
35. Smith S, et al. Defects in cytoskeletal signaling pathways, arrhythmia, and sudden cardiac death. *Front Physiol* 2012;**3**:122.
36. Colley BS, et al. Neurotrophin B receptor kinase increases Kv subfamily member 1.3 (Kv1.3) ion channel half-life and surface expression. *Neuroscience* 2007;**144**(2):531–46.
37. Chien AJ, et al. Roles of a membrane-localized beta subunit in the formation and targeting of functional L-type Ca^{2+} channels. *J Biol Chem* 1995;**270**(50):30036–44.
38. Di Biase V, et al. Surface traffic of dendritic CaV1.2 calcium channels in hippocampal neurons. *J Neurosci* 2011;**31**(38):13682–94.
39. Egger M, et al. Rapid turnover of the "functional" Na(+)-Ca(2+) exchanger in cardiac myocytes revealed by an antisense oligodeoxynucleotide approach. *Cell Calcium* 2005;**37**(3):233–43.
40. Maltsev VA, et al. Molecular identity of the late sodium current in adult dog cardiomyocytes identified by $Na_v1.5$ antisense inhibition. *Am J Physiol Heart Circ Physiol* 2008;**295**(2):H667–76.
41. Petitprez S, et al. SAP97 and dystrophin macromolecular complexes determine two pools of cardiac sodium channels $Na_v1.5$ in cardiomyocytes. *Circ Res* 2011;**108**(3):294–304.
42. Hamm-Alvarez SF, Sheetz MP. Microtubule-dependent vesicle transport: modulation of channel and transporter activity in liver and kidney. *Physiol Rev* 1998;**78**(4):1109–29.
43. Gaietta G, et al. Multicolor and electron microscopic imaging of connexin trafficking. *Science* 2002;**296**(5567):503–7.
44. Johnson RG, et al. Gap junctions assemble in the presence of cytoskeletal inhibitors, but enhanced assembly requires microtubules. *Exp Cell Res* 2002;**275**(1):67–80.

45. Lauf U, et al. Dynamic trafficking and delivery of connexons to the plasma membrane and accretion to gap junctions in living cells. *Proc Natl Acad Sci USA* 2002;**99**(16):10446–51.
46. Zadeh AD, et al. Kif5b is an essential forward trafficking motor for the Kv1.5 cardiac potassium channel. *J Physiol* 2009;**587**(Pt 19):4565–74.
47. Rhett JM, et al. Cx43 associates with Na(v)1.5 in the cardiomyocyte perinexus. *J Membr Biol* 2012;**245**(7):411–22.
48. Cusdin FS, Clare JJ, Jackson AP. Trafficking and cellular distribution of voltage-gated sodium channels. *Traffic* 2008;**9**(1):17–26.
49. Karunasekara Y, Dulhunty AF, Casarotto MG. The voltage-gated calcium-channel beta subunit: more than just an accessory. *Eur Biophys J* 2009;**39**(1):75–81.
50. Shikano S, et al. 14-3-3 proteins: regulation of endoplasmic reticulum localization and surface expression of membrane proteins. *Trends Cell Biol* 2006;**16**(7):370–5.
51. Watson P, et al. Coupling of ER exit to microtubules through direct interaction of COPII with dynactin. *Nat Cell Biol* 2005;**7**(1):48–55.
52. Glick BS, Nakano A. Membrane traffic within the Golgi apparatus. *Annu Rev Cell Dev Biol* 2009;**25**:113–32.
53. De Matteis MA, Luini A. Exiting the Golgi complex. *Nat Rev Mol Cell Biol* 2008;**9**(4):273–84.
54. Duffield A, Caplan MJ, Muth TR. Protein trafficking in polarized cells. *Int Rev Cell Mol Biol* 2008;**270**:145–79.
55. Verhey KJ, Hammond JW. Traffic control: regulation of kinesin motors. *Nat Rev Mol Cell Biol* 2009;**10**(11):765–77.
56. Lansbergen G, Akhmanova A. Microtubule plus end: a hub of cellular activities. *Traffic* 2006;**7**(5):499–507.
57. Lanzetti L. Actin in membrane trafficking. *Curr Opin Cell Biol* 2007;**19**(4):453–8.
58. Ross JL, Ali MY, Warshaw DM. Cargo transport: molecular motors navigate a complex cytoskeleton. *Curr Opin Cell Biol* 2008;**20**(1):41–7.
59. Jaiswal JK, Rivera VM, Simon SM. Exocytosis of post-Golgi vesicles is regulated by components of the endocytic machinery. *Cell* 2009;**137**(7):1308–19.
60. Hong TT, et al. BIN1 is reduced and Cav1.2 trafficking is impaired in human failing cardiomyocytes. *Heart Rhythm* 2012;**9**(5):812–20.
61. Laird DW. Life cycle of connexins in health and disease. *Biochem J* 2006;**394**(Pt 3):527–43.
62. Ligon LA, Holzbaur EL. Microtubules tethered at epithelial cell junctions by dynein facilitate efficient junction assembly. *Traffic* 2007;**8**(7):808–19.
63. Levy JR, Holzbaur EL. Special delivery: dynamic targeting via cortical capture of microtubules. *Dev Cell* 2007;**12**(3):320–2.
64. Hendricks AG, et al. Dynein tethers and stabilizes dynamic microtubule plus ends. *Curr Biol* 2012;**22**(7):632–7.
65. Chkourko HS, et al. Remodeling of mechanical junctions and of microtubule-associated proteins accompany cardiac connexin43 lateralization. *Heart Rhythm* 2012;**9**(7):1133–40.e6.
66. Zhang SS, et al. A micropatterning approach for imaging dynamic Cx43 trafficking to cell-cell borders. *FEBS Lett* 2014;**588**(8):1439–45.
67. Balijepalli RC, et al. Localization of cardiac L-type Ca(2+) channels to a caveolar macromolecular signaling complex is required for beta(2)-adrenergic regulation. *Proc Natl Acad Sci USA* 2006;**103**(19):7500–5.
68. Lowe JS, et al. Voltage-gated Na_v channel targeting in the heart requires an ankyrin-G dependent cellular pathway. *J Cell Biol* 2008;**180**(1):173–86.
69. Dun W, et al. Ankyrin-G participates in INa remodeling in myocytes from the border zones of infarcted canine heart. *PLoS One* 2013;**8**(10):e78087.
70. Cunha SR, Mohler PJ. Ankyrin-based cellular pathways for cardiac ion channel and transporter targeting and regulation. *Semin Cell Dev Biol* 2011;**22**(2):166–70.
71. Mohler PJ, Davis JQ, Bennett V. Ankyrin-B coordinates the Na/K ATPase, Na/Ca exchanger, and InsP3 receptor in a cardiac T-tubule/SR microdomain. *PLoS Biol* 2005;**3**(12):e423.

72. Hennessey JA, Wei EQ, Pitt GS. Fibroblast growth factor homologous factors modulate cardiac calcium channels. *Circ Res* 2013;**113**(4):381–8.

73. Wang C, et al. Fibroblast growth factor homologous factor 13 regulates Na⁺ channels and conduction velocity in murine hearts. *Circ Res* 2011;**109**(7):775–82.

74. Casini S, et al. Tubulin polymerization modifies cardiac sodium channel expression and gating. *Cardiovasc Res* 2010;**85**(4):691–700.

75. Meunier B, et al. The membrane-tubulating potential of amphiphysin 2/BIN1 is dependent on the microtubule-binding cytoplasmic linker protein 170 (CLIP-170). *Eur J Cell Biol* 2009;**88**(2):91–102.

76. Bartolini F, Ramalingam N, Gundersen GG. Actin-capping protein promotes microtubule stability by antagonizing the actin activity of mDia1. *Mol Biol Cell* 2012;**23**(20):4032–40.

77. Wittmann T, Bokoch GM, Waterman-Storer CM. Regulation of leading edge microtubule and actin dynamics downstream of Rac1. *J Cell Biol* 2003;**161**(5):845–51.

78. Smyth JW, Shaw RM. Autoregulation of connexin43 gap junction formation by internally translated isoforms. *Cell Rep* 2013;**5**(3):611–8.

79. Ul-Hussain M, et al. Internal ribosomal entry site (IRES) activity generates endogenous carboxyl-terminal domains of Cx43 and is responsive to hypoxic conditions. *J Biol Chem* 2014;**289**(30):20979–90.

80. Salat-Canela C, et al. Internal translation of the connexin 43 transcript. *Cell Commun Signal* 2014;**12**:31.

81. Meissner M, et al. Moderate calcium channel dysfunction in adult mice with inducible cardiomyocyte-specific excision of the cacnb2 gene. *J Biol Chem* 2011;**286**(18):15875–82.

82. Foell JD, et al. Molecular heterogeneity of calcium channel beta-subunits in canine and human heart: evidence for differential subcellular localization. *Physiol Genomics* 2004;**17**(2):183–200.

83. Lyon AR, et al. Plasticity of surface structures and beta2-adrenergic receptor localization in failing ventricular cardiomyocytes during recovery from heart failure. *Circ Heart Fail* 2012;**5**(3):357–65.

84. Caldwell JL, et al. Dependence of cardiac transverse tubules on the BAR domain protein amphiphysin II (BIN-1). *Circ Res* 2014;**115**(12):986–96.

85. Schumacher-Bass SM, et al. A role for myosin V motor proteins in the selective delivery of Kv channel isoforms to the membrane surface of cardiac myocytes. *Circ Res* 2014;**114**(6):982–92.

86. Rhett JM, Jourdan J, Gourdie RG. Connexin 43 connexon to gap junction transition is regulated by zonula occludens-1. *Mol Biol Cell* 2011;**22**(9):1516–28.

87. Oxford EM, et al. Ultrastructural changes in cardiac myocytes from Boxer dogs with arrhythmogenic right ventricular cardiomyopathy. *J Vet Cardiol* 2011;**13**(2):101–13.

88. Delmar M, Liang FX. Connexin43 and the regulation of intercalated disc function. *Heart Rhythm* 2012;**9**(5):835–8.

89. Kieken F, et al. Structural and molecular mechanisms of gap junction remodeling in epicardial border zone myocytes following myocardial infarction. *Circ Res* 2009;**104**(9):1103–12.

90. Heinzel FR, et al. Remodeling of T-tubules and reduced synchrony of Ca²⁺ release in myocytes from chronically ischemic myocardium. *Circ Res* 2008;**102**(3):338–46.

91. Wei S, et al. T-tubule remodeling during transition from hypertrophy to heart failure. *Circ Res* 2010;**107**(4):520–31.

92. Wagner E, et al. Stimulated emission depletion live-cell super-resolution imaging shows proliferative remodeling of T-tubule membrane structures after myocardial infarction. *Circ Res* 2012;**111**(4):402–14.

93. Zhang C, et al. Microtubule-mediated defects in junctophilin-2 trafficking contribute to myocyte transverse-tubule remodeling and Ca²⁺ handling dysfunction in heart failure. *Circulation* 2014;**129**(17):1742–50.

94. Leach RN, Desai JC, Orchard CH. Effect of cytoskeleton disruptors on L-type Ca channel distribution in rat ventricular myocytes. *Cell Calcium* 2005;**38**(5):515–26.

95. Tian Q, et al. Functional and morphological preservation of adult ventricular myocytes in culture by sub-micromolar cytochalasin D supplement. *J Mol Cell Cardiol* 2012;**52**(1):113–24.

96. Remo BF, et al. Phosphatase-resistant gap junctions inhibit pathological remodeling and prevent arrhythmias. *Circ Res* 2011;**108**(12):1459–66.

97. Leithe E, Rivedal E. Ubiquitination and down-regulation of gap junction protein connexin-43 in response to 12-O-tetradecanoylphorbol 13-acetate treatment. *J Biol Chem* 2004;**279**(48):50089–96.

98. Hesketh GG, et al. Ultrastructure and regulation of lateralized connexin43 in the failing heart. *Circ Res* 2010;**106**(6):1153–63.

99. Boassa D, et al. Trafficking and recycling of the connexin43 gap junction protein during mitosis. *Traffic* 2010;**11**(11):1471–86.

100. Gonzalez-Gutierrez G, et al. The Src homology 3 domain of the beta-subunit of voltage-gated calcium channels promotes endocytosis via dynamin interaction. *J Biol Chem* 2007;**282**(4):2156–62.

101. Green EM, et al. The tumor suppressor eIF3e mediates calcium-dependent internalization of the L-type calcium channel CaV1.2. *Neuron* 2007;**55**(4):615–32.

102. Best JM, et al. Small GTPase Rab11b regulates degradation of surface membrane L-type Cav1.2 channels. *Am J Physiol Cell Physiol* 2011;**300**(5):C1023–33.

103. Beardslee MA, et al. Dephosphorylation and intracellular redistribution of ventricular connexin43 during electrical uncoupling induced by ischemia. *Circ Res* 2000;**87**(8):656–62.

104. Marquez-Rosado L, et al. Connexin43 phosphorylation in brain, cardiac, endothelial and epithelial tissues. *Biochim Biophys Acta* 2012;**1818**(8):1985–92.

105. Majoul IV, et al. Limiting transport steps and novel interactions of Connexin-43 along the secretory pathway. *Histochem Cell Biol* 2009;**132**(3):263–80.

106. Batra N, et al. 14-3-3theta facilitates plasma membrane delivery and function of mechanosensitive connexin 43 hemichannels. *J Cell Sci* 2014;**127**(Pt 1):137–46.

107. Giepmans BN. Gap junctions and connexin-interacting proteins. *Cardiovasc Res* 2004;**62**(2):233–45.

108. Giepmans BN, Moolenaar WH. The gap junction protein connexin43 interacts with the second PDZ domain of the zona occludens-1 protein. *Curr Biol* 1998;**8**(16):931–4.

109. Laing JG, Chou BC, Steinberg TH. ZO-1 alters the plasma membrane localization and function of Cx43 in osteoblastic cells. *J Cell Sci* 2005;**118**(Pt 10):2167–76.

110. Hunter AW, et al. Zonula occludens-1 alters connexin43 gap junction size and organization by influencing channel accretion. *Mol Biol Cell* 2005;**16**(12):5686–98.

111. Hunter AW, Jourdan J, Gourdie RG. Fusion of GFP to the carboxyl terminus of connexin43 increases gap junction size in HeLa cells. *Cell Commun Adhes* 2003;**10**(4–6):211–4.

112. Chen J, et al. Domain-swapped dimerization of ZO-1 PDZ2 generates specific and regulatory connexin43-binding sites. *EMBO J* 2008;**27**(15):2113–23.

113. Bruce AF, et al. Gap junction remodelling in human heart failure is associated with increased interaction of connexin43 with ZO-1. *Cardiovasc Res* 2008;**77**(4):757–65.

Ryanodine Receptor Channelopathies in Skeletal and Cardiac Muscle

A.D. Hanna, L.J. Sharp, S.L. Hamilton

Baylor College of Medicine, Houston, TX, United States

INTRODUCTION TO RYANODINE RECEPTORS

RyR1 and RyR2 structure: Ryanodine receptors (RyRs) are the largest known ion channels and are composed of four identical subunits, each greater than 500 kDa in size, which assemble into a mushroom shaped homotetramer. RyR1 (5038 amino acids) is the predominant form in skeletal muscle, but RyR3 is expressed in low levels in some skeletal muscles. RyR2 (4967 amino acids) is the predominant form in cardiac muscle. However, both RyR1 and RyR2 are also expressed in B-lymphocytes and some areas of the brain, such that RyR mutations that are primarily associated with skeletal or cardiac muscle dysfunction could also impact immune and/or cognitive function.[1–4] RyRs have high interspecies homology with 88% homology between rabbit, pig, and human RyR1, and 98.6% homology between human and rabbit RyR2.[5] Although some regions of the three RyR isoforms have up to 90% homology, the overall homology is only 67–70%, due to three regions of diversity. In RyR1 these diverse regions are located between amino acids 4254–4631 (Divergent Region 1), 1342–1403 (Divergent Region 2), and 1872–1923 (Divergent Region 3) (Fig. 3.1).[6]

Structures of RyR1 were at less than 6.1 Å in three separate laboratories using cryo-electron microscopy. Each RyR protomer contains six transmembrane helices which include a pseudovoltage sensor domain and a pore domain.[7–9] The pore domain features an elongated sixth helix that, together with the selectivity filter, constitutes the ion conduction pathway.[8] This sixth helix interacts with an 80-residue C-terminal domain that

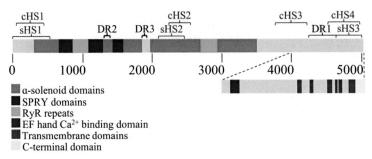

FIGURE 3.1 **Hot spots and divergent regions of RyR.** Functional domains, divergent regions, and mutation hot spots are depicted, in reference to the sequence of rabbit RyR1. Location of α-solenoid, SPRY, and RyR repeats is from Efremov[7] and of the transmembrane domains, C-terminal domain, and EF hand is from Yan.[8] The location of mutation hot spots for cardiac (cHS) and skeletal (sHS) RyR1 and divergent regions (DR) are from Priori & Chen[164] and MacLennan & Zvaritch.[61]

interfaces with an α-solenoid domain containing the EF hand Ca^{2+} binding motif.[7–9] The channel pore is surrounded by a large highly interconnected scaffold of α-solenoid repeats[7,9] which serve to confer flexibility to the structure.[9] This flexibility along with the numerous domain–domain interfaces (mediated by Van der Waals contacts[8]) may facilitate allosteric modulation of RyR1 activity by coupling conformational changes induced by modulator binding or posttranslational modification into changes in RyR1 gating.[7,9] Several peripheral domains were identified including three SPRY domains, thought to maintain structural integrity of the α-solenoid domains,[7] and the 1-2 Repeat and 3-4 Repeat regions, the last of which contains the S2843 PKA phosphorylation site.[7,9] The FKBP12 binding site was also identified in the periphery, at an interface common to the SPRY1, SPRY2, and α-solenoid domains, possibly allowing FKBPs to stabilize the closed state of RyR1.[9] Efremov et al.[7] compared the open and closed conformations of RyR1 and identified a conformational change in the RyR1 core that originated from the EF hand. For RyR opening, a relatively small conformational change in the EF hand domain is amplified by conformational changes of the α-solenoid helices, leading to an allosteric rearrangement of the C-terminal domain and membrane helices via the elongated S6 helix.[7]

A high-resolution structure for RyR2 is not currently available due to its instability under the conditions required for cryo-electron microscopy. Three-dimensional reconstruction of RyR2 at 41 Å revealed an overall high similarity to RyR1 with some localized regions of high variability.[10] A high-resolution crystal structure of the N-terminus of RyR2 has been obtained at 2.0 Å. These first 547 amino acids form three subdomains—the A, B, and C subdomains which are stabilized by a central chloride anion. This region was a site for 29 disease mutations that severely destabilized the subdomain interactions.[11]

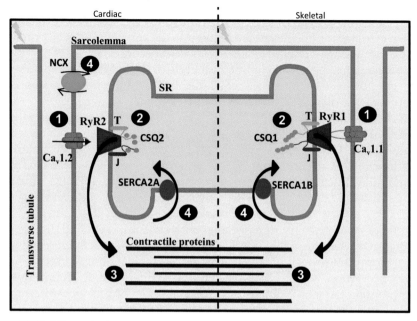

FIGURE 3.2 **Excitation–Contraction Coupling in Striated Muscle.** Schematic of striated muscle showing major Ca^{2+} currents that contribute to cardiac (left panel) and skeletal (right panel) muscle excitation–contraction coupling. (1) L-type Ca^{2+} channels are activated by depolarization of the t-tubule membrane. In cardiac muscle, the channel pore opens and Ca^{2+} enters the myocyte and activates RyR2. Skeletal muscle LTCCs undergo voltage-dependent conformational change and directly interact with RyR1 to activate it. (2) In skeletal and cardiac muscle, activation of RyR releases Ca^{2+} from the cytoplasm into the sarcoplasmic reticulum (SR) elevating the cytoplasmic Ca^{2+} concentration. (3) Cytoplasmic Ca^{2+}, together with ATP, bind to the contractile machinery and facilitate muscle contraction. (4) Muscle relaxation requires lowering of the cytoplasmic Ca^{2+} levels. In cardiac muscle this is achieved by SR Ca^{2+} ATPase (SERCA) and Na^+/Ca^{2+} exchanger (NCX). The contribution of NCX ranges from 5 to 30% depending on the species. In skeletal muscle there is no substantial efflux of Ca^{2+} during muscle relaxation, and the Ca^{2+} released via RyR1 is recovered into the SR by the SERCA pump. *CSQ*, Calsequestrin.

RyR function: RyRs play a critical role in excitation–contraction (EC) coupling, the process linking excitation of the cellular surface membrane (sarcolemma) to the mechanical contraction of intracellular muscle fibers. Depolarization reaches the interior of the muscle cell by a system of sarcolemmal invaginations called transverse tubules (t-tubules). Depolarization of the t-tubule activates voltage-gated, L-type Ca^{2+} channels (LTCC, $Ca_v1.1$ in skeletal muscle, and $Ca_v1.2$ in cardiac muscle) which then activate RyR channels positioned in the junctional face of the sarcoplasmic reticulum (SR) Ca^{2+} store (Fig. 3.2). In skeletal muscle, activation of RyR1 by $Ca_v1.1$ involves direct mechanical coupling[12] while in cardiac muscle, activation of RyR2 by $Ca_v1.2$ is by Ca^{2+}-induced Ca^{2+} release (CICR).[13] RyR activation allows bulk release of Ca^{2+} from the SR, which raises the cytoplasmic Ca^{2+}

from nanomolar to micromolar concentrations and, along with ATP, activates actin and myosin contractile filaments leading to muscle contraction. When cytoplasmic Ca^{2+} levels are reduced to the appropriate level, primarily by the SR Ca^{2+} ATPase (SERCA), relaxation occurs. The Ca^{2+} refilling of the SR prepares the muscle for the next cycle of Ca^{2+} release. Some Ca^{2+} is also extruded across the sarcolemmal membrane. In-depth reviews on EC coupling in cardiac and skeletal muscle are available elsewhere.[14–16]

RyR regulation: RyRs are the hubs of massive macromolecular complexes of proteins and enzymes that regulate RyR channel activity to modulate SR Ca^{2+} release. Some, including calsequestrin (CSQ), triadin, and junctin,[17–21] are primarily located in the SR lumen. Others, including calmodulin, junctophilin, and the FK506 binding proteins (FKBPs), bind to the cytoplasmic domains of RyR.[22–26] The importance of these proteins in regulating muscle function is evidenced by the findings that mutations in several of these proteins are associated with genetic or acquired disease in humans.[27–31] Adding to the complexity of RyR regulation is the susceptibility of the protein to posttranslational modifications including oxidation,[32–35] phosphorylation,[36–38] and *S*-nitrosylation.[39–41] Changes in RyR1 and RyR2 posttranslational modifications contribute to several forms of cardiac and skeletal muscle pathophysiology in animals[39,42–44] and humans.[45–47]

The multileveled regulation of RyRs ensures that SR Ca^{2+} release is fine-tuned to maintain optimal functional and metabolic output. In healthy muscle a combination of these regulatory factors (modulatory proteins and posttranslational modifications), as well as the innate sensitivity of the channel to activation/inhibition by Ca^{2+} and Mg^{2+} ensure maximal Ca^{2+} release during contraction/systole and minimal Ca^{2+} release during relaxation/diastole. Since striated muscle is an excitable tissue, ion fluxes are finely controlled by numerous proteins on both the sarcolemmal and SR membranes. RyR activity is modulated by several of these ionic species, the best documented being Ca^{2+} and Mg^{2+}. Therefore, dysregulated RyR function influences, and is influenced by, sarcolemmal and SR ion channels, pumps, and transporters. As detailed later, alterations in muscle Ca^{2+} homeostasis is a common feature of RyR-linked disease in both cardiac and skeletal muscle.

RYR1 MYOPATHIES

Acute or Evoked RyR1 Channelopathies: Malignant Hyperthermia, Enhanced Sensitivity to Heat, and Exercise-Induced Rhabdomyolysis

Malignant hyperthermia: Malignant hyperthermia (MH, MIM# 145600) is a potentially fatal hypermetabolic response to inhalation anesthetics alone or in combination with depolarizing muscle relaxants.

The hypermetabolic response manifests as hyperthermia, muscle rigidity, compartment syndrome (caused by pressure buildup within a muscle), rhabdomyolysis with myoglobinuria, hypercapnia, acidosis, hyperkalemia, and cardiac arrhythmias. MH can occur at any time during anesthesia or in the early postoperative period. Patients first develop tachycardia, elevated end-expired carbon dioxide levels, and muscle rigidity. Hyperthermia, which may develop later, occurs at a rate of 1–2°C every 5 min and can result in disseminated intravascular coagulation and widespread organ dysfunction. The same patient may or may not develop MH in response to a particular anesthetic agent at different times.[48] The estimated predisposing phenotype has been estimated to occur in up to 1 in 2000 individuals, though the actual incidence of MH is rare, occurring in only 1:100,000 surgical patients following anesthesia.[49,50]

Diagnosis of MH in individuals with concerning symptoms or a positive family history can be made either by in vitro contracture testing (IVCT) or by genetic testing for mutations in *RYR1* or *CACNA1S*. The in vitro contracture test exposes biopsied muscle tissue to increasing concentrations of either caffeine or halothane and measures the subsequent muscle contracture. However, this test is available only at certain limited centers and the IVCT must be performed within few hours of harvesting the muscle.

The treatment of episodes of MH consists of cessation of offending agents, hyperventilation to treat respiratory acidosis, and intravenous dantrolene given repeatedly until cardiac and respiratory stabilization. Dantrolene is currently the only available treatment for MH and functions by reducing RyR1 sensitivity to activation by Ca^{2+}[51] and caffeine.[52] Additional symptomatic treatment can include sodium bicarbonate for metabolic acidosis, cooling with chilled saline or ice packs, and amiodarone or beta-blockers to treat arrhythmias.[53] Future procedures requiring anesthesia should only take place in a hospital and patients should be encouraged to wear medical alert bracelets indicating MH susceptibility. Nearly 10% of episodes of MH result in death, with inadequate temperature monitoring during anesthesia being associated with increased risk of death. Clinicians should monitor core body temperatures intraoperatively whenever patients are undergoing general anesthesia for 30 min or more. Patients are encouraged to speak to family members about adverse reactions to anesthesia prior to surgery.[54]

Enhanced sensitivity to heat (EHS) and exercise-induced rhabdomyolysis (ER): MH or MH-like events have been reported in children following exposure to heat, who were later found to carry *RYR1* mutations (though in some cases the subject had survived an earlier, anesthetic-induced MH event).[55] Groom et al.[56] described two unrelated children with episodes of muscle rigidity following infection or heat, later found to have the same mutation in *RYR1*, which had previously been reported in patients with MH susceptibility.

RYR1 mutations are also associated with myalgia and rhabdomyolysis associated with exercise.[57–59] In one study, 24 individuals from 14 families were found to have *RYR1* mutations from a cohort of 39 families with unexplained exertional rhabdomyolysis (ER) or myalgia. In this cohort, symptomatic individuals were more often male, rhabdomyolysis occurred at an interval after exercise, and the first episode of rhabdomyolysis occurred anywhere from early childhood to 45 years of age. Muscle histology showed nonspecific findings such as fiber size variability, internalized nuclei, or core-like structures on oxidative staining. Additionally, in some patients, muscle MRI showed relative sparing of the rectus femoris muscle, as seen in some patients with *RYR1*-associated congenital myopathy or muscular dystrophy.[58]

Location of mutations in *RYR1* associated with MH/ESH/ER: The European Malignant Hyperthermia Group (EMHG) lists 34 causative *RYR1* mutations (Fig. 3.1), and hundreds more that are associated with MH susceptibility (reviewed in Refs. 60 and 61). Mutations in *RYR1* have been found in approximately 50–80% of cases of MH.[59] Rare mutations in *CACNA1S* have also been identified in MH susceptible individuals[62–64]; however, in up to 40% of MH cases, no mutations in either *RYR1* or *CACNA1S* are evident.[65] Although many *RYR1* mutations cluster within three "hot spot" regions, there are in fact mutations scattered throughout the entire amino acid sequence. MH/ESH/ER mutations, however, tend to cluster in hot spot regions 1 and 2, corresponding to the N-terminus and central regions (Fig. 3.1). These regions are primarily regulatory domains. MH-associated mutations are predicted to modify RyR1 gating by causing incorrect folding of subdomains and destabilizing domain–domain interactions either at the interfaces between RyR subunits or at the subdomain interfaces within the N-terminal domain.[66] Additionally, several MH mutations increase the thermal sensitivity of RyR1.[11,39]

Functional consequences of MH/ESH/ER mutations: Dysregulated Ca^{2+} release has long been recognized as a hallmark of MH. Research in pigs and mice with *RYR1* mutations has revealed critical details of the cellular mechanisms underlying this aberrant Ca^{2+} release. Even prior to the discovery of MH-associated mutations in *RYR1* and *CACNA1S*, it was hypothesized that triggering agents caused a pathological Ca^{2+} release into the muscle cytoplasm[67] and, indeed, muscles from MH patients and swine have often been found to contain a higher basal Ca^{2+} concentration.[68–70] MH mutations increased RyR1 sensitivity to activation by pharmacological agonists including anesthetics,[71,72] calcium,[73,74] caffeine,[71,75,76] 4-chloro-*m*-cresol (4-CMC)[76,77], and simvastatin.[78] An increase in voltage-dependent Ca^{2+} release was also identified in myotubes from porcine[79] and murine MH models[75,80,81] and in dyspedic myotubes suggesting an altered interaction with the $Ca_V1.1$, the voltage dependent Ca^{2+} channel.[76]

Analyses of Ca^{2+} handling in myotubes and muscle fibers expressing mutant RyR1 have generally (but not always[75,82]) found elevated cytoplasmic Ca^{2+} levels in MH compared to wild-type (WT).[39,74,82,83] The extent of the increase varied depending on the mutation[82,83] and temperature, with higher temperatures resulting in more extreme differences between control and MH resting Ca^{2+} levels.[39] Higher myoplasmic Ca^{2+} levels have been attributed to excessive SR Ca^{2+} release[39,75] and increased Ca^{2+} influx via sarcolemmal ion channels.[84]

SR Ca^{2+} load also regulates Ca^{2+} release and is finely controlled by the balance between Ca^{2+} release via RyR, Ca^{2+} uptake via SERCA, and buffering by the Ca^{2+} binding protein CSQ.[85] SR load in MH models, as indicated by the magnitude of Ca^{2+} release triggered by RyR1 agonists (usually caffeine or 4-CMC), has yielded variable results in MH muscle including reduced load,[82] or no change[75,82] compared to controls. Direct measurement of the SR load using SR-targeted biosensors in the Y522S MH mutant revealed a significant reduction in SR $[Ca^{2+}]$ at rest and during prolonged depolarization.[81] As with cytoplasmic Ca^{2+} levels, the apparent effect of an MH-associated mutation on SR Ca^{2+} store load is likely to vary with mutation, methodology, and the temperature at which experiments are conducted.[39,82]

Mutant RyRs have increased tendency to open at rest and are more sensitive to agonists (see previous discussion), but these channels are also more resistant to inactivation. RyR1 from MH susceptible humans,[72,86,87] swine,[88,89] and MH myotubes[76] were less sensitive to inhibition by both cytoplasmic Ca^{2+} (which at high micromolar concentrations normally inhibits RyR1) and Mg^{2+}, possibly due to reduced affinity for divalent cations in the MH channels.[88] Feng and colleagues, however, found no difference in Mg^{2+} sensitivity between WT and R163C RyR1.[74] Lowered Mg^{2+} inhibition has been associated with prolongation of Ca^{2+} transients and Ca^{2+} waves that propagate through the muscle fiber.[90] Reduced SR buffering in the Y522S MH model was also measured by Manno et al.[81] Thus, lower sensitivity to inhibition by Mg^{2+} may not only increase sensitivity of RyR1 to activation by triggers, but also may help facilitate the baseline elevation of cytoplasmic Ca^{2+} present in MH muscle.

Alterations in posttranslational modifications of RyR1 with MH mutations: Posttranslational modifications including phosphorylation, oxidation, and S-nitrosylation of RyR1 alter subunit/domain interactions, ionic sensitivity, and protein–protein interactions. Muscles of mice expressing RyR1 with MH-associated mutations display oxidative and nitrosative stress that increases with temperature.[39,91,92] In mice carrying the Y522S mutation, the triggering of an MH crisis is preceded by elevated levels of cytoplasmic Ca^{2+} and reactive oxygen/nitrogen species (ROS/RNS), leading to oxidative and nitrosative modifications of RyR1. Such modifications increase RyR1 Ca^{2+} leak, leading to a vicious cycle where increased leak leads to increased ROS/RNS production, causing further modification of RyR1 and increasing its sensitivity to triggers.[39]

Phosphorylation of RyR1 may also increase the sensitivity of mutant RyR1s to MH triggers (reviewed in Ref. 93). Catecholamine levels were found to be significantly elevated within 10–20 min of halothane exposure in MH swine,[55,94,95] and the increase was partially ameliorated by pretreatment with the β-blocker propranolol. Propranolol also blocked the MH-induced increases in oxygen consumption and the rise in body temperature.[55] Others have suggested that catecholamines did not themselves induce an MH episode, but rather prolong the hypermetabolic response.[96] A role for β-adrenergic stimulation in MH is also supported by the occurrence of MH crises in the absence of a pharmacological trigger in humans (see previous discussion) and animals[39,97] carrying MH mutations. In the R163C MH mouse model, RyR1 showed elevated phosphorylation at S2844; however, the functional significance of the increased phosphorylation is not known.[74] A role for adrenergic stimulation in MH pathogenesis would further align the etiologies of this disease with catecholaminergic polymorphic ventricular tachycardia (CPVT), which is triggered by stress and exercise-induced catecholamine elevation and is primarily associated with *RYR2* mutations (see later discussion). Less is known about the role of β-adrenergic stimulation in modulating Ca^{2+} handling is skeletal muscle.[98]

Congenital *RYR1* Myopathies

Location of mutations associated with congenital *RYR1* myopathies: Although most dominant central core disease (CCD) mutations tend to cluster in hot spot 3, which includes the channel pore (Fig. 3.1), mutations predicted to be hypomorphic (ie, resulting in only partial loss of gene function) are scattered throughout the gene. Nonhypomorphic mutations involved in recessive *RYR1* associated congenital myopathies appear in all three mutation hot spots.[99] However, there is strong association between recessive CCD and at least one mutation in hot spot 3.[99] Several aspects of altered Ca^{2+} handling (such as SR load, RyR sensitivity, and resting Ca^{2+} levels) exhibit mutation-dependent differences in severity and/or presentation. For example, a mutation in the peripheral N-terminal domain may manifest differently than a mutation in the transmembrane domain. The recent availability of the full-length RyR1 structure in high resolution[7–9] will be invaluable in determining how mutations in specific structural domain relate to alterations in RyR function and specific RyR1-linked congenital myopathies and may aid in the development of mutation-specific drug design.

Central core disease and other *RYR1*-linked congenital myopathies: Although MH, ESH, and ER are addressed separately from congenital RyR1 myopathies in this chapter, it should be recognized that MH, ESH, and ER can present with or without an associated congenital myopathy. *RYR1* channelopathies represent a continuum of not clearly distinct myopathies. Affected individuals with the same *RYR1* mutation can present with central cores, multiminicores, or even a myopathy without cores.[100] Congenital myopathies are a group of hereditary muscle diseases, which typically present in the neonatal or early childhood period with hypotonia and delayed motor development and have been historically thought to be nonprogressive or only slowly progressive. Further classification has been based on muscle histology. The most common congenital myopathy associated with mutations in *RYR1* is CCD, so named for areas of central clearing in Type I muscle fibers on oxidative stains (Fig. 3.3). Mutations in *RYR1* have also been associated with Congenital Fiber-Type Disproportion

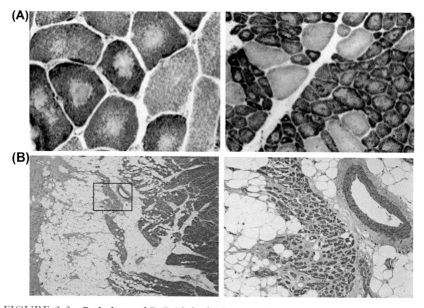

FIGURE 3.3   **Pathology of RyR-Linked Diseases.** (A) Muscle cross-sections from a CCD patient. Histology slides from the muscle of CCD patient. Images were obtained by Dr. Yadollah Harati, Baylor College of Medicine (B) H&E staining from an ARVC patient. Left panel is transmural view of the right ventricle, oriented with the epicardial surface to the left. Staining shows relative preservation of subendocardial myocardium with small islands of myocytes in the midwall. The fat is adipose tissue replacing myocardium. Right panel is higher magnification of the indicated region showing myocyte hypertrophy and interstitial fibrosis (the paler pink surrounding the myocytes). *ARVC images were obtained by Dr. Debra L. Kearney, Baylor College of Medicine.*

(MIM# 255310),[101] Type I Fiber Predominance,[102] Multiminicore disease (MIM# 255320), and Centronuclear myopathy (MIM# 160150).

Congenital myopathies are rare disorders, and the estimated prevalence has been based on epidemiological studies of small geographic areas. The estimated pediatric point prevalence has been estimated at 1:22,480 in western Sweden and 1:26,000 in southeastern Michigan (where the population mimics the overall demographics for the United States).[99,103] In Northern Ireland, the prevalence of adult and pediatric cases has been estimated at 1:28,600.[104] Mutations in *RYR1* are the most frequent cause of congenital myopathies, responsible for 44.4–59% of all cases.[105,106] CCD accounts for approximately 53% of congenital myopathies, and *RYR1* is also the most common gene responsible for CCD (>90% of cases).[50]

Initial descriptions of *RYR1*-associated congenital myopathies described dominantly inherited, mild, nonprogressive proximal and axial weakness, and histology consistent with central core disease.[107] However, the clinical spectrum has dramatically expanded to include both autosomal dominant and recessive disease with phenotypes ranging from mild to severe, and highly variable histologic findings on muscle biopsy. The most severe form of the *RYR1* congenital myopathies has been described to include decreased fetal movement, multiple skeletal abnormalities (kyphoscoliosis, arthrogryposis), severe weakness with patients never achieving independent ambulation, ophthalmoparesis, respiratory impairment, and feeding difficulties.[99,108] The severe phenotype can be associated with both dominant and recessive forms of the disease, though dominant disease tends to have a milder phenotype.[108] There is increased disease severity with compound heterozygous mutations with at least one hypomorphic allele and reduced RyR1 protein levels.[99] Additionally, compound heterozygous mutations are less likely to be associated with central cores and more likely to show more diverse muscle histology (commonly increased internalized nuclei, Type I fiber predominance, and hypertrophic Type II fibers).[99,109,110] However, even homozygous mutations can result in variable phenotype, as evidenced by the report of a consanguineous family affecting multiple family members and phenotype ranging from severe neonatal onset never achieving ambulation, to slowly progressive weakness with ambulation.[111] Overlap with MH has also been described. Carpenter[112] reported a consanguineous family with the homozygous *RYR1* mutation (R3772Q) which, when heterozygous, resulted in MH, but caused myopathy as well as MH when homozygous.

King-Denborough syndrome (MIM# 145600) is a rare disease associated with *RYR1* mutations and is characterized by dysmorphic features, MH, and myopathy. Dysmorphic features can include high arched palate, low set ears, ptosis, down-slanting palpebral fissures, micrognathia, malar hypoplasia, clinodactyly, and variable skeletal abnormalities. Histological changes are again nonspecific and may include fiber size variation, internalized nuclei, Type I predominance, or central cores. Western blot analysis may show reduced RyR1 protein levels.[113]

Diagnosis is typically made after muscle biopsy and confirmatory genetic testing. Muscle imaging with MRI or ultrasound may show relative sparing of the rectus femoris muscle compared to surrounding muscles, though this finding is less prevalent in patients with autosomal recessive disease and ophthalmoplegia.[114,115]

Patients are treated symptomatically with physical and occupational therapy and assistive devices as needed. Monitoring should include exams for scoliosis and respiratory insufficiency. Because the risk of MH in all genetic variants of *RYR1*-associated congenital myopathy is unknown, any patient with an *RYR1*-associated myopathy should be counseled to avoid volatile anesthetics and succinylcholine. Antioxidant treatment has been shown to improve muscle phenotype in a mouse model,[39] a zebra fish model of *RYR1* myopathy and myotubes derived from patients with *RYR1*-associated myopathies.[97,116] Stabilization of potassium homeostasis was also found to ameliorate the myopathy in mice with an *RYR1* mutation.[117]

The broad spectrum of myopathies associated with *RYR1* mutations and the distribution of disease-associated mutations in the primary sequence of RyR1 suggests the *RYR1* channelopathies are a wide continuum of muscle diseases with both overlapping and distinct features. In addition, the same mutation within a family can produce very different pathologies, suggesting genetic/environmental modifiers may exist, that affect disease presentation. These findings raise questions as to whether interventions for *RYR1* myopathies need to be individually tailored to the specific *RYR1* mutation, especially given that distinct mutations produce very different functional effects on RyR1 activity (as discussed later).

Alterations in RyR1 associated with CCD mutations: *RYR1* mutations associated with CCD fall into two classes: mutations that *increase* Ca^{2+} leak and, paradoxically, pore blocking mutations that *decrease* Ca^{2+} release. As with MH mutations, CCD mutants impact EC coupling and Ca^{2+} handling in skeletal muscle. Early functional analysis of CCD mutations made use of expression systems such as HEK-293 cells,[71,118,119] B-lymphocytes (which express the RyR1 protein),[120] and dyspedic myotubes[121] to characterize Ca^{2+}-handling parameters. CCD[71,118–120] and MH/CCD mutations[122] were associated with decreased SR stores and increased cytoplasmic Ca^{2+}. However, other studies found no change in resting Ca^{2+} or SR stores associated with the pore blocking CCD mutations.[121] Homozygous expression of CCD mutations induced a more severe phenotype then heterozygous mutation (as occurs in CCD patients).[118,119,122] In voltage-clamped myotubes, homozygous expression resulted in uncoupling of EC while in HEK cells, homozygous CCD mutations were more leaky[119] and had little or no caffeine induced Ca^{2+} release.[118] Several studies directly compared the effects of CCD and MH mutations on RyR1 function. N-terminal (hot spots 1 and 2) and C-terminal (hot spot 3) mutations elicited distinct alterations of RyR function (as assessed using dyspedic myotubes). N-terminal CCD mutations displayed similar characteristics to MH mutations (ie, enhanced SR leak and decreased SR store),[122] while the C-terminal mutations were more likely to have stable leak and SR store levels, with a deficiency in voltage-gated SR Ca^{2+} release.[121] Thus it was proposed that at least two distinct mechanisms of RyR dysfunction may account for the muscle weakness observed in CCD patients: one involving EC uncoupling and another involving leaky RyR1 channels, as seen in MH.[121]

Mice homozygous for the I4898T (IT) knock-in mutation exhibited delayed embryogenesis (from E15.5 onward), were born paralyzed, unable to breathe, and died shortly after birth.[123] Despite some ultrastructural abnormalities in the IT mice, interaction between DHPR and RyR1 channels was not significantly different from WT muscle. Furthermore, analysis of Ca^{2+}-handling parameters from myotubes of these animals found intact retrograde coupling between RyR1 and the DHPR and maintenance of SR store load. However, the homozygous IT channels lacked any voltage-dependent Ca^{2+} release and had severely reduced agonist induced Ca^{2+} release,[123] similar to the results obtained when IT channels were expressed in dyspedic myotubes.[121] Another outcome of this study was the confirmation of a role for RyR1-mediated Ca^{2+} release in late stage embryogenesis, including myogenesis (which had been suggested by earlier studies[124,125]), cardiogenesis, dermatogenesis, and ossification.[123]

Mice heterozygous for the knock-in IT mutation[126–128] had normal embryogenesis. After transient hypotonia and respiratory difficulties as neonates, mice developed progressive skeletal muscle weakness ranging from mild weakness to complete paralysis, though a minority remained asymptomatic. Affected mice were found to have progressive disorganization of muscle ultrastructure which included the formation of minicores, central cores, and rods.[126] A mouse model with the IT mutation on a mixed genetic background had a substantially milder phenotype with regard to muscle function but did have muscle ultrastructural abnormalities that were reminiscent of premature aging.[127] Analysis of EC coupling and Ca^{2+} handling in fibers from IT mice showed that mice had reduced electrical[127,128] and agonist evoked-Ca^{2+} release, due to a decrease in Ca^{2+} ion permeation via RyR1.[128] Previous studies of the IT mutation concluded that there was a decrease in SR store load, as assessed by caffeine or 4-CMC,[77,118,120] suggesting the reduced Ca^{2+} release in the IT mice was due to SR store depletion. However, Loy et al[128] found that the reduction in electrical and ligand-induced Ca^{2+} release in the IT mice was due to mutation-induced pore blockade rather than decreased SR stores.[128]

Further analysis of IT muscle found selective and differential ultrastructural modification in the muscles of IT mice.[127] In this model, there was selective ultrastructural modification of the Type I and Type IIb/IIx fibers. These modifications were specific to the affected fiber type, manifesting as core regions deficient in mitochondrial activity in the Type I fibers, and as dilation of the free SR and altered triad formation in the Type IIb/IIx fibers.[127] The authors also documented an increase in CSQ accumulation within the swollen SR and an age-related progression from minicores to central core and rod presentation. The simultaneous presence of cores and rods has been reported in other instances of clinical congenital myopathy.[129] Whether these ultrastructural changes directly correlate with functional changes is difficult to discern. It has been suggested that cores decrease the spatial and temporal homogeneous Ca^{2+} release resulting in regions of localized force production which, over time, modify the muscle ultrastructure, thereby inhibiting muscle function.[126]

RYR2-LINKED MYOPATHIES

Catecholaminergic Polymorphic Ventricular Tachycardia

CPVT is a rare condition associated with recurrent episodes of syncope or cardiac arrest due to ventricular arrhythmias occurring in response to emotion or exercise. Patients typically present with dizziness, palpitations, and syncope associated with exercise in the first decade of life. The electrocardiogram at rest may be normal or show only bradycardia. Heart structure is normal. Diagnosis is made by observing characteristic arrhythmias with exercise testing (which occur at threshold heart rates of 120–130) or isoproterenol injection.[29,130,131]

CPVT is now estimated to occur at a rate of 1:10,000; however, the asymptomatic nature of the disease in the absence of triggering conditions suggests that this is an under estimate.[132] Due to its rarity and a lack of distinct diagnostic criteria, little progress was made in terms of characterization of CPVT until the mid-1990s when Leenhardt et al.[130] characterized ECG traits and other clinical features in a cohort of CPVT-diagnosed children over a seven-year period. Features of CPVT differentiating it from other forms of arrhythmia include the occurrence of exercise or emotion-induced biventricular and polymorphic ventricular tachycardia and the absence of any structural or resting ECG abnormalities in individuals <40 years of age.[132] Initial clinical presentations of CPVT include syncope, palpitations, and sudden cardiac death. Syncope episodes may be accompanied by convulsions and urinary and fecal incontinence, leading to a false diagnosis of epilepsy. These factors contribute to an average delay in the correct diagnosis for 2–9 years.[130,133,134] Age of onset can vary substantially but typically occurs at ≤20 years.[31,130,135,136] In one study of 101 CPVT patients, 60% were symptomatic before diagnosis, with 74%

of those reporting syncope, 18% reporting aborted cardiac arrest, and 8% reporting palpitations or near syncope.[135] Mortality approaches 31% by age 30 and is higher with younger onset of symptoms and initial presentation as near-fatal events (cardiac arrest) and lack of treatment with β-blockers.[130,135]

There are two forms of CPVT which are distinguished by the affected gene. Type I (CPVT1, OMIM# 604772) is caused by mutations in *RYR2*, accounts for approximately 50–60% of cases of CPVT, and follows an autosomal dominant inheritance[31,137,138] whereas Type II (CPVT2, OMIM# 611938) is linked to mutations in the *CASQ2* gene and is inherited in an autosomal recessive pattern.[139] To date, 148 mutations in *RYR2* and 15 mutations in *CASQ2* have been identified in CPVT1 and 2, respectively, with the vast majority being single, missense mutations (Human Gene Mutation Database). Together, mutations in RyR2 and CSQ2 account for ~60% of all CPVT mutations, implying the existence of other causative genes. Indeed, CPVT has now been linked to mutations in the genes for triadin (autosomal recessive[140]) and calmodulin (autosomal dominant[28]), both proteins that associate directly with RyR2 and regulate SR Ca^{2+} release.

A comprehensive analysis of all 105 RyR2 exons revealed that CPVT1 mutations were localized to 45 exons and that 65% of these mutations were confined to only 16 exons.[136] Many of these mutations cluster in three regions of the RyR2 protein synonymous with the three regions on the RyR1 protein where MH/CCD mutations cluster (Fig. 3.1): an observation of potential significance in relating mutation localization to functional disturbances in RyR1/2. However, Jabbari[141] analyzed the frequency of previously reported genetic variants associated with CPVT against the Exome Sequencing Project database containing exome data from 6503 individuals and determined that the estimated prevalence of CPVT would be 1 in 150. Given the rarity of the disease, they concluded that the reported variants are not the monogenic cause of the disease.

Avoidance of strenuous exercise, competitive sports, and stressful environments are recommended in all patients diagnosed with CPVT.[132] β-blockers, especially the long acting β-blocker, nadolol, have been the preferred treatment for CPVT and have been shown to decrease frequency of arrhythmias and reduced mortality.[135] Patients should be placed on the highest tolerated dose. Periodic Holter monitoring should be performed to evaluate for continued arrhythmias. Verapamil was also effective in preventing arrhythmias when used in combination with β-blockers via inhibition of $Ca_v1.2$.[142] In patients where CPVT is refractory to β-blockade, flecainide, a class I antiarrhythmic that inhibits Na^+ channels, has been effective in reducing exercise-induced arrhythmias. Flecainide was proposed to prevent CPVT by directly inhibiting RyR2.[143–145] This idea has been challenged by others though who argue that the effects of flecainide on RyR2 are less important in the intact cell than effects of the drug on Na$^+$ channels.[146,147] Implantable cardiac defibrillators (ICD) should be offered only to patients with a history of cardiac arrest, poor medication compliance, or breakthrough arrhythmias on adequate doses of beta-blockers.[148] Left cardiac sympathetic denervation has also reduced the frequency of arrhythmias and need for ICD discharges for sustained ventricular tachycardia.[149,150] Future treatments for CPVT could make use of gene therapy. Delivery of an adeno-associated viral (AAV) construct to transfer the CASQ2 gene to mice carrying the R33Q-CASQ2 mutation was able to prevent ultrastructural changes and arrhythmogenesis when injected in newborn mice, while delivery in adult CPVT mice was able to reverse the phenotypic changes. These effects were accomplished with only ~40% of cardiomyocytes being infected with the AAV-CASQ2 construct.[151]

Autoregulation of Ca^{2+} handling in cardiomyocytes: Definition of the molecular mechanisms underlying cardiac Ca^{2+} channelopathies requires understanding whole cell Ca^{2+} homeostasis. There are two Ca^{2+} cycles in cardiomyocytes, one that moves Ca^{2+} across the sarcolemma and another that involves Ca^{2+} movement in and out of the SR.[152] Under normal conditions, a balance is maintained so Ca^{2+} that enters via Ca$_v$1.2 upon depolarization, is removed via the Na$^+$/Ca^{2+} exchanger (NCX) during diastole. Similarly, the amount of Ca^{2+} that is released from the SR by RyR2 during systole is equivalent to what it had taken up via SERCA during diastole.[153] Sarcolemmal ion channels also influence the activity of SR ion channels (for example, in CICR) and vice versa. This is evident in the way that SR Ca^{2+} release and uptake is "autoregulated," such that alteration in Ca^{2+} flux through one channel or transporter leads to compensatory changes in other Ca^{2+} fluxes to maintain overall Ca^{2+} flux balance.[154] However, conditions that alter multiple targets involved in Ca^{2+} signaling alter the set point at which Ca^{2+} flux balance occurs (reviewed in Ref. 155). Some of the effects of altered Ca^{2+} homeostasis in pathological conditions are discussed later.

Structural implications of CPVT mutations: At present only the N-terminal domain of RyR2 has been solved in high resolution and several CPVT mutations have been mapped to this region which is stabilized an anion binding site. Mutations in the N-terminal region altered subdomain orientations, weakening the interactions between these interfaces.[11] The decreased intramolecular stability increases channel opening by lowering the energy required to break intersubunit connections (reviewed in Ref. 156). Several mutations within the disease hot spots of RyR2 are located at interfaces of the cytoplasmic and N-terminal domains that are thought to participate in open/close conformational changes.[11,157] Interaction between the N-terminal and central domains via a "zipping" mechanism has been suggested to stabilize the closed conformation of RyR2.[158] CPVT mutations at these interfaces weaken the interactions between the domains, in effect "unzipping" the domain–domain interface and facilitating aberrant, diastolic Ca^{2+} release.[157,159] Therefore, increased RyR2 sensitivity associated with CPVT (see later discussion) may be due to the location of mutations at interfaces between RyR2 domains and subdomains.

Arrhythmogenesis in CPVT: The electrophysiological changes that occur in CPVT are associated with triggered activity (ie, extra heart beats), rather than a reentry mechanism or altered pacemaker activity. Triggered activity in CPVT is thought to be induced by afterdepolarizations (depolarizing oscillations of the membrane potential that occur during or after a sinoatrial node generated action potential). Afterdepolarizations are present in heart failure,[160–162] atrial fibrillation,[163,164] and digitalis toxicity[165] and may be the primary arrhythmogenic substrate in CPVT[166] (reviewed in Ref. 167). Early afterdepolarizations (EADs) occur during phase 2 or phase 3 of the ventricular/atrial action potential—before the membrane potential returns to baseline. Delayed afterdepolarizations (DADs) on the other hand, occur in phase 4—after the preceding action potential has ended and membrane potential has returned to baseline, but before the subsequent action potential is triggered by the sinoatrial node.[168] DADs occur as a result of increased cytoplasmic Ca^{2+} during diastole[169–171] which activates the forward mode NCX to extrude the excess Ca^{2+}.[161,171,172] If the accompanying influx of Na^+ is large enough, the ensuing depolarization of the surface membrane could exceed the threshold for Na^+ channel activation, triggering an action potential.[171] Conversely, EADs are more often associated with reactivation of the L-type Ca^{2+} current[173] and are likely to occur clinically when there is prolongation of the action potential duration or bradycardia and thus EADs have not always been considered as a likely source of triggered activity in CPVT (reviewed in Refs.14 and 174). Certain types of EADs have been attributed to NCX and Ca^{2+} overload mechanisms.[163,170,175] Cumulatively these studies suggest that both EADs and DADs should be considered in the pathogenesis of CPVT.

Mechanisms of arrhythmogenic Ca^{2+} release in CPVT: Current hypotheses of afterdepolarization generation in CPVT revolve around dysregulated RyR2 mediated Ca^{2+} release. CPVT mutations increase the sensitivity of RyR2 to activators, possibly as a result of weakened domain–domain interactions (see previous discussion). Increased sensitivity to activation lowered the threshold for Ca^{2+} release, leading to excess diastolic Ca^{2+} release.[176–180] Excess diastolic Ca^{2+} release could raise the cytoplasmic Ca^{2+} sufficiently to activate NCX and cause afterdepolarizations (see previous discussion). However, in the intact cell, there is a steep relationship between SR load and RyR2 Ca^{2+} release.[181–183] Hence, altered RyR2 activity does not generally cause such drastic Ca^{2+} release, since the SR soon becomes depleted.[153,154,180,184] A key event in CPVT pathogenesis is accelerated refilling of the SR.[177,180] In CPVT, rapid refilling of the store is accomplished by exercise or stress-induced β-adrenergic stimulation which phosphorylates $Ca_v1.2$ and phospholamban causing increased Ca^{2+} influx and increased SERCA2A activity. The combination of increased Ca^{2+} entry into the myocyte and increased SERCA2A activity greatly enhances Ca^{2+} uptake into the SR.[177,180] RyR2 phosphorylation with[43,178,185] or without[157,186] subsequent FKBP12.6 dissociation may contribute to increased sensitivity of the RyR2 to activation in CPVT. However, others have failed to detect a functional role for RyR2 phosphorylation in CPVT or a change in FKBP12.6 interaction with RyR2.[187,188] Regardless, in CPVT, rapid refilling of the SR in combination with the lowered threshold for RyR2 activation greatly facilitates excessive diastolic Ca^{2+} release leading to afterdepolarizations and triggered activity[176,177,180]; for a review see Ref. 167. This molecular pathway explains the requirement for β-adrenergic stimulation for triggered activity[184,189] and correlates well with the asymptomatic nature of CPVT in the absence of adrenergic stimulation.

Not all CPVT mutations induce a gain-of-function. The A4860G mutation decreased channel sensitivity to activation by luminal Ca^{2+}, and mutant cells were resistant to spontaneous Ca^{2+} oscillations, suggesting a higher threshold for activation.[190] Rather than being protective, the A4860G mutation in patients is associated with catecholaminergic idiopathic ventricular tachycardia,[134] suggesting that this mutation exerts its phenotypic effects by a cellular mechanism distinct from gain-of-function mutations (see previous discussion). Ca^{2+} alternans could be an arrhythmogenic substrate in loss-of-function CPVT mutations. Loss-of-function mutations may promote arrhythmia via a Ca^{2+} alternans–dependent model,[179,190–192] where there is beat to beat variation in Ca^{2+} transient amplitude, causing spatial desynchronization of SR Ca^{2+} release.[193]

CPVT mutations in RyR2 modulatory proteins: While not strictly an *RYR* channelopathy, mutations in the gene for CSQ2, a modulator of RyR-mediated Ca^{2+} release, are associated with an autosomal recessive form of CPVT (CPVT2).[29,139] Several mutations in *CASQ2* induce premature stop codons, resulting in deletion of the protein.[29] Mouse models of missense mutations in *CASQ2* also result in a substantial decrease in CSQ2 protein levels.[194,195] *CASQ2*-linked CPVT mice commonly exhibit exercise or stress-induced arrhythmias,[194–196] similar to *RYR2*-linked CPVT. At the cellular level, this arrhythmogenesis was associated with drastic changes in Ca^{2+} handling including reduced SR load and catecholamine-induced spontaneous Ca^{2+} oscillations with subsequent afterdepolarizations.[194–197] Most of these alterations mirror changes that occurred in CSQ2 knockout mice.[19,198] The deleterious effects of CSQ2 deletion/mutation have been attributed to several mechanisms (reviewed in Refs. 199 and 200). CSQ2 is the primary SR Ca^{2+} buffer and as such, is an important determinant of SR Ca^{2+} load.[19,194,198] Loss of SR buffering capacity accelerated refilling of the SR store,[201] which was enhanced with catecholamine treatment.[19] Another aspect of CSQ2 function is its influence on RyR2 refractoriness. CSQ2 deletion[201]/mutation[196,202] shortened the time in which RyR2 remained in a refractory state, allowing temporal synchronization of diastolic Ca^{2+} release. This synchronization facilitated the induction of triggered activity in the whole heart.[202] The absence of CSQ2 was also associated with an increase in RyR2 sensitivity to both luminal and cytoplasmic Ca^{2+} as assessed in CSQ-null myocytes and single channel experiments.[198,203–205] This increase in RyR2 sensitivity, in combination with accelerated SR refilling, would be expected to greatly increase the propensity for diastolic Ca^{2+} release as occurs in *RYR2*-linked CPVT mutations.

CPVT is also associated with mutations in two other RyR2 modulatory proteins, calmodulin and triadin. CPVT mutations in the gene for calmodulin (*CALM1*) (CPVT 4, OMIM# 614916) reduced the interaction of the protein with a calmodulin-binding domain peptide from RyR2 and altered the Ca^{2+} binding properties of calmodulin.[28] However, calmodulin is encoded by multiple genes in humans and also regulates several multiple sarcolemmal and intracellular ion channels (reviewed in Ref. 206) complicating the interpretation of these studies. Although there is currently no functional analysis, three CPVT-associated triadin mutations (CPVT 5, OMIM# 615441) cause either deletion of the protein or promote its degradation.[140] Knockout of triadin caused detrimental changes to the SR ultrastructure and EC coupling.[207] A confounding factor in considering the effect of triadin or CSQ2 deletion/mutation on SR Ca^{2+} handling is the coexisting ultrastructural changes and altered expression of other SR proteins.[194,196,198,207] Therefore, whether the aberrant Ca^{2+} release/propagation is due to CSQ or triadin absence per se, or to the accompanying changes in protein expression or SR architecture is important to consider. Whether or not the expression of these other SR proteins is altered in the setting of triadin or CSQ2-related CPVT in humans is unknown.

The current literature suggests that the CPVT subtypes resulting from mutation of calmodulin, CASQ2, triadin, or RyR2 are based on aberrant RyR2 Ca^{2+} release or Ca^{2+} leak demonstrating the importance of an intact, RyR2 macromolecular structure in maintaining cardiac health. This suggests that mutations in (at least) four different proteins could have a similar disease presentations. Although the field has made tremendous progress in identifying factors that increase Ca^{2+} leak, including altered domain interactions, posttranslational modifications, reduced refractoriness, and muscle ultrastructural changes, it remains to be determined what the relative importance of each of these and as yet unidentified factors is in the clinical pathogenesis of each CPVT subtype. Whether each distinct mechanism contributes differently to each type of CPVT, or even to different mutations within a type, will be vital knowledge in the design of targeted, effective medications. It must also be considered that the combination of all identified mutations still does not account for 100% of CPVT cases.

Arrhythmogenic Right Ventricular Cardiomyopathy

Arrhythmogenic right ventricular cardiomyopathy (ARVC) is a dominantly inherited progressive cardiomyopathy characterized by fatty-fibrous replacement of the right ventricular myocardium (Fig. 3.3) leading to ventricular arrhythmias, which can result in palpitations, syncope, and sudden cardiac death, particularly in young people. ARVC is genetically heterogeneous, though approximately 60% of cases result from mutations in cardiac desmosomal genes.[208] There are currently 13 known forms of ARVC, but only Type II (ARVC2, MIM# 600996) is linked to mutations in *RYR2*.[30,209–213]

In contrast to other genetic causes, *RYR2*-associated ARVC has an equal male to female ratio and higher penetrance and is associated with effort-induced ventricular arrhythmias.[213] Some patients described to have ARVD/C with *RYR2* mutations have normal right ventricular structure by echocardiogram, though fibrofatty infiltration has been found histologically.[30,209,212,214] Because of this, it has been suggested that *RYR2* mutations are in fact a phenotypic copy of ARVC and should be classified as CPVT instead.[215]

Diagnosis of ARVC is made based on 2009 revised diagnostic criteria which includes right ventricular morphologic imaging abnormalities, EKG changes, myocardial tissue histology, and family history.[216] Treatment includes avoidance of strenuous physical activity, antiarrhythmic drugs, and ICD placement.[215]

Arrhythmogenesis in ARVC: While both CPVT and general ARVC present with polymorphic/bidirectional ventricular tachycardia, the underlying mechanisms that lead to the arrhythmia differ. Replacement of regions of the right ventricular myocardium with fibrotic and fatty tissue leads to heterogeneity of the ventricular tissue and subsequent defects in electrical conduction. Additionally, several forms of ARVC are caused by mutations in genes that comprise the cardiac desmosome, a specialized adhesive structure that spans the intercellular space between two adjacent cardiomyocytes. The presence of desmosomes is thought to stabilize the three-dimensional structure of the myocardium and help to resist the mechanical stress imposed by contraction of the muscle wall (reviewed in Ref. 217). Dysfunction of these proteins is thought to cause conduction abnormalities which can manifest as arrhythmia and sudden cardiac death.[218] The specific mechanisms of arrhythmogenesis in ARVC2 are currently unknown but fibrofatty replacement of the myocardium is documented in this setting,[30,209,214] suggesting a role for conduction defects in the pathogenesis of ARVC2.

Ca^{2+} handling abnormalities in ARVC2: The occurrence of stress/exercise-induced ventricular tachycardia in ARVC2 raises the possibility that RyR2-linked ARVC mutations elicit changes in channel function similar to those that occur in CPVT and other forms of RyR2-related sudden cardiac death.[185,214,219] The discovery of ARVC2 mutations in the disease hot spots of RyR2 in close proximity to CPVT mutations led to the hypothesis that ARVC2 mutations elicit gain-of-function in RyR2.[213] This hypothesis was supported by finding of ARVC2 mutation-associated increases in spontaneous Ca^{2+} oscillations and channel sensitivity to activation, especially at diastolic Ca^{2+} concentrations.[176,192,210,220] Others reported differential sensitivity of the ARVC2 mutations to activation by agonists and noted that one mutation (L433P) did not exhibit a gain-of-function phenotype, suggesting some heterogeneity in the molecular basis of ARVC2 arrhythmogenesis.[192] In accordance with these findings, in the R176Q mouse model of ARVC2, mice were susceptible to isoproterenol-induced ventricular arrhythmias, and cardiomyocytes displayed isoproterenol-induced spontaneous Ca^{2+} oscillations. Although this mutation has been identified in ARVC2 patients, the mouse model had no histological abnormalities typical of ARVC, though it did have mild cardiomyopathy.[189]

Conclusions from functional analysis of ARVC2 mutations in RyR2 are confounded by the heterogeneous effects of the mutations and by the substantial crossover with CPVT mutations. While the ARVC2 mutations do alter RyR2 function, the mechanism by which alterations in RyR2 function lead to ARVC2 associated fibrofatty deposits is not known. One possibility is that RyR2 leak drives Ca^{2+} dependent apoptosis, leading to the histological changes.[213] Whether or not sudden cardiac death in patients carrying ARVC2 mutations results from a triggered activity mechanism, similar to that hypothesized for CPVT, or to histological changes in right ventricular wall remains to be assessed.

SUMMARY/CONCLUSION

Considering the complex mechanisms that exist in cardiac and skeletal muscle to regulate Ca^{2+} homeostasis, it is remarkable that a single point mutation in a protein of the size of RyR (~5000 amino acids) can elicit such devastating effects on the heart and skeletal muscle function. The association of RyR mutations with these diseases demonstrates the fundamental role of RyR in striated muscle function. While the multiple and complex regulatory mechanisms in place in both cardiac and skeletal muscle, for the most part, prevent many mutations from altering function under basal conditions, stress in the form of β-adrenergic stimulation, oxidative stress, or temperature overwhelm these control mechanisms allowing for aberrant Ca^{2+} release. Interventions should therefore be tailored to either alleviate the stress factors orstabilize RyR. The high variability in presentation of *RYR* channelopathies, huge range of mutation-specific changes in RyR activities, and the functional consequences of this altered RyR activity suggest the interventions need to be tailored to the specific RyR mutations. Identification of patient-specific *RYR* mutations and development of individualized interventions directed at the pathways altered by specific *RYR* mutations will be essential in the design of effective treatments for *RYR*-linked disease.

References

1. Santiago JA, Talbott GC, Lorenzon NM. Effects of a Y522S-RyR1 mutation on cerebellar Purkinje cell function. *Biophys J* 2010;**98**(3, Supplement 1):511a.
2. Hakamata Y, Nakai J, Takeshima H, Imoto K. Primary structure and distribution of a novel ryanodine receptor/calcium release channel from rabbit brain. *FEBS Lett* 1992;**312**(2–3):229–35.
3. Nakai J, Imagawa T, Hakamat Y, Shigekawa M, Takeshima H, Numa S. Primary structure and functional expression from cDNA of the cardiac ryanodine receptor/calcium release channel. *FEBS Lett* 1990;**271**(1–2):169–77.
4. Ohashi R, Sakata S, Naito A, Hirashima N, Tanaka M. Dendritic differentiation of cerebellar Purkinje cells is promoted by ryanodine receptors expressed by Purkinje and granule cells. *Dev Neurobiol* 2013;**74**(4):467–80.
5. Tunwell RE, Wickenden C, Bertrand BM, et al. The human cardiac muscle ryanodine receptor-calcium release channel: identification, primary structure and topological analysis. *Biochem J* 1996;**318**(Pt 2):477–87.
6. Rossi D, Sorrentino V. Molecular genetics of ryanodine receptors Ca^{2+}-release channels. *Cell Calcium* 2002;**32**(5–6):307–19.
7. Efremov RG, Leitner A, Aebersold R, Raunser S. Architecture and conformational switch mechanism of the ryanodine receptor. *Nature* 2015;**517**(7532):39–43.
8. Yan Z, Bai XC, Yan C, et al. Structure of the rabbit ryanodine receptor RyR1 at near-atomic resolution. *Nature* 2015;**517**(7532):50–5.
9. Zalk R, Clarke OB, des Georges A, et al. Structure of a mammalian ryanodine receptor. *Nature* 2015;**517**(7532):44–9.
10. Sharma MR, Penczek P, Grassucci R, Xin H-B, Fleischer S, Wagenknecht T. Cryoelectron microscopy and image analysis of the cardiac ryanodine receptor. *J Biol Chem* 1998:18429–34.
11. Kimlicka L, Tung CC, Carlsson AC, Lobo PA, Yuchi Z, Van Petegem F. The cardiac ryanodine receptor N-terminal region contains an anion binding site that is targeted by disease mutations. *Structure* 2013;**21**(8):1440–9.
12. Rios E, Brum G. Involvement of dihydropyridine receptors in excitation-contraction coupling in skeletal muscle. *Nature* 1987;**325**(6106):717–20.
13. Fabiato A, Fabiato F. Contractions induced by a calcium-triggered release of calcium from the sarcoplasmic reticulum of single skinned cardiac cells. *J Physiol* 1975;**249**(3):469–95.
14. Bers DM. Calcium cycling and signaling in cardiac myocytes. *Annu Rev Physiol* 2008;**70**:23–49.
15. Rebbeck RT, Karunasekara Y, Board PG, Beard NA, Casarotto MG, Dulhunty AF. Skeletal muscle excitation-contraction coupling: who are the dancing partners? *Int J Biochem Cell Biol* 2013;**48**:28–38.
16. Lamb GD. Excitation-contraction coupling in skeletal muscle: comparisons with cardiac muscle. *Clin Exp Pharmacol Physiol* 2000;**27**(3):216–24.

17. Altschafl BA, Arvanitis DA, Fuentes O, Yuan Q, Kranias EG, Valdivia HH. Dual role of junctin in the regulation of ryanodine receptors and calcium release in cardiac ventricular myocytes. *J Physiol* 2011;**589**(Pt 24):6063–80.

18. Beard NA, Sakowska MM, Dulhunty AF, Laver DR. Calsequestrin is an inhibitor of skeletal muscle ryanodine receptor calcium release channels. *Biophys J* 2002;**82**(1 Pt 1):310–20.

19. Terentyev D, Viatchenko-Karpinski S, Gyorke I, Volpe P, Williams SC, Gyorke S. Calsequestrin determines the functional size and stability of cardiac intracellular calcium stores: mechanism for hereditary arrhythmia. *Proc Natl Acad Sci USA* 2003;**100**(20):11759–64.

20. Wium E, Dulhunty AF, Beard NA. A skeletal muscle ryanodine receptor interaction domain in triadin. *PLoS One* 2012;**7**(8):e43817.

21. Zhang L, Kelley J, Schmeisser G, Kobayashi YM, Jones LR. Complex formation between junctin, triadin, calsequestrin, and the ryanodine receptor. Proteins of the cardiac junctional sarcoplasmic reticulum membrane. *J Biol Chem* 1997;**272**(37):23389–97.

22. Ahern GP, Junankar PR, Dulhunty AF. Single channel activity of the ryanodine receptor calcium release channel is modulated by FK-506. *FEBS Lett* 1994;**352**(3):369–74.

23. Hamilton SL, Serysheva I, Strasburg GM. Calmodulin and excitation-contraction coupling. *News Physiol Sci* 2000;**15**:281–4.

24. Ito K, Komazaki S, Sasamoto K, et al. Deficiency of triad junction and contraction in mutant skeletal muscle lacking junctophilin type 1. *J Cell Biol* 2001;**154**(5):1059–67.

25. Takeshima H, Komazaki S, Nishi M, Iino M, Kangawa K. Junctophilins: a novel family of junctional membrane complex proteins. *Mol Cell* 2000;**6**(1):11–22.

26. Timerman AP, Ogunbumni E, Freund E, Wiederrecht G, Marks AR, Fleischer S. The calcium release channel of sarcoplasmic reticulum is modulated by FK-506-binding protein. Dissociation and reconstitution of FKBP-12 to the calcium release channel of skeletal muscle sarcoplasmic reticulum. *J Biol Chem* 1993;**268**(31):22992–9.

27. Landstrom AP, Weisleder N, Batalden KB, et al. Mutations in JPH2-encoded junctophilin-2 associated with hypertrophic cardiomyopathy in humans. *J Mol Cell Cardiol* 2007;**42**(6):1026–35.

28. Nyegaard M, Overgaard MT, Sondergaard MT, et al. Mutations in calmodulin cause ventricular tachycardia and sudden cardiac death. *Am J Hum Genet* 2012;**91**(4):703–12.

29. Postma AV, Denjoy I, Hoorntje TM, et al. Absence of calsequestrin 2 causes severe forms of catecholaminergic polymorphic ventricular tachycardia. *Circ Res* 2002;**91**(8):e21–6.

30. Roux-Buisson N, Gandjbakhch E, Donal E, et al. Prevalence and significance of rare RYR2 variants in arrhythmogenic right ventricular cardiomyopathy/dysplasia: results of a systematic screening. *Heart Rhythm* 2014;**11**(11):1999–2009.

31. Priori SG, Napolitano C, Tiso N, et al. Mutations in the cardiac ryanodine receptor gene (hRyR2) underlie catecholaminergic polymorphic ventricular tachycardia. *Circulation* 2001;**103**(2):196–200.

32. Abramson JJ, Salama G. Critical sulfhydryls regulate calcium release from sarcoplasmic reticulum. *J Bioenerg Biomembr* 1989;**21**(2):283–94.

33. Aracena-Parks P, Goonasekera SA, Gilman CP, Dirksen RT, Hidalgo C, Hamilton SL. Identification of cysteines involved in S-nitrosylation, S-glutathionylation, and oxidation to disulfides in ryanodine receptor type 1. *J Biol Chem* 2006;**281**(52):40354–68.

34. Hidalgo C, Aracena P, Sanchez G, Donoso P. Redox regulation of calcium release in skeletal and cardiac muscle. *Biol Res* 2002;**35**(2):183–93.

35. Zable AC, Favero TG, Abramson JJ. Glutathione modulates ryanodine receptor from skeletal muscle sarcoplasmic reticulum. Evidence for redox regulation of the Ca^{2+} release mechanism. *J Biol Chem* 1997;**272**(11):7069–77.

36. Lokuta AJ, Rogers TB, Lederer WJ, Valdivia HH. Modulation of cardiac ryanodine receptors of swine and rabbit by a phosphorylation-dephosphorylation mechanism. *J Physiol* 1995;**487**(Pt 3):609–22.

37. Reiken S, Lacampagne A, Zhou H, et al. PKA phosphorylation activates the calcium release channel (ryanodine receptor) in skeletal muscle: defective regulation in heart failure. *J Cell Biol* 2003;**160**(6):919–28.

38. Suko J, Maurer-Fogy I, Plank B, et al. Phosphorylation of serine 2843 in ryanodine receptor-calcium release channel of skeletal muscle by cAMP-, cGMP- and CaM-dependent protein kinase. *Biochim Biophys Acta* 1993;**1175**(2):193–206.

39. Durham WJ, Aracena-Parks P, Long C, et al. RyR1 S-nitrosylation underlies environmental heat stroke and sudden death in Y522S RyR1 knockin mice. *Cell* 2008;**133**(1):53–65.

40. Eu JP, Xu L, Stamler JS, Meissner G. Regulation of ryanodine receptors by reactive nitrogen species. *Biochem Pharmacol* 1999;**57**(10):1079–84.

41. Stoyanovsky D, Murphy T, Anno PR, Kim YM, Salama G. Nitric oxide activates skeletal and cardiac ryanodine receptors. *Cell Calcium* 1997;**21**(1):19–29.

42. Gonzalez DR, Beigi F, Treuer AV, Hare JM. Deficient ryanodine receptor S-nitrosylation increases sarcoplasmic reticulum calcium leak and arrhythmogenesis in cardiomyocytes. *Proc Natl Acad Sci USA* 2007;**104**(51):20612–7.

43. Shan J, Xie W, Betzenhauser M, et al. Calcium leak through ryanodine receptors leads to atrial fibrillation in 3 mouse models of catecholaminergic polymorphic ventricular tachycardia. *Circ Res* 2012;**111**(6):708–17.

44. Terentyev D, Gyorke I, Belevych AE, et al. Redox modification of ryanodine receptors contributes to sarcoplasmic reticulum Ca^{2+} leak in chronic heart failure. *Circ Res* 2008;**103**(12):1466–72.

45. Marx SO, Reiken S, Hisamatsu Y, et al. PKA phosphorylation dissociates FKBP12.6 from the calcium release channel (ryanodine receptor): defective regulation in failing hearts. *Cell* 2000;**101**(4):365–76.

46. Fischer TH, Herting J, Tirilomis T, et al. Ca^{2+}/calmodulin-dependent protein kinase II and protein kinase A differentially regulate sarcoplasmic reticulum Ca^{2+} leak in human cardiac pathology. *Circulation* 2013;**128**(9):970–81.

47. Rullman E, Andersson DC, Melin M, et al. Modifications of skeletal muscle ryanodine receptor type 1 and exercise intolerance in heart failure. *J Heart Lung Transpl* 2013;**32**(9):925–9.

48. Hogan K. The anesthetic myopathies and malignant hyperthermias. *Curr Opin Neurol* 1998;**11**(5):469–76.

49. Brady JE, Sun LS, Rosenberg H, Li G. Prevalence of malignant hyperthermia due to anesthesia in New York State, 2001–2005. *Anesth Analg* 2009;**109**(4):1162–6.

50. Wu S, Ibarra MC, Malicdan MC, et al. Central core disease is due to RYR1 mutations in more than 90% of patients. *Brain* 2006;**129**(Pt 6):1470–80.

51. Fruen BR, Mickelson JR, Louis CF. Dantrolene inhibition of sarcoplasmic reticulum Ca^{2+} release by direct and specific action at skeletal muscle ryanodine receptors. *J Biol Chem* 1997;**272**(43):26965–71.

52. Zhao F, Li P, Chen SR, Louis CF, Fruen BR. Dantrolene inhibition of ryanodine receptor Ca^{2+} release channels. Molecular mechanism and isoform selectivity. *J Biol Chem* 2001;**276**(17):13810–6.

53. Glahn KP, Ellis FR, Halsall PJ, et al. Recognizing and managing a malignant hyperthermia crisis: guidelines from the European Malignant Hyperthermia Group. *Br J Anaesth* 2010;**105**(4):417–20.

54. Larach MG, Brandom BW, Allen GC, Gronert GA, Lehman EB. Malignant hyperthermia deaths related to inadequate temperature monitoring, 2007–2012: a report from the North American malignant hyperthermia registry of the malignant hyperthermia association of the United States. *Anesth Analg* 2014;**119**(6):1359–66.

55. Gronert GA, Tobin JR, Muldoon S. Malignant hyperthermia: human stress triggering. *Biochim Biophys Acta* 2011;**1813**(12):2191–2. author reply 3–4.

56. Groom L, Muldoon SM, Tang ZZ, et al. Identical de novo mutation in the type 1 ryanodine receptor gene associated with fatal, stress-induced malignant hyperthermia in two unrelated families. *Anesthesiology* 2011;**115**(5):938–45.

57. Davis M, Brown R, Dickson A, et al. Malignant hyperthermia associated with exercise-induced rhabdomyolysis or congenital abnormalities and a novel RYR1 mutation in New Zealand and Australian pedigrees. *Br J Anaesth* 2002;**88**(4):508–15.

58. Dlamini N, Voermans NC, Lillis S, et al. Mutations in RYR1 are a common cause of exertional myalgia and rhabdomyolysis. *Neuromuscul Disord* 2013;**23**(7):540–8.

59. Sambuughin N, Holley H, Muldoon S, et al. Screening of the entire ryanodine receptor type 1 coding region for sequence variants associated with malignant hyperthermia susceptibility in the North American population. *Anesthesiology* 2005;**102**(3):515–21.

60. Hirshey Dirksen SJ, Larach MG, Rosenberg H, et al. Special article: Future directions in malignant hyperthermia research and patient care. *Anesth Analg* 2011;**113**(5):1108–19.

61. Maclennan DH, Zvaritch E. Mechanistic models for muscle diseases and disorders originating in the sarcoplasmic reticulum. *Biochim Biophys Acta* 2011;**1813**(5):948–64.

62. Monnier N, Procaccio V, Stieglitz P, Lunardi J. Malignant-hyperthermia susceptibility is associated with a mutation of the alpha 1-subunit of the human dihydropyridine-sensitive L-type voltage-dependent calcium-channel receptor in skeletal muscle. *Am J Hum Genet* 1997;**60**(6):1316–25.

63. Toppin PJ, Chandy TT, Ghanekar A, Kraeva N, Beattie WS, Riazi S. A report of fulminant malignant hyperthermia in a patient with a novel mutation of the CACNA1S gene. *Can J Anaesth* 2010;**57**(7):689–93.

64. Carpenter D, Ringrose C, Leo V, et al. The role of CACNA1S in predisposition to malignant hyperthermia. *BMC Med Genet* 2009;**10**:104.

65. Robinson R, Carpenter D, Shaw MA, Halsall J, Hopkins P. Mutations in RYR1 in malignant hyperthermia and central core disease. *Hum Mutat* 2006;**27**(10):977–89.

66. Tung CC, Lobo PA, Kimlicka L, Van Petegem F. The amino-terminal disease hotspot of ryanodine receptors forms a cytoplasmic vestibule. *Nature* 2010;**468**(7323):585–8.

67. Britt BA, Kalow W, Endrenyi L. Effects of halothane and methoxyflurane on rat skeletal muscle mitochondria. *Biochem Pharmacol* 1972;**21**(8):1159–69.

68. Lopez JR, Alamo L, Caputo C, Wikinski J, Ledezma D. Intracellular ionized calcium concentration in muscles from humans with malignant hyperthermia. *Muscle Nerve* 1985;**8**(5):355–8.

69. Lopez JR, Gerardi A, Lopez MJ, Allen PD. Effects of dantrolene on myoplasmic free [Ca^{2+}] measured in vivo in patients susceptible to malignant hyperthermia. *Anesthesiology* 1992;**76**(5):711–9.

70. Nelson TE. Abnormality in calcium release from skeletal sarcoplasmic reticulum of pigs susceptible to malignant hyperthermia. *J Clin Invest* 1983;**72**(3):862–70.

71. Tong J, Oyamada H, Demaurex N, Grinstein S, McCarthy TV, MacLennan DH. Caffeine and halothane sensitivity of intracellular Ca^{2+} release is altered by 15 calcium release channel (ryanodine receptor) mutations associated with malignant hyperthermia and/or central core disease. *J Biol Chem* 1997;**272**(42):26332–9.

72. Duke AM, Hopkins PM, Halsall PJ, Steele DS. Mg^{2+} dependence of Ca^{2+} release from the sarcoplasmic reticulum induced by sevoflurane or halothane in skeletal muscle from humans susceptible to malignant hyperthermia. *Br J Anaesth* 2006;**97**(3):320–8.

73. Ohnishi ST, Taylor S, Gronert GA. Calcium-induced Ca^{2+} release from sarcoplasmic reticulum of pigs susceptible to malignant hyperthermia. The effects of halothane and dantrolene. *FEBS Lett* 1983;**161**(1):103–7.

74. Feng W, Barrientos GC, Cherednichenko G, et al. Functional and biochemical properties of ryanodine receptor type 1 channels from heterozygous R163C malignant hyperthermia-susceptible mice. *Mol Pharmacol* 2011;**79**(3):420–31.

75. Chelu MG, Goonasekera SA, Durham WJ, et al. Heat- and anesthesia-induced malignant hyperthermia in an RyR1 knock-in mouse. *Faseb J* 2006;**20**(2):329–30.

76. Yang T, Ta TA, Pessah IN, Allen PD. Functional defects in six ryanodine receptor isoform-1 (RyR1) mutations associated with malignant hyperthermia and their impact on skeletal excitation-contraction coupling. *J Biol Chem* 2003;**278**(28):25722–30.

77. Ducreux S, Zorzato F, Muller C, et al. Effect of ryanodine receptor mutations on inter-leukin-6 release and intracellular calcium homeostasis in human myotubes from malignant hyperthermia-susceptible individuals and patients affected by central core disease. *J Biol Chem* 2004;**279**(42):43838–46.

78. Knoblauch M, Dagnino-Acosta A, Hamilton SL. Mice with RyR1 mutation (Y524S) undergo hypermetabolic response to simvastatin. *Skelet Muscle* 2013;**3**(1):22.

79. Dietze B, Henke J, Eichinger HM, Lehmann-Horn F, Melzer W. Malignant hyperther-mia mutation Arg615Cys in the porcine ryanodine receptor alters voltage dependence of Ca^{2+} release. *J Physiol* 2000;**526**(Pt 3):507–14.

80. Andronache Z, Hamilton SL, Dirksen RT, Melzer W. A retrograde signal from RyR1 alters DHP receptor inactivation and limits window Ca^{2+} release in muscle fibers of Y522S RyR1 knock-in mice. *Proc Natl Acad Sci USA* 2009;**106**(11):4531–6.

81. Manno C, Figueroa L, Royer L, et al. Altered Ca^{2+} concentration, permeability and buff-ering in the myofibre Ca^{2+} store of a mouse model of malignant hyperthermia. *J Physiol* 2013;**591**(Pt 18):4439–57.

82. Dirksen RT, Avila G. Distinct effects on Ca^{2+} handling caused by malignant hyperther-mia and central core disease mutations in RyR1. *Biophys J* 2004;**87**(5):3193–204.

83. Yang T, Esteve E, Pessah IN, Molinski TF, Allen PD, Lopez JR. Elevated resting [Ca(2+)](i) in myotubes expressing malignant hyperthermia RyR1 cDNAs is partially restored by modulation of passive calcium leak from the SR. *Am J Physiol Cell Physiol* 2007;**292**(5):C1591–8.

84. Eltit JM, Ding X, Pessah IN, Allen PD, Lopez JR. Nonspecific sarcolemmal cation channels are critical for the pathogenesis of malignant hyperthermia. *Faseb J* 2013;**27**(3):991–1000.

85. Manno C, Sztretye M, Figueroa L, Allen PD, Rios E. Dynamic measurement of the cal-cium buffering properties of the sarcoplasmic reticulum in mouse skeletal muscle. *J Physiol* 2012;**591**(Pt 2):423–42.

86. Duke AM, Hopkins PM, Halsal JP, Steele DS. Mg^{2+} dependence of halothane-induced Ca^{2+} release from the sarcoplasmic reticulum in skeletal muscle from humans suscep-tible to malignant hyperthermia. *Anesthesiology* 2004;**101**(6):1339–46.

87. Duke AM, Hopkins PM, Steele DS. Effects of Mg(2+) and SR luminal Ca(2+) on caffeine-induced Ca(2+) release in skeletal muscle from humans susceptible to malignant hyper-thermia. *J Physiol* 2002;**544**(Pt 1):85–95.

88. Laver DR, Owen VJ, Junankar PR, Taske NL, Dulhunty AF, Lamb GD. Reduced inhibi-tory effect of Mg^{2+} on ryanodine receptor-Ca^{2+} release channels in malignant hyperther-mia. *Biophys J* 1997;**73**(4):1913–24.

89. Owen VJ, Taske NL, Lamb GD. Reduced Mg^{2+} inhibition of Ca^{2+} release in muscle fibers of pigs susceptible to malignant hyperthermia. *Am J Physiol* 1997;**272**(1 Pt 1):C203–11.

90. Cully TR, Edwards JN, Launikonis BS. Activation and propagation of Ca^{2+} release from inside the sarcoplasmic reticulum network of mammalian skeletal muscle. *J Physiol* 2014;**592**(Pt 17):3727–46.

91. Allen DG, Lamb GD, Westerblad H. Skeletal muscle fatigue: cellular mechanisms. *Physiol Rev* 2008;**88**(1):287–332.

92. Reid MB. Free radicals and muscle fatigue: of ROS, canaries, and the IOC. *Free Radic Biol Med* 2008;**44**(2):169–79.

93. Perry SM. Chapter 7 Investigations on the relationship between the autonomic nervous system and the triggering of malignant hyperthermia: a state-of-the-science review. *Annu Rev Nurs Res* 2014;**32**(1):135–54.

94. Haggendal J, Jonsson L, Johansson G, Bjurstrom S, Carlsten J. Disordered catechol-amine release in pigs susceptible to malignant hyperthermia. *Pharmacol Toxicol* 1988;**63**(4):257–61.

95. Roewer N, Dziadzka A, Greim CA, Kraas E, Schulte am Esch J. Cardiovascular and metabolic responses to anesthetic-induced malignant hyperthermia in swine. *Anesthesi-ology* 1995;**83**(1):141–59.

96. Haggendal J, Jonsson L, Carlsten J. The role of sympathetic activity in initiating malignant hyperthermia. *Acta Anaesthesiol Scand* 1990;**34**(8):677–82.

97. Lanner JT, Georgiou DK, Dagnino-Acosta A, et al. AICAR prevents heat-induced sudden death in RyR1 mutant mice independent of AMPK activation. *Nat Med* 2012;**18**(2):244–51.

98. Lynch GS, Ryall JG. Role of beta-adrenoceptor signaling in skeletal muscle: implications for muscle wasting and disease. *Physiol Rev* 2008;**88**(2):729–67.

99. Amburgey K, McNamara N, Bennett LR, McCormick ME, Acsadi G, Dowling JJ. Prevalence of congenital myopathies in a representative pediatric United States population. *Ann Neurol* 2011;**70**(4):662–5.

100. Ravenscroft G, Laing NG, Bonnemann CG. Pathophysiological concepts in the congenital myopathies: blurring the boundaries, sharpening the focus. *Brain* 2015;**138**(Pt 2):246–68.

101. Clarke NF, Waddell LB, Cooper ST, et al. Recessive mutations in RYR1 are a common cause of congenital fiber type disproportion. *Hum Mutat* 2010;**31**(7):E1544–50.

102. Sato I, Wu S, Ibarra MC, et al. Congenital neuromuscular disease with uniform type 1 fiber and RYR1 mutation. *Neurology* 2008;**70**(2):114–22.

103. Darin N, Tulinius M. Neuromuscular disorders in childhood: a descriptive epidemiological study from western Sweden. *Neuromuscul Disord* 2000;**10**(1):1–9.

104. Hughes MI, Hicks EM, Nevin NC, Patterson VH. The prevalence of inherited neuromuscular disease in Northern Ireland. *Neuromuscul Disord* 1996;**6**(1):69–73.

105. Colombo I, Scoto M, Manzur AY, et al. Congenital myopathies: natural history of a large pediatric cohort. *Neurology* 2015;**84**(1):28–35.

106. Maggi L, Scoto M, Cirak S, et al. Congenital myopathies–clinical features and frequency of individual subtypes diagnosed over a 5-year period in the United Kingdom. *Neuromuscul Disord* 2013;**23**(3):195–205.

107. Byrne E, Blumbergs PC, Hallpike JF. Central core disease. Study of a family with five affected generations. *J Neurol Sci* 1982;**53**(1):77–83.

108. Klein A, Lillis S, Munteanu I, et al. Clinical and genetic findings in a large cohort of patients with ryanodine receptor 1 gene-associated myopathies. *Hum Mutat* 2012;**33**(6):981–8.

109. Bharucha-Goebel DX, Santi M, Medne L, et al. Severe congenital RYR1-associated myopathy: the expanding clinicopathologic and genetic spectrum. *Neurology* 2013;**80**(17):1584–9.

110. Wilmshurst JM, Lillis S, Zhou H, et al. RYR1 mutations are a common cause of congenital myopathies with central nuclei. *Ann Neurol* 2010;**68**(5):717–26.

111. Attali R, Aharoni S, Treves S, et al. Variable myopathic presentation in a single family with novel skeletal RYR1 mutation. *PLoS One* 2013;**8**(7):e69296.

112. Carpenter D, Ismail A, Robinson RL, et al. A RYR1 mutation associated with recessive congenital myopathy and dominant malignant hyperthermia in Asian families. *Muscle Nerve* 2009;**40**(4):633–9.

113. Dowling JJ, Lillis S, Amburgey K, et al. King-Denborough syndrome with and without mutations in the skeletal muscle ryanodine receptor (RYR1) gene. *Neuromuscul Disord* 2011;**21**(6):420–7.

114. Goebel HH, Stenzel W. Practical application of electron microscopy to neuromuscular diseases. *Ultrastruct Pathol* 2013;**37**(1):15–8.

115. Klein A, Jungbluth H, Clement E, et al. Muscle magnetic resonance imaging in congenital myopathies due to ryanodine receptor type 1 gene mutations. *Arch Neurol* 2011;**68**(9):1171–9.

116. Dowling JJ, Arbogast S, Hur J, et al. Oxidative stress and successful antioxidant treatment in models of RYR1-related myopathy. *Brain* 2012;**135**(Pt 4):1115–27.

117. Hanson MG, Wilde JJ, Moreno RL, Minic AD, Niswander L. Potassium dependent rescue of a myopathy with core-like structures in mouse. *Elife* 2015;**4**.

118. Lynch PJ, Tong J, Lehane M, et al. A mutation in the transmembrane/luminal domain of the ryanodine receptor is associated with abnormal Ca^{2+} release channel function and severe central core disease. *Proc Natl Acad Sci USA* 1999;**96**(7):4164–9.

119. Tong J, McCarthy TV, MacLennan DH. Measurement of resting cytosolic Ca^{2+} concentrations and Ca^{2+} store size in HEK-293 cells transfected with malignant hyperthermia or central core disease mutant Ca^{2+} release channels. *J Biol Chem* 1999;**274**(2):693–702.

120. Tilgen N, Zorzato F, Halliger-Keller B, et al. Identification of four novel mutations in the C-terminal membrane spanning domain of the ryanodine receptor 1: association with central core disease and alteration of calcium homeostasis. *Hum Mol Genet* 2001;**10**(25):2879–87.

121. Avila G, O'Brien JJ, Dirksen RT. Excitation–contraction uncoupling by a human central core disease mutation in the ryanodine receptor. *Proc Natl Acad Sci USA* 2001;**98**(7):4215–20.

122. Avila G, Dirksen RT. Functional effects of central core disease mutations in the cytoplasmic region of the skeletal muscle ryanodine receptor. *J Gen Physiol* 2001;**118**(3):277–90.

123. Zvaritch E, Depreux F, Kraeva N, et al. An Ryr1I4895T mutation abolishes Ca^{2+} release channel function and delays development in homozygous offspring of a mutant mouse line. *Proc Natl Acad Sci USA* 2007;**104**(47):18537–42.

124. Pisaniello A, Serra C, Rossi D, et al. The block of ryanodine receptors selectively inhibits fetal myoblast differentiation. *J Cell Sci* 2003;**116**(Pt 8):1589–97.

125. Takeshima H, Iino M, Takekura H, et al. Excitation-contraction uncoupling and muscular degeneration in mice lacking functional skeletal muscle ryanodine-receptor gene. *Nature* 1994;**369**(6481):556–9.

126. Zvaritch E, Kraeva N, Bombardier E, et al. Ca^{2+} dysregulation in Ryr1(I4895T/wt) mice causes congenital myopathy with progressive formation of minicores, cores, and nemaline rods. *Proc Natl Acad Sci USA* 2009;**106**(51):21813–8.

127. Boncompagni S, Loy RE, Dirksen RT, Franzini-Armstrong C. The I4895T mutation in the type 1 ryanodine receptor induces fiber-type specific alterations in skeletal muscle that mimic premature aging. *Aging Cell* 2010;**9**(6):958–70.

128. Loy RE, Orynbayev M, Xu L, et al. Muscle weakness in Ryr1I4895T/WT knock-in mice as a result of reduced ryanodine receptor Ca^{2+} ion permeation and release from the sarcoplasmic reticulum. *J Gen Physiol* 2010;**137**(1):43–57.

129. Scacheri PC, Hoffman EP, Fratkin JD, et al. A novel ryanodine receptor gene mutation causing both cores and rods in congenital myopathy. *Neurology* 2000;**55**(11):1689–96.

130. Leenhardt A, Lucet V, Denjoy I, Grau F, Ngoc DD, Coumel P. Catecholaminergic polymorphic ventricular tachycardia in children. A 7-year follow-up of 21 patients. *Circulation* 1995;**91**(5):1512–9.

131. Sumitomo N, Harada K, Nagashima M, et al. Catecholaminergic polymorphic ventricular tachycardia: electrocardiographic characteristics and optimal therapeutic strategies to prevent sudden death. *Heart* 2003;**89**(1):66–70.

132. Priori SG, Wilde AA, Horie M, et al. HRS/EHRA/APHRS expert consensus statement on the diagnosis and management of patients with inherited primary arrhythmia syndromes: document endorsed by HRS, EHRA, and APHRS in May 2013 and by ACCF, AHA, PACES, and AEPC in June 2013. p. 1932.

133. Kozlovski J, Ingles J, Connell V, et al. Delay to diagnosis amongst patients with catecholaminergic polymorphic ventricular tachycardia. *Int J Cardiol* 2014;**176**(3):1402–4.

134. Priori SG, Napolitano C, Memmi M, et al. Clinical and molecular characterization of patients with catecholaminergic polymorphic ventricular tachycardia. *Circulation* 2002;**106**(1):69–74.

135. Hayashi M, Denjoy I, Extramiana F, et al. Incidence and risk factors of arrhythmic events in catecholaminergic polymorphic ventricular tachycardia. *Circulation* 2009;**119**(18):2426–34.
136. Medeiros-Domingo A, Bhuiyan ZA, Tester DJ, et al. The RYR2-encoded ryanodine receptor/calcium release channel in patients diagnosed previously with either catecholaminergic polymorphic ventricular tachycardia or genotype negative, exercise-induced long QT syndrome: a comprehensive open reading frame mutational analysis. *J Am Coll Cardiol* 2009;**54**(22):2065–74.
137. Laitinen PJ, Brown KM, Piippo K, et al. Mutations of the cardiac ryanodine receptor (RyR2) gene in familial polymorphic ventricular tachycardia. *Circulation* 2001;**103**(4):485–90.
138. Kawamura M, Ohno S, Naiki N, et al. Genetic background of catecholaminergic polymorphic ventricular tachycardia in Japan. *Circ J* 2013;**77**(7):1705–13.
139. Lahat H, Pras E, Olender T, et al. A missense mutation in a highly conserved region of CASQ2 is associated with autosomal recessive catecholamine-induced polymorphic ventricular tachycardia in Bedouin families from Israel. *Am J Hum Genet* 2001;**69**(6):1378–84.
140. Roux-Buisson N, Cacheux M, Fourest-Lieuvin A, et al. Absence of triadin, a protein of the calcium release complex, is responsible for cardiac arrhythmia with sudden death in human. *Hum Mol Genet* 2012;**21**(12):2759–67.
141. Jabbari J, Jabbari R, Nielsen MW, et al. New exome data question the pathogenicity of genetic variants previously associated with catecholaminergic polymorphic ventricular tachycardia. *Circ Cardiovasc Genet* 2013;**6**(5):481–9.
142. Rosso R, Kalman JM, Rogowski O, et al. Calcium channel blockers and beta-blockers versus beta-blockers alone for preventing exercise-induced arrhythmias in catecholaminergic polymorphic ventricular tachycardia. *Heart Rhythm* 2007;**4**(9):1149–54.
143. Hwang HS, Hasdemir C, Laver D, et al. Inhibition of cardiac Ca(2+) release channels (RyR2) determines efficacy of class I antiarrhythmic drugs in catecholaminergic polymorphic ventricular tachycardia. *Circ Arrhythm Electrophysiol* 2011;**4**(2):128–35.
144. van der Werf C, Kannankeril PJ, Sacher F, et al. Flecainide therapy reduces exercise-induced ventricular arrhythmias in patients with catecholaminergic polymorphic ventricular tachycardia. *J Am Coll Cardiol* 2011;**57**(22):2244–54.
145. Watanabe H, Chopra N, Laver D, et al. Flecainide prevents catecholaminergic polymorphic ventricular tachycardia in mice and humans. *Nat Med* 2009;**15**(4):380–3.
146. Bannister ML, Thomas NL, Sikkel MB, et al. The mechanism of flecainide action in CPVT does not involve a direct effect on RyR2. *Circ Res* 2015;**116**(8):1324–35.
147. Liu N, Denegri M, Ruan Y, et al. Short communication: flecainide exerts an antiarrhythmic effect in a mouse model of catecholaminergic polymorphic ventricular tachycardia by increasing the threshold for triggered activity. *Circ Res* 2011;**109**(3):291–5.
148. Leenhardt A, Denjoy I, Guicheney P. Catecholaminergic polymorphic ventricular tachycardia. *Circ Arrhythm Electrophysiol* 2012;**5**(5):1044–52.
149. Hofferberth SC, Cecchin F, Loberman D, Fynn-Thompson F. Left thoracoscopic sympathectomy for cardiac denervation in patients with life-threatening ventricular arrhythmias. *J Thorac Cardiovasc Surg* 2013;**147**(1):404–9.
150. Schneider HE, Steinmetz M, Krause U, Kriebel T, Ruschewski W, Paul T. Left cardiac sympathetic denervation for the management of life-threatening ventricular tachyarrhythmias in young patients with catecholaminergic polymorphic ventricular tachycardia and long QT syndrome. *Clin Res Cardiol* 2012;**102**(1):33–42.
151. Denegri M, Bongianino R, Lodola F, et al. Single delivery of an adeno-associated viral construct to transfer the CASQ2 gene to knock-in mice affected by catecholaminergic polymorphic ventricular tachycardia is able to cure the disease from birth to advanced age. *Circulation* 2014;**129**(25):2673–81.

152. Katz AM. *Physiology of the heart*. Lippincott Williams & Wilkins; 2006.
153. Eisner DA, Choi HS, Diaz ME, O'Neill SC, Trafford AW. Integrative analysis of calcium cycling in cardiac muscle. *Circ Res* 2000;**87**(12):1087–94.
154. Dibb KM, Graham HK, Venetucci LA, Eisner DA, Trafford AW. Analysis of cellular calcium fluxes in cardiac muscle to understand calcium homeostasis in the heart. *Cell Calcium* 2007;**42**(4–5):503–12.
155. Eisner D, Bode E, Venetucci L, Trafford A. Calcium flux balance in the heart. *J Mol Cell Cardiol* 2013;**58**:110–7.
156. Van Petegem F. Ryanodine receptors: allosteric ion channel giants. *J Mol Biol* 2015 ;**427**(1):31–53.
157. Uchinoumi H, Yano M, Suetomi T, et al. Catecholaminergic polymorphic ventricular tachycardia is caused by mutation-linked defective conformational regulation of the ryanodine receptor. *Circ Res* 2010;**106**(8):1413–24.
158. Yamamoto T, El-Hayek R, Ikemoto N. Postulated role of interdomain interaction within the ryanodine receptor in Ca(2+) channel regulation. *J Biol Chem* 2000;**275**(16):11618–25.
159. Yang Z, Ikemoto N, Lamb GD, Steele DS. The RyR2 central domain peptide DPc10 lowers the threshold for spontaneous Ca^{2+} release in permeabilized cardiomyocytes. *Cardiovasc Res* 2006;**70**(3):475–85.
160. Pogwizd SM, Qi M, Yuan W, Samarel AM, Bers DM. Upregulation of Na(+)/Ca(2+) exchanger expression and function in an arrhythmogenic rabbit model of heart failure. *Circ Res* 1999;**85**(11):1009–19.
161. Pogwizd SM, Schlotthauer K, Li L, Yuan W, Bers DM. Arrhythmogenesis and contractile dysfunction in heart failure: roles of sodium-calcium exchange, inward rectifier potassium current, and residual beta-adrenergic responsiveness. *Circ Res* 2001;**88**(11):1159–67.
162. Tomaselli GF, Beuckelmann DJ, Calkins HG, et al. Sudden cardiac death in heart failure. The role of abnormal repolarization. *Circulation* 1994;**90**(5):2534–9.
163. Burashnikov A, Antzelevitch C. Reinduction of atrial fibrillation immediately after termination of the arrhythmia is mediated by late phase 3 early afterdepolarization-induced triggered activity. *Circulation* 2003;**107**(18):2355–60.
164. Voigt N, Li N, Wang Q, et al. Enhanced sarcoplasmic reticulum Ca^{2+} leak and increased Na+-Ca^{2+} exchanger function underlie delayed afterdepolarizations in patients with chronic atrial fibrillation. *Circulation* 2012;**125**(17):2059–70.
165. Demiryurek AT, Demiryurek S. Cardiotoxicity of digitalis glycosides: roles of autonomic pathways, autacoids and ion channels. *Auton Autacoid Pharmacol* 2005;**25**(2):35–52.
166. Nakajima T, Kaneko Y, Taniguchi Y, et al. The mechanism of catecholaminergic polymorphic ventricular tachycardia may be triggered activity due to delayed afterdepolarization. *Eur Heart J* 1997;**18**(3):530–1.
167. Priori SG, Chen SR. Inherited dysfunction of sarcoplasmic reticulum Ca^{2+} handling and arrhythmogenesis. *Circ Res* 2011;**108**(7):871–83.
168. Dumotier BM. A straightforward guide to the basic science behind arrhythmogenesis. *Heart* 2014;**100**(24):1907–15.
169. Capogrossi MC, Houser SR, Bahinski A, Lakatta EG. Synchronous occurrence of spontaneous localized calcium release from the sarcoplasmic reticulum generates action potentials in rat cardiac ventricular myocytes at normal resting membrane potential. *Circ Res* 1987;**61**(4):498–503.
170. Luo CH, Rudy Y. A dynamic model of the cardiac ventricular action potential. II. Afterdepolarizations, triggered activity, and potentiation. *Circ Res* 1994;**74**(6):1097–113.
171. Schlotthauer K, Bers DM. Sarcoplasmic reticulum Ca(2+) release causes myocyte depolarization. Underlying mechanism and threshold for triggered action potentials. *Circ Res* 2000;**87**(9):774–80.
172. Verkerk AO, Veldkamp MW, Baartscheer A, et al. Ionic mechanism of delayed afterdepolarizations in ventricular cells isolated from human end-stage failing hearts. *Circulation* 2001;**104**(22):2728–33.

173. Xie Y, Sato D, Garfinkel A, Qu Z, Weiss JN. So little source, so much sink: requirements for afterdepolarizations to propagate in tissue. *Biophys J* 2010;**99**(5):1408–15.
174. Priori SG, Napolitano C. Cardiac and skeletal muscle disorders caused by mutations in the intracellular Ca^{2+} release channels. *J Clin Invest* 2005;**115**(8):2033–8.
175. Maruyama M, Xiao J, Zhou Q, et al. Carvedilol analogue inhibits triggered activities evoked by both early and delayed afterdepolarizations. *Heart Rhythm* 2012;**10**(1):101–7.
176. Jiang D, Wang R, Xiao B, et al. Enhanced store overload-induced Ca^{2+} release and channel sensitivity to luminal Ca^{2+} activation are common defects of RyR2 mutations linked to ventricular tachycardia and sudden death. *Circ Res* 2005;**97**(11):1173–81.
177. Jiang D, Xiao B, Yang D, et al. RyR2 mutations linked to ventricular tachycardia and sudden death reduce the threshold for store-overload-induced Ca^{2+} release (SOICR). *Proc Natl Acad Sci USA* 2004;**101**(35):13062–7.
178. Lehnart SE, Wehrens XH, Laitinen PJ, et al. Sudden death in familial polymorphic ventricular tachycardia associated with calcium release channel (ryanodine receptor) leak. *Circulation* 2004;**109**(25):3208–14.
179. Thomas NL, George CH, Lai FA. Functional heterogeneity of ryanodine receptor mutations associated with sudden cardiac death. *Cardiovasc Res* 2004;**64**(1):52–60.
180. Kashimura T, Briston SJ, Trafford AW, et al. In the RyR2(R4496C) mouse model of CPVT, beta-adrenergic stimulation induces Ca waves by increasing SR Ca content and not by decreasing the threshold for Ca waves. *Circ Res* 2010;**107**(12):1483–9.
181. Bassani JW, Yuan W, Bers DM. Fractional SR Ca release is regulated by trigger Ca and SR Ca content in cardiac myocytes. *Am J Physiol* 1995;**268**(5 Pt 1):C1313–9.
182. Shannon TR, Ginsburg KS, Bers DM. Potentiation of fractional sarcoplasmic reticulum calcium release by total and free intra-sarcoplasmic reticulum calcium concentration. *Biophys J* 2000;**78**(1):334–43.
183. Shannon TR, Ginsburg KS, Bers DM. Quantitative assessment of the SR Ca^{2+} leak-load relationship. *Circ Res* 2002;**91**(7):594–600.
184. Venetucci LA, Trafford AW, Eisner DA. Increasing ryanodine receptor open probability alone does not produce arrhythmogenic calcium waves: threshold sarcoplasmic reticulum calcium content is required. *Circ Res* 2007;**100**(1):105–11.
185. Wehrens XH, Lehnart SE, Huang F, et al. FKBP12.6 deficiency and defective calcium release channel (ryanodine receptor) function linked to exercise-induced sudden cardiac death. *Cell* 2003;**113**(7):829–40.
186. Xiao B, Tian X, Xie W, et al. Functional consequence of protein kinase A-dependent phosphorylation of the cardiac ryanodine receptor: sensitization of store overload-induced Ca^{2+} release. *J Biol Chem* 2007;**282**(41):30256–64.
187. Liu N, Colombi B, Memmi M, et al. Arrhythmogenesis in catecholaminergic polymorphic ventricular tachycardia: insights from a RyR2 R4496C knock-in mouse model. *Circ Res* 2006;**99**(3):292–8.
188. George CH, Higgs GV, Lai FA. Ryanodine receptor mutations associated with stress-induced ventricular tachycardia mediate increased calcium release in stimulated cardiomyocytes. *Circ Res* 2003;**93**(6):531–40.
189. Kannankeril PJ, Mitchell BM, Goonasekera SA, et al. Mice with the R176Q cardiac ryanodine receptor mutation exhibit catecholamine-induced ventricular tachycardia and cardiomyopathy. *Proc Natl Acad Sci USA* 2006;**103**(32):12179–84.
190. Jiang D, Chen W, Wang R, Zhang L, Chen SR. Loss of luminal Ca^{2+} activation in the cardiac ryanodine receptor is associated with ventricular fibrillation and sudden death. *Proc Natl Acad Sci USA* 2007;**104**(46):18309–14.
191. Bround MJ, Asghari P, Wambolt RB, et al. Cardiac ryanodine receptors control heart rate and rhythmicity in adult mice. *Cardiovasc Res* 2012;**96**(3):372–80.
192. Thomas NL, Lai FA, George CH. Differential Ca^{2+} sensitivity of RyR2 mutations reveals distinct mechanisms of channel dysfunction in sudden cardiac death. *Biochem Biophys Res Commun* 2005;**331**(1):231–8.

193. Diaz ME, Eisner DA, O'Neill SC. Depressed ryanodine receptor activity increases variability and duration of the systolic Ca^{2+} transient in rat ventricular myocytes. *Circ Res* 2002;**91**(7):585–93.

194. Rizzi N, Liu N, Napolitano C, et al. Unexpected structural and functional consequences of the R33Q homozygous mutation in cardiac calsequestrin: a complex arrhythmogenic cascade in a knock in mouse model. *Circ Res* 2008;**103**(3):298–306.

195. Song L, Alcalai R, Arad M, et al. Calsequestrin 2 (CASQ2) mutations increase expression of calreticulin and ryanodine receptors, causing catecholaminergic polymorphic ventricular tachycardia. *J Clin Invest* 2007;**117**(7):1814–23.

196. Liu N, Denegri M, Dun W, et al. Abnormal propagation of calcium waves and ultrastructural remodeling in recessive catecholaminergic polymorphic ventricular tachycardia. *Circ Res* 2013;**113**(2):142–52.

197. Viatchenko-Karpinski S, Terentyev D, Gyorke I, et al. Abnormal calcium signaling and sudden cardiac death associated with mutation of calsequestrin. *Circ Res* 2004;**94**(4):471–7.

198. Knollmann BC, Chopra N, Hlaing T, et al. Casq2 deletion causes sarcoplasmic reticulum volume increase, premature Ca^{2+} release, and catecholaminergic polymorphic ventricular tachycardia. *J Clin Invest* 2006;**116**(9):2510–20.

199. Faggioni M, Knollmann BC. Calsequestrin 2 and arrhythmias. *Am J Physiol Heart Circ Physiol* 2012;**302**(6):H1250–60.

200. Radwanski PB, Belevych AE, Brunello L, Carnes CA, Gyorke S. Store-dependent deactivation: cooling the chain-reaction of myocardial calcium signaling. *J Mol Cell Cardiol* 2012;**58**:77–83.

201. Kornyeyev D, Petrosky AD, Zepeda B, Ferreiro M, Knollmann B, Escobar AL. Calsequestrin 2 deletion shortens the refractoriness of Ca(2)(+) release and reduces rate-dependent Ca(2)(+)-alternans in intact mouse hearts. *J Mol Cell Cardiol* 2011;**52**(1):21–31.

202. Brunello L, Slabaugh JL, Radwanski PB, et al. Decreased RyR2 refractoriness determines myocardial synchronization of aberrant Ca^{2+} release in a genetic model of arrhythmia. *Proc Natl Acad Sci USA* 2013;**110**(25):10312–7.

203. Chen H, Valle G, Furlan S, et al. Mechanism of calsequestrin regulation of single cardiac ryanodine receptor in normal and pathological conditions. *J Gen Physiol* 2013;**142**(2):127–36.

204. Gyorke I, Hester N, Jones LR, Gyorke S. The role of calsequestrin, triadin, and junctin in conferring cardiac ryanodine receptor responsiveness to luminal calcium. *Biophys J* 2004;**86**(4):2121–8.

205. Qin J, Valle G, Nani A, et al. Luminal Ca^{2+} regulation of single cardiac ryanodine receptors: insights provided by calsequestrin and its mutants. *J Gen Physiol* 2008;**131**(4):325–34.

206. Saucerman JJ, Bers DM. Calmodulin binding proteins provide domains of local Ca^{2+} signaling in cardiac myocytes. *J Mol Cell Cardiol* 2011;**52**(2):312–6.

207. Chopra N, Yang T, Asghari P, et al. Ablation of triadin causes loss of cardiac Ca^{2+} release units, impaired excitation-contraction coupling, and cardiac arrhythmias. *Proc Natl Acad Sci USA* 2009;**106**(18):7636–41.

208. Quarta G, Syrris P, Ashworth M, et al. Mutations in the Lamin A/C gene mimic arrhythmogenic right ventricular cardiomyopathy. *Eur Heart J* 2011;**33**(9):1128–36.

209. Bauce B, Nava A, Rampazzo A, et al. Familial effort polymorphic ventricular arrhythmias in arrhythmogenic right ventricular cardiomyopathy map to chromosome 1q42-43. *Am J Cardiol* 2000;**85**(5):573–9.

210. Milting H, Lukas N, Klauke B, et al. Composite polymorphisms in the ryanodine receptor 2 gene associated with arrhythmogenic right ventricular cardiomyopathy. *Cardiovasc Res* 2006;**71**(3):496–505.

211. Nava A, Canciani B, Daliento L, et al. Juvenile sudden death and effort ventricular tachycardias in a family with right ventricular cardiomyopathy. *Int J Cardiol* 1988;**21**(2):111–126.

212. Rampazzo A, Nava A, Erne P, et al. A new locus for arrhythmogenic right ventricular cardiomyopathy (ARVD2) maps to chromosome 1q42-q43. *Hum Mol Genet* 1995;**4**(11):2151–4.

213. Tiso N, Stephan DA, Nava A, et al. Identification of mutations in the cardiac ryanodine receptor gene in families affected with arrhythmogenic right ventricular cardiomyopathy type 2 (ARVD2). *Hum Mol Genet* 2001;**10**(3):189–94.

214. d'Amati G, Bagattin A, Bauce B, et al. Juvenile sudden death in a family with polymorphic ventricular arrhythmias caused by a novel RyR2 gene mutation: evidence of specific morphological substrates. *Hum Pathol* 2005;**36**(7):761–7.

215. Basso C, Bauce B, Corrado D, Thiene G. Pathophysiology of arrhythmogenic cardiomyopathy. *Nat Rev Cardiol* 2011;**9**(4):223–33.

216. Marcus FI, Zareba W, Calkins H, et al. Arrhythmogenic right ventricular cardiomyopathy/dysplasia clinical presentation and diagnostic evaluation: results from the North American Multidisciplinary Study. *Heart Rhythm* 2009;**6**(7):984–92.

217. Delmar M, McKenna WJ. The cardiac desmosome and arrhythmogenic cardiomyopathies: from gene to disease. *Circ Res* 2010;**107**(6):700–14.

218. Basso C, Corrado D, Marcus FI, Nava A, Thiene G. Arrhythmogenic right ventricular cardiomyopathy. *Lancet* 2009;**373**(9671):1289–300.

219. Terentyev D, Nori A, Santoro M, et al. Abnormal interactions of calsequestrin with the ryanodine receptor calcium release channel complex linked to exercise-induced sudden cardiac death. *Circ Res* 2006;**98**(9):1151–8.

220. Koop A, Goldmann P, Chen SR, Thieleczek R, Varsanyi M. ARVC-related mutations in divergent region 3 alter functional properties of the cardiac ryanodine receptor. *Biophys J* 2008;**94**(12):4668–77.

Dravet Syndrome: A Sodium Channel Interneuronopathy

W.A. Catterall

University of Washington, Seattle, WA, United States

INTRODUCTION

Neurological and psychiatric syndromes often have multiple disease traits, yet it is unknown how such multifaceted deficits arise from single mutations. Haploinsufficiency of the voltage-gated sodium channel $Na_V1.1$ causes Dravet syndrome (DS), an intractable childhood-onset epilepsy with ataxia, hyperactivity, sleep disorder, cognitive deficit, autistic-like behaviors, and premature death.[1,2] The gene encoding $Na_V1.1$ channels, *SCN1A*, is the gene most highly associated with epilepsy.[3] It is also closely associated with autism spectrum disorders in genome-wide association studies and in deep DNA-sequencing of genetic variants.[4,5] This chapter reviews research on mouse genetic models of DS that have given valuable insights into pathophysiology and potential therapies for this devastating childhood disease.

VOLTAGE-GATED SODIUM CHANNELS

Voltage-gated Na^+ channels in the brain are complexes of a 260-kD α-subunit in association with auxiliary β-subunits (β1–β4) of 33–36 kD.[6] The α subunit contains the voltage sensors and the ion-conducting pore in four internally repeated domains (I–IV), which each consists of six α-helical transmembrane segments (S1–S6) and a pore loop connecting S5 and S6.[6] The β-subunits modify the kinetics and voltage dependence of gating and serve as cell adhesion molecules interacting with extracellular matrix, other cell adhesion molecules, and the cytoskeleton.[7,8] The voltage-gated ion channels are among the most ancient and conserved gene families, with sequence identity of >50% in the transmembrane domains of human sodium channel α subunits and those of the simplest multicellular eukaryotes. The mammalian genome contains nine functional voltage-gated sodium channel α subunits, which differ in patterns of tissue expression and biophysical properties. The $Na_V1.1$, $Na_V1.2$, $Na_V1.3$, and $Na_V1.6$ channel subtypes, encoded by the *SCN1A*, *SCN2A*, *SCN3A*, and *SCN8A* genes, are the primary sodium channels in the central nervous system.[6,9–11] $Na_V1.1$ and $Na_V1.3$ channels are primarily localized in cell bodies and axon initial segments,[12–15] $Na_V1.2$ channels in unmyelinated or premyelinated axons and dendrites,[12,13] and $Na_V1.6$ channels in myelinated axons and in dendrites.[16,17] These channels participate in generation of both somatodendritic and axonal action potentials (APs).[18–22]

$Na_V1.1$ CHANNELS AND INHERITED EPILEPSY

Screening of human patients with inherited epilepsy first led to the identification of mutations of $Na_V1.1$ channels encoded by the *SCN1A* gene in families with Generalized Epilepsy with Febrile Seizures Plus (GEFS+, OMIM 604233).[23] This inherited epilepsy syndrome causes febrile seizures and spontaneous seizures, but they are usually well controlled by conventional antiepileptic drug therapy. More than 20 different mutations have been identified in GEFS+ patients and their family members. GEFS+ is caused by inherited missense mutations that alter the functional properties of the primary pore-forming α subunit of the channel.[24] Moreover, a mutation in the $Na_V\beta1$ subunit also causes GEFS+ epilepsy, very likely by impairing expression and function of $Na_V1.1$ channels.[25]

Unexpectedly, subsequent work showed that children with Dravet syndrome (also called Severe Myoclonic Epilepsy, OMIM 607208)[26] carry de novo mutations in one allele of the *SCN1A* gene, leading to haploinsufficiency of $Na_V1.1$ channels.[26–31] DS is a much more severe form of epilepsy than GEFS+, and it is not well controlled with conventional antiepileptic drug therapy. More than 600 de novo mutations in the coding sequences of the *SCN1A* gene have been identified, accounting for more than 80% of DS cases.[24,26,29,32–34] More than half of the DS mutations cause loss-of-function due to stop codons or deletions, demonstrating that haploinsufficiency of *SCN1A* is pathogenic. Since only coding regions of the gene are sequenced, it is plausible that many of the remaining 20% of DS patients harbor mutations in regulatory regions of the gene outside of the coding sequences that impair or prevent channel expression. Studies show that homozygous loss-of-function mutations in the $Na_V\beta1$ subunits also cause DS, probably by impairing expression of $Na_V1.1$ channels on the cell surface.[35]

DRAVET SYNDROME

DS begins during the first year of life, with seizures often associated with elevated body temperature due to fever or bathing, and progresses to prolonged, clustered, or continuous seizures and to status epilepticus.[36,37] After the second year of life, patients develop comorbidities including psychomotor delay, ataxia, and cognitive impairment. Medically refractory seizures, including frequent and prolonged episodes of status epilepticus, contribute to substantial sudden unexpected death in epilepsy (SUDEP) in the time period of 1–4 years of age, and there is an unfavorable long-term outcome for the children who survive to sexual maturity.[37,38] Patients who reach their teenage years typically have IQs of 50 and require life-long care. It is a surprise that haploinsufficiency of a Na_V channel causes such a severe epilepsy syndrome, because reduced sodium current should lead to hypoexcitability rather than hyperexcitability. To understand the mechanistic basis for hyperexcitability, epilepsy, and comorbidities in DS, animal models were generated by targeted deletion or mutation of the *Scn1a* gene in mice.[39,40]

EPILEPSY AND PREMATURE DEATH IN DRAVET SYNDROME

Timing of DS and $Na_V1.1$ expression in mouse and man: Individuals with DS develop normally from birth until the time of initial seizure presentation at 5–11 months.[1] Similarly, DS mice are indistinguishable from their wild-type (WT) littermates at birth and develop normally through the third week of life. During the fourth and fifth postnatal weeks, DS mice become susceptible to thermally induced and spontaneous seizures as well as premature death.[39–41] Susceptibility to thermally induced seizures, spontaneous seizures, and SUDEP begins near the time of weaning in both mouse (P21) and human (6–9 months), suggesting that disease onset occurs at a similar stage of development in both species. The timing of premature death in mouse and human is illustrated in Fig. 4.1 for mice described in Yu et al. (2006)[39] and humans in a well-studied Japanese cohort.[42] The similarity in timing of disease onset and progression to SUDEP in human and mouse DS suggests that there is a common underlying molecular event triggering disease onset in the two species and that the mouse model is an accurate phenocopy of the human disease.

In rodents, $Na_V1.1$, $Na_V1.2$, and $Na_V1.6$ channel expression increases progressively from birth, while $Na_V1.3$ channel expression is highest just prior to birth and declines in early postnatal life.[43–46] In mice, the decline in $Na_V1.3$ channel protein reaches its lowest levels at P21, and the rise in

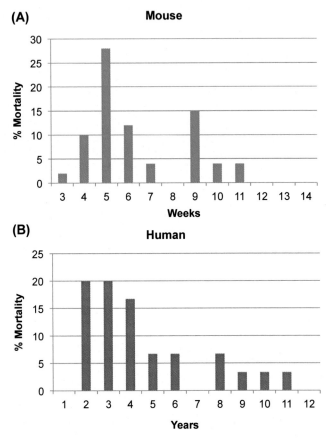

FIGURE 4.1   Timing of premature death in mouse (A) and human (B) Dravet syndrome. y-axis, % mortality for deaths from SUDEP; x-axis, time in weeks or years during which the indicated number of deaths occurred. *Mortality data are replotted from (A) Yu FH, Mantegazza M, Westenbroek RE, et al. Reduced sodium current in GABAergic interneurons in a mouse model of severe myoclonic epilepsy in infancy.* Nat Neurosci 2006;**9**(9):1142–9 *and (B) Sakauchi M, Oguni H, Kato I, et al. Mortality in Dravet syndrome: search for risk factors in Japanese patients.* Epilepsia 2011;**52**(Suppl. 2):50–4.

$Na_V1.1$ channels begins at P10 and continues through P30. Susceptibility to thermally induced seizures begins at P20, and there is a sharp increase in the frequency of spontaneous seizures and death from P20 to P25 (Fig. 4.1).[41,47] Children diagnosed with DS have an average age of seizure onset of 5.7 months.[1] DS progresses over 1–2 years to multiple forms of spontaneous seizures, including generalized tonic-clonic seizures, resistance to drug therapy, developmental delay, and permanent mental and physical disabilities.[1,2] Following the onset of increased seizure incidence, the risk of SUDEP is also elevated with the majority of unexpected deaths occurring during the first 4 years of life (Fig. 4.1).[42] Immunoblotting with subtype-specific antibodies showed that Na_V protein expression in human brain parallels that in rodent brain, with levels of $Na_V1.1$ low at birth and steadily increasing to peak values by 20 months while $Na_V1.3$ protein expression is high at birth and decreases to a low level at 6 months of age.[48] The time at which the decline of $Na_V1.3$ and the rise of $Na_V1.1$ expression cross at 5–6 months correlates closely with seizure onset in patients with DS. These data in human and mouse suggest that the timing of DS onset is a result of the failure of $Na_V1.1$ channels to fully replace the normal developmental loss of embryonic $Na_V1.3$ channels. The time of crossover of declining levels of $Na_V1.3$ channels and rising levels of $Na_V1.1$ channels may define the tipping point for development of DS, as the mutations characteristic of this disease prevent the normal increase in functional $Na_V1.1$ channels in forebrain GABAergic inhibitory neurons, leading to disinhibition of neural circuits, epilepsy, developmental delay, comorbidities, and SUDEP.

Selective loss of excitability of GABAergic interneurons and hyperexcitability in DS mice: Homozygous null $Na_V1.1(-/-)$ mice developed ataxia and died on P15.[39,40] Heterozygous $Na_V1.1(+/-)$ mice exhibited spontaneous seizures and sporadic deaths beginning after P21, with a striking dependence on genetic background.[39] Loss of $Na_V1.1$ did not change voltage-dependent activation or inactivation of sodium channels in hippocampal neurons.[39] However, the sodium current density was substantially reduced in inhibitory interneurons of $Na_V1.1(+/-)$ and $Na_V1.1(-/-)$ mice, but not in their excitatory pyramidal neurons. This reduction in sodium current caused a loss of sustained high-frequency firing of APs in hippocampal and cortical interneurons,[39,40] thereby impairing their inhibitory function that depends on generation of high-frequency bursts of APs. These results suggested that reduced sodium currents in GABAergic inhibitory interneurons in $Na_V1.1(+/-)$ heterozygotes may cause the hyperexcitability that leads to epilepsy in patients with DS.

Thermally induced seizures in DS mice: Children with DS frequently have seizures with elevated body temperature as their first symptom.[49] Experiments with our mouse model of DS demonstrated that haploinsufficiency of $Na_V1.1$ channels is sufficient for induction of seizures by elevated body temperature.[41] P17-18 mice with DS did not have thermally induced seizures, but nearly all P20-22 and P30-46 mice with DS had myoclonic seizures followed by generalized seizures with elevated core body temperature. Spontaneous seizures were only observed in mice older than P21, indicating that mice with DS become susceptible to temperature-induced seizures before spontaneous seizures, similar to children with DS. Interictal spike activity was seen at normal body temperature in most P30-46 mice with DS but not in P20-22 or P17-18 mice, suggesting that interictal epileptic activity correlates with seizure susceptibility. These results define a critical developmental transition for susceptibility to seizures in a mouse model of DS and reveal a close correspondence between human and mouse DS in the striking temperature- and age-dependence of DS onset and progression.

Selective gene deletion in forebrain interneurons replicates DS in mice: Our studies of DS mice revealed reduced sodium currents and impaired excitability in GABAergic interneurons in the hippocampus, leading to the hypothesis that impaired excitability of GABAergic inhibitory neurons is the cause of epilepsy and premature death in DS. However, other classes of GABAergic interneurons are less impaired, so the direct cause of hyperexcitability, epilepsy, and premature death remained unresolved. We generated a floxed $Scn1a$ mouse line and used the Cre-Lox method driven by an enhancer from the $Dlx1,2$ locus for conditional deletion of $Scn1a$ in forebrain GABAergic neurons.[50] Immunocytochemical studies demonstrated selective loss of $Na_V1.1$ channels in GABAergic interneurons in cerebral cortex and hippocampus.[50] Mice with this deletion were equally susceptible to thermal induction of seizures as mice with global deletion of $Scn1a$ and had similar seizure severity (Fig. 4.2A and B).[50] Specifically deleted mice also died prematurely following generalized tonic-clonic seizures with a dependence on age that was similar to DS mice with global deletion of $Na_V1.1$ channels (Fig. 4.2C).[50] All premature deaths followed a series of seizures that increased in severity to Racine 5 shortly before death, as illustrated for a typical DS mouse in Fig. 4.2D.[50] Evidently, loss of $Na_V1.1$ channels in forebrain GABAergic interneurons is both necessary and sufficient to cause epilepsy and premature death in DS mice.

Mechanism of SUDEP in DS mice: SUDEP is the most common cause of death in intractable epilepsies, but physiological mechanisms that lead to SUDEP are unknown. We studied the mechanism of premature death in $Scn1a$ heterozygous knockout mice and conditional brain- and cardiac-specific knockouts during the period of frequent SUDEP in postnatal week 4.[47] Video monitoring demonstrated that SUDEP occurred immediately following generalized tonic-clonic seizures, as shown for one exemplar forebrain-specific knockout mouse in Fig. 4.2D. A history of multiple seizures was a strong risk factor for SUDEP.[47] Combined video-electroencephalography-electrocardiography revealed suppressed interictal resting heart-rate variability and episodes of ictal bradycardia associated with the tonic phases of generalized tonic-clonic seizures.[47] Prolonged atropine-sensitive ictal bradycardia preceded SUDEP. Similar studies in conditional knockout mice demonstrated that brain, but not cardiac, knockout of $Scn1a$ produced similar cardiac and SUDEP phenotypes as in DS mice.[47] Atropine or N-methyl scopolamine treatment reduced the incidence of ictal bradycardia and SUDEP in DS mice.[47] These findings suggest that SUDEP is caused by parasympathetic hyperactivity immediately following tonic-clonic seizures in DS mice, which leads to lethal bradycardia and electrical dysfunction of the ventricle.

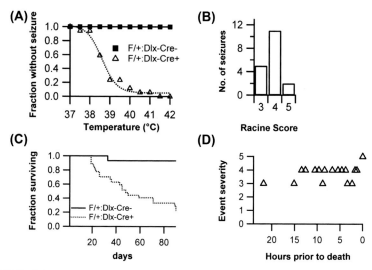

FIGURE 4.2 Conditional heterozygotes experience excess death and both thermally induced and spontaneous seizures. (A) Thermal induction of seizures in F/+:Dlx-Cre- and F/+:Dlx-Cre+ mice. (B) Severity of spontaneous seizures in F/+:Dlx-Cre+ mice. (C) Survival of F/+:Dlx-Cre- and F/+:Dlx-Cre+ mice. The fraction of each genotype surviving is plotted versus postnatal day (F/+:Dlx-Cre- $n=45$, F/+:Dlx-Cre+ $n=27$). (D) Representative example of spontaneous seizure progression in an F/+:Dlx-Cre+ mouse. The Racine score of each seizure is plotted as a function of the time before death ($t=0$ at time of death). *Adapted from Cheah CS, Yu FH, Westenbroek RE, et al. Specific deletion of Na$_V$1.1 sodium channels in inhibitory interneurons causes seizures and premature death in a mouse model of Dravet syndrome. Proc Natl Acad Sci USA 2012;109(36):14646–51.*

COMORBIDITIES IN A MOUSE MODEL OF DRAVET SYNDROME

Loss of excitability of Purkinje neurons and ataxia in DS mice: Ataxia, spasticity, and failure of motor coordination contribute substantially to the developmental delay and functional impairments of DS patients.[1] Cerebellar Purkinje cells are GABAergic inhibitory neurons that serve as the output pathway for information on movement, coordination, and balance from the cerebellar cortex. Degeneration of Purkinje neurons and abnormal expression of voltage-gated ion channels in them are associated with ataxia.[51–54] Behavioral assessment indicated severe motor deficits in homozygous Na$_V$1.1 knockout mice.[39,55] A milder impairment of normal gait, resulting in failure of proper foot placement, was observed in the heterozygotes after P21.[55] The amplitudes of whole-cell peak, persistent, and resurgent sodium currents in Purkinje neurons were reduced by 58–69%, without detectable change in the kinetics or voltage dependence of channel activation or inactivation. Current-clamp recordings revealed that the firing rates of Purkinje neurons from mutant mice were substantially reduced, with no effect on threshold for AP generation.[55] The results show that Na$_V$1.1 channels play a crucial role in the excitability of cerebellar Purkinje neurons, with major contributions to peak, persistent, and resurgent forms of sodium current and to sustained AP firing. Loss of these channels in Purkinje neurons of mutant mice and DS patients may be sufficient to cause their ataxia and related functional deficits (Table 4.1).

TABLE 4.1   Causes of Comorbidities in Dravet Syndrome

Comorbidity	Symptoms in DS Mice	Physiological Correlates	Causal Evidence for Mechanism
Ataxia	Abnormal foot placement in walking[55]	Failure of action potential firing in GABAergic cerebellar Purkinje neurons[55]	
Circadian rhythm	Long circadian cycle; weak light-induced phase shift; increased negative masking[57]	Failure of action potential firing in GABAergic neurons in the suprachiasmatic nucleus[57, 58]	Not observed in forebrain-specific knockout mouse[58]; reversed by clonazepam[57]
Sleep impairment	Reduced nonrapid-eye-movement sleep, delta wave power, sleep spindles[58]	Failure of rebound action potential firing in GABAergic neurons in the reticular nucleus of the thalamus[58]	Observed in forebrain-specific knockout mouse[58]
Cognitive deficit	Failure of spatial learning and memory[63]	Failure of action potential firing by GABAergic neurons in hippocampus and cerebral cortex[63]	Observed in forebrain-specific knockout mouse[63]; reversed by clonazepam[63]
Autistic-like behavior	Impaired social interaction; repetitive behaviors[63]	Failure of action potential firing by PV-expressing GABAergic neurons in hippocampus and cerebral cortex[63, 92]	Observed in forebrain-specific and PV-specific knock-out mouse; reversed by clonazepam[63, 92]
Hyperactivity	Increased distance explored in open field[63, 92]	Failure of action potential firing by SST-expressing GABAergic neurons in hippocampus and cerebral cortex[63, 92]	Observed in forebrain-specific and SST-specific knock-out mouse; reversed by clonazepam[63, 92]

AP, action potential; *SCN*, suprachiasmatic nucleus of the hypothalamus; *RNT*, reticular nucleus of the thalamus; *REM*, rapid-eye-movement.

Circadian rhythm defects and sleep disturbance in DS mice: Sleep defects are common in DS and are often severe.[1,56] Children fail to develop a normal 24-h sleep cycle and often wake during the night, suggesting impairment of both circadian rhythm and sleep quality. DS mice have a striking failure of regulation of circadian rhythm.[57] $Na_V1.1$ is the primary voltage-gated Na^+ channel in the suprachiasmatic nucleus (SCN) of the hypothalamus, which is composed entirely of GABAergic neurons. DS mice have longer circadian period than WT mice and lack light-induced phase shifts.[57] In contrast, Scn1a(+/−) mice have exaggerated light-induced negative-masking behavior and normal electroretinogram, suggesting an intact retina light response.[57] Electrical stimulation of the optic chiasm elicits reduced calcium transients and impaired ventrodorsal communication in SCN neurons from Scn1a(+/−) mice, and this communication is barely detectable in the homozygous gene knockout.[57] These results demonstrate that the $Na_V1.1$ channel mutation and its associated impairment of interneuronal communication between GABAergic neurons in the dorsal and ventral SCN lead to major deficits in the function of the master circadian pacemaker (Table 4.1).

Poor quality of sleep is common in DS,[56] but the physiological mechanism has been unknown. In DS mice, electroencephalographic studies revealed abnormal sleep, including reduced delta wave power, reduced sleep spindles, increased brief wakes, and numerous interictal spikes in nonrapid-eye-movement sleep.[58] Theta power was reduced in rapid-eye-movement sleep.[58] Mice with $Na_V1.1$ deleted specifically in forebrain interneurons exhibited similar sleep pathology to DS mice, but without changes in circadian rhythm.[58] These results confirm that the poor quality of sleep in DS mice is indeed due to failure of sleep rhythms in the brain rather than the SCN. Sleep architecture depends on oscillatory activity in the thalamocortical network generated by excitatory neurons in the ventrobasal nucleus and inhibitory GABAergic neurons in the reticular nucleus of the thalamus. Whole-cell Na_V current was reduced in GABAergic neurons of the reticular nucleus but not in excitatory neurons in the ventrobasal nucleus.[58] Rebound firing of APs following hyperpolarization, the firing signature of reticular nucleus neurons during sleep, was also reduced.[58] These results demonstrate imbalance of excitatory versus inhibitory neurons in this circuit. As predicted from this functional impairment, homeostatic rebound of slow wave activity following sleep deprivation was also impaired.[58] Our results show that impairment of Na_V currents and excitability of GABAergic reticular nucleus of the thalamus (RNT) neurons are correlated with impaired sleep quality and homeostasis in DS mice (Table 4.1).

Cognitive and behavioral deficits in DS mice: Unlike other generalized epilepsy disorders, DS is accompanied by characteristic neuropsychiatric comorbidities, including hyperactivity, attention deficit, delayed psychomotor development, anxiety-like behaviors, impaired social interactions, restricted interests, and severe cognitive deficits.[2,59–62] These comorbidities in DS overlap with symptoms of autism-spectrum disorders, and clinical studies suggest that DS patients have autism-spectrum behaviors.[59] DS mice display hyperactivity, stereotyped behaviors, social interaction deficits, and impaired context-dependent spatial memory.[63] Olfactory sensitivity is retained, but novel food odors and social odors are aversive to DS mice.[63] As for epilepsy and premature death, selective deletion of $Na_V1.1$ channels in forebrain interneurons is sufficient to cause these behavioral and cognitive impairments.[63] These results demonstrate a critical role for $Na_V1.1$ channels in forebrain GABAergic inhibitory neurons in the cognitive deficit, hyperactivity, and autistic-like behaviors of DS mice (Table 4.1).

GENETIC AND PHARMACOLOGICAL TREATMENT OF DRAVET SYNDROME

Balancing excitation and inhibition with genetic compensation: If DS is caused by loss of sodium current and failure of firing of GABAergic interneurons,[39,55] it may be compensated by mutations that reduce the sodium current and AP firing of excitatory neurons and thereby rebalance excitation and inhibition in the brain. Such genetic compensation can be studied by mating mouse lines having different well-defined genetic deficiencies. $Na_V1.6$ channels encoded by the *Scn8a* gene are highly expressed in excitatory neurons, and their functional properties are well suited to driving repetitive firing.[52,64,65] Double heterozygous mice with haploinsufficiency for both *Scn1a* and *Scn8a* did indeed have reduced susceptibility to drug-induced seizures and improved lifespan compared to $Na_V1.1$ heterozygotes.[66] These results support the concept that loss-of-function mutations in $Na_V1.1$ channels in DS cause an imbalance of excitation over inhibition in the brain and that this imbalance can be partially compensated by a corresponding reduction in expression of $Na_V1.6$ channels.

Balancing excitation and inhibition with pharmacological treatment: The imbalance of excitation and inhibition could also be corrected by drug treatment. Reducing sodium channel activity by treatment of patients with sodium channel blocking antiepileptic drugs exacerbates seizure frequency and intensity in DS,[67] consistent with the hypothesis that failure of AP firing due to reduced sodium current in inhibitory neurons is the underlying mechanism of the disease. However, increasing GABAergic neurotransmission by pharmacological treatment is a viable treatment approach. The benzodiazepines are positive allosteric modulators of GABA activation of the postsynaptic $GABA_A$ receptors. Therefore, they would increase the postsynaptic response to GABA released in response to APs. However, the reduced frequency of APs in GABAergic inhibitory neurons in DS decreases phasic release of GABA and impairs inhibitory neurotransmission, even in the presence of a benzodiazepine. Drugs such as tiagabine increase the concentration of GABA in the synaptic cleft by inhibiting its reuptake into nerve terminals and glia. Using thermally induced seizures in a mouse model of DS as a test system,[41] we found that combinations of tiagabine and clonazepam are effective in completely preventing thermally induced myoclonic and generalized tonic-clonic seizures.[68] The effects of the drugs are synergistic in preventing thermally induced seizures, and their synergy of action allows lower doses to be used to reduce sedative side effects, as measured in the rotarod test of balance and motor coordination.[68] These encouraging results suggest that similar combination drug therapies may be useful for control of seizures in children with DS.

Low-dose benzodiazepine treatment rescues cognitive and behavioral deficits in DS mice: Enhancing inhibitory neurotransmission might also be effective in treatment of the cognitive and behavioral deficits that are comorbidities of DS if these conditions are caused by the impairment of inhibitory neurotransmission itself rather than by neuronal damage from seizures. However, treatment is restricted to low dose levels of benzodiazepines that do not engage the sedative effects of these drugs. Remarkably, treatment with low-dose clonazepam, a classic benzodiazepine positive allosteric modulator of $GABA_A$ receptors, completely rescued the abnormal social behaviors and deficits in context-dependent fear memory in DS mice, demonstrating that they are caused by impaired GABAergic neurotransmission and not by neuronal damage from recurrent seizures.[63] Moreover, similar beneficial effects were observed in studies of the BTBR mouse, a well-validated model of idiopathic autism,[69] suggesting the potential for use of this therapeutic strategy in autism as well as DS. A clinical trial of low-dose therapy of autism with benzodiazepine-related drugs is in progress based on these results (http://clinicaltrials.gov/show/NCT01966679). Unfortunately, almost all DS patients are treated with high doses of benzodiazepines for control of their epilepsy; therefore, therapy of cognitive deficits and autistic-like behaviors with low-dose benzodiazepines would only be beneficial for the small subset of DS patients who are not receiving high-dose benzodiazepine therapy for seizure prevention.

GENETIC BACKGROUND EFFECTS IN DRAVET SYNDROME

Individual patients with apparently complete loss-of-function mutations demonstrate a broad range of disease severity and a wide-ranging combination of symptoms, suggesting that genetic background can strongly influence the severity and clinical presentation of DS.[1] Similarly, the severity of DS in mice has a striking dependence on genetic background. While DS mice in 129/SvJ genetic background (DS/129) survive nearly as well as WT, up to 80% of DS mice harboring the same mutation in C57BL/6 genetic background (DS/B6) die by 12 weeks of age.[39] Moreover, spontaneous generalized tonic-clonic (GTC) seizures are rare in DS/129, whereas they are frequent in DS/B6.[39,70] DS/129 mice survive similarly to WT mice,[39,70] and no signs of behavioral seizures were detected in video recordings during P21–P28. At P21, elevated temperature induced GTC seizures in both genetic backgrounds, but with different sensitivity (Fig. 4.3A).[41,71] Similar to DS patients,[72] DS/B6 mice had thermally induced GTC seizures just above body temperature, with a mean temperature for seizure induction of $37.9 \pm 0.3°C$.[71] In contrast, no DS/129 mice had GTC seizures below 38.5°C, but all had GTC seizures at higher temperatures, with a mean temperature for seizure induction of $39.1 \pm 0.2°C$.[71] Another striking difference between the genetic backgrounds became evident through analysis of myoclonic seizures. This type of seizure is characteristic in DS but is less common in GEFS+.[73] Similar to DS patients, all DS/B6 mice had myoclonic seizures preceding their thermally induced GTC seizures (Fig. 4.3B).[41,71] In contrast, 30% of DS/129 mice did not have myoclonic seizures (Fig. 4.3B).[71] Thus, in parallel with their lesser susceptibility to spontaneous death, the seizure phenotype is much less severe in DS/129 mice than it is in DS/B6 mice.

Since developmental delay, cognitive impairment, and social interaction deficits are observed in most DS patients,[74,75] we compared the cognitive and social abilities of DS/B6 and DS/129 mice. In contrast to DS/B6 mice,[63] DS/129 mice did not show impaired context-dependent fear memory, and their performance was indistinguishable from WT/129 (Fig. 4.3C).[71] On the other hand, similar to DS/B6 mice,[63] DS/129 mice had social interaction deficits in the three-chamber test for social interaction (Fig. 4.3D).[71] In this test, mice are placed in the central compartment of a three-chambered plexiglass box and allowed to interact with an object (a small empty cage) in the left side chamber or with a stranger mouse (in an identical small cage) in the right side chamber. Whereas WT/129 mice spent far more time interacting with a stranger mouse than with an inanimate object, DS/129 mice did not show a significant preference (Fig. 4.3D).[71] Together, these pathophysiological and behavioral data suggest that, while DS/B6 mice are an accurate phenocopy of human DS, the milder epilepsy, lower number of myoclonic seizures, and lack of cognitive deficit in DS/129 mice are more consistent with the milder clinical manifestations of human GEFS+.

To assess the physiological effects of genetic background, we studied excitability of pyramidal neurons and GABAergic interneurons in hippocampal slices, in which cellular morphology and circuit properties

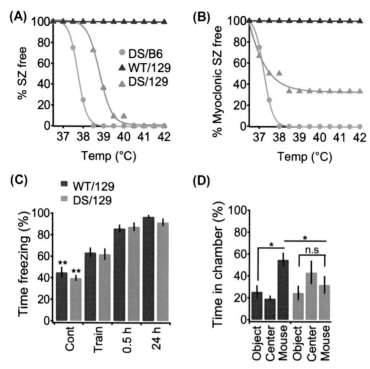

FIGURE 4.3 Epilepsy, spatial learning, and social interaction in DS/B6 and DS/129 mice. (A) Induction of seizures by increasing body core temperature. Percentage of P21 mice remaining free of behavioral seizures (SZ) versus body core temperature, fitted with a sigmoid line. Wild-type mice of either strain do not have seizures at the tested temperatures. DS/129, $n = 11$; DS/B6, $n = 4$. (B) Percentage of mice remaining free of myoclonic seizures. (C) Contextual fear conditioning test. WT/129 and DS/129 display similar freezing behavior during training and when reintroduced to the chamber 0.5 and 24 h later. $n = 14$ for each genotype. (D) Three-chamber experiment. WT/129 mice spend more time in the chamber housing a stranger mouse (mouse) than the chamber housing an empty cage (object) or in the connecting chamber (center). DS/129 mice have no preference for either chamber. $n = 10$ for each genotype. AP, action potential. *Adapted from Rubinstein M, Westenbroek RE, Yu FH, Jones CJ, Scheuer T, Catterall WA. Genetic background modulates impaired excitability of inhibitory neurons in a mouse model of Dravet syndrome. Neurobiol Dis 2015;73:106–17.*

are preserved. Deletion of Na$_V$1.1 had no effect on firing of CA1 pyramidal cells of either DS/B6 or DS/129 mice. We did not detect any difference in threshold, rheobase, or average firing frequency of APs in response to 1 s of depolarizing current injection. We also tested AP firing in response to a short (10 ms) depolarizing current injection and found no differences (Fig. 4.4A and B).[71] These data are consistent with the view that disinhibition caused by loss-of-function in interneurons, and not hyperexcitability of pyramidal cells, is the neuronal basis of epilepsy in DS mice.[39,40,50,76]

To compare the effects of the DS mutation on a specific class of inhibitory interneurons, we examined the excitability and function of CA1 stratum oriens (SO) interneurons. Horizontal SO interneurons are activated by recurrent collaterals of pyramidal cells, mediate feedback inhibition of neighboring pyramidal cells,[77–79] and regulate the strength of depolarization of CA1 principal cells by the temporoammonic pathway.[80,81] Injection of depolarizing current induced APs in CA1 pyramidal cells and SO interneurons of all genotypes, but the output of APs was diminished for SO interneurons in DS mice (Fig. 4.4C and D).[71] At a stimulus frequency of 50 Hz, SO interneurons in DS/B6 mice showed a substantial impairment

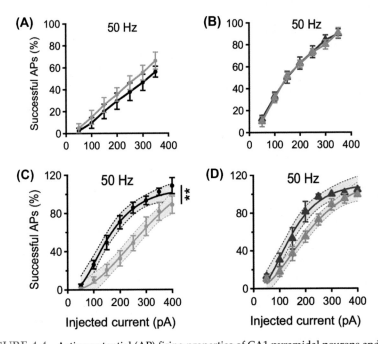

FIGURE 4.4   Action potential (AP) firing properties of CA1 pyramidal neurons and hippocampal interneurons in DS/129 and DS/B6 mice. Percent successful APs in response to 10 ms, depolarizing current pulses at 50 Hz. 100% indicates one successful AP in every trial. (A) CA1 pyramidal cells. *Black*, WT/B6; *gray*, DS/B6. (B) CA1 pyramidal cells. *Dark blue*, WT/129; *light blue*, DS/129. (C) Hippocampal stratum oriens (SO) interneurons. *Black*, WT/B6; *gray*, DS/B6. (D) Hippocampal SO interneurons. *Dark blue*, WT/129; *light blue*, DS/129. *Solid lines* in panels C and D are fits of an empirical equation $(A[1 - \exp(-I \times D)])^3$ to the data where A is the amplitude, I is the current, and D is a constant. The shaded area is the 95% confidence interval of the fit, and the nonoverlapping areas are statistically significant ($p < 0.05$). Statistical significance was also calculated using area under the curve. DS/B6 fire less AP compared to WT/B6 at 50 Hz, $p < 0.01$. AP firing of 129/DS was similar to that of 129/WT at tested frequencies $p > 0.05$. $n = 5–7$ (A and B) and $n = 8–11$ (C and D) for each genotype. *Adapted from Rubinstein M, Westenbroek RE, Yu FH, Jones CJ, Scheuer T, Catterall WA. Genetic background modulates impaired excitability of inhibitory neurons in a mouse model of Dravet syndrome. Neurobiol Dis 2015;73:106–17.*

of AP firing (Fig. 4.4C) and DS/129 mice showed a milder impairment in this experimental protocol (Fig. 4.4D). Similar deficits were observed at 1 and 10 Hz stimulus frequencies.[71] We also observed characteristic differences in threshold, rheobase, and boosting of synaptic signals by subthreshold sodium currents in SO interneurons in hippocampal slices from DS/B6 and DS/129 mice, which were consistent with a milder loss of electrical excitability in the DS/129 mice.[71] As OL-M interneurons usually fire at theta frequencies (5–10 Hz),[82–84] the reduced ability of DS/B6 interneurons to fire APs in the frequency range from 1 to 50 Hz suggests significant impairment of their normal physiological function. The lesser impairment of AP firing in DS/129 mice would contribute to their milder epilepsy, lack of premature death, and lack of cognitive deficit.

In a complementary study, Mistry et al.[85] examined a different genetic model of DS in C57BL/6J and 129S6/SvEvTac mice using the whole-cell voltage clamp method to record sodium currents in acutely dissociated neurons at P21. Consistent with our results outlined above, they found a larger reduction in sodium current in dissociated hippocampal interneurons in the C57BL/6J genetic background than in the 129S6/SvEvTac background. However, they also found a strain-dependent increase in sodium currents in the cell bodies of acutely dissociated hippocampal pyramidal cells, in contrast to the lack of effect we observed on AP firing of pyramidal cells in acutely dissected hippocampal slices. This difference in results may reflect differences between the DS mutations studied, the exact genetic backgrounds of the mouse strains, or the electrophysiological recordings in dissociated cells versus brain slices. Nevertheless, in both studies, a greater loss of sodium current and cellular excitability was observed in inhibitory interneurons in the B6 mouse genetic background.

GENETIC DISSECTION OF PHENOTYPES IN DRAVET SYNDROME

In DS mice, epilepsy, premature death, and comorbidities are all caused by deletion of $Na_V1.1$ channels in GABAergic interneurons, but the roles of different classes of interneurons in the many phenotypes of DS remain incompletely defined. We have used specific gene deletion with the Cre-Lox method and $Na_V1.1$ floxed mice[50] to dissect the functional roles of different classes of interneurons in DS. Parvalbumin (PV)-expressing and somatostatin (SST)-expressing interneurons account for a large fraction of total interneurons. For instance, in Layer V of the cerebral cortex these two classes include >80% of interneurons.[86] These interneurons have complementary roles in control of pyramidal cell excitability. PV cells form synapses on the cell body and proximal dendrites and oppose AP initiation, whereas SST cells form synapses on the distal dendrites and oppose incoming excitatory postsynaptic currents. Previous studies showed that deletion of both *Scn1a* alleles in PV-expressing interneurons, a more severe genetic ablation than DS, caused seizures and premature death.[76] We investigated specific contributions of PV- and SST-expressing interneurons to the physiological and behavioral consequences of DS using mice with selective heterozygous and homozygous deletion of $Na_V1.1$ in PV-expressing interneurons, SST-expressing interneurons, or both.

Physiological effects of selective deletion of $Na_V1.1$ channels: Layer V pyramidal cells are the principal output neurons of the cerebral cortex, and their activity is tightly controlled by Layer V PV-expressing, fast-spiking interneurons and SST-expressing Martinotti cells.[86] Electrophysiological recordings in cortical slices from mice with global deletion of $Na_V1.1$ demonstrated reduced excitability of both PV-expressing and SST-expressing interneurons, including increased threshold, increased rheobase, reduced output of APs in response to depolarizing stimuli, and increased frequency of failures of firing during trains of APs.[87] Spread of electrical excitation along the cortical layers is controlled by disynaptic inhibition, in which firing of one pyramidal cell excites a neighboring SST-expressing GABAergic Martinotti cell, which in turn makes inhibitory synaptic input on neighboring pyramidal cells.[88] In addition to the direct effects on interneuron excitability, we found that deletion of $Na_V1.1$ channels in SST-expressing Martinotti cells substantially reduces disynaptic inhibition.[87] Disruption of this local inhibitory circuit would greatly facilitate the spread of excitability across the cerebral cortex.

Similar brain-slice recordings revealed comparable deficits in interneuron excitability and AP generation in mice with functional haploinsufficiency of $Na_V1.1$ after specific Cre-dependent deletion in PV-expressing and SST-expressing forebrain interneurons only.[50] For example, the output of APs generated by injection of depolarizing current was reduced for both interneuron types across a broad range of stimulus intensities (Fig. 4.5). These results demonstrate that the Cre-Lox method was effective in creating functional haploinsufficiency of $Na_V1.1$ channels in both PV and SST interneurons.

Interneurons in the hippocampus also express PV and SST, and they are involved in seizure generation in DS.[89] As a complement to our studies of cortical interneurons, we tested the functional consequence of selective deletion of $Na_V1.1$ in CA1 horizontal stratum oriens lacunosum-moleculare cells, which are critical for regulation of AP firing of CA1 pyramidal cells and exert important inhibitory control of circuit function.[77,80,81] These cells represent a distinct subtype of interneurons that express both SST and PV.[90,91] Their excitability was reduced in DS mice with global deletion of $Na_V1.1$,[92] and a similar deficit was observed in mice with haploinsufficiency of $Na_V1.1$ channels in both PV and SST neurons.[92] Both threshold and rheobase for AP generation in response to injection of depolarizing

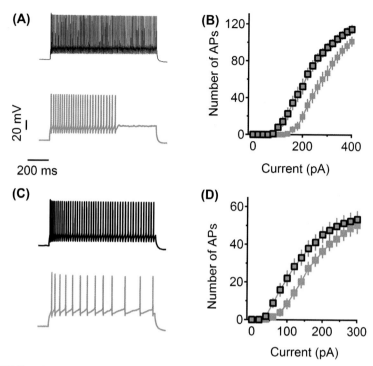

FIGURE 4.5   Deletion of $Na_V1.1$ results in reduced excitability in Layer V cortical inter-neurons. (A) Sample traces of whole-cell current-clamp recordings of action potentials (APs) in parvalbumin-expressing neurons in response to injection of 240-pA current. (B) Average number of APs in response to 1-s depolarizing current injection at the indicated intensity ($n = 15$–19). (C) Sample traces of whole-cell current-clamp recordings of APs in SST-expressing neurons in response to injection of 160 pA current ($n = 15$–17). (D) Average number of APs in response to 1-s depolarizing current injection at the indicated intensity. *Black squares, WT; blue, PV-Cre: yellow, SST-Cre. Adapted from Rubinstein M, Han S, Tai C, et al. Dissecting the phenotypes of Dravet syndrome by gene deletion.* Brain 2015;**138**(8):2219.

current were increased, and the threshold for generation of APs in response to synaptic input was also increased.[92] Together, these physiological defi-cits would greatly impair the function of these interneurons in controlling electrical excitability in hippocampal circuits.

   Epilepsy in mice with selective deletion of $Na_V1.1$ channels: We tested mice with selective deletion of $Na_V1.1$ in PV-expressing or SST-expressing interneurons for their susceptibility to thermally induced seizures at two different ages: P21, the age at which DS mice with global $Na_V1.1$ deletion first experience spontaneous seizures,[39,41] and P35 following the period of most intense spontaneous seizures.[47] None of the WT $Scn1a^{+/+}$ mice expe-rienced seizures up to 42°C. With mice that were heterozygous for deletion of $Scn1a$ but homozygous for expression of PV-Cre ($PV^{Cre/Cre}$, $Scn1a^{fl/+}$), thermal induction caused seizures in 50% of mice at 41°C on P21, and all mice had seizures by 42°C (Fig. 4.6A, blue).[92] Moreover, at P35, all of these mice had thermally induced seizures (Fig. 4.6B, blue).[92] Loss-of-function of Na$_V$1.1 in SST neurons was sufficient to induce thermal seizures in 50% of mice at 40.5°C at P35, and all of these mice had seizures by 42°C (Fig. 4.6B, yellow).[92] In contrast, no seizures were induced in mice at P21 (Fig. 4.6A, yellow).[92] As PV-expressing and SST-expressing neurons both contributed to epilepsy in DS, we tested for synergistic effects. While SST-DS mice were not susceptible to thermally induced seizures at P21, all PV-DS and PV&SST-DS mice had thermally induced seizures (Fig. 4.6A, green).[92] Moreover, at P35, the double deletion was more effective in thermal induction of seizures than single deletions (Fig. 4.6B), and seizure duration in PV&SST-DS mice was approximately twofold longer than in PV-DS mice and

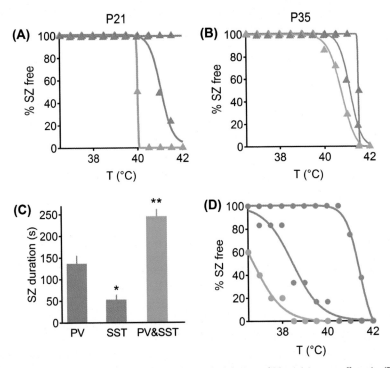

FIGURE 4.6   Epileptic phenotype of mice with deletion of Na$_V$1.1 in parvalbumin (PV)- and somatostatin (SST)-expressing interneurons. (A and B) Heterozygous deletion of Na$_V$1.1 (Scn1a$^{fl/+}$). Mean values±SEM for the percentage of mice remaining free of behavioral seizures (SZ) at the indicated body core temperatures at P21 (A) and P35 (B). (C) Mean values±SEM for seizure duration at P35. Statistical analysis was done using one-way ANOVA, **$p < 0.01$, *$p < 0.05$. PV$^{Cre/Cre}$-DS, $n = 9$; SST$^{Cre/Cre}$-DS, $n = 5$; PV$^{Cre/Cre}$&SST-DS, $n = 8$. Scn1a$^{+/+}$ mice of either cohort did not have seizures at the tested temperatures ($n = 3-7$ for each genotype). (D) Homozygous deletion of Na$_V$1.1 (Scn1a$^{fl/fl}$). Percentage of mice remaining free of behavioral seizures (SZ) at the indicated body core temperatures at P21. Blue, PV-Cre; yellow, SST-Cre; green, PV-Cre&SST-Cre. Adapted from Rubinstein M, Han S, Tai C, et al. Dissecting the phenotypes of Dravet syndrome by gene deletion. Brain 2015;138(8):2219.

approximately fivefold longer than in SST-DS mice (Fig. 4.6C).[92] Thus, heterozygous deletion of $Na_V1.1$ simultaneously in both types of interneurons results in earlier onset of seizure susceptibility and longer duration of thermally induced seizures, and those effects are greater than additive.

Mice with homozygous deletion of *Scn1a* have severe ataxia, frequent seizures, and all die by P14.[39,40,55] Homozygous deletion of *Scn1a* in PV-expressing neurons resulted in severe ataxia, spontaneous convulsive seizures, and high sensitivity to thermally induced seizures (Fig. 4.6D, blue), and these mice often died prematurely.[92] In contrast, the epileptic phenotype of mice with complete deletion of *Scn1a* in SST-expressing neurons (SST^{Cre}, *Scn1a*$^{fl/fl}$) was milder. These mice had seizures at high temperature (Fig. 4.6D, yellow), and no premature death was observed.[92] Mice with complete deletion of $Na_V1.1$ in both PV and SST interneurons (PV^{Cre}, SST^{Cre}, *Scn1a*$^{fl/fl}$) exhibited ataxia, frequent spontaneous convulsive seizures, and thermally induced seizures at low temperatures (Fig. 4.6D, green). Premature death was frequent.[92] These results with homozygous deletion of $Na_V1.1$ in PV&SST-DS mice support the conclusion that deletion in both PV- and SST-expressing neurons has synergistic effects on epilepsy and premature death.

Differential effects of deletion of $Na_V1.1$ in PV- and SST-expressing interneurons on hyperactivity and autistic-like behaviors: DS mice with global deletion of *Scn1a* display behavioral traits that are found in DS patients, including hyperactivity and autistic-like social interaction deficits.[63] *Scn1a*$^{fl/+}$ mice with specific heterozygous deletion of $Na_V1.1$ in forebrain interneurons displayed all of the DS-associated behavioral deficits.[63] To examine the functional role of deletion of $Na_V1.1$ channels in different interneuron classes, we performed behavioral tests with PV-DS and SST-DS mice. PV-DS mice displayed normal locomotor activity, whereas SST-DS mice were hyperactive and traveled significantly longer distances.[92] To examine autistic-like behaviors, we tested the social behavior of these mice with the three-chamber test of social interaction. *Scn1a*$^{+/+}$ mice spent approximately twofold more time interacting with a stranger mouse than with the empty cage (Fig. 4.7A and B, control bars outlined in black).[92] In contrast, PV-DS mice displayed a social interaction deficit and exhibited no preference for interaction with a stranger mouse compared to a small empty mouse cage (Fig. 4.7A and B).[92] No social deficit was apparent for SST-DS mice, which displayed a strong preference for interaction with the stranger mouse (Fig. 4.7C and D).[92] These data provide evidence that reduced electrical excitability in PV-expressing interneurons contributes specifically to social interaction deficits in DS, but impairment of excitability of SST-expressing interneurons does not.

Synergistic effects of deletion of $Na_V1.1$ in PV-expressing and SST-expressing interneurons on spatial learning and memory: Cognitive impairment is characteristic in DS patients. Similarly, DS mice with

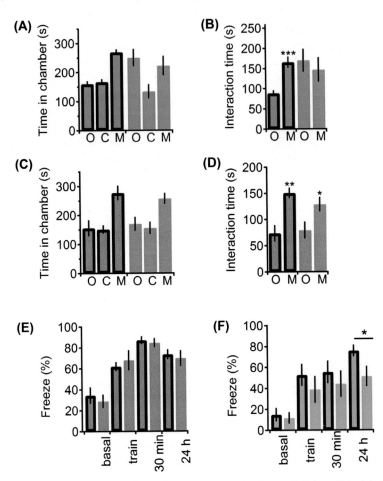

FIGURE 4.7 Social and cognitive deficits in mice with selective deletion of $Na_V1.1$. (A–D) Three-chamber test of social interaction for PV (A, B, $n = 8–9$) and SST (C, D, $n = 11–14$). Time in each chamber (A, C) and total interaction time (B, D). O, object (an empty small mouse cage); C, center; M, mouse (a stranger mouse in an identical small cage). Statistical analysis was done using Student's t-test: $*p < 0.05$, $**p < 0.01$, $***p < 0.001$. (E and F) Context-dependent fear conditioning test in SST (E, $n = 10$) and PV$^{Cre/Cre}$ & SST mice (F, $n = 7–9$). Basal, freezing time (%) during exploration of a cage containing an electrical grid as a floor and markings for identification; training, freezing time (%) during presentation of a 0.6 mA, 2-s long footshock; 30 min, freezing time (%) upon return to the same cage after 30 min but without shock; 24 h, freezing time (%) upon return to the same cage after 24 h but without shock. *Adapted from Rubinstein M, Han S, Tai C, et al. Dissecting the phenotypes of Dravet syndrome by gene deletion. Brain 2015;**138**(8):2219.*

global or specific heterozygous deletion of *Scn1a* in forebrain interneurons, have deficits in short-term and long-term context-dependent fear memory and in spatial memory.[63] In contrast, PV-DS mice (not shown) and SST-DS mice (Fig. 4.7E) did not demonstrate deficits in context-dependent fear-conditioning.[92] All of these mice displayed robust freezing behavior when they returned to the fearful spatial context of a footshock 30 min and 24 h later. In contrast, double mutant PV&SST-DS mice demonstrated normal freezing behavior immediately following a footshock and 30 min later, but their freezing behavior was significantly reduced after 24 h (Fig. 4.7F).[92] These results indicate that simultaneous deletion of $Na_V1.1$ in both PV- and SST-expressing neurons causes impairment of long-term context-dependent fear memory but not short-term memory.

DRAVET SYNDROME AS A SODIUM CHANNEL INTERNEURONOPATHY

Our studies with a mouse model of DS lead to the hypothesis that this disease is a sodium channel interneuronopathy. Physiological studies of dissociated neurons and intact neurons in hippocampal and cortical brain slices show that heterozygous deletion of $Na_V1.1$ channels causes reduced sodium currents and/or impaired AP firing in GABAergic interneurons in the hippocampus (dissociated interneurons and OL-M cells in slices),[39,71] cerebellum (Purkinje neurons),[55] SCN (ventral SCN neurons),[57] thalamus (RNT neurons),[58] and cerebral cortex (Layer V PV- and SST-expressing interneurons).[87] In contrast, we have not detected deficits in sodium currents or AP firing in excitatory neurons from hippocampus,[39,71] cerebral cortex (Layer V),[87] or thalamus (ventrobasal nucleus).[58] Similarly, studies of spontaneous synaptic activity in brain slices reveal a deficit in frequency of inhibitory neurotransmission, consistent with reduced AP firing in interneurons and a resulting increase in excitatory postsynaptic events caused as a secondary consequence of reduced inhibition.[63,87] These extensive studies show that the primary physiological deficit in these neuronal classes in DS mice is impaired AP firing in GABAergic interneurons. Extensive genetic and pharmacological studies also support this conclusion. Specific deletion of $Na_V1.1$ channels in forebrain GABAergic interneurons recapitulates the epilepsy, premature death, reduced sleep quality, hyperactivity, cognitive deficit, and autistic-like behaviors of DS mice.[50,63] Deletion in PV- and SST-expressing interneurons cleanly dissects these phenotypes.[92] Pharmacological treatments with GABA-enhancing drugs prevent seizures and improve circadian rhythms,[57,68] and treatment with low doses of the classical benzodiazepine clonazepam reverses hyperactivity, cognitive deficit, and autistic-like behaviors.[63] Altogether, the physiological, genetic, and pharmacological studies of DS mice lead to the conclusion that this multifaceted disease and its many comorbidities arise from failure of firing of GABAergic interneurons in many or all brain circuits, leading to disinhibition of individual neural circuits and resulting impairment of specific physiological, cognitive, and behavioral functions. This unified hypothesis for epilepsy and comorbidities in DS is encouraging with respect to therapeutic approaches, because enhancement of GABAergic neurotransmission in relevant circuits by pharmacologic or genetic means may be broadly beneficial for patients. Many neurological and psychiatric diseases have multifaceted symptoms that may arise from a single genetic or other pathologic cause. Our results provide the first example, to our knowledge, in which the multifaceted symptoms of a complex neuropsychiatric disease can be traced to the same cellular mechanism, ie, impaired firing of GABAergic inhibitory neurons occurring in different neural circuits in functionally distinct parts of the brain.

CONTRASTING VIEWS FROM STUDIES OF HUMAN-INDUCED PLURIPOTENT STEM CELLS

One of the most promising research approaches for the future is modeling human disease processes in human induced pluripotent stem cells that have been stimulated to mature into cell types of pathophysiological interest. In general, this approach is closest to providing key insights into the disease process for systems in which the pathophysiology is cell-autonomous and does not depend on complex cell–cell and cell–tissue interactions. In this context, diseases of the central nervous system will likely be among the most difficult to model using induced pluripotent stem cells because of the complexity of cell–cell and cell–environment interactions in the brain. Three studies using induced pluripotent stem cells from DS patients have been reported to date.[93–95] Different methods have been used to induce differentiation in the three studies, but in each case the level of sodium currents and cellular excitability achieved after differentiation in vitro are far less than observed in brain neurons isolated from experimental animals, the ratio of excitatory to inhibitory neurons is different among the studies and different from that observed in major brain areas in vivo, and the level of synapse formation and functional synaptic interaction is much less than observed in vivo or in cultures of explanted neurons. Therefore, caution is required in interpreting the results from these pioneering, but still early stage, studies with stem cells. In two of these studies,[93,94] increases in the excitability of excitatory pyramidal neurons were observed, which must result from indirect compensatory changes since these cells have loss-of-function mutations in $Na_V1.1$ channels. In one study,[95] loss of excitability of GABAergic interneurons was observed, which might reflect direct functional effects of the loss-of-function mutations in $Na_V1.1$ channels in these cells. These studies highlight the need to examine the noncell-autonomous effects of loss-of-function mutations in sodium channels in these differentiating neuronal cultures to understand how these mutations can increase sodium currents and electrical excitability of developing pyramidal neurons in vitro. They also highlight the need to improve the level of differentiation in cultures of neurons induced from pluripotent stem cells to achieve levels of electrical excitability and synaptic activity that are comparable to those in vivo. Nevertheless, even when mature levels of electrical excitability and synaptic activity are achieved in vitro, these induced neural cells will lack the strict three-dimensional organization of the brain, so continuous interplay between studies of the intact brain of experimental animals in vivo and studies of human cells differentiated in cell culture in vitro will be necessary to define the mechanisms of pathophysiology and test novel modes of therapy for DS and other complex diseases of the central nervous system.

Acknowledgments

Research in the author's laboratory was supported by research grants from the National Institutes of Health (R01 NS25704), the McKnight Foundation, and the SFARI Program of the Simons Foundation. I thank Dr. Christine Cheah for valuable comments on this manuscript.

References

1. Dravet C, Bureau M, Oguni H, Fukuyama Y, Cokar O. Severe myoclonic epilepsy in infancy: Dravet syndrome. *Adv Neurol* 2005;**95**:71–102.
2. Wolff M, Casse-Perrot C, Dravet C. Severe myoclonic epilepsy of infants (Dravet syndrome): natural history and neuropsychological findings. *Epilepsia* 2006;**47**(Suppl. 2): 45–8.
3. Catterall WA, Kalume F, Oakley JC. Na$_V$1.1 channels and epilepsy. *J Physiol* 2010;**588**(Pt 11): 1849–59.
4. O'Roak BJ, Deriziotis P, Lee C, et al. Exome sequencing in sporadic autism spectrum disorders identifies severe de novo mutations. *Nat Genet* 2011;**43**(6):585–9.
5. O'Roak BJ, Vives L, Girirajan S, et al. Sporadic autism exomes reveal a highly interconnected protein network of de novo mutations. *Nature* 2012;**485**(7397):246–50.
6. Catterall WA. From ionic currents to molecular mechanisms: the structure and function of voltage-gated sodium channels. *Neuron* 2000;**26**(1):13–25.
7. Isom LL. The role of sodium channels in cell adhesion. *Front Biosci* 2002;**7**:12–23.
8. Isom LL, Ragsdale DS, De Jongh KS, et al. Structure and function of the β2 subunit of brain sodium channels, a transmembrane glycoprotein with a CAM motif. *Cell* 1995;**83**(3):433–42.
9. Goldin AL. Resurgence of sodium channel research. *Annu Rev Physiol* 2001;**63**:871–94.
10. Goldin AL, Barchi RL, Caldwell JH, et al. Nomenclature of voltage-gated sodium channels. *Neuron* 2000;**28**(2):365–8.
11. Trimmer JS, Rhodes KJ. Localization of voltage-gated ion channels in mammalian brain. *Annu Rev Physiol* 2004;**66**:477–519.
12. Westenbroek RE, Merrick DK, Catterall WA. Differential subcellular localization of the RI and RII Na$^+$ channel subtypes in central neurons. *Neuron* 1989;**3**(6):695–704.
13. Westenbroek RE, Noebels JL, Catterall WA. Elevated expression of type II Na$^+$ channels in hypomyelinated axons of shiverer mouse brain. *J Neurosci* 1992;**12**(6):2259–67.
14. Lorincz A, Nusser Z. Cell-type-dependent molecular composition of the axon initial segment. *J Neurosci* 2008;**28**(53):14329–40.
15. Vacher H, Mohapatra DP, Trimmer JS. Localization and targeting of voltage-dependent ion channels in mammalian central neurons. *Physiol Rev* 2008;**88**(4):1407–47.

16. Krzemien DM, Schaller KL, Levinson SR, Caldwell JH. Immunolocalization of sodium channel isoform NaCh6 in the nervous system. *J Comp Neurol* 2000;**420**(1):70–83.

17. Jenkins SM, Bennett V. Ankyrin-G coordinates assembly of the spectrin-based membrane skeleton, voltage-gated sodium channels, and L1 CAMs at Purkinje neuron initial segments. *J Cell Biol* 2001;**155**(5):739–46.

18. Stuart GJ, Sakmann B. Active propagation of somatic action potentials into neocortical pyramidal cell dendrites. *Nature* 1994;**367**:69–72.

19. Johnston D, Magee JC, Colbert CM, Christie BR. Active properties of neuronal dendrites. *Annu Rev Neurosci* 1996;**19**:165–86.

20. Callaway JC, Ross WN. Spatial distribution of synaptically activated sodium concentration changes in cerebellar Purkinje neurons. *J Neurophysiol* 1997;**77**(1):145–52.

21. Raman IM, Bean BP. Properties of sodium currents and action potential firing in isolated cerebellar Purkinje neurons. *Ann NY Acad Sci* 1999;**868**:93–6.

22. Khaliq ZM, Raman IM. Relative contributions of axonal and somatic Na channels to action potential initiation in cerebellar Purkinje neurons. *J Neurosci* 2006;**26**(7):1935–44.

23. Escayg A, MacDonald BT, Meisler MH, et al. Mutations of SCN1A, encoding a neuronal sodium channel, in two families with GEFS+2. *Nat Genet* 2000;**24**(4):343–5.

24. Meisler MH, Kearney JA. Sodium channel mutations in epilepsy and other neurological disorders. *J Clin Invest* 2005;**115**(8):2010–7.

25. Wallace RH, Wang DW, Singh R, et al. Febrile seizures and generalized epilepsy associated with a mutation in the sodium channel β1 subunit gene SCN1B. *Nat Genet* 1998;**19**:366–70.

26. Claes L, Del-Favero J, Ceulemans B, Lagae L, Van Broeckhoven C, De Jonghe P. De novo mutations in the sodium-channel gene SCN1A cause severe myoclonic epilepsy of infancy. *Am J Hum Genet* 2001;**68**(6):1327–32.

27. Ohmori I, Ouchida M, Ohtsuka Y, Oka E, Shimizu K. Significant correlation of the SCN1A mutations and severe myoclonic epilepsy in infancy. *Biochem Biophys Res Commun* 2002;**295**(1):17–23.

28. Fujiwara T, Sugawara T, Mazaki-Miyazaki E, et al. Mutations of sodium channel α subunit type 1 (*SCN1A*) in intractable childhood epilepsies with frequent generalized tonic-clonic seizures. *Brain* 2003;**126**(Pt 3):531–46.

29. Claes L, Ceulemans B, Audenaert D, et al. De novo SCN1A mutations are a major cause of severe myoclonic epilepsy of infancy. *Hum Mutat* 2003;**21**(6):615–21.

30. Sugawara T, Mazaki-Miyazaki E, Fukushima K, et al. Frequent mutations of SCN1A in severe myoclonic epilepsy in infancy. *Neurology* 2002;**58**(7):1122–4.

31. Fukuma G, Oguni H, Shirasaka Y, et al. Mutations of neuronal voltage-gated Na+ channel α1 subunit gene SCN1A in core severe myoclonic epilepsy in infancy (SMEI) and in borderline SMEI (SMEB). *Epilepsia* 2004;**45**(2):140–8.

32. Depienne C, Trouillard O, Gourfinkel-An I, et al. Mechanisms for variable expressivity of inherited SCN1A mutations causing Dravet syndrome. *J Med Genet* 2010;**47**(6):404–10.

33. Suls A, Velizarova R, Yordanova I, et al. Four generations of epilepsy caused by an inherited microdeletion of the SCN1A gene. *Neurology* 2010;**75**(1):72–6.

34. Marini C, Scheffer IE, Nabbout R, et al. The genetics of Dravet syndrome. *Epilepsia* 2011;**52**(Suppl. 2):24–9.

35. Patino GA, Claes LR, Lopez-Santiago LF, et al. A functional null mutation of SCN1B in a patient with Dravet syndrome. *J Neurosci* 2009;**29**(34):10764–78.

36. Engel Jr J, International League Against Epilepsy. A proposed diagnostic scheme for people with epileptic seizures and with epilepsy: report of the ILAE Task Force on Classification and Terminology. *Epilepsia* 2001;**42**(6):796–803.

37. Dravet C, Bureau M, Guerrini R, Giraud N, Roger J. Severe myoclonic epilepsy in infants. In: Roger J, Dravet C, Bureau M, Dreifus FE, Perret A, Wolf P, editors. *Epileptic syndromes in infancy, childhood and adolescence*. 2nd ed. London: John Libbey; 1992. p. 75–102.

38. Oguni H, Hayashi K, Awaya Y, Fukuyama Y, Osawa M. Severe myoclonic epilepsy in infants–a review based on the Tokyo Women's Medical University series of 84 cases. *Brain Dev* 2001;**23**(7):736–48.
39. Yu FH, Mantegazza M, Westenbroek RE, et al. Reduced sodium current in GABAergic interneurons in a mouse model of severe myoclonic epilepsy in infancy. *Nat Neurosci* 2006;**9**(9):1142–9.
40. Ogiwara I, Miyamoto H, Morita N, et al. Na$_V$1.1 localizes to axons of parvalbumin-positive inhibitory interneurons: a circuit basis for epileptic seizures in mice carrying an Scn1a gene mutation. *J Neurosci* 2007;**27**(22):5903–14.
41. Oakley JC, Kalume F, Yu FH, Scheuer T, Catterall WA. Temperature- and age-dependent seizures in a mouse model of severe myoclonic epilepsy in infancy. *Proc Natl Acad Sci USA* 2009;**106**(10):3994–9.
42. Sakauchi M, Oguni H, Kato I, et al. Mortality in Dravet syndrome: search for risk factors in Japanese patients. *Epilepsia* 2011;**52**(Suppl. 2):50–4.
43. Gordon D, Merrick D, Auld V, et al. Tissue-specific expression of the RI and RII sodium channel subtypes. *Proc Natl Acad Sci USA* 1987;**84**:8682–6.
44. Beckh S, Noda M, Lübbert H, Numa S. Differential regulation of three sodium channel messenger RNAs in the rat central nervous system during development. *EMBO J* 1989;**8**:3611–6.
45. Felts PA, Yokoyama S, Dib-Hajj S, Black JA, Waxman SG. Sodium channel α-subunit mRNAs I, II, III, NaG, Na6 and hNE (PN1): different expression patterns in developing rat nervous system. *Brain Res Mol Brain Res* 1997;**45**(1):71–82.
46. Gong B, Rhodes KJ, Bekele-Arcuri Z, Trimmer JS. Type I and type II Na$^+$ channel α-subunit polypeptides exhibit distinct spatial and temporal patterning, and association with auxiliary subunits in rat brain. *J Comp Neurol* 1999;**412**(2):342–52.
47. Kalume F, Westenbroek RE, Cheah CS, et al. Sudden unexpected death in a mouse model of Dravet syndrome. *J Clin Invest* 2013;**123**(4):1798–808.
48. Cheah CS, Westenbroek RE, Roden WH, et al. Correlations in timing of sodium channel expression, epilepsy, and sudden death in Dravet syndrome. *Channels (Austin)* 2013;**7**(6):468–72.
49. Oguni H, Hayashi K, Osawa M, et al. Severe myoclonic epilepsy in infancy: clinical analysis and relation to SCN1A mutations in a Japanese cohort. *Adv Neurol* 2005;**95**:103–17.
50. Cheah CS, Yu FH, Westenbroek RE, et al. Specific deletion of Na$_V$1.1 sodium channels in inhibitory interneurons causes seizures and premature death in a mouse model of Dravet syndrome. *Proc Natl Acad Sci USA* 2012;**109**(36):14646–51.
51. Grusser-Cornehls U, Baurle J. Mutant mice as a model for cerebellar ataxia. *Prog Neurobiol* 2001;**63**(5):489–540.
52. Raman IM, Bean BP. Resurgent sodium current and action potential formation in dissociated cerebellar Purkinje neurons. *J Neurosci* 1997;**17**(12):4517–26.
53. Sausbier M, Hu H, Arntz C, et al. Cerebellar ataxia and Purkinje cell dysfunction caused by Ca^{2+}-activated K$^+$ channel deficiency. *Proc Natl Acad Sci USA* 2004;**101**(25):9474–8.
54. Fletcher CF, Lutz CM, O'Sullivan TN, et al. Absence epilepsy in tottering mutant mice is associated with calcium channel defects. *Cell* 1996;**87**(4):607–17.
55. Kalume F, Yu FH, Westenbroek RE, Scheuer T, Catterall WA. Reduced sodium current in Purkinje neurons from Na$_V$1.1 mutant mice: implications for ataxia in severe myoclonic epilepsy in infancy. *J Neurosci* 2007;**27**(41):11065–74.
56. Nolan KJ, Camfield CS, Camfield PR. Coping with Dravet syndrome: parental experiences with a catastrophic epilepsy. *Dev Med Child Neurol* 2006;**48**(9):761–5.
57. Han S, Yu FH, Schwartz MD, et al. Na$_V$1.1 channels are critical for intercellular communication in the suprachiasmatic nucleus and for normal circadian rhythms. *Proc Natl Acad Sci USA* 2012;**109**(6):E368–77.
58. Kalume F, Oakley JC, Westenbroek RE, et al. Sleep impairment and reduced interneuron excitability in a mouse model of Dravet Syndrome. *Neurobiol Dis* 2015;**77**:141–54.

59. Genton P, Velizarova R, Dravet C. Dravet syndrome: the long-term outcome. *Epilepsia* 2011;**52**(Suppl. 2):44–9.
60. Brunklaus A, Dorris L, Zuberi SM. Comorbidities and predictors of health-related quality of life in Dravet syndrome. *Epilepsia* 2011;**52**(8):1476–82.
61. Besag FM. Behavioral aspects of pediatric epilepsy syndromes. *Epilepsy Behav* 2004; **5**(Suppl. 1):S3–13.
62. Mahoney K, Moore SJ, Buckley D, et al. Variable neurologic phenotype in a GEFS+ family with a novel mutation in SCN1A. *Seizure* 2009;**18**(7):492–7.
63. Han S, Tai C, Westenbroek RE, et al. Autistic-like behaviour in *Scn1a*$^{+/-}$ mice and rescue by enhanced GABA-mediated neurotransmission. *Nature* 2012;**489**(7416):385–90.
64. Raman IM, Bean BP. Ionic currents underlying spontaneous action potentials in isolated cerebellar Purkinje neurons. *J Neurosci* 1999;**19**:1664–74.
65. Chen Y, Yu FH, Sharp EM, Beacham D, Scheuer T, Catterall WA. Functional properties and differential neuromodulation of Na$_V$1.6 channels. *Mol Cell Neurosci* 2008;**38**(4):607–15.
66. Martin MS, Tang B, Papale LA, Yu FH, Catterall WA, Escayg A. The voltage-gated sodium channel Scn8a is a genetic modifier of severe myoclonic epilepsy of infancy. *Hum Mol Genet* 2007;**16**(23):2892–9.
67. Guerrini R, Dravet C, Genton P, Belmonte A, Kaminska A, Dulac O. Lamotrigine and seizure aggravation in severe myoclonic epilepsy. *Epilepsia* 1998;**39**(5):508–12.
68. Oakley JC, Cho AR, Cheah CS, Scheuer T, Catterall WA. Synergistic GABA-enhancing therapy against seizures in a mouse model of Dravet syndrome. *J Pharmacol Exp Ther* 2013;**345**(2):215–24.
69. Han S, Tai C, Jones CJ, Scheuer T, Catterall WA. Enhancement of inhibitory neurotransmission by GABA$_A$ receptors having α2,3-subunits ameliorates behavioral deficits in a mouse model of autism. *Neuron* 2014;**81**(6):1282–9.
70. Miller AR, Hawkins NA, McCollom CE, Kearney JA. Mapping genetic modifiers of survival in a mouse model of Dravet syndrome. *Genes Brain Behav* 2013;**13**(2):163–72.
71. Rubinstein M, Westenbroek RE, Yu FH, Jones CJ, Scheuer T, Catterall WA. Genetic background modulates impaired excitability of inhibitory neurons in a mouse model of Dravet syndrome. *Neurobiol Dis* 2015;**73**:106–17.
72. Verbeek NE, van der Maas NAT, Jansen FE, van Kempen MJA, Lindhout D, Brilstra EH. Prevalence of SCN1A related Dravet syndrome among children reported with seizures following vaccination: a population-based ten-year cohort study. *PLoS One* 2013;**8**(6):e65758.
73. Gambardella A, Marini C. Clinical spectrum of SCN1A mutations. *Epilepsia* 2009; **50**(Suppl. 5):20–3.
74. Guzzetta F. Cognitive and behavioral characteristics of children with Dravet syndrome: an overview. *Epilepsia* 2011;**52**(Suppl. 2):35–8.
75. Buoni S, Orrico A, Galli L, et al. SCN1A (2528delG) novel truncating mutation with benign outcome of severe myoclonic epilepsy of infancy. *Neurology* 2006;**66**(4):606–7.
76. Ogiwara I, Iwasato T, Miyamoto H, et al. Na$_V$1.1 haploinsufficiency in excitatory neurons ameliorates seizure-associated sudden death in a mouse model of Dravet syndrome. *Hum Mol Genet* 2013;**22**(23):4784–804.
77. Pouille F, Scanziani M. Routing of spike series by dynamic circuits in the hippocampus. *Nature* 2004;**429**(6993):717–23.
78. Klausberger T, Somogyi P. Neuronal diversity and temporal dynamics: the unity of hippocampal circuit operations. *Science* 2008;**321**(5885):53–7.
79. Blasco-Ibanez JM, Freund TF. Synaptic input of horizontal interneurons in stratum oriens of the hippocampal CA1 subfield: structural basis of feed-back activation. *Eur J Neurosci* 1995;**7**(10):2170–80.
80. Coulter DA, Yue C, Ang CW, et al. Hippocampal microcircuit dynamics probed using optical imaging approaches. *J Physiol* 2011;**589**(8):1893–903.

81. Maccaferri G, McBain CJ. Passive propagation of LTD to stratum oriens-alveus inhibitory neurons modulates the temporoammonic input to the hippocampal CA1 region. *Neuron* 1995;**15**(1):137–45.

82. Maccaferri G. Stratum oriens horizontal interneurone diversity and hippocampal network dynamics. *J Physiol* 2005;**562**(Pt 1):73–80.

83. Maccaferri G, McBain CJ. The hyperpolarization-activated current (Ih) and its contribution to pacemaker activity in rat CA1 hippocampal stratum oriens-alveus interneurones. *J Physiol* 1996;**497**(Pt 1):119–30.

84. Klausberger T, Magill PJ, Marton LF, et al. Brain-state- and cell-type-specific firing of hippocampal interneurons in vivo. *Nature* 2003;**421**(6925):844–8.

85. Mistry AM, Thompson CH, Miller AR, Vanoye CG, George Jr AL, Kearney JA. Strain- and age-dependent hippocampal neuron sodium currents correlate with epilepsy severity in Dravet syndrome mice. *Neurobiol Dis* 2014;**65**:1–11.

86. Rudy B, Fishell G, Lee S, Hjerling-Leffler J. Three groups of interneurons account for nearly 100% of neocortical GABAergic neurons. *Dev Neurobiol* 2011;**71**(1):45–61.

87. Tai C, Abe Y, Westenbroek RE, Scheuer T, Catterall WA. Impaired excitability of somatostatin- and parvalbumin-expressing cortical interneurons in a mouse model of Dravet syndrome. *Proc Natl Acad Sci USA* 2014;**111**(30):E3139–48.

88. Silberberg G, Markram H. Disynaptic inhibition between neocortical pyramidal cells mediated by Martinotti cells. *Neuron* 2007;**53**(5):735–46.

89. Liautard C, Scalmani P, Carriero G, de Curtis M, Franceschetti S, Mantegazza M. Hippocampal hyperexcitability and specific epileptiform activity in a mouse model of Dravet syndrome. *Epilepsia* 2013;**54**(7):1251–61.

90. Maccaferri G, Roberts JD, Szucs P, Cottingham CA, Somogyi P. Cell surface domain specific postsynaptic currents evoked by identified GABAergic neurones in rat hippocampus in vitro. *J Physiol* 2000;**524**(Pt 1):91–116.

91. Ferraguti F, Cobden P, Pollard M, et al. Immunolocalization of metabotropic glutamate receptor 1α (mGluR1α) in distinct classes of interneuron in the CA1 region of the rat hippocampus. *Hippocampus* 2004;**14**(2):193–215.

92. Rubinstein M, Han S, Tai C, et al. Dissecting the phenotypes of Dravet syndrome by gene deletion. *Brain* 2015;**138**(8):2219.

93. Liu Y, Lopez-Santiago LF, Yuan Y, et al. Dravet syndrome patient-derived neurons suggest a novel epilepsy mechanism. *Ann Neurol* 2013;**74**(1):128–39.

94. Jiao J, Yang Y, Shi Y, et al. Modeling Dravet syndrome using induced pluripotent stem cells (iPSCs) and directly converted neurons. *Hum Mol Genet* 2013;**22**(21):4241–52.

95. Higurashi N, Uchida T, Lossin C, et al. A human Dravet syndrome model from patient induced pluripotent stem cells. *Mol Brain* 2013;**6**:19.

Diagnosis, Treatment, and Mechanisms of Long QT Syndrome

E. Wan, S.O. Marx

Columbia University, New York, NY, United States

INTRODUCTION

Long QT syndrome (LQTS) is characterized by the prolongation of the QT interval on the surface electrocardiogram (ECG). Congenital LQTS is a hereditary disorder in which family members have delayed cardiac repolarization manifested on the surface ECG as QT interval prolongation. The disorder has variable penetrance and is associated with an increased incidence of syncope, ventricular tachyarrhythmias, including torsade de pointes and sudden cardiac death (SCD). The estimated prevalence of the disease is approximately 1:2000.[1]

In 1957, Jervell and Lange-Nielsen described a family with LQTS.[2] The family consisted of six children, four of whom were deaf with frequent fainting spells precipitated by emotion and exercise. Three of these children died suddenly while playing. An ECG of the surviving deaf child revealed a markedly prolonged QT interval. The parents were unaffected. In 1958, Levine and Woodworth reported the sudden death of a 13-year-old deaf boy with syncope and QT prolongation.[3] These case reports of congenital deafness, syncope, prolonged repolarization manifest on ECG, and sudden death suggested an autosomal recessive mode of inheritance. In 1963, Romano[4] and in 1964, Ward[5] reported families with QT prolongation in one parent and several children with recurrent syncope, sudden death, but normal hearing. These case reports suggested an autosomal dominant model of inheritance. Subsequent to these reports, additional cases were identified and the International LQTS Registry was established in 1979.[6] About 85% of the reported cases are inherited from one parent, with the remaining 15% of affected patients having de novo mutations.

In seminal studies, mutations in the *KCNQ1*, *KCNH2*, and *SCN5A* genes, encoding the α subunit of ion channels that conduct the potassium (K^+) delayed rectifier currents (I_{Ks} and I_{Kr}) and the sodium (Na^+) current (I_{Na}) were identified, referred to as LQT1, LQT2, and LQT3 respectively.[7–12] In general, loss-of-function mutations of repolarizing currents (K^+ channels) and gain-of-function mutations of inward depolarizing currents (Na^+ channels) are responsible for the altered repolarization. Subsequent studies have led to the identification of additional genes and gene mutations causing LQTS, including other pore-forming α subunits of ion channels ($Ca_V1.2$—LQT8,[13] Kir2.1—LQT7,[14,15] Kir3.4—LQT13[16]), ion channel β subunits (KCNE1—LQT5,[17] KCNE2—LQT6,[18] Na_v β4—LQT10[19]), scaffolding proteins (ankyrin-B—LQT4,[20] caveolin-3—LQT9,[21] yotiao—AKAP9,[22] α-1 syntrophin—LQT12[23,24]), and regulatory proteins (calmodulin—LQT14[25]). LQT1, LQT2, and LQT3 comprise more than 95% of the genotype-positive LQTS and 65–75% of the gene mutations, while each of the other susceptibility genes accounts for <1%, leaving approximately 20–25% of patients with undefined genotypes.[26–29]

DIAGNOSIS OF LONG QT SYNDROME

An inherent challenge in the diagnosis of LQTS is that mutations often have incomplete penetrance and variable expressivity. Thus, the hallmark of the syndrome, a prolonged ventricular repolarization manifested by a long QT interval, may not be observed on a single ECG, if at all. The measurement of the QT interval can vary according to physician and LQTS expertise,[30] and an expert interpretation of the QT interval and ST-T wave morphology is critically important.[31]

Approximately 50% of mutation carriers never have symptoms. Most others have one or many episodes of syncope but do not die suddenly.[32] Of the patients who become symptomatic, 50% experience their first cardiac event by age 12 and 90% by the age of 40.[33] According to the International LQTS Registry, sudden death occurs in only about 4% of affected persons.

A diagnostic scoring system factoring in presence of prolonged QT interval on ECG, presence of symptoms, and family history, was created before the identification of the 14 LQTS disease-causing genes and has been subsequently modified. Patients with a Schwartz score of ≥3.5 in the absence of a secondary cause of prolonged QT interval (for instance drugs, acquired cardiac conditions, electrolyte imbalance, and unbalanced diets[34]) are diagnosed with LQTS. The Schwartz criteria cannot be used to identify silent mutation carriers, which are relatively common, accounting for 36% of LQT1 patients, 19% of LQT2 patients, and 10% of LQT3 patients.[27] The Heart Rhythm Society/European Heart Rhythm Association (EHRA)/Asia Pacific Heart Rhythm Society (APHRS) published in 2013 expert consensus statements on the diagnosis of LQTS.[34]

Patients at the highest and lowest risk for the development of SCD are relatively easy to identify. For the majority of patients, however, risk stratification is difficult. For most genetic mutations, specific locations (transmembrane vs. intracellular domain), type of mutation (substitution vs. missense), and degree of dysfunction are associated with different risks. In general, Jervell and Lange-Nielsen syndrome and Timothy syndrome (LQT8) are highly malignant. Very high-risk markers are patients who have suffered a cardiac arrest and received cardiopulmonary resuscitation and patients with spontaneous torsade de pointes.[28] A patient is of high risk when the QTc > 500 ms and is extremely high when QTc > 600 ms. Overt T-wave alternans is also a sign of electrical instability.

Comprehensive genetic testing is now available to identify the mutations responsible for LQT.[35] It is recommended for patients with a strong suspicion of LQTS based on ECG testing, stress testing with exercise or catecholamine testing, clinical history and/or family history, for any asymptomatic patient with QTc > 480 ms (prepubertal) or >500 ms (adults) in the absence of other clinical conditions, and for family members and other relatives subsequent to identification of the LQTS causative mutation in the index case. Importantly, a negative genetic test in a patient with the phenotype of LQTS does not remove the diagnosis.

CLINICAL COURSE

The clinical course of patients with LQTS is variable, influenced by age, genotype, gender, environmental factors, therapy, and incomplete penetrance with variable expressivity, which may be caused by genetic and environmental factors.[28,31] Modifier genes are common genetic variants and have less effect on the QT interval than the causal gene, but by summing the effects of multiple genes, they may have an important effect on the phenotype. For example, genome-wide association studies have shown that single nucleotide polymorphisms (SNPs) in NOS1AP are modulators of the QT interval in patients with LQTS as well as the general population.[36–38] SNPs in noncoding regions (intron, 3'UTR) of KCNQ1 are also important modifiers of the QT interval.[39,40]

Locati et al. showed that the male gender is independently associated with an increased risk of cardiac events before aged 15 years, whereas a gender risk reversal is observed in females after 14 years old. Zareba et al. showed that during childhood, boys with LQT1 had a 71% increase in risk of a first cardiac event compared to girls, whereas no significant gender difference was observed in patients with LQT2 and LQT3.[41] In more recent studies from the International LQTS Registry, the onset of gender risk reversal for aborted cardiac arrest or SCD occurred at a later age, after 20 years. Hormonal factors may be one of the major causes associated with the increased risk associated with female gender in the postadolescence period. Estrogens modify the expression of K^+ channels and may inhibit I_{Ks}.[42]

The largest difference in risk factors for cardiac events is between patients with LQT3 syndrome patients (*SCN5A* mutations) and patients with LQT1 syndrome (*KCNQ1* mutations) or LQT2 syndrome (*KCNH2* mutations). In patients with LQT3, the risk of cardiac events is greatest during rest (bradycardia). In contrast, cardiac events in patients with LQT2 mutations were associated with arousal and/or conditions in which patients were startled, whereas in patients with LQT1 mutations, the greatest risk of cardiac events occurred during exercise or conditions of elevated sympathetic nerve activity.[43,44]

GENES RESPONSIBLE FOR LONG QT SYNDROME (SEE TABLE 5.1)

KCNQ1 ($K_V7.1$, K_VLQT1) (LQT1)

LQT1 is the most common type of inherited LQT syndrome, accounting for ~35% of all patients and >50% of genotyped patients.[28] Four KCNQ1 α subunits assemble to form the I_{Ks} channel. Although a tetramer of KCNQ1 subunits alone forms a voltage-gated channel, only KCNQ1 and KCNE1 together form a channel with the slow activation and deactivation kinetics and the minimal inactivation characteristics of I_{Ks}.[45,46] Furthermore, KCNE1 is necessary for sympathetic modulation of I_{Ks}.[47]

Exercise is the major trigger for cardiac arrhythmias in patients with mutations of KCNQ1. Phosphorylated KCNQ1 increases I_{Ks} current and the shortening of the action potential duration that occurs allows for sufficient diastolic intervals in the face of increased heart rate. Sympathetic nervous system regulation of cardiac action potential duration is mediated by β-adrenergic receptor activation, which increases I_{Ks} by the phosphorylation of Ser^{27} on the N-terminus of KCNQ1.[48] To facilitate PKA phosphorylation, KCNQ1 must also assemble with the A-kinase anchoring protein (AKAP) Yotiao.[48] Mutations in KCNQ1 or Yotiao (AKAP9) that disrupt this complex give rise to LQT1[48] and LQT11 variants.[22] AKAP9 also assembles other modulators of I_{Ks} including protein phosphatase 1 (PP1),[48] adenylyl cyclase,[49] and phosphodiesterase PDE4D3.[50] The auxiliary subunit KCNE1 may play a role in transducing protein phosphorylation into channel function. For instance, the naturally occurring LQT5 mutation D76N prevented regulation of I_{Ks} by cAMP.[47] Future studies

TABLE 5.1 Cardiac Sodium Channelopathies

	Gene	Protein	Effect	Phenotype	References
CLINICAL DIAGNOSES					
LQT1	KCNQ1	I_{KS} α subunit (K_vLQT1)	↓K^+ outward current	↑QTc	Wang et al.[8]
LQT2	KCNH2	I_{Kr} (HERG)	↓K^+ outward current	↑QTc	Curran et al.[67]
LQT3	SCN5A	$Na_v1.5$	↑I_{Na}	↑QTc	Wang et al.[7]
LQT4	ANK2	Ankyrin-B	↓Na^+ and Ca^{2+} channels	↑QTc Mohler	Mohler et al.[23]
LQT5	KCNE1	I_{KS} β subunit (MinK)	↓K^+ outward current	↑QTc, AF	Barhanin et al.[43]
LQT6	KCNE2	I_{Kr} (MiRP)	↓K^+ outward current	↑QTc	Abbott et al.[21]
LQT7 (Andersen–Tawil syndrome)	KCNJ2	I_{K1} (Kir2.1)	↓K^+ outward current	↑QTc, periodic paralysis, low set ears, hypoplastic mandible	Plaster et al.[17]
LQT8 (Timothy syndrome)	CACNA1c	$Ca_v1.2$	↑I_{Na} due to altered kinetic gating	↑QTc, syndactyly, septal defect, Patent foramen ovale, autism, developmental disorders	Splawski et al.[16]
LQT9	CAV3	Caveolin	↑I_{Na} due to altered kinetic gating	↑QTc	Vatta et al.[24]
LQT10	SCN4B	Sodium channel β4	↓β4, ↑I_{Na}	↑QTc	Medeiro-Domingo et al.[22]
LQT11	AKAP9	Yotiao	↑I_{Ks}	↑QTc	Chen et al.[25]
LQT12	SNTA1	α-1 syntrophin	↑I_{Na}	↑QTc	Ueda et al.[26]
LQT13	KCNJ5	Kir3.4 (GIRK4)	↓$I_{K,ACh}$	↑QTc, atrial arrhythmias	Wang et al.[68]
LQT14	CALM	Calmodulin	Ca^{2+} signaling abnormalities	↑QTc, ventricular arrhythmias, seizures, developmental disorders	Crotti et al.[69]

may use human-induced pluripotent stem cell cardiomyocytes (hIPSC-CM) to investigate the mechanistic basis for heritable cardiac arrhythmias, as was shown for hiPSC-CM derived from LQT1 patients.[51]

KCNH2 ($K_V11.1$, hERG) (LQT2)

The *KCNH2* gene encodes the human ether-a-go-go-related protein (hERG1), which is responsible for the rapid component of the delayed rectifier repolarization current (I_{Kr}) in the heart. I_{Kr} is a major determinant of the plateau phase duration and the rate of phases 2 and 3 repolarization of action potentials in human cardiomyocytes. These channels rapidly activate and inactivate when the cardiomyocyte is depolarized. The channels remain closed during the plateau phase, but during phase 3 repolarization, they rapidly recover from inactivation and reenter an open state. Like many K_v channels, phosphatidylinositol-4,5-biphosphate (PIP2) increases current magnitude and slows deactivation.[52]

Mutations of *KCNH2* are responsible for LQT2.[10,12] The majority of mutations disrupt folding or trafficking of channels to the plasma membrane.[53] Patients with pore mutations have more severe clinical manifestations and experience a higher frequency of arrhythmia-related cardiac events at an earlier age than do subjects with nonpore mutations.[54]

Hypokalemia is an important risk factor for patients with LQTS. Cardiac action potentials are shorter at higher extracellular K^+ and longer at lower extracellular K^+ concentrations.[55] These effects have been linked to the effects of extracellular K^+ on I_{Kr}, based upon the findings that a reduction in extracellular K^+ can: (1) enhance inactivation[56]; (2) increase inhibition by Na^+, since K^+ and Na^+ can compete for binding to the external pore[57]; and (3) acutely decrease the amplitude of I_{Kr} in cardiomyocytes.[58]

Acquired LQTS is most often due to specific drugs, hypokalemia, or hypomagnesemia that may precipitate torsade de pointes and cause SCD. The hallmark mechanism of acquired LQTS and torsade de pointes is the blockade of I_{Kr} by specific drugs.[59] The unusual susceptibility of I_{Kr} channels to various drugs are accounted by: (1) aromatic amino acids (Tyr[652] and Phe[656]) with side chains directed toward the large central cavity of the pore, which provides high-affinity binding sites for many drugs; (2) while most K^+ channels contain two Pro residues in the helix that forms the part of the pore that restricts access to the drug binding site, these two Pro residues are absent in KCNH2.[60,61]

SCN5A ($Na_V1.5$) (LQT3)

The *SCN5A* gene encodes the cardiac Na^+ channel α subunit, also known as $Na_v1.5$. The Na^+ current underlies initiation of the cardiac action potential and plays a prominent role in Na^+ homeostasis by loading cardiomyocytes with Na^+ ions.[43] After opening briefly during the upstroke of the action potential, individual Na^+ channels usually inactivate and remain inactivated until repolarization is completed. Openings after phase 0 of wild-type (normal) Na^+ channels create a small persistent (also termed "late" or "sustained") current, normally <0.1% of peak Na^+ current, that persists throughout the plateau phase of the cardiac action potential. An increased persistent current in patients with LQT3 is the basis for the prolonged action potential duration and QT interval. In normal ventricular myocardium at a rate of 60 beats/min, persistent Na^+ current during phase 2 of the cardiac action potential plateau is estimated to be 30% of the total Na^+ influx via Na^+ channels, but this can be increased several fold when persistent Na^+ current is increased. For instance, enhancement of persistent Na^+ current by fivefold during the action potential plateau may double the total Na^+ influx into the myocyte during a cardiac cycle.[62] Na^+ channel mutations, pathological conditions, such as heart failure, hypoxia, inflammation, oxidative stress, and pharmacological agents can either delay or destabilize Na^+ channel inactivation, thereby increasing persistent current.[63] Ranolazine, a specific inhibitor of persistent Na^+ current, markedly reduces the persistent current.[64,65]

Unlike LQT1 and LQT2 patients who usually experience increase cardiac risks during physical or emotional stress, patients afflicted with LQT3 usually experience arrhythmias at rest or during sleep. This is because the amplitude of the persistent Na^+ current is greater at slow than at fast heart rates because an increased rate of Na^+ channel opening increases channel inactivation.

Transgenic mouse models with mutations of the $Na_v1.5$ channel, ΔKPQ and N1325S mutants, have prolonged action potential duration and QT interval, spontaneous ventricular tachycardia and ventricular fibrillation, and early afterdepolarizations.[43,66] Induced pluripotent stem cell derived cardiomyocytes (iPSCs) from patients with LQT3 also offer an opportunity to investigate the pharmacology of disease processes in therapeutically and genetically relevant primary cell types *in vitro*.[67,68]

ANK2 (LQT4)

LQT4 was mapped to chromosome 4q25–27 in a large French family with autosomal dominant LQTS associated with bradycardia and atrial fibrillation.[69] Mohler and colleagues identified a loss-of-function mutation in *ANK2* (ankyrin-B) as the gene responsible for LQT4, establishing ankyrin-B gene as the first non-ion channel gene involved in LQTS.[20] Ankyrins are a family of adapter proteins that link integral membrane proteins to spectrin-based cytoskeleton. Ankyrins bind to several ion channel proteins, including the anion exchanger ($Cl^-/HCO3^-$ exchanger),

Na^+/K^+-ATPase, $Na_v1.5$, and Na_V^+/Ca^{2+} exchanger (NCX), and the intracellular Ca^{2+} release channels, inositol trisphosphate receptors and ryanodine receptors.

KCNE1 (LQT5)

As discussed earlier, I_{Ks} is composed of the pore-forming KCNQ1 α subunit and the auxiliary β subunit (KCNE1). In the absence of KCNE1, the currents produced by a tetramer of KCNQ1 α subunits are smaller and more rapidly activating and inactivating. When α and β subunits are coexpressed, activation and deactivation kinetics are markedly slowed, inactivation is absent, and the voltage-dependence of activation is shifted rightward toward more depolarized potentials.[45,46] Usually, LQT5 mutations in KCNE1 cause reductions in I_{Ks} current density, thereby prolonging the action potential duration. Although not extensively studied, KCNE1 mutations have been associated with defective trafficking of I_{Ks} channel[70,71] and defective I_{Ks} complex assembly.[72]

KCNE2 (LQT6)

Initially, KCNE2 was thought to be part of the hERG (KCNH2) macromolecular complex.[18] KCNE2 has subsequently been shown to associate, when heterologously expressed, with several K^+ channels including KCNQ1,[73,74] $K_v1.4$, $K_v2.1$,[75] $K_v3.1$ and $K_v3.2$,[76] $K_v4.2$ and $K_v4.3$,[77] and several HCN channels[78,79] (see Ref. 80 for review). It remains to be established whether human KCNE2 associates *in vivo* with these channels in the heart as well as the mechanism(s) by which mutant KCNE2 alters cardiac repolarization leading to LQTS.[80]

KCNJ2 (Kir2.1) (LQT7)

Andersen–Tawil syndrome (ATS), also known as LQT7, is due in 60% of cases to a missense mutation in the gene *KCNJ2*, which encodes for the inward K^+ rectifier channel Kir2.1 (I_{K1}). Mutant subunits alter the function and/or insertion in the membrane. Many mutations prevent the interaction of phosphatidylinositol-4,5-biphosphate (PIP2) and the channel, which is needed for channel activation.[81,82]

ATS was first reported by Dr. Ellen Andersen in 1971 after a series of patients were found to have periodic skeletal muscle paralysis, ventricular ectopy, and dysmorphic features.[83] Dr. Rabi Tawil elucidated the genetic abnormalities.[84] The diagnosis of ATS is based on abnormal cranial, facial, and skeletal morphologies, along with electrocardiographic criteria. ATS must have two of the three specific features: (1) structural dysmorphia, (2) periodic paralysis, and (3) premature ventricular complexes, bidirectional ventricular tachycardia, or polymorphic ventricular tachycardia. In the case where only one of the above features is present, a family history is necessary to make the diagnosis. The ventricular tachycardia is usually nonsustained, with rates ≤ 150 beats/min, although the tachycardia burden is often significant enough to cause a tachycardia-induced cardiomyopathy.[81,82]

CACNA1 ($Ca_V1.2$) (LQT8)

Timothy syndrome is an inherited primary arrhythmia with a high mortality and a complex phenotype that involves the cardiac, endocrine, and central nervous systems.[13,85] Timothy syndrome is a mutation in the *CACNA1* gene, which encodes the voltage-dependent L-type Ca^{2+} channel, $Ca_V1.2$. Ca^{2+} influx through $Ca_V1.2$ channels initiates excitation–contraction coupling by activating ryanodine receptors on the sarcoplasmic reticulum and contributes to the plateau phase of the action potential. Identified as a novel cardiac arrhythmia syndrome associated with syndactyly and dysmorphic facial features, the genetic basis for Timothy syndrome was discovered to be a gain-of-function mutation (G406R) in *CACNA1C*. The mutation greatly slows channel inactivation and thereby prolongs cellular repolarization in cardiac myocytes.[13] The high mortality of Timothy syndrome is due to ventricular tachycardia or fibrillation. Due to the low prevalence, Timothy syndrome has not been well studied and therapies are not supported by specific data. Induced human pluripotent stem cells differentiated into cardiomyocytes have been used to identify potential treatment strategies for Timothy syndrome. Dolmetsch and colleagues reported that roscovitine, a compound that increases the voltage-dependent inactivation of $Ca_V1.2$, restored the electrical and Ca^{2+} signaling properties of cardiomyocytes from Timothy syndrome patients.[86]

CAV3 (LQT9)

LQT9 is a mutation of the *CAV3* gene, which encodes caveolin-3, the major scaffolding protein present in caveolae in the heart.[21] Caveolae serve as platform to organize and regulate signal transduction pathways. Ion channels and transporters localized to caveolae include $Na_V1.5$, $K_V1.5$, Kir2.1, $Ca_V1.2$, and the $NCX._V$[20] Mutations in caveolin-3 directly modify $Na_V1.5$ kinetics, resulting in a two- to threefold increase in the persistent Na_V^+ current, similar to the increased persistent Na^+ current observed in LQT3 mutations.[21] Mutant caveolin-3 may also decrease the cell surface expression of Kir2.1, which may be additive to the persistent Na^+ current, prolonging repolarization and leading to arrhythmia generation in LQT9.[87]

SCN4B (LQT10)

SCN4B (LQT10) is a rare LQTS-susceptibility gene that encodes a cardiac Na$^+$ channel β subunit. An L179F (C535T) missense mutation in *SCN4B* caused an eightfold (compared with *SCN5A* alone) and threefold (compared with *SCN5A* + WT β4) increase in persistent Na$^+$ current.[19] This is consistent with the molecular/electrophysiological phenotype previously discussed for LQT3 and LQT9 phenotypes.

AKAP9 (LQT11)

LQT11 is a mutation of Yotiao, an A-kinase anchoring protein (AKAP9) shown to associate with KCNQ1 and facilitate both the phosphorylation of Ser^{27} on the N-terminus of KCNQ1 and the increased I_{Ks} in response to β-adrenergic stimulation.[48] Yotiao binds to KCNQ1 by a leucine zipper motif, which is disrupted by an LQTS mutation (KCNQ1-G589D).[48,88] Similarly, a mutation of *AKAP9*, S1570L, causes disruption, but not ablation, of the interaction between KCNQ1 and Yotiao, reduction of PKA-dependent phosphorylation of KCNQ1, and marked inhibition of the functional response of I_{Ks} current to cAMP stimulation.[22]

SNTA1 (LQT12)

LQT12 is due a missense mutation of the α-1 syntrophin gene (A390V-SNTA1) found in a patient with recurrent syncope and markedly prolonged QT interval.[23] α-1 syntrophin links neuronal nitric oxide synthase (nNOS) to the nNOS inhibitor plasma membrane Ca^{2+}-ATPase subtype 4b (PMCA4b) and also associates with SCN5A. The mutant protein A390V-SNTA1, however, selectively disrupted the association of PMCA4b with this complex and increased nitrosylation of SCN5A.[23] Using heterologous expression, mutant A390V-SNTA1 was shown to increase peak and persistent Na$^+$ current compared with WT-SNTA1. Expression of A390V-SNTA1 in cardiomyocytes also increased persistent Na$^+$ current, consistent with an LQT3 phenotype.[23]

KCNJ5 (Kir3.4) (LQT13)

LQT13 is caused by a loss-of-function of the G-protein coupled inward rectifier K$^+$ channel subtype 4, encoded by *KCNJ5*.[16,89] Heterologous expression studies with the mutant Kir3.4-Gly387Arg revealed a loss-of-function electrophysiological phenotype resulting from reduced plasma membrane expression.

CALM (LQT14)

Exome sequencing of two unrelated infants presenting with recurrent cardiac arrest identified missense mutations in calmodulin that are associated with severe early-onset LQTS.[25] Expression of calmodulin mutants in adult guinea pig ventricular myocytes induced action potential prolongation and strongly suppressed Ca^{2+}/calmodulin-dependent inactivation (CDI) of $Ca_v1.2$.[90] Thus, similar to the reduced voltage-dependent inactivation of $Ca_v1.2$ in Timothy syndrome (LQT8), inappropriate Ca^{2+}-calmodulin-dependent inactivation of $Ca_v1.2$ is a potential molecular mechanism of LQT14.

ACQUIRED LONG QT

Specific drugs and electrolyte abnormalities, such as hypokalemia and hypomagnesemia, can cause prolonged QT interval on an ECG leading to torsade de pointes and sudden death. Women are more commonly affected by acquired LQTS than men in association with medications or electrolyte abnormalities.[91] As discussed earlier, the hallmark mechanism of acquired LQTS and torsade de pointes is the blockade of I_{Kr} by specific drugs.[59] Inhibitors of phosphoinositide-3-kinase (PI3K) or tyrosine kinases, which are used as anticancer drugs, delay repolarization, thereby causing QT prolongation.[92] These drugs decrease the delayed rectifier currents I_{Kr} and I_{Ks}, the L-type Ca^{2+} current, and the peak Na^+ current, but increased persistent Na^+ current, identifying the molecular mechanism by which some tyrosine kinase inhibitors in clinical use are associated with increased risk of malignant ventricular arrhythmias.[92]

TREATMENT STRATEGIES

Therapeutic interventions for LQTS include medical, device, and surgical therapies. Due to the relatively low prevalence of LQTS, however, the assessment of their efficacies are based predominantly on observational long-term registry studies. The HRS/EHRA Expert Consensus proposes as a class I indication that patients with LQTS should avoid QT prolonging drugs (www.qtdrugs.org), and there should be careful correction of any electrolyte abnormalities, such as hypokalemia, which may occur during diarrhea or vomiting and subsequently cause QT interval lengthening.[35]

All patients who are asymptomatic with QTc≥470ms and/or symptomatic with syncope or documented ventricular tachycardia or ventricular fibrillation should receive β-blockers.[93] β-blockers should also be prescribed for patients with LQT1, LQT2, or LQT3 with intermediate or high risk, and possibly those with low risk, which should be considered on a personalized basis.[93] Propranolol (2–4 mg/kg/day) or nadolol (1–1.5 mg/kg/day) is the preferred therapy.[31] β-blocker therapy has been shown to have significant reduction in mortality in LQT1 and LQT2. Since the beginning of genotype–phenotype correlation studies, β-blocker therapy often was not even started in LQT3 patients based on the assumption that β-blockers were ineffective. In a large study of LQT3 patients ($n = 403$), among those without events in the first year of life, mortality on β-blockers is just ~3%.[94–96] Thus, β-blockers are now also recommended in patients with LQT3.[94]

Clinical trials have shown that implantation of implantable cardiac defibrillators (ICDs) significantly improved mortality in high-risk LQTS patients. In high-risk LQTS patients who underwent ICD implantation, there was 1 death in 73 patients compared to 26 deaths in 161 patients who had not received ICD implantation.[97] An ICD is recommended for patients who have frequent syncopal episodes despite maximal doses of β-blocker therapy or at high risk of SCD such as patients with history of SCD or, infant cases of less than 1 year of age, or those with Jervell Lange-Nielsen syndrome. Implantation of an ICD is a class I indication in LQTS patients who are survivors of a cardiac arrest.[31,35,97]

Genetic testing is now readily available and can be helpful in delineating usage of adjunctive therapy. For example, K^+ supplementation has been used in LQT2 because the hERG channel is sensitive to extracellular K^+ levels.[61] Na^+ channel blockers, such as flecainide, mexiletine, and ranolazine, can shorten the QT interval and normalize repolarization.[98–100] Clinical studies are ongoing to explore the efficacy of ranolazine in LQT3 patients (NCT01648205).

Patients with recurrent syncope despite medical treatment with β-blockers, and repeated ICD shocks for ventricular arrhythmias, should be considered for surgical left cervicothoracic sympathetic denervation (LCSD). The purpose of LCSD is to directly decrease adrenergic stimulation by severing the sympathetic fibers that innervate the myocardium and the coronary arteries.[101]

CONCLUSIONS

Despite LQTS being a rare disease, important insights have been derived from basic and clinical studies exploring its genetics and unique electrophysiology manifestations. These studies have identified unique gene-specific approaches to treat diseases associated with abnormal cardiac repolarization, providing some of the first examples of personalized medicine. Ongoing investigations are laying the foundation for more personalized approaches, from mutation-specific risk stratification to mutation-based treatments.

References

1. Schwartz PJ, Stramba-Badiale M, Crotti L, Pedrazzini M, Besana A, Bosi G, et al. Prevalence of the congenital long-QT syndrome. *Circulation* 2009;**120**(18):1761–7.
2. Jervell A, Lange-Nielsen F. Congenital deaf-mutism, functional heart disease with prolongation of the Q-T interval and sudden death. *Am Heart J* 1957;**54**(1):59–68.
3. Levine SA, Woodworth CR. Congenital deaf-mutism, prolonged QT interval, syncopal attacks and sudden death. *N Engl J Med* 1958;**259**(9):412–7.
4. Romano C, Gemme G, Pongiglione R. Rare cardiac arrythmias of the pediatric age. II. Syncopal attacks due to paroxysmal ventricular fibrillation. (Presentation of 1st case in Italian pediatric literature). *Clin Pediatr (Bologna)* 1963;**45**:656–83.
5. Ward OC. A new familial cardiac syndrome in children. *J Ir Med Assoc* 1964;**54**:103–6.
6. Moss AJ, Schwartz PJ, Crampton RS, Locati E, Carleen E. The long QT syndrome: a prospective international study. *Circulation* 1985;**71**(1):17–21.
7. Wang Q, Shen J, Splawski I, Atkinson D, Li Z, Robinson JL, et al. SCN5A mutations associated with an inherited cardiac arrhythmia, long QT syndrome. *Cell* 1995;**80**(5):805–11.
8. Wang Q, Curran ME, Splawski I, Burn TC, Millholland JM, VanRaay TJ, et al. Positional cloning of a novel potassium channel gene: K$_V$LQT1 mutations cause cardiac arrhythmias. *Nat Genet* 1996;**12**(1):17–23.
9. Splawski I, Timothy KW, Vincent GM, Atkinson DL, Keating MT. Molecular basis of the long-QT syndrome associated with deafness. *N Engl J Med* 1997;**336**(22):1562–7.
10. Sanguinetti MC, Jiang C, Curran ME, Keating MT. A mechanistic link between an inherited and an acquired cardiac arrhythmia: HERG encodes the I_{Kr} potassium channel. *Cell* 1995;**81**(2):299–307.
11. Bennett PB, Yazawa K, Makita N, George Jr AL. Molecular mechanism for an inherited cardiac arrhythmia. *Nature* 1995;**376**(6542):683–5.
12. Curran ME, Splawski I, Timothy KW, Vincent GM, Green ED, Keating MT. A molecular basis for cardiac arrhythmia: HERG mutations cause long QT syndrome. *Cell* 1995;**80**(5):795–803.
13. Splawski I, Timothy KW, Sharpe LM, Decher N, Kumar P, Bloise R, et al. Ca(V)1.2 calcium channel dysfunction causes a multisystem disorder including arrhythmia and autism. *Cell* 2004;**119**(1):19–31.
14. Plaster NM, Tawil R, Tristani-Firouzi M, Canun S, Bendahhou S, Tsunoda A, et al. Mutations in Kir2.1 cause the developmental and episodic electrical phenotypes of Andersen's syndrome. *Cell* 2001;**105**(4):511–9.
15. Ai T, Fujiwara Y, Tsuji K, Otani H, Nakano S, Kubo Y, et al. Novel KCNJ2 mutation in familial periodic paralysis with ventricular dysrhythmia. *Circulation* 2002;**105**(22):2592–4.
16. Yang Y, Yang Y, Liang B, Liu J, Li J, Grunnet M, et al. Identification of a Kir3.4 mutation in congenital long QT syndrome. *Am J Hum Genet* 2010;**86**(6):872–80.
17. Splawski I, Tristani-Firouzi M, Lehmann MH, Sanguinetti MC, Keating MT. Mutations in the hminK gene cause long QT syndrome and suppress I_{Ks} function. *Nat Genet* 1997;**17**(3):338–40.
18. Abbott GW, Sesti F, Splawski I, Buck ME, Lehmann MH, Timothy KW, et al. MiRP1 forms I_{Kr} potassium channels with HERG and is associated with cardiac arrhythmia. *Cell* 1999;**97**(2):175–87.
19. Medeiros-Domingo A, Kaku T, Tester DJ, Iturralde-Torres P, Itty A, Ye B, et al. SCN4B-encoded sodium channel beta4 subunit in congenital long-QT syndrome. *Circulation* 2007;**116**(2):134–42.
20. Mohler PJ, Schott JJ, Gramolini AO, Dilly KW, Guatimosim S, duBell WH, et al. Ankyrin-B mutation causes type 4 long-QT cardiac arrhythmia and sudden cardiac death. *Nature* 2003;**421**(6923):634–9.

21. Vatta M, Ackerman MJ, Ye B, Makielski JC, Ughanze EE, Taylor EW, et al. Mutant caveolin-3 induces persistent late sodium current and is associated with long-QT syndrome. *Circulation* 2006;**114**(20):2104–12.

22. Chen L, Marquardt ML, Tester DJ, Sampson KJ, Ackerman MJ, Kass RS. Mutation of an A-kinase-anchoring protein causes long-QT syndrome. *Proc Natl Acad Sci USA* 2007;**104**(52):20990–5.

23. Ueda K, Valdivia C, Medeiros-Domingo A, Tester DJ, Vatta M, Farrugia G, et al. Syntrophin mutation associated with long QT syndrome through activation of the nNOS-SCN5A macromolecular complex. *Proc Natl Acad Sci USA* 2008;**105**(27):9355–60.

24. Wu G, Ai T, Kim JJ, Mohapatra B, Xi Y, Li Z, et al. Alpha-1-syntrophin mutation and the long-QT syndrome: a disease of sodium channel disruption. *Circ Arrhythm Electrophysiol* 2008;**1**(3):193–201.

25. Crotti L, Johnson CN, Graf E, De Ferrari GM, Cuneo BF, Ovadia M, et al. Calmodulin mutations associated with recurrent cardiac arrest in infants. *Circulation* 2013;**127**(9):1009–17.

26. Splawski I, Shen J, Timothy KW, Lehmann MH, Priori S, Robinson JL, et al. Spectrum of mutations in long-QT syndrome genes. K$_V$LQT1, HERG, SCN5A, KCNE1, and KCNE2. *Circulation* 2000;**102**(10):1178–85.

27. Priori SG, Schwartz PJ, Napolitano C, Bloise R, Ronchetti E, Grillo M, et al. Risk stratification in the long-QT syndrome. *N Engl J Med* 2003;**348**(19):1866–74.

28. Goldenberg I, Moss AJ. Long QT syndrome. *J Am Coll Cardiol* 2008;**51**(24):2291–300.

29. Ackerman MJ, Mohler PJ. Defining a new paradigm for human arrhythmia syndromes: phenotypic manifestations of gene mutations in ion channel- and transporter-associated proteins. *Circ Res* 2010;**107**(4):457–65.

30. Viskin S. The QT interval: too long, too short or just right. *Heart Rhythm* 2009;**6**(5):711–5.

31. Mizusawa Y, Horie M, Wilde AA. Genetic and clinical advances in congenital long QT syndrome. *Circ J* 2014;**78**(12):2827–33.

32. Vincent GM. The long-QT syndrome–bedside to bench to bedside. *N Engl J Med* 2003;**348**(19):1837–8.

33. Moss AJ, Schwartz PJ, Crampton RS, Tzivoni D, Locati EH, MacCluer J, et al. The long QT syndrome. Prospective longitudinal study of 328 families. *Circulation* 1991;**84**(3):1136–44.

34. Priori SG, Wilde AA, Horie M, Cho Y, Behr ER, Berul C, et al. Executive summary: HRS/EHRA/APHRS expert consensus statement on the diagnosis and management of patients with inherited primary arrhythmia syndromes. *Europace* 2013;**15**(10):1389–406.

35. Priori SG, Wilde AA, Horie M, Cho Y, Behr ER, Berul C, et al. HRS/EHRA/APHRS expert consensus statement on the diagnosis and management of patients with inherited primary arrhythmia syndromes: document endorsed by HRS, EHRA, and APHRS in May 2013 and by ACCF, AHA, PACES, and AEPC in June 2013. *Heart Rhythm* 2013;**10**(12):1932–63.

36. Crotti L, Monti MC, Insolia R, Peljto A, Goosen A, Brink PA, et al. NOS1AP is a genetic modifier of the long-QT syndrome. *Circulation* 2009;**120**(17):1657–63.

37. Tomas M, Napolitano C, De Giuli L, Bloise R, Subirana I, Malovini A, et al. Polymorphisms in the NOS1AP gene modulate QT interval duration and risk of arrhythmias in the long QT syndrome. *J Am Coll Cardiol* 2010;**55**(24):2745–52.

38. Arking DE, Pfeufer A, Post W, Kao WH, Newton-Cheh C, Ikeda M, et al. A common genetic variant in the NOS1 regulator NOS1AP modulates cardiac repolarization. *Nat Genet* 2006;**38**(6):644–51.

39. Duchatelet S, Crotti L, Peat RA, Denjoy I, Itoh H, Berthet M, et al. Identification of a KCNQ1 polymorphism acting as a protective modifier against arrhythmic risk in long-QT syndrome. *Circ Cardiovasc Genet* 2013;**6**(4):354–61.

40. Amin AS, Giudicessi JR, Tijsen AJ, Spanjaart AM, Reckman YJ, Klemens CA, et al. Variants in the 3' untranslated region of the KCNQ1-encoded K$_v$7.1 potassium channel modify disease severity in patients with type 1 long QT syndrome in an allele-specific manner. *Eur Heart J* 2012;**33**(6):714–23.

41. Zareba W, Moss AJ, Locati EH, Lehmann MH, Peterson DR, Hall WJ, et al. Modulating effects of age and gender on the clinical course of long QT syndrome by genotype. *J Am Coll Cardiol* 2003;**42**(1):103–9.

42. Boyle MB, MacLusky NJ, Naftolin F, Kaczmarek LK. Hormonal regulation of K$^+$-channel messenger RNA in rat myometrium during oestrus cycle and in pregnancy. *Nature* 1987;**330**(6146):373–5.

43. Moss AJ, Kass RS. Long QT syndrome: from channels to cardiac arrhythmias. *J Clin Invest* 2005;**115**(8):2018–24.

44. Schwartz PJ, Priori SG, Spazzolini C, Moss AJ, Vincent GM, Napolitano C, et al. Genotype-phenotype correlation in the long-QT syndrome: gene-specific triggers for life-threatening arrhythmias. *Circulation* 2001;**103**(1):89–95.

45. Sanguinetti MC, Curran ME, Zou A, Shen J, Spector PS, Atkinson DL, et al. Coassembly of K(V)LQT1 and minK (IsK) proteins to form cardiac I(Ks) potassium channel. *Nature* 1996;**384**(6604):80–3.

46. Barhanin J, Lesage F, Guillemare E, Fink M, Lazdunski M, Romey G. K(V)LQT1 and lsK (minK) proteins associate to form the I(Ks) cardiac potassium current. *Nature* 1996;**384**(6604):78–80.

47. Kurokawa J, Chen L, Kass RS. Requirement of subunit expression for cAMP-mediated regulation of a heart potassium channel. *Proc Natl Acad Sci USA* 2003;**100**(4):2122–7.

48. Marx SO, Kurokawa J, Reiken S, Motoike H, D'Armiento J, Marks AR, et al. Requirement of a macromolecular signaling complex for beta adrenergic receptor modulation of the KCNQ1-KCNE1 potassium channel. *Science* 2002;**295**(5554):496–9.

49. Li Y, Chen L, Kass RS, Dessauer CW. The A-kinase anchoring protein Yotiao facilitates complex formation between adenylyl cyclase type 9 and the I_{Ks} potassium channel in heart. *J Biol Chem* 2012;**287**(35):29815–24.

50. Terrenoire C, Houslay MD, Baillie GS, Kass RS. The cardiac I_{Ks} potassium channel macromolecular complex includes the phosphodiesterase PDE4D3. *J Biol Chem* 2009;**284**(14):9140–6.

51. Moretti A, Bellin M, Welling A, Jung CB, Lam JT, Bott-Flugel L, et al. Patient-specific induced pluripotent stem-cell models for long-QT syndrome. *N Engl J Med* 2010;**363**(15):1397–409.

52. Rodriguez N, Amarouch MY, Montnach J, Piron J, Labro AJ, Charpentier F, et al. Phosphatidylinositol-4,5-bisphosphate (PIP(2)) stabilizes the open pore conformation of the K$_V$11.1 (hERG) channel. *Biophys J* 2010;**99**(4):1110–8.

53. Delisle BP, Anson BD, Rajamani S, January CT. Biology of cardiac arrhythmias: ion channel protein trafficking. *Circ Res* 2004;**94**(11):1418–28.

54. Shimizu W, Moss AJ, Wilde AA, Towbin JA, Ackerman MJ, January CT, et al. Genotype-phenotype aspects of type 2 long QT syndrome. *J Am Coll Cardiol* 2009;**54**(22):2052–62.

55. Roden DM, Hoffman BF. Action potential prolongation and induction of abnormal automaticity by low quinidine concentrations in canine Purkinje fibers. Relationship to potassium and cycle length. *Circ Res* 1985;**56**(6):857–67.

56. Yang T, Snyders DJ, Roden DM. Rapid inactivation determines the rectification and [K$^+$]o dependence of the rapid component of the delayed rectifier K$^+$ current in cardiac cells. *Circ Res* 1997;**80**(6):782–9.

57. Mullins FM, Stepanovic SZ, Desai RR, George Jr AL, Balser JR. Extracellular sodium interacts with the HERG channel at an outer pore site. *J Gen Physiol* 2002;**120**(4):517–37.

58. Guo J, Massaeli H, Xu J, Jia Z, Wigle JT, Mesaeli N, et al. Extracellular K$^+$ concentration controls cell surface density of I_{Kr} in rabbit hearts and of the HERG channel in human cell lines. *J Clin Invest* 2009;**119**(9):2745–57.

59. Roden DM, Viswanathan PC. Genetics of acquired long QT syndrome. *J Clin Invest* 2005;**115**(8):2025–32.
60. Fernandez D, Ghanta A, Kauffman GW, Sanguinetti MC. Physicochemical features of the HERG channel drug binding site. *J Biol Chem* 2004;**279**(11):10120–7.
61. Sanguinetti MC, Tristani-Firouzi M. hERG potassium channels and cardiac arrhythmia. *Nature* 2006;**440**(7083):463–9.
62. Makielski JC, Farley AL. Na(+) current in human ventricle: implications for sodium loading and homeostasis. *J Cardiovasc Electrophysiol* 2006;**17**(Suppl. 1):S15–20.
63. Toischer K, Hartmann N, Wagner S, Fischer TH, Herting J, Danner BC, et al. Role of late sodium current as a potential arrhythmogenic mechanism in the progression of pressure-induced heart disease. *J Mol Cell Cardiol* 2013;**61**:111–22.
64. Song Y, Shryock JC, Wu L, Belardinelli L. Antagonism by ranolazine of the pro-arrhythmic effects of increasing late I_{Na} in guinea pig ventricular myocytes. *J Cardiovasc Pharmacol* 2004;**44**(2):192–9.
65. Wu L, Shryock JC, Song Y, Li Y, Antzelevitch C, Belardinelli L. Antiarrhythmic effects of ranolazine in a guinea pig in vitro model of long-QT syndrome. *J Pharmacol Exp Ther* 2004;**310**(2):599–605.
66. Tian XL, Yong SL, Wan X, Wu L, Chung MK, Tchou PJ, et al. Mechanisms by which SCN5A mutation N1325S causes cardiac arrhythmias and sudden death in vivo. *Cardiovasc Res* 2004;**61**(2):256–67.
67. Terrenoire C, Wang K, Tung KW, Chung WK, Pass RH, Lu JT, et al. Induced pluripotent stem cells used to reveal drug actions in a long QT syndrome family with complex genetics. *J Gen Physiol* 2013;**141**(1):61–72.
68. Ma D, Wei H, Zhao Y, Lu J, Li G, Sahib NB, et al. Modeling type 3 long QT syndrome with cardiomyocytes derived from patient-specific induced pluripotent stem cells. *Int J Cardiol* 2013;**168**(6):5277–86.
69. Schott JJ, Charpentier F, Peltier S, Foley P, Drouin E, Bouhour JB, et al. Mapping of a gene for long QT syndrome to chromosome 4q25-27. *Am J Hum Genet* 1995;**57**(5):1114–22.
70. Bianchi L, Shen Z, Dennis AT, Priori SG, Napolitano C, Ronchetti E, et al. Cellular dysfunction of LQT5-minK mutants: abnormalities of I_{Ks}, I_{Kr} and trafficking in long QT syndrome. *Hum Mol Genet* 1999;**8**(8):1499–507.
71. Krumerman A, Gao X, Bian JS, Melman YF, Kagan A, McDonald TV. An LQT mutant minK alters K_VLQT1 trafficking. *Am J Physiol Cell Physiol* 2004;**286**(6):C1453–63.
72. Harmer SC, Wilson AJ, Aldridge R, Tinker A. Mechanisms of disease pathogenesis in long QT syndrome type 5. *Am J Physiol Cell Physiol* 2010;**298**(2):C263–73.
73. Toyoda F, Ueyama H, Ding WG, Matsuura H. Modulation of functional properties of KCNQ1 channel by association of KCNE1 and KCNE2. *Biochem Biophys Res Commun* 2006;**344**(3):814–20.
74. Jiang M, Xu X, Wang Y, Toyoda F, Liu XS, Zhang M, et al. Dynamic partnership between KCNQ1 and KCNE1 and influence on cardiac I_{Ks} current amplitude by KCNE2. *J Biol Chem* 2009;**284**(24):16452–62.
75. McCrossan ZA, Roepke TK, Lewis A, Panaghie G, Abbott GW. Regulation of the Kv2.1 potassium channel by MinK and MiRP1. *J Membr Biol* 2009;**228**(1):1–14.
76. Lewis A, McCrossan ZA, Abbott GW. MinK, MiRP1, and MiRP2 diversify Kv3.1 and Kv3.2 potassium channel gating. *J Biol Chem* 2004;**279**(9):7884–92.
77. Zhang M, Jiang M, Tseng GN. minK-related peptide 1 associates with Kv4.2 and modulates its gating function: potential role as beta subunit of cardiac transient outward channel? *Circ Res* 2001;**88**(10):1012–9.
78. Yu H, Wu J, Potapova I, Wymore RT, Holmes B, Zuckerman J, et al. MinK-related peptide 1: a beta subunit for the HCN ion channel subunit family enhances expression and speeds activation. *Circ Res* 2001;**88**(12):E84–7.

79. Decher N, Bundis F, Vajna R, Steinmeyer K. KCNE2 modulates current amplitudes and activation kinetics of HCN4: influence of KCNE family members on HCN4 currents. *Pflugers Arch* 2003;**446**(6):633–40.

80. Eldstrom J, Fedida D. The voltage-gated channel accessory protein KCNE2: multiple ion channel partners, multiple ways to long QT syndrome. *Expert Rev Mol Med* 2011; **13**:e38.

81. Tristani-Firouzi M, Etheridge SP. Kir 2.1 channelopathies: the Andersen-Tawil syndrome. *Pflugers Arch* 2010;**460**(2):289–94.

82. Hsiao PY, Tien HC, Lo CP, Juang JM, Wang YH, Sung RJ. Gene mutations in cardiac arrhythmias: a review of recent evidence in ion channelopathies. *Appl Clin Genet* 2013;**6**:1–13.

83. Andersen ED, Krasilnikoff PA, Overvad H. Intermittent muscular weakness, extrasystoles, and multiple developmental anomalies. A new syndrome? *Acta Paediatr Scand* 1971;**60**(5):559–64.

84. Tawil R, Ptacek LJ, Pavlakis SG, DeVivo DC, Penn AS, Ozdemir C, et al. Andersen's syndrome: potassium-sensitive periodic paralysis, ventricular ectopy, and dysmorphic features. *Ann Neurol* 1994;**35**(3):326–30.

85. Marks ML, Whisler SL, Clericuzio C, Keating M. A new form of long QT syndrome associated with syndactyly. *J Am Coll Cardiol* 1995;**25**(1):59–64.

86. Yazawa M, Hsueh B, Jia X, Pasca AM, Bernstein JA, Hallmayer J, et al. Using induced pluripotent stem cells to investigate cardiac phenotypes in Timothy syndrome. *Nature* 2011;**471**(7337):230–4.

87. Vaidyanathan R, Vega AL, Song C, Zhou Q, Tan BH, Berger S, et al. The interaction of caveolin 3 protein with the potassium inward rectifier channel Kir2.1: physiology and pathology related to long QT syndrome 9 (LQT9). *J Biol Chem* 2013;**288**(24): 17472–80.

88. Piippo K, Swan H, Pasternack M, Chapman H, Paavonen K, Viitasalo M, et al. A founder mutation of the potassium channel KCNQ1 in long QT syndrome: implications for estimation of disease prevalence and molecular diagnostics. *J Am Coll Cardiol* 2001;**37**(2):562–8.

89. Wang F, Liu J, Hong L, Liang B, Graff C, Yang Y, et al. The phenotype characteristics of type 13 long QT syndrome with mutation in KCNJ5 (Kir3.4-G387R). *Heart Rhythm* 2013;**10**(10):1500–6.

90. Limpitikul WB, Dick IE, Joshi-Mukherjee R, Overgaard MT, George Jr AL, Yue DT. Calmodulin mutations associated with long QT syndrome prevent inactivation of cardiac L-type Ca(2+) currents and promote proarrhythmic behavior in ventricular myocytes. *J Mol Cell Cardiol* 2014;**74**:115–24.

91. Zipes DP, Camm AJ, Borggrefe M, Buxton AE, Chaitman B, Fromer M, et al. ACC/ AHA/ESC 2006 guidelines for management of patients with ventricular arrhythmias and the prevention of sudden cardiac death: a report of the American College of Cardiology/American Heart Association Task Force and the European Society of Cardiology Committee for Practice Guidelines (writing committee to develop guidelines for management of patients with ventricular arrhythmias and the prevention of sudden cardiac death): developed in collaboration with the European Heart Rhythm Association and the Heart Rhythm Society. *Circulation* 2006;**114**(10):e385–484.

92. Lu Z, Wu CY, Jiang YP, Ballou LM, Clausen C, Cohen IS, et al. Suppression of phosphoinositide 3-kinase signaling and alteration of multiple ion currents in drug-induced long QT syndrome. *Sci Transl Med* 2012;**4**(131):131ra50.

93. Priori SG, Wilde AA, Horie M, Cho Y, Behr ER, Berul C, et al. Executive summary: HRS/EHRA/APHRS expert consensus statement on the diagnosis and management of patients with inherited primary arrhythmia syndromes. *Heart Rhythm* 2013;**10**(12):e85–108.

94. Calvillo L, Spazzolini C, Vullo E, Insolia R, Crotti L, Schwartz PJ. Propranolol prevents life-threatening arrhythmias in LQT3 transgenic mice: implications for the clinical management of LQT3 patients. *Heart Rhythm* 2014;**11**(1):126–32.

95. Wilde AA, Kaufman ES, Shimizu W. Sodium channel mutations, risk of cardiac events, and efficacy of beta-blocker therapy in type 3 long QT syndrome. *Heart Rhythm* 2012;**9**(Suppl.):S321.

96. Schwartz PJ, Spazzolini C, Crotti L. All LQT3 patients need an ICD: true or false? *Heart Rhythm* 2009;**6**(1):113–20.

97. Zareba W, Moss AJ, Daubert JP, Hall WJ, Robinson JL, Andrews M. Implantable cardioverter defibrillator in high-risk long QT syndrome patients. *J Cardiovasc Electrophysiol* 2003;**14**(4):337–41.

98. Moss AJ, Zareba W, Schwarz KQ, Rosero S, McNitt S, Robinson JL. Ranolazine shortens repolarization in patients with sustained inward sodium current due to type-3 long-QT syndrome. *J Cardiovasc Electrophysiol* 2008;**19**(12):1289–93.

99. Windle JR, Geletka RC, Moss AJ, Zareba W, Atkins DL. Normalization of ventricular repolarization with flecainide in long QT syndrome patients with SCN5A: DeltaKPQ mutation. *Ann Noninvasive Electrocardiol* 2001;**6**(2):153–8.

100. Moss AJ, Windle JR, Hall WJ, Zareba W, Robinson JL, McNitt S, et al. Safety and efficacy of flecainide in subjects with Long QT-3 syndrome (DeltaKPQ mutation): a randomized, double-blind, placebo-controlled clinical trial. *Ann Noninvasive Electrocardiol* 2005;**10**(Suppl. 4):59–66.

101. Schwartz PJ. Cardiac sympathetic denervation to prevent life-threatening arrhythmias. *Nat Rev Cardiol* 2014;**11**(6):346–53.

Ion Channels in Cancer

W.J. Brackenbury

University of York, York, United Kingdom

INTRODUCTION

Ion channels are integral membrane spanning proteins that permit passive flow of ions, based on switching between closed and open conformational states. Gating of ion channels between conformational states is typically regulated by changes in transmembrane voltage (V_m), specific extracellular or intracellular ligands, or local mechanical changes.[1] A multiplicity of ion channel types, with various degrees of selectivity to different ions, gating, and kinetics, generates considerable flexibility, regulating a number of key physiological processes, including, but not limited to, sensory transduction, electrical activity, synaptic function, cell volume, exocytosis, muscle contraction, transepithelial ion flux, cell division, and cellular migration.[1–3] As a result of these critical roles, it is perhaps no surprise therefore that ion channels are widely expressed in cancer cells, where they have been shown to regulate various pathophysiological behaviors that contribute to tumor progression.[4] The purpose of this chapter is to introduce the types of ion channels that have been identified in cancer cells, to discuss our understanding of the contribution of these ion channels to tumorigenesis and cancer progression, and finally to consider their potential utility as novel therapeutic targets in oncology.[5]

ION CHANNELS IDENTIFIED IN CANCERS

A number of reviews have described the expression and function of different classes of ion channels in tumors in considerable detail.[6–13] For simplicity, here, the key ion channels that have been identified in tumors are grouped broadly according to ion selectivity and channel classification, rather than by tumor type (Table 6.1; Fig. 6.1). For an alternative approach,

TABLE 6.1   Ion Channel Expression in Various Cancers

Ion Channel	Cancer	Change in Expression Relative to Normal Tissue	References
$K_v11.1$	Colon, esophageal, leukemia	Up	14–16
$K_v1.3$	Breast, lymphoma, weakly metastatic prostate	Up	17–21
$K_{Ca}3.1$	Breast, lymphoma	Up	21,22
$K_v10.1$	Colon, gastric	Up	23–25
$K_v1.5$	Breast, glioma, lung	Down	26,27
$K_{ir}3.1$	Breast	Up	28
$K_{2P}9.1$	Breast, colon	Up	29,30
$K_{2P}2.1$	Prostate	Up	31
BK_{Ca}	Breast, glioma, leukemia, prostate	Up	32–36
$Na_v1.5$	Breast, colon, lymphoma	Up	37–39
$Na_v1.7$	Prostate	Up	40
Voltage-gated Na^+ channel $\beta1$	Breast, prostate, cervical	Up	41–43
Voltage-gated Na^+ channel $\beta2$	Prostate	–	44
ENaC/degenerin	Adenoid, colon, glioma, melanoma	–	45–48
$Ca_v3.1$	Breast, prostate	–	49,50
$Ca_v3.2$	Prostate	–	51,52
$Ca_v2.3$	Nephroblastoma	Up	53
SOC (Orai1/STIM)	Breast, prostate	–	54,55
Transient receptor potential canonical	Glioma, prostate	–	56,57
TRPM8	Prostate	Up	58
TRPV6	Prostate	Up	59
$\alpha7$-containing nicotinic acetylcholine receptor	Lung	Up	60
NMDA receptor	Pancreatic	Up	61
P2X7 receptor	Cervical	–	62

TABLE 6.1   Ion Channel Expression in Various Cancers—cont'd

Ion Channel	Cancer	Change in Expression Relative to Normal Tissue	References
ClC-3	Glioma	Up	63
ClC-4	Breast, osteosarcoma	Both	64,65
CLCA2	Breast, colon	Down	66,67
GABA$_A$ receptor	Breast	Up	68

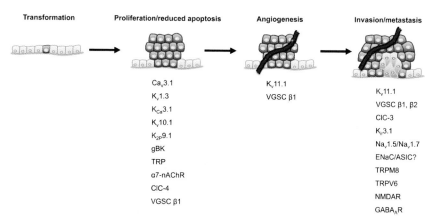

FIGURE 6.1   Ion channel expression during tumor progression. Various ion channels are expressed in transformed cancer cells, contributing to their proliferation. In response to signaling interactions between cancer cells and heterogeneous local tumor microenvironment, expression of other ion channels, including K$_V$11.1 and voltage-gated Na$^+$ channels (VGSCs), is proposed to be upregulated with concomitant downregulation in expression of other (mainly K$^+$) channels. The changing profile of ion channels promotes angiogenesis, invasion, and metastasis. *NMDAR*, N-methyl-D-aspartate receptor. Figure was produced using Science Slides software.

outlining current understanding of the full ion channel complement of different tumors, the reader is referred to other excellent reviews.[5,69] In addition, while the main focus of this chapter is on ion channels expressed predominantly at the plasma membrane, it should be noted that a growing body of evidence suggests that a number of intracellular ion channels also play key roles in cancer, reviewed in detail elsewhere.[70,71]

K$^+$ Channels

A number of different types of K$^+$ channels have been identified in a range of cancers.[11,72] One of the best-characterized examples is the rapid delayed rectifier voltage-gated $K_v11.1$ (hERG1) channel, which is upregulated in acute myeloid leukemia cells, compared with normal lymphocytes.[14] Furthermore, $K_v11.1$ expression correlates with increased risk of cancer relapse and reduced overall survival of cancer patients.[73] $K_v11.1$ is also upregulated in esophageal and colon tumors.[15,16] Other delayed rectifier K^+ channels have been detected in cancer cells. For example, delayed rectifier K^+ channels are expressed in THP-1 human monocytic leukemia cells.[74] In addition, $K_v1.3$ is upregulated in breast tumor biopsies, compared to normal breast tissue.[17] Interestingly, in prostate cancer cells, $K_v1.3$ is mainly expressed earlier in tumor progression and is downregulated in metastatic tumors and cell lines.[18–20] Along with $K_v1.3$, $K_v1.5$, $K_v3.1$, and $K_v10.1$ have been detected in colon carcinoma samples,[23,24] and $K_v10.1$ expression associates with metastasis in gastric tumors[25] and poor prognosis in colon cancers.[23] Interestingly, however, $K_v1.5$ has been shown to be downregulated in a number of cancer types, and its reexpression promotes apoptosis.[26,27]

Other classes of K^+ channels have also been detected in tumor cells. For example, inwardly rectifying $K_{ir}3.1$ channels are overexpressed in breast tumors, correlating with lymph node metastasis.[28] The two-pore TWIK-related acid-sensitive $K_{2P}9.1$ channel is amplified in colon and breast tumors and may play a role in promoting resistance to hypoxia and serum deprivation in the latter,[29,30] whereas $K_{2P}2.1$ is upregulated in prostate cancer, compared to normal prostate tissues and cell lines.[31] The large conductance (BK_{Ca}) Ca^{2+}-activated K^+ channel $K_{Ca}1.1$ is expressed in Friend erythroleukemia cells in response to adhesion to fibronectin.[32,33] Similarly, the main K^+ channel expressed in gliomas is the $K_{Ca}1.1$ variant, gBK, which is upregulated via the erbB2/neuregulin signaling axis.[34,35] $K_{Ca}1.1$ channels have also been reported in prostate and breast cancer cells.[36]

Na+ Channels

Voltage-gated Na^+ channels (VGSCs), classically responsible for action potential initiation and conduction in excitable cells,[1] are also widely expressed in cells that are considered "nonexcitable," including fibroblasts, immune cells, glia, and cancer cells.[75] VGSCs are composed of pore-forming α subunits and nonpore-forming β subunits, which modulate channel gating, and are cell adhesion molecules.[76,77] VGSC α subunits have been identified in cell lines and biopsy tissue from breast cancer,[37,78,79] cervical cancer,[41,80] colon cancer,[38] glioma,[81,82] mesothelioma,[83] lung cancer,[84–87] ovarian cancer,[88] and prostate cancer.[40,89–92] They have also been detected in lymphoma,[39] leukemia,[93] melanoma,[94,95] and neuroblastoma cell lines.[96] Although a range of different α subunit isoforms are expressed across different tumor types, of particular interest is the fact that the cardiac isoform, $Na_v1.5$, is most highly expressed α subunit in lymphoma, breast and colon cancer cells,[37–39] whereas the peripheral nerve isotype, $Na_v1.7$, is predominant in prostate cancer cells.[40] $Na_v1.5$ expression correlates with metastatic capability and poor clinical outcome in breast cancer.[37,79] $Na_v1.7$ expression and Na^+ current also correlate with metastatic potential in prostate cancer.[40,89,91,92,97] A similar pattern has been reported for ovarian[88] and colon cancers,[38] whereas there is no correlation with metastatic potential in lung cancer cell lines,[84,86] and in glioma, the correlation appears to be reversed.[82]

VGSC β subunits have been detected in breast,[42,98] prostate,[43,44] nonsmall cell lung,[86] and cervical cancers.[41] As with the α subunits, β subunit expression appears to vary across different tumor types,[6] and β1 is the most abundant β subunit in breast, prostate, and cervical cancer cells. In summary, VGSC α and β subunits appear to be expressed at varying levels in different cancer types and may perform diverse functions in these tissues. Non-VGSCs are also expressed in tumor cells.[99] Epithelial Na^+ channel (ENaC) and acid-sensitive ion channel (ASIC) variants of the ENaC/degenerin family have been detected in melanoma,[45] adenoid cystic carcinoma,[46] colon cancer,[47] and glioma.[48]

Ca^{2+} Channels

Both voltage-gated Ca^{2+} channels and other types of Ca^{2+} channels have been reported in tumor cells. For example, $Ca_v3.x$ (T-type) voltage-gated Ca^{2+} channels have been detected in MCF-7 breast cancer cells and LNCaP prostate cancer cells, and, in the latter, are activated by androgens.[49,50] Other Ca^{2+} channel types, including store-operated Ca^{2+} (SOC) channels have been reported in breast cancer cells,[54] and various transient receptor potential canonical (TRPC) channels have also been detected in gliomas and prostate cancer cells.[56,57] In addition, TRPM8 and TRPV6 are expressed in prostate cancers, correlating with high grade and metastatic potential.[58,59]

α7-Containing nicotinic acetylcholine receptors, which are permeable to Ca^{2+}, are upregulated in nonsmall cell lung cancer cells and are thought to mediate the mitogenic effects of nicotine in lung cancer.[60] The N-methyl-D-aspartate glutamate receptor is upregulated at the invasive fronts of pancreatic neuroendocrine tumors and is thought to be activated by fluid flow due to high interstitial pressure, promoting glutamate-dependent malignancy in tumor cells.[61] However, these studies relating to ligand-gated cation channels, while interesting, should be interpreted with a degree of caution, given that channel activity has not been determined directly by electrophysiological methods.

Cl⁻ Channels

Glioma cells have been shown to express the ClC-3 channel, typically found in membrane vesicles, at the plasma membrane.[63] In osteosarcoma cells, expression of the intracellular Cl⁻ channel ClC-4 is upregulated following DNA damage, promoting cell survival.[64] In contrast, ClC-4 is downregulated in breast cancer cells, but upregulated in stroma, in line with malignant progression,[65] suggesting a tumor- or cell-specific function of this particular channel. CLCA2, a Ca^{2+}-activated Cl⁻ channel and/or accessory subunit, has been shown to be downregulated in breast and colon cancer cells and may function as a tumor suppressor.[66,67] Finally, the Cl⁻-permeant $GABA_A$ receptor is upregulated in breast cancer metastases to brain, in an adaptive response of the tumor cells to the neural niche.[68]

REGULATION OF ION CHANNEL EXPRESSION IN CANCER

The expression profile of ion channels in cancer is complex. Increasing evidence suggests that a multiplicity of channel types is expressed across different cancers, and, as yet, no single unifying pattern has been identified. Furthermore, the mechanisms by which ion channels are up- or downregulated in tumors appear to be diverse. For example, several channels have been shown to be overexpressed in tumors as a result of gene amplification, including $K_{2P}9.1$ in breast and colon cancers and $Ca_v2.3$ in Wilms' kidney tumors.[29,30,53] Channel expression may also be regulated by epigenetic mechanisms in certain cases. For example, several channel-related genes, including *CLCA2*, *KCNH5*, and *CACNA1G*, encoding the Ca^{2+}-activated Cl⁻ channel/regulator 2, $K_v10.2$, and $Ca_v3.1$, respectively, are downregulated by methylation in various tumors.[100–102] In addition, recent evidence suggests that microRNAs may also regulate channel expression in cancer cells, eg, miR-34a and miR-296-3p both suppress $K_v10.1$ expression.[103,104]

Channel expression in tumors is regulated by alternative splicing. Expression of alternative splice variants of gBK, $K_v11.1$, and $Na_v1.5$ has been reported in various tumor types.[37,63,73] In the case of the latter, upregulation of the neonatal splice variant of $Na_v1.5$ in breast tumors has been suggested to be an example of oncofetal expression, that is, aberrant reexpression of genes expressed early in development may be a contributing factor to tumor progression.[37] However, the broader significance of alternative splicing of ion channels to tumorigenesis and cancer progression is still unclear.

For some tumor-expressed ion channels, expression and activity are regulated by a number of serum factors. For example, in ML-1 myeloblastic leukemia cells, serum, and epidermal growth factor (EGF) upregulate K^+ currents via phosphorylation.[105] In addition, both $K_v1.3$ and $K_{Ca}3.1$ are upregulated by serum factors in Daudi B-lymphoma cells.[21] Similarly, in glioma cells, gBK can be activated by growth factors and neurotransmitters.[35] In breast cancer cells, erbB2 stimulates the functional expression of SOC channels.[54] Growth factors, including insulin-like growth factor 1 (IGF1), increase the expression of both $K_v10.1$ and $K_{Ca}3.1$ in MCF-7 breast cancer cells.[22] In prostate cancer cells, expression of $Na_v1.7$ is promoted by both EGF and nerve growth factor.[106–108] Steroid hormones may also play a role. For example, in MDA-MB-231 cells, estrogen has been shown to increase Na^+ current carried by $Na_v1.5$, via the G-protein-coupled estrogen receptor, GPR30.[109] In prostate cancer cells, androgens increase expression of TRPM8[85], but decrease expression of VGSC $\beta1$ subunits.[43] Finally, under specific conditions, channel expression may be autoregulated. For example, in prostate and breast cancer cell lines, VGSC α subunit expression is activity dependent: Na^+ current has been shown to activate PKA and thereby promote trafficking of $Na_v1.5/Na_v1.7$ to the plasma membrane, in a positive feedback loop.[110,111]

ION CHANNEL FUNCTION IN CANCER CELLS

Aberrant expression of ion channels in tumor cells has been shown to regulate a multiplicity of cellular functions, a number of which are thought to contribute to various critical aspects of tumor progression, discussed later (Table 6.2). In particular, there is now strong evidence that ion channels regulate V_m and intracellular signaling cascades, the latter through both conducting (dependent on ion flux) and nonconducting (independent of ion flux) mechanisms. The V_m of cancer cells is generally relatively depolarized compared to terminally differentiated cells.[112] This depolarized V_m (in the range ~−5 to −55 mV, dependent on tumor cell type[112]) may have a major impact on the conductance of voltage-gated ion channels. For example, VGSCs typically inactivate within a few milliseconds of opening following sustained depolarization, and remain inactivated until the membrane repolarizes. However, several subtypes, including $Na_v1.5$, which is expressed in cells from a number of tumor types, do not inactivate completely and carry a small persistent Na^+ current.[149,150] In cancer cells with a depolarized V_m, it is likely that the majority of VGSCs are in the inactivated state, and therefore it is likely the persistent Na^+ current that is responsible for the tumor-promoting behavior of these channels.[6]

TABLE 6.2   Key Ion Channel Functional Effects in Cancer

Functional Role	Ion Channel	Mechanism	References
Proliferation	$K_v1.3$, $K_v10.1$, $K_v11.1$, $K_{ir}2.2$, $K_{ir}3.1$, $K_{ir}3.4$, K_{Ca}, $K_{2P}2.1$, $Ca_v3.2$, Orai3, transient receptor potential canonical, TRPV6/7, TRPM7/8	Control of V_m, intracellular Ca^{2+}, pH	31,55–59,112–126
Invasion	$K_v11.1$, $Na_v1.5$, $Na_v1.7$, voltage-gated Na^+ channel β1	Adhesion complexes, activation of cathepsin proteases	37,73,78,91,98,127,128
Apoptosis resistance	Orai1/Orai3, TRPV6, P2X7, $K_v1.5$, ClC-3, voltage-gated Na^+ channel	Altered Ca^{2+} signaling to evade Ca^{2+}-mediated death pathways	26,55,62,98,129–135
Cell volume regulation	ClC-3, K_{Ca}	Co-ordinated regulation of osmolarity	136–138
Angiogenesis	TRPC1/3/4/6, $K_{Ca}1.1/3.1$, $K_v11.1$, $K_v10.1$, voltage-gated Na^+ channel β1	Ca^{2+}-dependent signaling, nonconducting mechanisms, hypoxia-inducible factor 1 activation	98,139–148

Membrane Potential Regulation

The V_m has been shown to be functionally instructive in regulating a number of cellular processes in both normal and transformed cells, including proliferation, migration, and differentiation.[151,152] Changes in V_m also regulate wound healing, patterning, tissue development, and regeneration.[153–156] Furthermore, depolarization itself may be sufficient to induce neoplastic transformation.[151] The relationship between V_m, transformation and cellular behavior is complex because the V_m itself changes in individual cells as they progress through the cell cycle.[9,154] Thus, hyperpolarization is required for S phase initiation at the G1/S checkpoint,[157] whereas G2/M transition requires a depolarized V_m.[154,158,159] These changes in V_m depend on alterations in the expression profiles of a number of ion channels. For example, in neuroblastoma and glioma cells, upregulation of $K_v11.1$ and downregulation of inward rectifier K^+ channels that are normally present in differentiated glial cells results in cellular depolarization.[63,113]

The notion that changes in V_m contribute to regulating cell cycle progression and thus proliferation is supported by the observation that inhibition of

K^+ currents reduces proliferation of cancer cells in a number of studies.[112,114] In particular, inhibition of $K_v10.1$ reduces proliferation in cancer cell lines, whereas overexpression of this channel in Chinese hamster ovary (CHO) cells induces tumors in mice.[115] $K_v10.1$ expression is cell cycle dependent,[160] and $K_v10.1$ activation promotes hyperpolarization at the G1/S checkpoint.[161] Similarly, $K_v11.1$ activity is cell cycle dependent, and K^+ current carried by $K_v11.1$ increases tumor cell proliferation.[113,116,117] Further, inward rectifier channels are also required for proliferation of tumor cells, including $K_{ir}2.2$, $K_{ir}3.1$, and $K_{ir}3.4$.[118–122] In contrast, in glioma cells, $K_{ir}4.1$ increases quiescence.[162] Other classes of K^+ channels have been implicated in regulating V_m and cell cycle progression in tumor cells. For example, in glioma cells, inhibition of K_{Ca} channels causes cell cycle arrest in S phase and leads to apoptosis.[123] In addition, in MCF-7 cells, inhibition of ATP-sensitive K_{ATP} channels reversibly traps cells in G_0/G_1.[163] $K_2P2.1$ increases proliferation of prostate cancer cells, and its overexpression in CHO cells increases the proportion of cells in S phase.[31] Finally, Cl^- conductance is also linked to cell cycle progression, and in glioma cells, ClC-3 conductance significantly increases during M phase.[164] However, at this stage, it is not clear whether Ca^{2+} or Na^+ channels also contribute to V_m alterations and cell cycle regulation in tumor cells. Thus, heterogeneity of expression of different classes of ion channels (mainly those carrying K^+), and resultant variability in voltage dependence, channel regulation, and downstream signaling, has diverse effects on V_m and cell cycle progression in different tumor cell types.

The mechanisms by which ion channels regulate proliferation appear complex, and the question of whether it is the ion channels themselves or ion channel–mediated changes in V_m that contribute to cell cycle progression remains a matter of debate. Clearly, V_m hyperpolarization would increase the electrochemical gradient for Ca^{2+} and may increase Ca^{2+} influx through transient receptor potential (TRP) channels. Ca^{2+} itself is required for G1/S transition,[165] and in turn, may upregulate various K^+ channels that promote sustained proliferation. For example, in breast cancer cells, upregulation of $K_v10.1$ promotes hyperpolarization, inducing G1/S transition, and hyperpolarization-induced Ca^{2+} influx then activates $K_{Ca}3.1$ and inhibits $K_v10.1$ via calmodulin.[166,167] Alternatively, ion channels may enhance proliferation via other nonconducting signaling mechanisms. For example, both wild-type and nonconducting mutant $K_v10.1$ channels promote proliferation when transfected into heterologous cells, and gating of $K_v10.1$ itself, rather than K^+ conductance, is required for p38 mitogen-activated protein kinase activation.[168]

Regulation of Cell Volume and Motility

A growing body of evidence indicates that a number of ion channels play a critical role in regulating cellular migration.[136] A compelling hypothesis to explain channel involvement in cancer cell migration is through regulation of hydrodynamic changes by ion fluxes. In this model, ion channels can regulate global cell volume or local swelling or shrinking of cellular protrusions/processes. Thus, channel-mediated ion influx can induce osmotic inflow of water through nearby aquaporins at the leading edge of a migrating cell, resulting in local volume expansion, or vice versa at the trailing edge, resulting in shrinkage.[2] The active cotransporter NKCC1 localizes to the leading edge of migrating human glioma cells, facilitating Cl^- accumulation, thus providing an electrochemical driving force for Cl^- efflux to osmotically regulate cytoplasmic water level and modulating cell volume/promoting migration.[169,170] In agreement with this, ClC-3, expressed in human glioma cells and nasopharyngeal carcinoma cells, promotes migration and invasion.[171,172] Volume-activated Cl^- currents carried by ClC-3 promote regulatory volume decrease, the process by which a cell decreases in volume after swelling, suggesting that ClC-3 channels contribute to changes in volume that are required as cells migrate through narrow spaces.[137]

K_{Ca} channels are proposed to provide cation conductance that would maintain electroneutrality following Cl^- channel activation in migrating cells.[136] In agreement with this, $K_{Ca}3.1$ promotes invasion of glioma cells into surrounding brain parenchyma.[173] Interestingly, this mechanism appears to replicate the process by which intermediate conductance Ca^{2+}-activated K^+ channels regulate cell volume changes and migration in nontransformed cells.[155] Ca^{2+} is thought to be a major regulator of volume-changing ion channels. For example, Ca^{2+}/calmodulin-dependent protein kinase II (CaMKII) phosphorylation regulates ClC-3-mediated migration of glioma cells.[174] In addition, ligands that increase intracellular Ca^{2+} concentration can activate K_{Ca} channels that contribute to volume changes and migration of glioma cells, including gBK.[138] These local increases in intracellular Ca^{2+} may be regulated, in part, by TRPC channels expressed in caveolar lipid rafts at the leading edge of migrating glioma cells, together with Ca^{2+}-permeant 3-hydroxy-5-methyl-4-isoxazole propionic acid (AMPA) receptors.[56,175] Other Ca^{2+}-permeant channels, including $Ca_v3.1$, TRPC1, TRPV1, and the SOC channel constituents Orai1 and STIM1, have also been shown to promote migration of various tumor cells, suggesting a general requirement for Ca^{2+} signaling.[12,176–179]

Na^+ current carried through VGSCs enhances a number of cellular behaviors associated with metastasis in cancer cell lines, including outgrowth of neurite-like processes, endocytosis, galvanotaxis, gene expression, vesicular patterning, detachment from substrate, migration, and invasion.[91,180–187] In particular, VGSCs enhance migration and/or invasion in a number of different cancer cell types, including breast cancer, colon cancer, prostate cancer, lung cancer, cervical cancer, lymphoma, mesothelioma, and ovarian cancer.[37–39,41,78,83,86,88,89,91,110,188] In several cancer types, specific α subunit isoforms have been shown to contribute to increased invasive capacity. For example, in metastatic breast cancer cells, the neonatal DI:S3 splice variant of $Na_v1.5$ potentiates migration and invasion,[127] whereas $Na_v1.6$ enhances invasion in cervical cancer cells,[41] and $Na_v1.6$ and $Na_v1.7$ increase invasion in prostate cancer cells.[189] However, expression of any VGSC subtype may be generally important for promoting invasion. In fact, overexpression of $Na_v1.4$ in weakly metastatic prostate cancer cells increases their invasiveness, suggesting that VGSC expression is necessary and sufficient to promote metastatic behavior.[97]

The mechanisms by which Na^+ current carried by VGSCs increase migration and invasion of cancer cells are not fully clear, although several models have been proposed in different cancer types. In breast cancer cells, Na^+ current carried by $Na_v1.5$ increases H^+ efflux through the Na^+/H^+ exchanger NHE1, resulting in extracellular perimembrane acidification.[128] This extracellular reduction in pH provides a favorable environment for the activation of various pH-dependent cathepsin proteases, which enhance the cells' invasiveness and invadopodia formation.[190,191] In colon cancer cells, $Na_v1.5$ has been shown to regulate a network of invasion-promoting genes,[38] suggesting that VGSC-mediated gene regulation may also play a role. An additional possibility is that VGSCs may regulate the intracellular Ca^{2+} level. For example, in melanoma cells and THP-1 macrophages, activation of VGSCs results in Ca^{2+} release from mitochondria, which increases invadopodia formation and invasion.[94] Finally, Na^+ current carried by other non-VGSCs may play a role, eg, in glioma cells, ENaC and ASIC1 channels potentiate cellular migration.[192] Further work is required to establish whether these models are specific to the cancer types in which they were identified, or, whether they are generally representative of other neoplastic cells expressing VGSCs/other Na^+ channels.

Nonconducting Functions

In addition to regulating various aspects of cancer progression through ion flux, ion channels may also participate in nonconducting signaling interactions, eg, adhesion interactions, regulation of enzymatic activity, or conformational coupling with other proteins. Indeed, it has become increasingly evident that ion channels function as macromolecular signaling complexes that can participate in various signaling mechanisms on multiple timescales.[193] Furthermore, nonconducting signaling may accompany, interact with, and feed back on the functional consequences of ion flow (Fig. 6.2(A) and (B)).

A well-established example of nonconducting signaling is the interaction between K^+ channels and β_1 integrin. Cell adhesion interactions between integrins and the extracellular matrix regulate diverse processes, including cell survival, migration, proliferation, and differentiation.

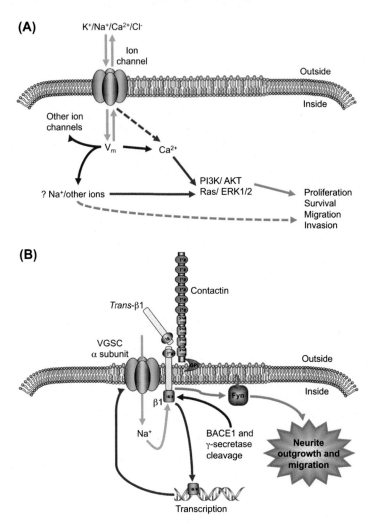

FIGURE 6.2 Ion channel modulation of intracellular signaling. (A) Ion flux alters membrane potential, V_m, which in turn activates intracellular Ca^{2+} release via multiple mechanisms, detailed in main text. Ca^{2+} and/or other ions, eg, Na^+, activate kinase cascades, eg, PI3K/AKT, Ras/ERK, which regulate cellular proliferation, survival/apoptosis, migration, and invasion. (B) Nonconducting signaling via the voltage-gated Na^+ channel (VGSC) macromolecular complex. Adhesion interactions between β1 on an adjacent cell and the VGSC macromolecular signaling complex initiate a signaling cascade through fyn kinase leading to neurite outgrowth and migration. In addition, cleavage of β1 by BACE1 and γ-secretase is proposed to release the β1 intracellular domain, which may enhance transcription of VGSC α subunit(s). Figure was produced using Science Slides software. *Panel B reproduced with permission from Brackenbury WJ, Djamgoz MB, Isom LL. An emerging role for voltage-gated Na^+ channels in cellular migration: regulation of central nervous system development and potentiation of invasive cancers.* Neuroscientist 2008;**14**(6):571–83.

Signaling cascades downstream from integrins through integrin-linked kinase and focal adhesion kinase (FAK) activate further kinases including phosphoinositide-3 kinase (PI3K).[194] In activated T lymphocytes and melanoma cells, $K_v1.3$ channels associate with and activate β_1 integrin.[195,196] Thus, macromolecular signaling complexes between ion channels and β_1 integrin can form at adhesive interaction sites, activating intracellular signaling cascades. Similarly, in neuroblastoma cells, complexes between $K_v11.1$ and β_1 integrin occur at adhesion sites, likely within lipid rafts and/or caveolae, recruiting FAK and Rac1.[197] Importantly, this process is dependent on ion flow through $K_v11.1$, thus suggesting that (nonconducting) cell adhesion–mediated signaling may also require channel conductance, although the mechanisms underlying this requirement are not yet clear. In acute myeloid leukemia cells, the $K_v11.1$-β_1 integrin complex also contains the vascular endothelial growth factor (VEGF) receptor 1, Flt-1, regulating Flt-1-dependent proliferation and migration.[73] In addition, heterotypic adhesion interactions between $K_v11.1$-β_1 integrin complexes on childhood acute B lymphoblastic leukemia cells and bone marrow–derived stromal cells activate signaling pathways that promote tumor cell survival and chemoresistance.[198]

A second important example of nonconducting signaling promoting cancer progression relates to VGSCs. VGSC β subunits modulate the biophysical properties of the α subunit, but are also members of the immunoglobulin superfamily of cell adhesion molecules.[77,193,199] In particular, the $\beta1$ subunit can interact homophilically, and also heterophilically with several other cell adhesion molecules and extracellular matrix proteins, including the $\beta2$ subunit, contactin, neurofascin-186, NrCAM, N-cadherin, and tenascin-R.[200–204] In addition, $\beta2$ can interact with tenascin-C and tenascin-R.[205] VGSC β subunits play a critical role in central nervous system development.[206] In cerebellar granule neurons, $\beta1$ regulates neurite outgrowth through *trans*-homophilic adhesion, via a mechanism involving fyn kinase and contactin (Fig. 6.2B).[207–209] Importantly, $\beta1$-mediated neurite outgrowth also requires the presence and function of the associated pore-forming α subunit, $Na_v1.6$, suggesting that a reciprocal relationship exists between VGSC α and β subunits, thus regulating intracellular signaling.[210] Proteolytic processing of VGSC β subunits by $\alpha/\beta/\gamma$-secretases releases the extracellular and intracellular domains.[211,212] In the case of $\beta2$, the released soluble intracellular domain has been shown to translocate to the nucleus and regulate gene expression.[213]

In breast cancer cells, $\beta1$ reduces apoptosis and increases adhesion, tumor growth, invasion, and metastatic dissemination.[42,98] $\beta1$ also promotes neurite-like process outgrowth in breast cancer cells cocultured with fibroblasts, thus promoting a mesenchymal-like, invasive phenotype. $\beta1$-mediated process outgrowth occurs via *trans*-homophilic adhesion and is dependent on fyn kinase activity and Na^+ current through the α subunit.[98]

Therefore, the mechanism by which VGSC complexes regulate neurite outgrowth and migration during nervous system development may be recapitulated in breast cancer cells during invasion and metastasis. Similar to β1, β2 induces an elongate bipolar morphology when expressed in prostate cancer cells, increasing adhesion and migration.[44] However, β2 overexpression reduces tumor take and growth in mice, suggesting that β2 may enhance invasion while also retarding tumor formation.[44] In support of this, β2 enhances association between prostate cancer cells and neurons in organotypic cultures, suggesting that it may promote perineural invasion and metastasis.[214]

A different situation occurs for the VGSC β3 subunit, which inhibits colony formation, and may function as a tumor suppressor.[215] The *SCN3B* gene (encoding β3) contains two response elements to the tumor suppressor p53 and is upregulated in p53-expressing fibroblasts in response to treatment with the cytotoxic chemotherapeutic drug doxorubicin.[215] β3 also promotes apoptosis in response to several anticancer drugs.[215] In summary, ion channel auxiliary subunits and/or other components of these macromolecular protein complexes appear to regulate a number of signaling cascades in cancer cells through nonconducting mechanisms, which, in turn, feed back on tumor progression.

ION CHANNEL INVOLVEMENT IN THE HALLMARKS OF CANCER

The ion channel expression profile changes during tumor progression. For example, an expression profiling study has shown that a molecular gene signature containing 30 ion channels reliably predicted clinical outcome in breast cancer patients.[216] In general, the K^+ channels are highly expressed relatively early in tumor progression, likely predominantly promoting proliferation (Fig. 6.1).[5] There is more variable ion channel expression in more advanced tumors and metastatic cells, perhaps reflecting the changes in the heterogeneous microenvironments faced by tumor cells as they start to disseminate. In both breast and prostate cancer cells, K^+ and Ca^{2+} channel expression appears to be replaced by VGSC expression in latter stages, likely promoting metastasis.[217] Similarly, in leukemias and gastrointestinal tumors, K^+ channel expression (eg, $K_v1.3$) is largely limited to early stages.[5] In contrast, $K_v11.1$ appears to play a significant role in later tumor stages, promoting transendothelial migration, adhesion, invasion, and drug resistance.[73,139,194,198] It is therefore possible that ion channel expression is finely tuned in different tumors/cancer stages and may be related to specific functional specializations during tumor development and acquisition of multidrug resistance.[218] In 2000, Hanahan and Weinberg proposed that tumor growth can be defined by six critical hallmarks: (1) self-sufficiency in growth signals, (2) insensitivity to anti-growth signals, (3) evasion of apoptosis, (4) limitless replicative potential, (5) sustained angiogenesis, and (6) tissue invasion and metastasis.[219] A decade later, in 2011, they proposed several further, emerging characteristics: deregulating cellular energetics, avoiding immune destruction, genetic instability, and tumor-promoting inflammation.[220]

Increasing evidence suggests that ion channels are involved in defining the various cancer hallmarks (Fig. 6.3), covered in detail in a number of excellent reviews.[4,114,222–224] Self-sufficiency in growth signals may be promoted by Ca^{2+} signaling regulated by altered ion channel expression. For example, in prostate cancer cells, $Ca_v3.2$ promotes neuroendocrine differentiation and secretion of mitogenic factors.[51,52] TRPM8 may also promote mitogen secretion in prostate cancer cells.[58] In addition, Ca_v3 channels are proposed to promote tumor cell proliferation via Ca^{2+}-mediated progression through the S phase of the cell cycle,[124] and various TRP channels, including TRPC6, TRPV6, TRPM7, and TRPM8, promote cancer cell proliferation.[57,58,125,126] Furthermore, as discussed earlier, through regulation of the V_m, various K^+ channels, including K_{ATP}, K_{ir}, K_{2P}, and K_v channels, play a critical role in promoting tumor cell proliferation.[11] Given that K^+ channels also regulate cell volume and pH,[11,225] the mechanisms by which they regulate cellular proliferation are likely to be diverse and dependent on cancer cell type and subcellular distribution.

Insensitivity to antigrowth signals in tumor cells may be enhanced by aberrant expression of Ca^{2+}-permeable channels leading to altered Ca^{2+}

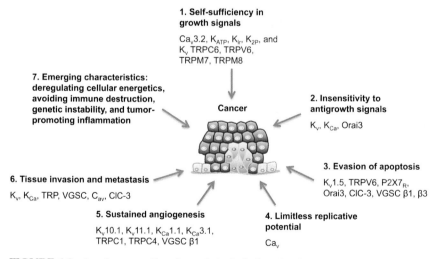

FIGURE 6.3 Involvement of ion channels in the hallmarks of cancer. Each of the six main hallmarks is listed, together with the emerging characteristics, defined by Hanahan and Weinberg.[219–221] The principal ion channels contributing to these features, described in the main text, are indicated.

homeostasis. For example, the SOC entry that accompanies antigrowth signaling is disrupted in cancer cells.[226] In prostate cancer cells, reduced SOC entry prevents Ca^{2+} overload, thus permitting evasion of apoptosis.[129] Recently, it has been shown that in prostate cancer, upregulation of Orai3 represents an oncogenic switch, in which Orai1/Orai3 heteromeric store-independent channels subsequently promote proliferation and apoptosis resistance.[55] In addition, reduction of Ca^{2+} influx via downregulation of other Ca^{2+}-permeant channels in cancer cells, including TRPV6 and ionotropic purinergic P2X7 receptors also enables apoptosis resistance.[62,130,131] In several cancer types, higher mitochondrial V_m and lower expression of the redox-sensitive $K_v1.5$ is proposed to reduce K^+ efflux and consequent collapse of the plasma membrane V_m, thereby enabling evasion of apoptosis and enhancing resistance to cytotoxic drugs.[26,132] Volume-regulated anion channels, including ClC-3, also promote apoptosis resistance via enhanced regulatory volume decrease in response to hypoosmotic stress.[133,134] Blockade of VGSCs in T-lymphoma cells prevents apoptosis, suggesting that Na^+ influx is also required.[135] On the other hand, overexpression of the VGSC β1 subunit in breast cancer cells reduces apoptosis, suggesting that VGSC-mediated adhesion may play a role.[98] Less is known about the potential involvement of ion channels in promoting the limitless replicative potential of cancer cells. However, it has been shown in ovarian cancer cells that localized Ca^{2+} influx via voltage-gated Ca^{2+} channels promotes telomerase activity, thus preventing telomere shortening and replicative senescence.[227] Further work is needed to establish the ion channels involved and the spatial/temporal distribution of Ca^{2+} signaling required.

Sustained angiogenesis requires Ca^{2+}-dependent signaling.[224] In particular, VEGF signaling, endothelial cell proliferation, and neovascularization require the activity of SOC channels, including TRPC1 and TRPC4.[140–142] The receptor-gated TRPC3 and TRPC6 channels are also required for VEGF-induced vascular permeability,[143] suggesting that Ca^{2+} influx through different classes of TRPs may have diverse contributions to various aspects of angiogenesis. $K_{Ca}1.1$ and $K_{Ca}3.1$ channels have been proposed to mediate V_m hyperpolarization in endothelial cells, thus providing an electrochemical driving force for Ca^{2+} entry, which would then promote vascular permeability and angiogenesis in colon tumors and brain metastases.[144,145] In addition, $K_v11.1$ promotes VEGF secretion in glioma cells and gastric tumors.[146,147] Interestingly, the mechanism for $K_v11.1$ involvement may be distinct from Ca^{2+}: in colon cancer, $K_v11.1$ activity modulates β1 integrin-mediated adhesion signaling giving rise to VEGF secretion.[139] Several lines of evidence suggest that ion channels may also promote angiogenesis by nonconducting mechanisms. For example, $K_v10.1$ enhances hypoxia-inducible factor 1 (HIF-1) activity and promotes VEGF secretion and angiogenesis, independent of its function as an ion channel.[148] In addition, the VGSC β1 subunit enhances VEGF secretion and angiogenesis in breast cancer, potentially via its function as a cell adhesion molecule.[98]

The sixth cancer hallmark, tissue invasion and metastasis, is also regulated, at least in part, by several ion channels including VGSCs, Ca^{2+}, K$^+$, and Cl$^-$ channels,[6,11,12,228] discussed in relation to cellular migration and invasion earlier in this chapter. Finally, with respect to the emerging hallmarks, including evasion of immune destruction and tumor-promoting inflammation, it is important to note the critical importance of the heterogeneous tumor microenvironment to cancer progression.[220] Heterotypic interactions between cancer, mesenchymal, endothelial, and immune cells contribute to this environment, and it is likely that ion channels play an important role, both in cancer cells and their counterparts in the reactive stroma.[223] This involvement of ion channels in the heterogeneous tumor microenvironment likely occurs through the various mechanisms discussed elsewhere in this chapter, including adhesion interactions, eg, K$_v$11.1/β_1 integrin signaling, VGSC β subunit signaling, as well as ion conductance in endothelial or recruited immune cells. Further work is required to establish the extent of involvement of ion channels in these emerging cancer hallmarks.

ION CHANNELS AS THERAPEUTIC TARGETS

Given the abundant and aberrant expression of various ion channels in tumors, and their diverse contributions to aspects of cancer progression, they may present ideal therapeutic targets (Fig. 6.4). Furthermore, ion channel–inhibiting drugs may be novel anticancer agents (Table 6.3). Indeed, plasma membrane ion channels are particularly attractive targets because many blockers act extracellularly and can be screened directly in tumor cell lines using electrophysiological methods, eg, patch clamping. However, relatively little progress has been made in harnessing the therapeutic potential of ion channels in tumors. One significant drawback with targeting various ion channels is that blockers may undesirably target the same channels expressed in other tissues. For example, drugs inhibiting K$_v$11.1 channels, which are widely expressed in tumors, slow cardiac repolarization and can cause the life-threatening ventricular arrhythmia *torsade de pointes*.[5] Nonetheless, K$_v$11.1-inhibiting drugs reduce proliferation, invasion, and secretory activity of various tumor cell lines in vitro and reduce tumor growth and angiogenesis in gastric tumors and leukemias in mouse models in vivo.[139,147,198] Thus, specifically targeting K$_v$11.1 channels in tumors, while avoiding cardiac K$_v$11.1 channels, may hold considerable therapeutic value.

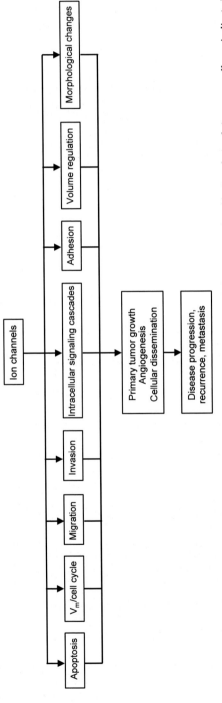

FIGURE 6.4   Multiple roles of ion channels in cancer progression. The key cellular behaviors regulated by ion channels in cancer cells are indicated. Collectively, these ion channel–regulated behaviors contribute to growth of the primary tumor, sustained angiogenesis, and metastatic dissemination, ultimately resulting in disease progression, recurrence, and metastasis.

TABLE 6.3   Ion Channel Blockers Effective in Cancer Cells

Drug	Target Channel/ Tumor	Effect	References
TEA, 4-AP	K_v channels	Inhibition of proliferation	7,229
TRAM-34, charybdotoxin	$K_{Ca}3.1$	Inhibition of proliferation	7,21
Iberiotoxin	$K_{Ca}1.1$	Inhibition of proliferation	7,230
Astemizole, terfenadine	$K_v10.1$	Inhibition of proliferation	229
E4031	$K_v11.1$	Inhibition of proliferation, invasion	14,147,198
TTX, phenytoin, ranolazine	Voltage-gated Na^+ channel	Inhibition of invasion and metastasis	37,231–233
Chlorotoxin	ClC-3	Inhibition of invasion	234

One possible approach to achieving higher specificity for ion channels expressed in tumors is to exploit specific functional states that are predominant in cancer cells. For example, $K_v11.1$ inhibitors that target channels in the open state, the typical situation in proliferating tumor cells with a V_m close to −40 mV, may have a beneficial effect by targeting tumors, without affecting cardiac $K_v11.1$ channels, which are mainly in the deactivated/inactivated states, dependent on the V_m during the cardiac action potential.[5]

State-dependent channel block may also be useful in the case of VGSCs. In particular, several Class Ib antiarrhythmic drugs are use dependent and preferentially bind to VGSCs in their inactivated state and thus are effective at selectively suppressing channel activity in hyperexcitable neurons.[235,236] Given that cancer cells have a relatively depolarized V_m, and VGSCs would be mainly inactivated, it is likely that the persistent Na^+ current is important for promoting metastatic behavior.[6] Therefore, drugs that selectively target inactivated VGSCs and inhibit the persistent current might be effective in reducing the invasiveness of metastatic tumor cells. In support of this notion, phenytoin, a widely prescribed anticonvulsant and Class Ib antiarrhythmic agent, inhibits Na^+ current, migration and invasion of metastatic breast cancer cells in vitro, and tumor growth and metastasis in vivo.[79,231] Phenytoin has also been shown to inhibit migration, and secretion of interleukin-6 and prostate-specific antigen, in prostate cancer cells.[183,237] In addition, phenytoin also inhibits endocytosis in lung cancer cells.[84] Other VGSC-inhibiting drugs, including carbamazepine, ranolazine and riluzole, have also been shown to inhibit metastatic cellular behaviors.[237–239] Ranolazine reduces lung metastases in mice.[232] Riluzole also reduces breast tumor growth in mice and is being explored as a novel therapeutic in patients with late-stage melanoma.[240,241] Interestingly, in addition to blocking VGSCs, riluzole also inhibits metabotropic glutamate receptor signaling, and it is the latter mechanism that is proposed for its potential efficacy as an anticancer agent.[242] Nonetheless, collectively, these data suggest that repurposing VGSC-targeting antiepileptic drugs such as phenytoin to cancer may improve clinical outcome.[243] Indeed, use of VGSC-inhibiting drugs during radical prostatectomy surgery associates with significantly reduced tumor recurrence and metastasis.[244]

Approaches aimed at improving the specificity of channel blockers have been applied in several cases.[5] For example, phenytoin analogs have been developed and tested with the aim of improving selectivity for VGSCs and reducing acute toxicity. These novel α-hydroxy-α-phenylamides are more potent VGSC inhibitors than phenytoin and inhibit prostate cancer cell growth to a greater extent.[245] In addition, novel phenytoin analogs have been developed with increased potency specifically against $Na_v1.5$.[246] Thus, it is possible that development of such molecules with increased specificity for specific channel subtypes and/or functional states may hold promise as novel therapeutics in the cancer setting. An alternative approach that has been proposed for targeting ion channels in cancer cells is to use peptide toxins, which are often very selective. For example, local injection of tetrodotoxin, a VGSC-blocking neurotoxin into prostate tumors in rats significantly reduces lung metastasis and improves over-all survival.[233] However, the acute toxicity of tetrodotoxin means that it could not be used as a systemic antimetastatic therapy. Nonetheless, mod-ification of such toxins to target specific subtypes/locations may be pos-sible. The scorpion toxin chlorotoxin inhibits ClC-3 channels by binding a membrane-bound metalloproteinase. Radiolabeled I^{121}-chlorotoxin has completed phase I and II clinical trials as a new route to deliver radiation to ClC-3-expressing gliomas for targeted radiotherapy.[136,234]

A further possible approach to target ion channel activity in tumor cells is through the use of function-blocking antibodies. This approach has the advantage of potentially targeting one channel subunit/isotype that is specific to the tumor and may thus provide improved selectiv-ity over homogeneous classes of inhibitors. For example, monoclonal antibodies targeting $K_v10.1$ specifically inhibit K^+ current through this channel and reduce tumor growth in mice.[247] Such antibodies may also have use as diagnostic tools and/or to deliver cytotoxic drugs directly to tumor cells.[5] In addition, a polyclonal antibody raised against the D1:S3/4 linker of the neonatal splice variant of $Na_v1.5$ inhibits Na^+ cur-rent, with a significantly higher potency for "neonatal" $Na_v1.5$ than "adult" $Na_v1.5$ channels.[248] Given that the neonatal $Na_v1.5$ splice form is highly expressed in breast tumors, it is possible that antibodies targeting this variant may have therapeutic value, with minimal offtarget toxicity, especially given the absence of the neonatal variant in other normal adult tissues.[248] Indeed, this antibody significantly reduces migration and invasion of metastatic breast cancer cells in vitro, to an extent equivalent to tetrodotoxin or siRNA.[127] Interestingly, other anticancer drugs and antibodies raised against other membrane targets may also inhibit, directly or indirectly, ion channels. For example, rituximab, which targets CD20 in B cell large lymphomas, also inhibits $K_{Ca}3.1$ channels,[21] and tamoxifen, which targets the estrogen receptor in breast cancers, also inhibits VGSCs.[249]

CONCLUDING REMARKS

Cancer morbidity and mortality are increasing as a result of an aging population, and while there have been considerable improvements in treatment and survival times for patients suffering from a number of different types of cancer, there is still an urgent need to identify and develop novel molecular targets and therapies with curative intent. In particular, metastasis remains the main cause of deaths from solid tumors, and yet there are no effective curative therapies in the clinic. Ion channels are widely expressed in tumors and play manifold critical roles in carcinogenesis, cancer cell survival, tumor progression, metastasis, and drug resistance. Channels expressed at the plasma membrane of cancer cells represent theoretically attractive novel therapeutic targets due to their accessibility from the extracellular space. Although much further work needs to be done to fully understand the integrated function of the multiplicity of ion channels present in tumors, current and future work will no doubt establish whether ion channel–targeting strategies have clinical value as novel therapeutic interventions.

Acknowledgment

This work was supported by the Medical Research Council [Fellowship G1000508].

References

1. Hille B. *Ionic channels of excitable membranes*. 2nd ed. Sunderland (Massachusetts): Sinauer Associates Inc.; 1992.
2. Schwab A, Fabian A, Hanley PJ, Stock C. Role of ion channels and transporters in cell migration. *Physiol Rev* 2012;**92**(4):1865–913.
3. Kaczmarek LK. Non-conducting functions of voltage-gated ion channels. *Nat Rev Neurosci* 2006;**7**(10):761–71.
4. Prevarskaya N, Skryma R, Shuba Y. Ion channels and the hallmarks of cancer. *Trends Mol Med* 2010;**16**(3):107–21.

5. Arcangeli A, Crociani O, Lastraioli E, Masi A, Pillozzi S, Becchetti A. Targeting ion channels in cancer: a novel frontier in antineoplastic therapy. *Curr Med Chem* 2009;**16**(1):66–93.
6. Brackenbury WJ. Voltage-gated sodium channels and metastatic disease. *Channels (Austin)* 2012;**6**(5):352–61.
7. Kunzelmann K. Ion channels and cancer. *J Membr Biol* 2005;**205**(3):159–73.
8. Pardo LA, Contreras-Jurado C, Zientkowska M, Alves F, Stuhmer W. Role of voltage-gated potassium channels in cancer. *J Membr Biol* 2005;**205**(3):115–24.
9. Wonderlin WF, Strobl JS. Potassium channels, proliferation and G1 progression. *J Membr Biol* 1996;**154**(2):91–107.
10. Fiske JL, Fomin VP, Brown ML, Duncan RL, Sikes RA. Voltage-sensitive ion channels and cancer. *Cancer Metastasis Rev* 2006;**25**(3):493–500.
11. Pardo LA, Stuhmer W. The roles of K(+) channels in cancer. *Nat Rev Cancer* 2014;**14**(1):39–48.
12. Prevarskaya N, Skryma R, Shuba Y. Calcium in tumour metastasis: new roles for known actors. *Nat Rev Cancer* 2011;**11**(8):609–18.
13. Fraser SP, Ozerlat-Gunduz I, Brackenbury WJ, et al. Regulation of voltage-gated sodium channel expression in cancer: hormones, growth factors and auto-regulation. *Philos Trans R Soc Lond B Biol Sci* 2014;**369**(1638):20130105.
14. Pillozzi S, Brizzi MF, Balzi M, et al. HERG potassium channels are constitutively expressed in primary human acute myeloid leukemias and regulate cell proliferation of normal and leukemic hemopoietic progenitors. *Leukemia* 2002;**16**(9):1791–8.
15. Lastraioli E, Guasti L, Crociani O, et al. herg1 gene and HERG1 protein are overexpressed in colorectal cancers and regulate cell invasion of tumor cells. *Cancer Res* 2004;**64**(2):606–11.
16. Lastraioli E, Taddei A, Messerini L, et al. hERG1 channels in human esophagus: evidence for their aberrant expression in the malignant progression of Barrett's esophagus. *J Cell Physiol* 2006;**209**(2):398–404.
17. Abdul M, Santo A, Hoosein N. Activity of potassium channel-blockers in breast cancer. *Anticancer Res* 2003;**23**(4):3347–51.
18. Abdul M, Hoosein N. Reduced Kv1.3 potassium channel expression in human prostate cancer. *J Membr Biol* 2006;**214**(2):99–102.
19. Laniado ME, Fraser SP, Djamgoz MB. Voltage-gated K(+) channel activity in human prostate cancer cell lines of markedly different metastatic potential: distinguishing characteristics of PC-3 and LNCaP cells. *Prostate* 2001;**46**(4):262–74.
20. Fraser SP, Grimes JA, Diss JK, Stewart D, Dolly JO, Djamgoz MB. Predominant expression of Kv1.3 voltage-gated K(+) channel subunit in rat prostate cancer cell lines: electrophysiological, pharmacological and molecular characterisation. *Pflugers Arch* 2003;**446**(5):559–71.
21. Wang J, Xu YQ, Liang YY, Gongora R, Warnock DG, Ma HP. An intermediate-conductance Ca(2+)-activated K(+) channel mediates B lymphoma cell cycle progression induced by serum. *Pflugers Arch* 2007;**454**(6):945–56.
22. Roy J, Vantol B, Cowley EA, Blay J, Linsdell P. Pharmacological separation of hEAG and hERG K+ channel function in the human mammary carcinoma cell line MCF-7. *Oncol Rep* 2008;**19**(6):1511–6.
23. Ousingsawat J, Spitzner M, Puntheeranurak S, et al. Expression of voltage-gated potassium channels in human and mouse colonic carcinoma. *Clin Cancer Res* 2007;**13**(3):824–31.
24. Hemmerlein B, Weseloh RM, Mello de Queiroz F, et al. Overexpression of Eag1 potassium channels in clinical tumours. *Mol Cancer* 2006;**5**:41.
25. Ding XW, Luo HS, Jin X, Yan JJ, Ai YW. Aberrant expression of Eag1 potassium channels in gastric cancer patients and cell lines. *Med Oncol* 2007;**24**(3):345–50.
26. Bonnet S, Archer SL, Allalunis-Turner J, et al. A mitochondria-K+ channel axis is suppressed in cancer and its normalization promotes apoptosis and inhibits cancer growth. *Cancer Cell* 2007;**11**(1):37–51.

27. Ryland KE, Svoboda LK, Vesely ED, et al. Polycomb-dependent repression of the potassium channel-encoding gene KCNA5 promotes cancer cell survival under conditions of stress. *Oncogene* 2014. http://dx.doi.org/10.1038/onc.2014.384.

28. Stringer BK, Cooper AG, Shepard SB. Overexpression of the G-protein inwardly rectifying potassium channel 1 (GIRK1) in primary breast carcinomas correlates with axillary lymph node metastasis. *Cancer Res* 2001;**61**(2):582–8.

29. Mu D, Chen L, Zhang X, et al. Genomic amplification and oncogenic properties of the KCNK9 potassium channel gene. *Cancer Cell* 2003;**3**(3):297–302.

30. Kim CJ, Cho YG, Jeong SW, et al. Altered expression of KCNK9 in colorectal cancers. *APMIS* 2004;**112**(9):588–94.

31. Voloshyna I, Besana A, Castillo M, et al. TREK-1 is a novel molecular target in prostate cancer. *Cancer Res* 2008;**68**(4):1197–203.

32. Arcangeli A, Wanke E, Olivotto M, Camagni S, Ferroni A. Three types of ion channels are present on the plasma membrane of Friend erythroleukemia cells. *Biochem Biophys Res Commun* 1987;**146**(3):1450–7.

33. Becchetti A, Arcangeli A, Del Bene MR, Olivotto M, Wanke E. Response to fibronectin-integrin interaction in leukaemia cells: delayed enhancing of a K$^+$ current. *Proc Biol Sci* 1992;**248**(1323):235–40.

34. Liu X, Chang Y, Reinhart PH, Sontheimer H. Cloning and characterization of glioma BK, a novel BK channel isoform highly expressed in human glioma cells. *J Neurosci* 2002;**22**(5):1840–9.

35. Olsen ML, Weaver AK, Ritch PS, Sontheimer H. Modulation of glioma BK channels via erbB2. *J Neurosci Res* 2005;**81**(2):179–89.

36. Gessner G, Schonherr K, Soom M, et al. BKCa channels activating at resting potential without calcium in LNCaP prostate cancer cells. *J Membr Biol* 2005;**208**(3):229–40.

37. Fraser SP, Diss JK, Chioni AM, et al. Voltage-gated sodium channel expression and potentiation of human breast cancer metastasis. *Clin Cancer Res* 2005;**11**:5381–9.

38. House CD, Vaske CJ, Schwartz A, et al. Voltage-gated Na$^+$ channel SCN5A is a key regulator of a gene transcriptional network that controls colon cancer invasion. *Cancer Res* 2010;**70**(17):6957–67.

39. Fraser SP, Diss JK, Lloyd LJ, et al. T-lymphocyte invasiveness: control by voltage-gated Na$^+$ channel activity. *FEBS Lett* 2004;**569**(1–3):191–4.

40. Diss JK, Archer SN, Hirano J, Fraser SP, Djamgoz MB. Expression profiles of voltage-gated Na$^+$ channel alpha-subunit genes in rat and human prostate cancer cell lines. *Prostate* 2001;**48**(3):165–78.

41. Hernandez-Plata E, Ortiz CS, Marquina-Castillo B, et al. Overexpression of Na(V) 1.6 channels is associated with the invasion capacity of human cervical cancer. *Int J Cancer* 2012;**130**:2013–23.

42. Chioni AM, Brackenbury WJ, Calhoun JD, Isom LL, Djamgoz MB. A novel adhesion molecule in human breast cancer cells: voltage-gated Na$^+$ channel beta1 subunit. *Int J Biochem Cell Biol* 2009;**41**(5):1216–27.

43. Diss JK, Fraser SP, Walker MM, Patel A, Latchman DS, Djamgoz MB. Beta-subunits of voltage-gated sodium channels in human prostate cancer: quantitative in vitro and in vivo analyses of mRNA expression. *Prostate Cancer Prostatic Dis* 2008;**11**(4):325–33.

44. Jansson KH, Lynch JE, Lepori-Bui N, Czymmek KJ, Duncan RL, Sikes RA. Overexpression of the VSSC-associated CAM, beta-2, enhances LNCaP cell metastasis associated behavior. *Prostate* 2012;**72**(10):1080–92.

45. Yamamura H, Ugawa S, Ueda T, Shimada S. Expression analysis of the epithelial Na$^+$ channel delta subunit in human melanoma G-361 cells. *Biochem Biophys Res Commun* 2008;**366**(2):489–92.

46. Ye JH, Gao J, Wu YN, Hu YJ, Zhang CP, Xu TL. Identification of acid-sensing ion channels in adenoid cystic carcinomas. *Biochem Biophys Res Commun* 2007;**355**(4):986–92.

47. Ousingsawat J, Spitzner M, Schreiber R, Kunzelmann K. Upregulation of colonic ion channels in APC(Min/+) mice. *Pflugers Arch* 2008;**456**(5):847–55.

48. Berdiev BK, Xia J, McLean LA, et al. Acid-sensing ion channels in malignant gliomas. *J Biol Chem* 2003;**278**(17):15023–34.

49. Bertolesi GE, Shi C, Elbaum L, et al. The Ca(2+) channel antagonists mibefradil and pimozide inhibit cell growth via different cytotoxic mechanisms. *Mol Pharmacol* 2002;**62**(2):210–9.

50. Sun YH, Gao X, Tang YJ, Xu CL, Wang LH. Androgens induce increases in intracellular calcium via a G protein-coupled receptor in LNCaP prostate cancer cells. *J Androl* 2006;**27**(5):671–8.

51. Mariot P, Vanoverberghe K, Lalevee N, Rossier MF, Prevarskaya N. Overexpression of an alpha 1H (Cav3.2) T-type calcium channel during neuroendocrine differentiation of human prostate cancer cells. *J Biol Chem* 2002;**277**(13):10824–33.

52. Gackiere F, Bidaux G, Delcourt P, et al. CaV3.2 T-type calcium channels are involved in calcium-dependent secretion of neuroendocrine prostate cancer cells. *J Biol Chem* 2008;**283**(15):10162–73.

53. Natrajan R, Little SE, Reis-Filho JS, et al. Amplification and overexpression of CACNA1E correlates with relapse in favorable histology Wilms' tumors. *Clin Cancer Res* 2006;**12**(24):7284–93.

54. Liao JY, Li LL, Wei Q, Yue JC. Heregulinbeta activates store-operated Ca^{2+} channels through c-erbB2 receptor level-dependent pathway in human breast cancer cells. *Arch Biochem Biophys* 2007;**458**(2):244–52.

55. Dubois C, Vanden Abeele F, Lehen'kyi V, et al. Remodeling of channel-forming ORAI proteins determines an oncogenic switch in prostate cancer. *Cancer Cell* 2014;**26**(1):19–32.

56. Bomben VC, Sontheimer HW. Inhibition of transient receptor potential canonical channels impairs cytokinesis in human malignant gliomas. *Cell Prolif* 2008;**41**(1):98–121.

57. Thebault S, Flourakis M, Vanoverberghe K, et al. Differential role of transient receptor potential channels in Ca^{2+} entry and proliferation of prostate cancer epithelial cells. *Cancer Res* 2006;**66**(4):2038–47.

58. Bidaux G, Flourakis M, Thebault S, et al. Prostate cell differentiation status determines transient receptor potential melastatin member 8 channel subcellular localization and function. *J Clin Invest* 2007;**117**(6):1647–57.

59. Lehen'kyi V, Flourakis M, Skryma R, Prevarskaya N. TRPV6 channel controls prostate cancer cell proliferation via Ca(2+)/NFAT-dependent pathways. *Oncogene* 2007;**26**(52):7380–5.

60. Egleton RD, Brown KC, Dasgupta P. Nicotinic acetylcholine receptors in cancer: multiple roles in proliferation and inhibition of apoptosis. *Trends Pharmacol Sci* 2008;**29**(3):151–8.

61. Li L, Hanahan D. Hijacking the neuronal NMDAR signaling circuit to promote tumor growth and invasion. *Cell* 2013;**153**(1):86–100.

62. Feng YH, Li X, Wang L, Zhou L, Gorodeski GI. A truncated P2X7 receptor variant (P2X7-j) endogenously expressed in cervical cancer cells antagonizes the full-length P2X7 receptor through hetero-oligomerization. *J Biol Chem* 2006;**281**(25):17228–37.

63. Olsen ML, Schade S, Lyons SA, Amaral MD, Sontheimer H. Expression of voltage-gated chloride channels in human glioma cells. *J Neurosci* 2003;**23**(13):5572–82.

64. Suh KS, Mutoh M, Gerdes M, et al. Antisense suppression of the chloride intracellular channel family induces apoptosis, enhances tumor necrosis factor (alpha)-induced apoptosis, and inhibits tumor growth. *Cancer Res* 2005;**65**(2):562–71.

65. Suh KS, Crutchley JM, Koochek A, et al. Reciprocal modifications of CLIC4 in tumor epithelium and stroma mark malignant progression of multiple human cancers. *Clin Cancer Res* 2007;**13**(1):121–31.

66. Gruber AD, Pauli BU. Tumorigenicity of human breast cancer is associated with loss of the Ca^{2+}-activated chloride channel CLCA2. *Cancer Res* 1999;**59**(21):5488–91.

67. Bustin SA, Li SR, Dorudi S. Expression of the Ca^{2+}-activated chloride channel genes CLCA1 and CLCA2 is downregulated in human colorectal cancer. *DNA Cell Biol* 2001;**20**(6):331–8.
68. Neman J, Termini J, Wilczynski S, et al. Human breast cancer metastases to the brain display GABAergic properties in the neural niche. *Proc Natl Acad Sci USA* 2014;**111**(3): 984–9.
69. Schonherr R. Clinical relevance of ion channels for diagnosis and therapy of cancer. *J Membr Biol* 2005;**205**(3):175–84.
70. Leanza L, Biasutto L, Manago A, Gulbins E, Zoratti M, Szabo I. Intracellular ion channels and cancer. *Front Physiol* 2013;**4**:227.
71. Leanza L, Zoratti M, Gulbins E, Szabo I. Mitochondrial ion channels as oncological targets. *Oncogene* 2014;**33**(49):5569–81.
72. Huang X, Jan LY. Targeting potassium channels in cancer. *J Cell Biol* 2014;**206**(2):151–62.
73. Pillozzi S, Brizzi MF, Bernabei PA, et al. VEGFR-1 (FLT-1), beta1 integrin, and hERG K^+ channel for a macromolecular signaling complex in acute myeloid leukemia: role in cell migration and clinical outcome. *Blood* 2007;**110**(4):1238–50.
74. DeCoursey TE, Kim SY, Silver MR, Quandt FN. Ion channel expression in PMA-differentiated human THP-1 macrophages. *J Membr Biol* 1996;**152**(2):141–57.
75. Black JA, Waxman SG. Noncanonical roles of voltage-gated sodium channels. *Neuron* 2013;**80**(2):280–91.
76. Catterall WA. From ionic currents to molecular mechanisms: the structure and function of voltage-gated sodium channels. *Neuron* 2000;**26**(1):13–25.
77. Brackenbury WJ, Isom LL. Na channel beta subunits: overachievers of the ion channel family. *Front Pharmacol* 2011;**2**:53.
78. Roger S, Besson P, Le Guennec JY. Involvement of a novel fast inward sodium current in the invasion capacity of a breast cancer cell line. *Biochim Biophys Acta* 2003;**1616**(2): 107–11.
79. Yang M, Kozminski DJ, Wold LA, et al. Therapeutic potential for phenytoin: targeting Na(v)1.5 sodium channels to reduce migration and invasion in metastatic breast cancer. *Breast Cancer Res Treat* 2012;**134**(2):603–15.
80. Diaz D, Delgadillo DM, Hernandez-Gallegos E, et al. Functional expression of voltage-gated sodium channels in primary cultures of human cervical cancer. *J Cell Physiol* 2007;**210**(2):469–78.
81. Joshi AD, Parsons DW, Velculescu VE, Riggins GJ. Sodium ion channel mutations in glioblastoma patients correlate with shorter survival. *Mol Cancer* 2011;**10**:17.
82. Schrey M, Codina C, Kraft R, et al. Molecular characterization of voltage-gated sodium channels in human gliomas. *Neuroreport* 2002;**13**(18):2493–8.
83. Fulgenzi G, Graciotti L, Faronato M, et al. Human neoplastic mesothelial cells express voltage-gated sodium channels involved in cell motility. *Int J Biochem Cell Biol* 2006;**38**(7):1146–59.
84. Onganer PU, Djamgoz MB. Small-cell lung cancer (human): potentiation of endocytic membrane activity by voltage-gated Na^+ channel expression in vitro. *J Membr Biol* 2005;**204**(2):67–75.
85. Blandino JK, Viglione MP, Bradley WA, Oie HK, Kim YI. Voltage-dependent sodium channels in human small-cell lung cancer cells: role in action potentials and inhibition by Lambert-Eaton syndrome IgG. *J Membr Biol* 1995;**143**(2):153–63.
86. Roger S, Rollin J, Barascu A, et al. Voltage-gated sodium channels potentiate the invasive capacities of human non-small-cell lung cancer cell lines. *Int J Biochem Cell Biol* 2007;**39**(4):774–86.
87. Onganer PU, Seckl MJ, Djamgoz MB. Neuronal characteristics of small-cell lung cancer. *Br J Cancer* 2005;**93**:1197–201.
88. Gao R, Shen Y, Cai J, Lei M, Wang Z. Expression of voltage-gated sodium channel alpha subunit in human ovarian cancer. *Oncol Rep* 2010;**23**(5):1293–9.

89. Laniado ME, Lalani EN, Fraser SP, et al. Expression and functional analysis of voltage-activated Na⁺ channels in human prostate cancer cell lines and their contribution to invasion in vitro. *Am J Pathol* 1997;**150**(4):1213–21.

90. Smith P, Rhodes NP, Shortland AP, et al. Sodium channel protein expression enhances the invasiveness of rat and human prostate cancer cells. *FEBS Lett* 1998;**423**(1): 19–24.

91. Grimes JA, Fraser SP, Stephens GJ, et al. Differential expression of voltage-activated Na⁺ currents in two prostatic tumour cell lines: contribution to invasiveness in vitro. *FEBS Lett* 1995;**369**(2–3):290–4.

92. Diss JK, Stewart D, Pani F, et al. A potential novel marker for human prostate cancer: voltage-gated sodium channel expression in vivo. *Prostate Cancer Prostatic Dis* 2005;**8**(3):266–73.

93. Yamashita N, Hamada H, Tsuruo T, Ogata E. Enhancement of voltage-gated Na⁺ channel current associated with multidrug resistance in human leukemia cells. *Cancer Res* 1987;**47**(14):3736–41.

94. Carrithers MD, Chatterjee G, Carrithers LM, et al. Regulation of podosome formation in macrophages by a novel splice variant of the sodium channel SCN8A. *J Biol Chem* 2009;**284**(12):8114–26.

95. Allen DH, Lepple-Wienhues A, Cahalan MD. Ion channel phenotype of melanoma cell lines. *J Membr Biol* 1997;**155**(1):27–34.

96. Ou SW, Kameyama A, Hao LY, et al. Tetrodotoxin-resistant Na⁺ channels in human neuroblastoma cells are encoded by new variants of Nav1.5/SCN5A. *Eur J Neurosci* 2005;**22**(4):793–801.

97. Bennett ES, Smith BA, Harper JM. Voltage-gated Na⁺ channels confer invasive properties on human prostate cancer cells. *Pflugers Arch* 2004;**447**(6):908–14.

98. Nelson M, Millican-Slater R, Forrest LC, Brackenbury WJ. The sodium channel beta1 subunit mediates outgrowth of neurite-like processes on breast cancer cells and promotes tumour growth and metastasis. *Int J Cancer* 2014;**135**(10):2338–51.

99. Qadri YJ, Rooj AK, Fuller CM. ENaCs and ASICs as therapeutic targets. *Am J Physiol Cell Physiol* 2012;**302**(7):C943–65.

100. Li X, Cowell JK, Sossey-Alaoui K. CLCA2 tumour suppressor gene in 1p31 is epigenetically regulated in breast cancer. *Oncogene* 2004;**23**(7):1474–80.

101. Feng Q, Hawes SE, Stern JE, et al. DNA methylation in tumor and matched normal tissues from non-small cell lung cancer patients. *Cancer Epidemiol Biomarkers Prev* 2008;**17**(3):645–54.

102. Toyota M, Ho C, Ohe-Toyota M, Baylin SB, Issa JP. Inactivation of CACNA1G, a T-type calcium channel gene, by aberrant methylation of its 5′ CpG island in human tumors. *Cancer Res* 1999;**59**(18):4535–41.

103. Lin H, Li Z, Chen C, et al. Transcriptional and post-transcriptional mechanisms for oncogenic overexpression of ether a go-go K⁺ channel. *PLoS One* 2011;**6**(5):e20362.

104. Bai Y, Liao H, Liu T, et al. MiR-296-3p regulates cell growth and multi-drug resistance of human glioblastoma by targeting ether-a-go-go (EAG1). *Eur J Cancer* 2013;**49**(3):710–24.

105. Xu D, Wang L, Dai W, Lu L. A requirement for K⁺-channel activity in growth factor-mediated extracellular signal-regulated kinase activation in human myeloblastic leukemia ML-1 cells. *Blood* 1999;**94**(1):139–45.

106. Brackenbury WJ, Djamgoz MB. Nerve growth factor enhances voltage-gated Na⁺ channel activity and Transwell migration in Mat-LyLu rat prostate cancer cell line. *J Cell Physiol* 2007;**210**(3):602–8.

107. Ding Y, Brackenbury WJ, Onganer PU, et al. Epidermal growth factor upregulates motility of Mat-LyLu rat prostate cancer cells partially via voltage-gated Na⁺ channel activity. *J Cell Physiol* 2008;**215**(1):77–81.

108. Onganer PU, Djamgoz MB. Epidermal growth factor potentiates in vitro metastatic behaviour of human prostate cancer PC-3M cells: involvement of voltage-gated sodium channel. *Mol Cancer* 2007;**6**(1):76.

109. Fraser SP, Ozerlat-Gunduz I, Onkal R, Diss JK, Latchman DS, Djamgoz MB. Estrogen and non-genomic upregulation of voltage-gated Na(+) channel activity in MDA-MB-231 human breast cancer cells: role in adhesion. *J Cell Physiol* 2010;**224**(2):527–39.

110. Brackenbury WJ, Djamgoz MB. Activity-dependent regulation of voltage-gated Na⁺ channel expression in Mat-LyLu rat prostate cancer cell line. *J Physiol* 2006;**573**(Pt 2):343–56.

111. Chioni AM, Shao D, Grose R, Djamgoz MB. Protein kinase A and regulation of neonatal Nav1.5 expression in human breast cancer cells: activity-dependent positive feedback and cellular migration. *Int J Biochem Cell Biol* 2010;**42**(2):346–58.

112. Yang M, Brackenbury WJ. Membrane potential and cancer progression. *Front Physiol* 2013;**4**:185.

113. Bianchi L, Wible B, Arcangeli A, et al. herg encodes a K⁺ current highly conserved in tumors of different histogenesis: a selective advantage for cancer cells? *Cancer Res* 1998;**58**(4):815–22.

114. Becchetti A. Ion channels and transporters in cancer. 1. Ion channels and cell proliferation in cancer. *Am J Physiol Cell Physiol* 2011;**301**(2):C255–65.

115. Pardo LA, del Camino D, Sanchez A, et al. Oncogenic potential of EAG K(+) channels. *EMBO J* 1999;**18**(20):5540–7.

116. Arcangeli A. Expression and role of hERG channels in cancer cells. *Novartis Found Symp* 2005;**266**:225–32. discussion 32-4.

117. Wang H, Zhang Y, Cao L, et al. HERG K⁺ channel, a regulator of tumor cell apoptosis and proliferation. *Cancer Res* 2002;**62**(17):4843–8.

118. Lee I, Park C, Kang WK. Knockdown of inwardly rectifying potassium channel Kir2.2 suppresses tumorigenesis by inducing reactive oxygen species-mediated cellular senescence. *Mol Cancer Ther* 2010;**9**(11):2951–9.

119. Plummer 3rd HK, Dhar MS, Cekanova M, Schuller HM. Expression of G-protein inwardly rectifying potassium channels (GIRKs) in lung cancer cell lines. *BMC Cancer* 2005;**5**:104.

120. Takanami I, Inoue Y, Gika M. G-protein inwardly rectifying potassium channel 1 (GIRK 1) gene expression correlates with tumor progression in non-small cell lung cancer. *BMC Cancer* 2004;**4**:79.

121. Plummer 3rd HK, Yu Q, Cakir Y, Schuller HM. Expression of inwardly rectifying potassium channels (GIRKs) and beta-adrenergic regulation of breast cancer cell lines. *BMC Cancer* 2004;**4**:93.

122. Wagner V, Stadelmeyer E, Riederer M, et al. Cloning and characterisation of GIRK1 variants resulting from alternative RNA editing of the KCNJ3 gene transcript in a human breast cancer cell line. *J Cell Biochem* 2010;**110**(3):598–608.

123. Weaver AK, Liu X, Sontheimer H. Role for calcium-activated potassium channels (BK) in growth control of human malignant glioma cells. *J Neurosci Res* 2004;**78**(2):224–34.

124. Lu F, Chen H, Zhou C, et al. T-type Ca²⁺ channel expression in human esophageal carcinomas: a functional role in proliferation. *Cell Calcium* 2008;**43**(1):49–58.

125. Bolanz KA, Hediger MA, Landowski CP. The role of TRPV6 in breast carcinogenesis. *Mol Cancer Ther* 2008;**7**(2):271–9.

126. Guilbert A, Gautier M, Dhennin-Duthille I, Haren N, Sevestre H, Ouadid-Ahidouch H. Evidence that TRPM7 is required for breast cancer cell proliferation. *Am J Physiol Cell Physiol* 2009;**297**(3):C493–502.

127. Brackenbury WJ, Chioni AM, Diss JK, Djamgoz MB. The neonatal splice variant of Nav1.5 potentiates in vitro metastatic behaviour of MDA-MB-231 human breast cancer cells. *Breast Cancer Res Treat* 2007;**101**(2):149–60.

128. Gillet L, Roger S, Besson P, et al. Voltage-gated sodium channel activity promotes cysteine cathepsin-dependent invasiveness and colony growth of human cancer cells. *J Biol Chem* 2009;**284**(13):8680–91.

129. Vanden Abeele F, Skryma R, Shuba Y, et al. Bcl-2-dependent modulation of Ca(2+) homeostasis and store-operated channels in prostate cancer cells. *Cancer Cell* 2002;**1**(2):169–79.

130. Chow J, Norng M, Zhang J, Chai J. TRPV6 mediates capsaicin-induced apoptosis in gastric cancer cells–Mechanisms behind a possible new "hot" cancer treatment. *Biochim Biophys Acta* 2007;**1773**(4):565–76.

131. Raphael M, Lehen'kyi V, Vandenberghe M, et al. TRPV6 calcium channel translocates to the plasma membrane via Orai1-mediated mechanism and controls cancer cell survival. *Proc Natl Acad Sci USA* 2014;**111**(37):E3870–9.

132. Han Y, Shi Y, Han Z, Sun L, Fan D. Detection of potassium currents and regulation of multidrug resistance by potassium channels in human gastric cancer cells. *Cell Biol Int* 2007;**31**(7):741–7.

133. Lemonnier L, Lazarenko R, Shuba Y, et al. Alterations in the regulatory volume decrease (RVD) and swelling-activated Cl-current associated with neuroendocrine differentiation of prostate cancer epithelial cells. *Endocr Relat Cancer* 2005;**12**(2):335–49.

134. Lemonnier L, Shuba Y, Crepin A, et al. Bcl-2-dependent modulation of swelling-activated Cl-current and ClC-3 expression in human prostate cancer epithelial cells. *Cancer Res* 2004;**64**(14):4841–8.

135. Bortner CD, Cidlowski JA. Uncoupling cell shrinkage from apoptosis reveals that Na+ influx is required for volume loss during programmed cell death. *J Biol Chem* 2003;**278**(40):39176–84.

136. Cuddapah VA, Sontheimer H. Ion channels and transporters in cancer. 2. Ion channels and the control of cancer cell migration. *Am J Physiol Cell Physiol* 2011;**301**(3):C541–9.

137. Watkins S, Sontheimer H. Hydrodynamic cellular volume changes enable glioma cell invasion. *J Neurosci* 2011;**31**(47):17250–9.

138. Seifert S, Sontheimer H. Bradykinin enhances invasion of malignant glioma into the brain parenchyma by inducing cells to undergo amoeboid migration. *J Physiol* 2014;**592**(Pt 22):5109–27.

139. Crociani O, Zanieri F, Pillozzi S, et al. hERG1 channels modulate integrin signaling to trigger angiogenesis and tumor progression in colorectal cancer. *Sci Rep* 2013;**3**:3308.

140. Faehling M, Kroll J, Fohr KJ, et al. Essential role of calcium in vascular endothelial growth factor A-induced signaling: mechanism of the antiangiogenic effect of carboxyamidotriazole. *FASEB J* 2002;**16**(13):1805–7.

141. Luzzi KJ, Varghese HJ, MacDonald IC, et al. Inhibition of angiogenesis in liver metastases by carboxyamidotriazole (CAI). *Angiogenesis* 1998;**2**(4):373–9.

142. Kwan HY, Huang Y, Yao X. TRP channels in endothelial function and dysfunction. *Biochim Biophys Acta* 2007;**1772**(8):907–14.

143. Cheng HW, James AF, Foster RR, Hancox JC, Bates DO. VEGF activates receptor-operated cation channels in human microvascular endothelial cells. *Arterioscler Thromb Vasc Biol* 2006;**26**(8):1768–76.

144. Kohler R, Degenhardt C, Kuhn M, Runkel N, Paul M, Hoyer J. Expression and function of endothelial Ca(2+)-activated K(+) channels in human mesenteric artery: a single-cell reverse transcriptase-polymerase chain reaction and electrophysiological study in situ. *Circ Res* 2000;**87**(6):496–503.

145. Hu J, Yuan X, Ko MK, et al. Calcium-activated potassium channels mediated blood–brain tumor barrier opening in a rat metastatic brain tumor model. *Mol Cancer* 2007;**6**:22.

146. Masi A, Becchetti A, Restano-Cassulini R, et al. hERG1 channels are overexpressed in glioblastoma multiforme and modulate VEGF secretion in glioblastoma cell lines. *Br J Cancer* 2005;**93**(7):781–92.

147. Crociani O, Lastraioli E, Boni L, et al. hERG1 channels regulate VEGF-A secretion in human gastric cancer: clinicopathological correlations and therapeutical implications. *Clin Cancer Res* 2014;**20**(6):1502–12.

148. Downie BR, Sanchez A, Knotgen H, et al. Eag1 expression interferes with hypoxia homeostasis and induces angiogenesis in tumors. *J Biol Chem* 2008;**283**(52):36234–40.

149. Ju YK, Saint DA, Gage PW. Hypoxia increases persistent sodium current in rat ventricular myocytes. *J Physiol* 1996;**497**(2):337–47.

150. Crill WE. Persistent sodium current in mammalian central neurons. *Annu Rev Physiol* 1996;**58**:349–62.

151. Lobikin M, Chernet B, Lobo D, Levin M. Resting potential, oncogene-induced tumorigenesis, and metastasis: the bioelectric basis of cancer in vivo. *Phys Biol* 2012;**9**(6):065002.

152. Chernet BT, Levin M. Transmembrane voltage potential of somatic cells controls oncogene-mediated tumorigenesis at long-range. *Oncotarget* 2014;**5**(10):3287–306.

153. Sundelacruz S, Levin M, Kaplan DL. Role of membrane potential in the regulation of cell proliferation and differentiation. *Stem cell Rev* 2009;**5**(3):231–46.

154. Blackiston DJ, McLaughlin KA, Levin M. Bioelectric controls of cell proliferation: ion channels, membrane voltage and the cell cycle. *Cell Cycle* 2009;**8**(21):3519–28.

155. Schwab A, Nechyporuk-Zloy V, Fabian A, Stock C. Cells move when ions and water flow. *Pflugers Arch* 2007;**453**(4):421–32.

156. Nuccitelli R. A role for endogenous electric fields in wound healing. *Curr Top Dev Biol* 2003;**58**:1–26.

157. Wonderlin WF, Woodfork KA, Strobl JS. Changes in membrane potential during the progression of MCF-7 human mammary tumor cells through the cell cycle. *J Cell Physiol* 1995;**165**(1):177–85.

158. Boonstra J, Mummery CL, Tertoolen LG, Van Der Saag PT, De Laat SW. Cation transport and growth regulation in neuroblastoma cells. Modulations of K$^+$ transport and electrical membrane properties during the cell cycle. *J Cell Physiol* 1981;**107**(1):75–83.

159. Sachs HG, Stambrook PJ, Ebert JD. Changes in membrane potential during the cell cycle. *Exp Cell Res* 1974;**83**(2):362–6.

160. Meyer R, Heinemann SH. Characterization of an eag-like potassium channel in human neuroblastoma cells. *J Physiol* 1998;**508**(Pt 1):49–56.

161. Ouadid-Ahidouch H, Le Bourhis X, Roudbaraki M, Toillon RA, Delcourt P, Prevarskaya N. Changes in the K$^+$ current-density of MCF-7 cells during progression through the cell cycle: possible involvement of a h-ether.a-gogo K$^+$ channel. *Recept Channels* 2001;**7**(5):345–56.

162. Higashimori H, Sontheimer H. Role of Kir4.1 channels in growth control of glia. *Glia* 2007;**55**(16):1668–79.

163. Woodfork KA, Wonderlin WF, Peterson VA, Strobl JS. Inhibition of ATP-sensitive potassium channels causes reversible cell-cycle arrest of human breast cancer cells in tissue culture. *J Cell Physiol* 1995;**162**(2):163–71.

164. Habela CW, Olsen ML, Sontheimer H. ClC3 is a critical regulator of the cell cycle in normal and malignant glial cells. *J Neurosci* 2008;**28**(37):9205–17.

165. Choi J, Chiang A, Taulier N, Gros R, Pirani A, Husain M. A calmodulin-binding site on cyclin E mediates Ca^{2+}-sensitive G1/s transitions in vascular smooth muscle cells. *Circ Res* 2006;**98**(10):1273–81.

166. Ziechner U, Schonherr R, Born AK, et al. Inhibition of human ether a go-go potassium channels by Ca^{2+}/calmodulin binding to the cytosolic N- and C-termini. *FEBS J* 2006;**273**(5):1074–86.

167. Khanna R, Chang MC, Joiner WJ, Kaczmarek LK, Schlichter LC. hSK4/hIK1, a calmodulin-binding KCa channel in human T lymphocytes. Roles in proliferation and volume regulation. *J Biol Chem* 1999;**274**(21):14838–49.

168. Hegle AP, Marble DD, Wilson GF. A voltage-driven switch for ion-independent signaling by ether-a-go-go K$^+$ channels. *Proc Natl Acad Sci USA* 2006;**103**(8):2886–91.

169. Haas BR, Sontheimer H. Inhibition of the sodium-potassium-chloride cotransporter isoform-1 reduces glioma invasion. *Cancer Res* 2010;**70**(13):5597–606.

170. Habela CW, Ernest NJ, Swindall AF, Sontheimer H. Chloride accumulation drives volume dynamics underlying cell proliferation and migration. *J Neurophysiol* 2009;**101**(2):750–7.

171. Lui VC, Lung SS, Pu JK, Hung KN, Leung GK. Invasion of human glioma cells is regulated by multiple chloride channels including ClC-3. *Anticancer Res* 2010;**30**(11):4515–24.

172. Mao J, Chen L, Xu B, et al. Suppression of ClC-3 channel expression reduces migration of nasopharyngeal carcinoma cells. *Biochem Pharmacol* 2008;**75**(9):1706–16.

173. Turner KL, Honasoge A, Robert SM, McFerrin MM, Sontheimer H. A proinvasive role for the Ca(2+)-activated K(+) channel KCa3.1 in malignant glioma. *Glia* 2014;**62**(6):971–81.

174. Cuddapah VA, Sontheimer H. Molecular interaction and functional regulation of ClC-3 by Ca^{2+}/calmodulin-dependent protein kinase II (CaMKII) in human malignant glioma. *J Biol Chem* 2010;**285**(15):11188–96.

175. Ishiuchi S, Tsuzuki K, Yoshida Y, et al. Blockage of Ca(2+)-permeable AMPA receptors suppresses migration and induces apoptosis in human glioblastoma cells. *Nat Med* 2002;**8**(9):971–8.

176. Huang JB, Kindzelskii AL, Clark AJ, Petty HR. Identification of channels promoting calcium spikes and waves in HT1080 tumor cells: their apparent roles in cell motility and invasion. *Cancer Res* 2004;**64**(7):2482–9.

177. Waning J, Vriens J, Owsianik G, et al. A novel function of capsaicin-sensitive TRPV1 channels: involvement in cell migration. *Cell Calcium* 2007;**42**(1):17–25.

178. Yang S, Zhang JJ, Huang XY. Orai1 and STIM1 are critical for breast tumor cell migration and metastasis. *Cancer Cell* 2009;**15**(2):124–34.

179. Lee JM, Davis FM, Roberts-Thomson SJ, Monteith GR. Ion channels and transporters in cancer. 4. Remodeling of Ca(2+) signaling in tumorigenesis: role of Ca(2+) transport. *Am J Physiol Cell Physiol* 2011;**301**(5):C969–76.

180. Mycielska ME, Fraser SP, Szatkowski M, Djamgoz MB. Contribution of functional voltage-gated Na+ channel expression to cell behaviors involved in the metastatic cascade in rat prostate cancer: II. Secretory membrane activity. *J Cell Physiol* 2003;**195**(3):461–9.

181. Djamgoz MBA, Mycielska M, Madeja Z, Fraser SP, Korohoda W. Directional movement of rat prostate cancer cells in direct-current electric field: involvement of voltage gated Na+ channel activity. *J Cell Sci* 2001;**114**(Pt 14):2697–705.

182. Mycielska ME, Palmer CP, Brackenbury WJ, Djamgoz MB. Expression of Na+-dependent citrate transport in a strongly metastatic human prostate cancer PC-3M cell line: regulation by voltage-gated Na+ channel activity. *J Physiol* 2005;**563**(Pt 2):393–408.

183. Fraser SP, Salvador V, Manning EA, et al. Contribution of functional voltage-gated Na+ channel expression to cell behaviors involved in the metastatic cascade in rat prostate cancer: I. Lateral motility. *J Cell Physiol* 2003;**195**(3):479–87.

184. Fraser SP, Ding Y, Liu A, Foster CS, Djamgoz MB. Tetrodotoxin suppresses morphological enhancement of the metastatic MAT-LyLu rat prostate cancer cell line. *Cell Tissue Res* 1999;**295**(3):505–12.

185. Krasowska M, Grzywna ZJ, Mycielska ME, Djamgoz MB. Patterning of endocytic vesicles and its control by voltage-gated Na+ channel activity in rat prostate cancer cells: fractal analyses. *Eur Biophys J* 2004;**33**(6):535–42.

186. Krasowska M, Grzywna ZJ, Mycielska ME, Djamgoz MB. Fractal analysis and ionic dependence of endocytotic membrane activity of human breast cancer cells. *Eur Biophys J* 2009;**38**(8):1115–25.

187. Palmer CP, Mycielska ME, Burcu H, et al. Single cell adhesion measuring apparatus (SCAMA): application to cancer cell lines of different metastatic potential and voltage-gated Na+ channel expression. *Eur Biophys J* 2008;**37**(4):359–68.

188. Campbell TM, Main MJ, Fitzgerald EM. Functional expression of the voltage-gated Na(+)-channel Nav1.7 is necessary for EGF-mediated invasion in human non-small cell lung cancer cells. *J Cell Sci* 2013;**126**(Pt 21):4939–49.

189. Nakajima T, Kubota N, Tsutsumi T, et al. Eicosapentaenoic acid inhibits voltage-gated sodium channels and invasiveness in prostate cancer cells. *Br J Pharmacol* 2009;**156**(3):420–31.

190. Brisson L, Driffort V, Benoist L, et al. NaV1.5 Na(+) channels allosterically regulate the NHE-1 exchanger and promote the activity of breast cancer cell invadopodia. *J Cell Sci* 2013;**126**(Pt 21):4835–42.

191. Brisson L, Gillet L, Calaghan S, et al. Na(V)1.5 enhances breast cancer cell invasiveness by increasing NHE1-dependent H(+) efflux in caveolae. *Oncogene* 2011;**30**(17):2070–6.

192. Kapoor N, Bartoszewski R, Qadri YJ, et al. Knockdown of ASIC1 and epithelial sodium channel subunits inhibits glioblastoma whole cell current and cell migration. *J Biol Chem* 2009;**284**(36):24526–41.

193. Brackenbury WJ, Djamgoz MB, Isom LL. An emerging role for voltage-gated Na+ channels in cellular migration: regulation of central nervous system development and potentiation of invasive cancers. *Neuroscientist* 2008;**14**(6):571–83.

194. Arcangeli A, Becchetti A. Complex functional interaction between integrin receptors and ion channels. *Trends Cell Biol* 2006;**16**(12):631–9.

195. Levite M, Cahalon L, Peretz A, et al. Extracellular K(+) and opening of voltage-gated potassium channels activate T cell integrin function: physical and functional association between Kv1.3 channels and beta1 integrins. *J Exp Med* 2000;**191**(7):1167–76.

196. Artym VV, Petty HR. Molecular proximity of Kv1.3 voltage-gated potassium channels and beta(1)-integrins on the plasma membrane of melanoma cells: effects of cell adherence and channel blockers. *J Gen Physiol* 2002;**120**(1):29–37.

197. Cherubini A, Hofmann G, Pillozzi S, et al. Human ether-a-go-go-related gene 1 channels are physically linked to beta1 integrins and modulate adhesion-dependent signaling. *Mol Biol Cell* 2005;**16**(6):2972–83.

198. Pillozzi S, Masselli M, De Lorenzo E, et al. Chemotherapy resistance in acute lymphoblastic leukemia requires hERG1 channels and is overcome by hERG1 blockers. *Blood* 2011;**117**(3):902–14.

199. Brackenbury WJ, Isom LL. Voltage-gated Na+ channels: potential for beta subunits as therapeutic targets. *Expert Opin Ther Targets* 2008;**12**(9):1191–203.

200. McEwen DP, Isom LL. Heterophilic interactions of sodium channel beta1 subunits with axonal and glial cell adhesion molecules. *J Biol Chem* 2004;**279**(50):52744–52.

201. Kazarinova-Noyes K, Malhotra JD, McEwen DP, et al. Contactin associates with Na+ channels and increases their functional expression. *J Neurosci* 2001;**21**:7517–25.

202. Ratcliffe CF, Westenbroek RE, Curtis R, Catterall WA. Sodium channel beta1 and beta3 subunits associate with neurofascin through their extracellular immunoglobulin-like domain. *J Cell Biol* 2001;**154**(2):427–34.

203. Malhotra JD, Kazen-Gillespie K, Hortsch M, Isom LL. Sodium channel β subunits mediate homophilic cell adhesion and recruit ankyrin to points of cell-cell contact. *J Biol Chem* 2000;**275**:11383–8.

204. Xiao ZC, Ragsdale DS, Malhotra JD, et al. Tenascin-R is a functional modulator of sodium channel beta subunits. *J Biol Chem* 1999;**274**(37):26511–7.

205. Srinivasan J, Schachner M, Catterall WA. Interaction of voltage-gated sodium channels with the extracellular matrix molecules tenascin-C and tenascin-R. *Proc Natl Acad Sci USA* 1998;**95**(26):15753–7.

206. Brackenbury WJ, Yuan Y, O'Malley HA, Parent JM, Isom LL. Abnormal neuronal patterning occurs during early postnatal brain development of Scn1b-null mice and precedes hyperexcitability. *Proc Natl Acad Sci USA* 2013;**110**(3):1089–94.

207. Brackenbury WJ, Davis TH, Chen C, et al. Voltage-gated Na+ channel β1 subunit-mediated neurite outgrowth requires fyn kinase and contributes to central nervous system development in vivo. *J Neurosci* 2008;**28**(12):3246–56.

208. Davis TH, Chen C, Isom LL. Sodium channel beta1 subunits promote neurite outgrowth in cerebellar granule neurons. *J Biol Chem* 2004;**279**(49):51424–32.

209. Patino GA, Brackenbury WJ, Bao YY, et al. Voltage-gated Na+ channel beta 1B: a secreted cell adhesion molecule involved in human epilepsy. *J Neurosci* 2011;**31**(41):14577–91.

210. Brackenbury WJ, Calhoun JD, Chen C, et al. Functional reciprocity between Na$^+$ channel Nav1.6 and β1 subunits in the coordinated regulation of excitability and neurite outgrowth. *Proc Natl Acad Sci USA* 2010;**107**(5):2283–8.

211. Kim DY, Mackenzie Ingano LA, Carey BW, Pettingell WP, Kovacs DM. Presenilin/gamma-secretase-mediated cleavage of the voltage-gated sodium channel beta 2 subunit regulates cell adhesion and migration. *J Biol Chem* 2005;**280**(24):23251–61.

212. Wong HK, Sakurai T, Oyama F, et al. Beta subunits of voltage-gated sodium channels are novel substrates of BACE1 and gamma-secretase. *J Biol Chem* 2005;**280**(24):23009–17.

213. Kim DY, Carey BW, Wang H, et al. BACE1 regulates voltage-gated sodium channels and neuronal activity. *Nat Cell Biol* 2007;**9**(7):755–64.

214. Jansson KH, Castillo DG, Morris JW, et al. Identification of beta-2 as a key cell adhesion molecule in PCa cell neurotropic behavior: a novel ex vivo and biophysical approach. *PLoS One* 2014;**9**(6):e98408.

215. Adachi K, Toyota M, Sasaki Y, et al. Identification of SCN3B as a novel p53-inducible proapoptotic gene. *Oncogene* 2004;**23**(47):7791–8.

216. Ko JH, Ko EA, Gu W, Lim I, Bang H, Zhou T. Expression profiling of ion channel genes predicts clinical outcome in breast cancer. *Mol Cancer* 2013;**12**(1):106.

217. Onkal R, Djamgoz MB. Molecular pharmacology of voltage-gated sodium channel expression in metastatic disease: clinical potential of neonatal Nav1.5 in breast cancer. *Eur J Pharmacol* 2009;**625**(1–3):206–19.

218. Hoffmann EK, Lambert IH. Ion channels and transporters in the development of drug resistance in cancer cells. *Philos Trans R Soc Lond B Biol Sci* 2014;**369**(1638):20130109.

219. Hanahan D, Weinberg RA. The hallmarks of cancer. *Cell* 2000;**100**(1):57–70.

220. Hanahan D, Weinberg RA. Hallmarks of cancer: the next generation. *Cell* 2011;**144**(5):646–74.

221. Fraser SP, Pardo LA. Ion channels: functional expression and therapeutic potential in cancer. Colloquium on Ion Channels and Cancer. *EMBO Rep* 2008;**9**(6):512–5.

222. Lehen'kyi V, Shapovalov G, Skryma R, Prevarskaya N. Ion channels and transporters in cancer. 5. Ion channels in control of cancer and cell apoptosis. *Am J Physiol Cell Physiol* 2011;**301**(6):C1281–9.

223. Arcangeli A. Ion channels and transporters in cancer. 3. Ion channels in the tumor cell-microenvironment cross talk. *Am J Physiol Cell Physiol* 2011;**301**(4):C762–71.

224. Fiorio Pla A, Avanzato D, Munaron L, Ambudkar IS. Ion channels and transporters in cancer. 6. Vascularizing the tumor: TRP channels as molecular targets. *Am J Physiol Cell Physiol* 2012;**302**(1):C9–15.

225. Rouzaire-Dubois B, Malo M, Milandri JB, Dubois JM. Cell size-proliferation relationship in rat glioma cells. *Glia* 2004;**45**(3):249–57.

226. Schofl C, Rossig L, Mader T, et al. Impairment of ATP-induced Ca^{2+}-signalling in human thyroid cancer cells. *Mol Cell Endocrinol* 1997;**133**(1):33–9.

227. Alfonso-De Matte MY, Moses-Soto H, Kruk PA. Calcium-mediated telomerase activity in ovarian epithelial cells. *Arch Biochem Biophys* 2002;**399**(2):239–44.

228. Cronin NB, O'Reilly A, Duclohier H, Wallace BA. Effects of deglycosylation of sodium channels on their structure and function. *Biochemistry* 2005;**44**(2):441–9.

229. Spitzner M, Ousingsawat J, Scheidt K, Kunzelmann K, Schreiber R. Voltage-gated K$^+$ channels support proliferation of colonic carcinoma cells. *FASEB J* 2007;**21**(1):35–44.

230. Bloch M, Ousingsawat J, Simon R, et al. KCNMA1 gene amplification promotes tumor cell proliferation in human prostate cancer. *Oncogene* 2007;**26**(17):2525–34.

231. Nelson M, Yang M, Dowle AA, Thomas JR, Brackenbury WJ. The sodium channel-blocking antiepileptic drug phenytoin inhibits breast tumour growth and metastasis. *Mol Cancer* 2015;**14**(1):13.

232. Driffort V, Gillet L, Bon E, et al. Ranolazine inhibits NaV1.5-mediated breast cancer cell invasiveness and lung colonization. *Mol Cancer* 2014;**13**(1):264.

233. Yildirim S, Altun S, Gumushan H, Patel A, Djamgoz MB. Voltage-gated sodium channel activity promotes prostate cancer metastasis in vivo. *Cancer Lett* 2012;**323**(1):58–61.

234. Mamelak AN, Rosenfeld S, Bucholz R, et al. Phase I single-dose study of intracavitary-administered iodine-131-TM-601 in adults with recurrent high-grade glioma. *J Clin Oncol* 2006;**24**(22):3644–50.

235. Mantegazza M, Curia G, Biagini G, Ragsdale DS, Avoli M. Voltage-gated sodium channels as therapeutic targets in epilepsy and other neurological disorders. *Lancet Neurol* 2010;**9**(4):413–24.

236. Ragsdale DS, Scheuer T, Catterall WA. Frequency and voltage-dependent inhibition of type IIA Na^+ channels, expressed in a mammalian cell line, by local anesthetic, antiarrhythmic, and anticonvulsant drugs. *Mol Pharmacol* 1991;**40**(5):756–65.

237. Abdul M, Hoosein N. Inhibition by anticonvulsants of prostate-specific antigen and interleukin-6 secretion by human prostate cancer cells. *Anticancer Res* 2001;**21**(3B): 2045–8.

238. Abdul M, Hoosein N. Voltage-gated sodium ion channels in prostate cancer: expression and activity. *Anticancer Res* 2002;**22**(3):1727–30.

239. Djamgoz MB, Onkal R. Persistent current blockers of voltage-gated sodium channels: a clinical opportunity for controlling metastatic disease. *Recent Pat Anticancer Drug Discov* 2013;**8**(1):66–84.

240. Speyer CL, Smith JS, Banda M, Devries JA, Mekani T, Gorski DH. Metabotropic glutamate receptor-1: a potential therapeutic target for the treatment of breast cancer. *Breast Cancer Res Treat* 2012;**132**(2):565–73.

241. Yip D, Le MN, Chan JL, et al. A phase 0 trial of riluzole in patients with resectable stage III and IV melanoma. *Clin Cancer Res* 2009;**15**(11):3896–902.

242. Wen Y, Li J, Koo J, et al. Activation of the glutamate receptor GRM1 enhances angiogenic signaling to drive melanoma progression. *Cancer Res* 2014;**74**(9):2499–509.

243. Fairhurst C, Watt I, Martin F, Bland M, Brackenbury WJ. Exposure to sodium channel-inhibiting drugs and cancer survival: protocol for a cohort study using the QResearch primary care database. *BMJ Open* 2014;**4**(11):e006604.

244. Biki B, Mascha E, Moriarty DC, Fitzpatrick JM, Sessler DI, Buggy DJ. Anesthetic technique for radical prostatectomy surgery affects cancer recurrence: a retrospective analysis. *Anesthesiology* 2008;**109**(2):180–7.

245. Anderson JD, Hansen TP, Lenkowski PW, et al. Voltage-gated sodium channel blockers as cytostatic inhibitors of the androgen-independent prostate cancer cell line PC-3. *Mol Cancer Ther* 2003;**2**(11):1149–54.

246. Lenkowski PW, Ko SH, Anderson JD, Brown ML, Patel MK. Block of human NaV1.5 sodium channels by novel alpha-hydroxyphenylamide analogues of phenytoin. *Eur J Pharm Sci* 2004;**21**(5):635–44.

247. Gomez-Varela D, Zwick-Wallasch E, Knotgen H, et al. Monoclonal antibody blockade of the human Eag1 potassium channel function exerts antitumor activity. *Cancer Res* 2007;**67**(15):7343–9.

248. Chioni AM, Fraser SP, Pani F, et al. A novel polyclonal antibody specific for the $Na_v1.5$ voltage-gated Na^+ channel 'neonatal' splice form. *J Neurosci Methods* 2005;**147**(2):88–98.

249. Smitherman KA, Sontheimer H. Inhibition of glial Na^+ and K^+ currents by tamoxifen. *J Membr Biol* 2001;**181**(2):125–35.

TMEM16 Membrane Proteins in Health and Disease

H. Yang, L.Y. Jan

University of California, San Francisco, CA, United States

Phospholipid molecules form the basic structure of membranes that isolate the cells from their external environment and create different subcellular compartments. The hydrophobic core of the lipid bilayer generates a huge energy barrier to hamper the diffusion of charged molecules across the cell membranes. Moreover, the exchange of phospholipids from one leaflet to another leaflet is rarely displayed in the cell membrane, which uses flippases and floppases to establish the asymmetry of lipid distribution in the two leaflets.[1] Cells have evolved a large variety of ion channel and transporter proteins to catalyze the movement of different substrates across membranes utilizing electrochemical gradients or energy from chemical reactions or physical stimuli (Fig. 7.1). About 800 out of the more than 6000 human membrane proteins have been associated with ion channel or transporter functions[2]; however, a significant fraction of the human membrane proteome has not been functionally characterized,[2,3] as in the case of the TransMEMbrane protein 16 (TMEM16) family of transmembrane (TM) proteins with unknown functions up until the past decade. Because of the collective efforts of scientists in recent years, it becomes clear that the TMEM16 family not only includes the long sought-after calcium-activated chloride channels (CaCCs),[4–6] but also the calcium-activated nonselective channels[7,8] and the mysterious calcium-activated lipid scramblases,[8–11] which can quickly move phospholipids from one leaflet to the other and destroy their asymmetric distribution in the cell membrane in a calcium-dependent manner. In this chapter we focus on the recent structure-function characterizations of TMEM16 proteins and discuss the current understanding of their physiological functions and potential pathological roles in human diseases. As to topics such as

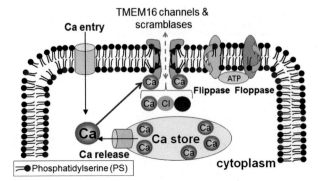

FIGURE 7.1 **Overview of main functions of the calcium-activated TransMEMbrane protein 16 (TMEM16) proteins.** The TMEM16 proteins on the plasma membrane can act as nonselective ion channels, lipid scramblases, or both in response to calcium entry through cell surface calcium-permeable channels or calcium release from intracellular calcium stores. By catalyzing the nonspecific membrane lipid transposition, the calcium-dependent lipid scramblases can rapidly destroy the lipid asymmetry across the lipid bilayer, which is dedicatedly maintained by the ATP-dependent lipid transporters, flippases and floppases.

posttranslational modification, alternative splicing, and pharmacological profiles of TMEM16 proteins that are not discussed here, readers are referred to other reviews for further information.[12–25]

THE DISCOVERIES CONCERNING THE TMEM16 FAMILY

The TMEM16 family of genes was first systematically described in 2003.[26] The cDNA clone, *FLJ10261* (corresponding to *TMEM16A* or *ANO1*), was identified during in silico characterization of the human chromosome 11q13 region that is frequently amplified in various tumor types such as head and neck tumors, parathyroid tumors, and ovarian tumors and in breast, pancreatic, gastric, and uterine cancers. The *TMEM16A* gene encodes a 986-amino-acid protein predicted to contain eight TM regions with the N- and C-terminal domains exposed to the cytosol. Other human TMEM16 family members (Fig. 7.2A) were subsequently identified based on sequence similarity through bioinformatic analysis.[26–30] In addition to bioinformatics, early genetic analysis and expression profiling also implicated the importance of TMEM16 proteins in human health and disease. For instance, TMEM16A (also known as DOG1, ORAOV2, or TAOS2) was overexpressed in gastrointestinal stromal tumors (GISTs)[31] and oral squamous cell carcinomas.[32] TMEM16G (also known as NGEP or D-TMPP) is highly expressed in normal and cancerous prostate,[33,34] whereas mutations of TMEM16E (also known as GDD1) were associated with gnathodiaphyseal dysplasia (GDD).[27,35]

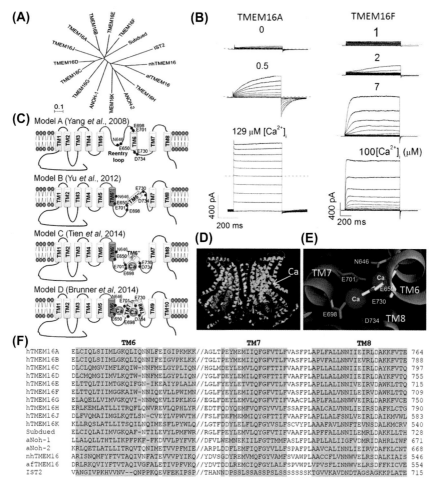

FIGURE 7.2 **Structure and function of TransMEMbrane protein 16 (TMEM16) proteins.** (A) Phylogeny of various human TMEM16 family proteins and their orthologues. Sequence alignment was performed using CLUSTALW2, and the phylogeny tree was built using SeaView. Scale bar represents 0.1 nucleotide substitutions per site. (B) Representative macroscopic current traces of TMEM16A (left) and TMEM16F (right) channels exposed to various intracellular calcium concentrations. Testing potentials were from −80 mV to +180 mV with 20-mV increments. (C) Evolution of our understanding on the membrane topology of TMEM16 proteins. Six calcium-binding residues identified in the fungal nhTMEM16 structure were mapped on each of these models. Numbering of these residues was based on the mouse TMEM16A (Uniprot ID: Q8BHY3-2). The locations of these residues are for demonstration purpose only. (D) Side view of the fungal nhTMEM16 structure (PDB ID: 4WIS) from within the membrane. Two subunits were shown in *cyan* and *green*. *Red color* labels the interaction interface of two TMEM16 subunits computed using the "InterfaceResidues" script in Pymol. Calcium ions are shown in *magenta*. (E) The calcium-binding pocket is in the middle of the nhTMEM16 transmembrane domain. (F) The calcium-binding residues are highly conserved among TMEM16 family members. Subdued is a *Drosophila* TMEM16 homologue, aNoh-1 and aNoh-2 are the two TMEM16 homologues in *Caenorhabditis elegans*. Fungal afTMEM16 and nhTMEM16 are the TMEM16 homologues in *Aspergillus fumigatus* and *Nectria haematococca*, respectively. Increased sodium tolerance protein 2 (Ist2) is the only TMEM16 in yeast.

An important leap in understanding the functions of the TMEM16 protein family was made in 2008 when three independent groups[4-6] simultaneously discovered that the TMEM16A is the long sought-after CaCC, the molecular identity of which remained elusive for one-quarter of a century.[36-38] By searching among uncharacterized membrane proteins with multiple TM segments, the Oh laboratory discovered TMEM16A as a bona fide CaCC that is activated by intracellular calcium and various calcium-mobilizing stimuli.[4] Heterologous expression of TMEM16A in different cell lines elicits a current that for the first time resembled the "classical" CaCCs, with hallmark features of anion selectivity ($NO_3^- > I^- > Br^- > Cl^- > F^-$), intracellular calcium-dependent activation, voltage-dependent activation at submicromolar calcium concentrations (Fig. 7.2B), and inhibition by the pharmacological agents such as niflumic acid and 5-nitro-2-(3-phenylpropylamino)-benzoic acid.[12,24] TMEM16A is expressed in various secretory epithelia, the retina, and sensory neurons. Knockdown of mouse TMEM16A markedly reduces native CaCC currents and saliva production in mice.[4] Sequence analysis and membrane topology predication indicate that TMEM16A has no apparent similarity to previously characterized membrane proteins. These authors named this protein family anoctamins (abbreviated Ano) to signify that TMEM16A is an anionic channel (*ano*) with eight (*oct*) TM segments and to imply that its family members have similar structure and function. As discussed later, this nomenclature seems no longer appropriate because (1) recent structural characterizations support a 10-TM segment topology for TMEM16 proteins (Fig. 7.2C and D)[11]; (2) some TMEM16 proteins are nonselective ion channels permeable to cations, and some TMEM16 proteins can scramble phospholipids with or without displaying detectable channel activity (Fig. 7.1)[7-9,11,39]; (3) Ano has a potentially embarrassing meaning in Spanish and Italian.

The Galietta laboratory identified TMEM16A using a functional genomics strategy.[6] It was known that prolonged treatment of bronchial epithelial cells with a proinflammatory cytokine [interleukin (IL)-4] causes increased CaCC activity, presumably by upregulating expression of the CaCC gene. Therefore the authors performed a global gene expression analysis to identify membrane proteins regulated by IL4. Transfection of human pancreatic and bronchial epithelial cells with specific small-interfering RNA against each of these proteins revealed that TMEM16A is associated with CaCC conductance, as measured with halide-sensitive fluorescent proteins, short-circuit current, and patch-clamp techniques. Further evidence showed that TMEM16A knockdown does not alter intracellular calcium signaling induced by calcium-mobilizing agents, indicating that TMEM16A is an intrinsic constituent of the CaCC.

Using a classical expression cloning approach, the Jan laboratory discovered that the TMEM16A orthologue in *Xenopus laevis* frogs are responsible for the endogenous CaCC currents in *Xenopus* oocytes,[5] which generate the fertilization potential to prevent fusion of multiple sperms to an egg, or polyspermia.[40] Noticing the fact that the oocytes of Axolotl salamander (*Ambystoma mexicanum*) are physiologically polyspermic, the authors found that Axolotl oocytes lack CaCC conductance. They then generated a cDNA library of *Xenopus* oocytes and injected mRNA fractions extracted from *Xenopus* oocytes into Axolotl oocytes. After many cycles of cDNA fractionation and functional screening in Axolotl oocytes, the authors narrowed down to the *Xenopus TMEM16A* gene, which gives rise to the classical CaCC currents observed in *Xenopus* oocytes. They further showed that heterologous expression of murine TMEM16A and TMEM16B in HEK-293 cells also generates CaCC currents, indicating that these two closely related paralogues in the TMEM16 family form the bona fide CaCCs.

Although the ion channel functions of other TMEM16 family members were being explored, Suzuki and colleagues reported a novel finding that TMEM16F can promote calcium-dependent scrambling of membrane phospholipids in 2010.[9] Phospholipids in the plasma membrane (PM) are asymmetrically distributed with phosphatidylcholine (PC) and sphingomyelin (SPH) predominantly in the outer leaflet, whereas phosphatidylethanolamine (PE) and phosphatidylserine (PS) are enriched in the inner leaflet (Fig. 7.1). This asymmetry is delicately maintained by the adenosine triphosphate (ATP)-dependent lipid transporters, flippases and floppases, which actively catalyze the *trans*-bilayer movement of PS/PE to the inner leaflet and PC/SPH to the outer leaflet, respectively.[1,41,42] Under certain physiological or pathological conditions, a third type of lipid transporting proteins called scramblases are activated in a calcium-dependent but energy-independent fashion. These scramblases rapidly mix the lipid species on both leaflets and collapse the lipid asymmetry, resulting in exposure of PS on the cell surface.[1,43,44] Through rounds of fluorescence-activated cell sorting to search for enhanced calcium-induced PS exposure in a B cell line, Suzuki and colleagues successfully established a subline (Ba/F3-PS19) that showed spontaneous PS exposure. Subsequent sequencing analysis demonstrated that a gain-of-function mutation (D409G) in TMEM16F is responsible for the PS exposure in this subline. Overexpression of TMEM16F in a B cell line greatly enhanced PS exposure whereas knockdown of TMEM16F attenuated PS exposure, further demonstrating that TMEM16F is critical for lipid scrambling. Consistent with this finding, loss-of-function mutations of TMEM16F were identified in patients with Scott syndrome,[9,45] an inherited bleeding disorder caused by defective scramblase activity in platelets.[46]

Interestingly, TMEM16F also forms a calcium-activated ion channel. TMEM16F has been proposed to form a volume-regulated anion channel,[47] a CaCC,[48] and an outwardly rectifying chloride channel.[49]

On the other hand, a comprehensive biophysical characterization of TMEM16F using inside-out patch clamp technique strongly suggests that TMEM16F forms a small conductance, calcium-activated non-selective cation (SCAN) channel.[7] The activation of TMEM16F channel requires synergistic action of both calcium and depolarization (Fig. 7.2B). Unlike TMEM16A-CaCC, which exhibits outward rectification at low internal calcium concentrations but has a linear current–voltage (I-V) relationship at higher calcium concentrations—a characteristic of classical CaCC—the TMEM16F currents show prominent outward rectification even at high calcium concentrations (Fig. 7.2B). On the basis of noise analysis, its single channel conductance was estimated to be approximately 0.5 pS,[7] several-fold smaller than that of the CaCCs (1–8 pS).[4,50,51] The most contentious aspect of the TMEM16F channels is their ion selectivity. On the basis of the measurement of reversal potential changes using whole-cell patch clamp of HEK-293 cells expressing TMEM16F, several groups reported that TMEM16F forms anion channels that have significant sodium conductance (relative permeation ratio $P_{Na}/P_{Cl} = \sim 0.3$).[52–55] In contrast, Yang et al. showed that TMEM16F channels have high cation conductance as evident in inside-out patches with the native TMEM16F in murine megakaryocytes and heterologously expressed TMEM16F in HEK-293 cells, *Xenopus* oocytes, and Axolotl oocytes ($P_{Na}/P_{Cl} = 7.0$ in HEK-293 cells).[7] To further prove that the TMEM16F channel prefers to permeate cations, the authors also identified a key residue (Q559) for ion selectivity in TM segment 5 (Fig. 7.3B). Swapping this residue between the TMEM16F channels and the TMEM16A-CaCC rendered the Q559K-TMEM16F channel more permeable to chloride and the K584Q-TMEM16A channel more permeable to sodium. As discussed later, the crystal structures of a fungal TMEM16 homologue (nhTMEM16) place the equivalent residue of Q559 in the center of the hydrophilic "subunit cavity" (Fig. 7.3B), where the putative ion conduction pore is proposed to be located,[11] further supporting the key role of this residue in determining the ion selectivity of the TMEM16 channels. The discrepancy about TMEM16F ion selectivity might be partially attributed to the differences between the inside-out and whole-cell patch recording conducted in different studies. In contrast to the inside-out patches where TMEM16F channels can be rapidly activated by elevating the internal calcium level and depolarizing the membrane potential,[7] the TMEM16F current under whole-cell patch clamp configuration cannot be observed until more than 10 min after cell break-in.[52] This long latency is much slower than the diffusion rate of calcium ions from patch pipette into the cell, suggesting that complex cellular events might have occurred before the TMEM16F channel can be activated. Although the exact reason for this latency is still unclear, the properties of TMEM16F channels, including ion

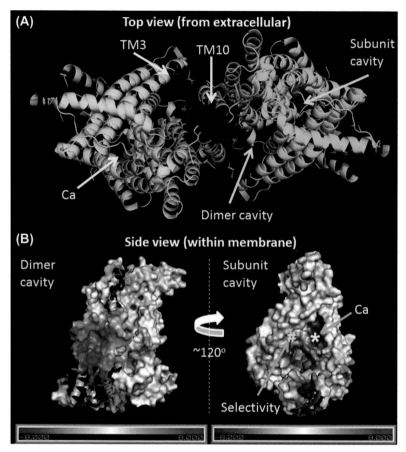

FIGURE 7.3 **The dimer cavity and the subunit cavity in TransMEMbrane protein 16 (TMEM16) proteins are important for their oligomerization and function.** (A) Top view of the fungal nhTMEM16 structure from the extracellular side. *Cyan arrows* point to the dimer cavity and the subunit cavity in the transmembrane domain. *Red color* labels the interaction interface of two TMEM16 subunits computed using the "InterfaceResidues" script in Pymol. Calcium ions are shown in *magenta*. (B) Side view of the dimer cavity and the subunit cavity from within the membrane. Another subunit in the dimer cavity was removed for clarity. Electrostatic surface potential representations of these two membrane cavities were calculated and plotted using PyMOL plugin "APBS" in ±8 kT/e scale.

selectivity, could have been modified during this long latency period. Moreover, it is well documented that HEK-293 cells have robustly expressed endogenous ion channels.[56] These endogenous currents, as well as membrane leak currents under the whole-cell configuration, may also confound the reversal potential measurement. Thus further characterization is needed to resolve the discrepancy regarding the ion selectivity of TMEM16F channels.

EVOLUTION AND FUNCTIONAL DIVERGENCE OF TMEM16 HOMOLOGUES

To better understand the TMEM16 proteins and their roles in human health and disease, it is important to gain insights from evolution and the functional divergence of TMEM16 proteins in different organisms. Comprehensive bioinformatics analysis of the TMEM16 gene family suggests that these are evolutionarily ancient proteins expressed in different eukaryotic organisms from fungi to invertebrates to mammals.[57–59] Increased sodium tolerance protein 2 (Ist2) is the only TMEM16 protein in the budding yeast *Saccharomyces cerevisiae* and shares only 10% sequence identity with TMEM16A CaCC (Fig. 7.2A). It was first identified through a mutant screen and was initially proposed to be a potential ion channel to regulate osmotic tolerance.[60] Recent studies discovered that, instead of being an ion channel on the PM, Ist2 resides in the membrane of the cortical endoplasmic reticulum (ER) and plays an important role in ER-PM tethering.[61,62] The electrostatic interaction between the highly basic C-terminus of Ist2 and the negatively charged phosphoinositides in the inner leaflet of the PM controls the formation of the ER-PM junction.[61,63] Deletion of Ist2 in yeast results in an increased distance between the ER and PM and accessibility of ribosomes to the space between the nucleus and the PM.[64] No calcium dependence of Ist2 function has been observed.[8]

Different from Ist2, two other fungal TMEM16F homologues clearly show calcium-dependent activity.[8,11] By purifying and reconstituting the *Aspergillus fumigatus* homologue afTMEM16 into lipid bilayer and liposomes, Malvezzi et al. elegantly demonstrated that afTMEM16 is a dual-function protein with both calcium-activated ion channel activity and calcium-activated lipid scrambling activity.[8] Similar to TMEM16F, the afTMEM16 channel is also gated by both calcium and voltage, with a minor cationic preference P_K/P_{Cl} of approximately 1.5. Its single-channel conductance is approximately 300 pS when reconstituted in artificial lipid bilayer, significantly larger than that of TMEM16A and TMEM16F channels (0.5–8 pS). In the presence of calcium, afTMEM16 can scramble lipids in liposomes at the rate of 10^4 lipids per second with a moderate selectivity of PC > PE > PS. Interestingly, its ion and lipid transport activity is controlled by a common calcium-binding site, which is highly conserved in the TMEM16 protein family (Fig. 7.3F). Mutating the two calcium-binding residues in TM7 greatly impairs both channel and lipid scramblase activity, further supporting that afTMEM16 is a dual-function calcium-activated membrane protein.

Most recently, the crystal structure and functional characterization of another fungal TMEM16F homologue (nhTMEM16) from *Nectria haematococca* have been reported.[11] Bearing 48% sequence identity with afTMEM16, nhTMEM16 also exhibits calcium-activated lipid scramblase activity in reconstituted liposomes. However, nhTMEM16 did not give rise to ion channel activity when reconstituted in lipid bilayer or expressed in mammalian cells. The crystal structure of nhTMEM16 and the insights it affords in our understanding about TMEM16 proteins will be discussed in detail in the following section.

The functions of TMEM16 proteins in invertebrates have also been explored. Two TMEM16 family members in *Caenorhabditis elegans*, ANOH-1 and ANOH-2, were recently characterized.[59] ANOH-1 is mainly expressed in intestinal cells and the cilia endings of amphid sensory neurons that detect external chemical and nociceptive cues. RNA interference silencing of *anoh-1* reduces the ability of worms to avoid high osmolarity without affecting amphid cilia development, chemotaxis, or withdrawal from noxious stimuli. ANOH-2 is expressed in mechanoreceptive neurons and the spermatheca, but loss of *anoh-2* does not result in a clear phenotype. There are five TMEM16 family members in *Drosophila melanogaster*. Axs is closely related to TMEM16H and TMEM16K,[59] is associated with the ER membrane, and regulates meiotic spindle assembly and chromosome segregation.[65] Mutation of Axs causes a high frequency of X chromosome nondisjunction during female meiosis.[66] It is unknown if Axs is associated with any channel activity or lipid scrambling activity. Another TMEM16 family member named Subdued, a *Drosophila* ortholog of mammalian TMEM16A and TMEM16B, displays characteristics of classic CaCCs, albeit with lower calcium sensitivity than the mammalian counterpart TMEM16A/16B.[39] Subdued channels are required for host defense[39] and mediate thermal nociception[67] in *Drosophila*.

In summary, the functions TMEM16 of proteins are diverse based on the current knowledge of the vertebrate TMEM16 family members and their homologues in lower organisms. A TMEM16 protein could be an intracellular membrane protein that regulates organelle interactions (yeast Ist2 and *Drosophila* Axs) whereas on the PM they can serve as an ion channel (TMEM16A and TMEM16B CaCCs), a lipid scramblase (fungal nhTMEM16), or both (TMEM16F and fungal afTMEM16) (Fig. 7.1), or they could possibly be an auxiliary subunit to augment the sodium sensitivity of the Slack sodium-activated potassium channel (TMEM16C).[68] These findings raise the possibility that the early ancestors of TMEM16 proteins might have been nonselective to their substrates, and under evolutionary pressure more and more strict selectivity was gradually acquired, resulting in the divergent TMEM16 family members with separate functional subtypes. As discussed in the following section, most of the TMEM16 proteins share the highly conserved calcium-binding residues (Fig. 7.2F), suggesting that these functionally divergent proteins might share a common calcium-dependent regulation mechanism that was acquired early during evolution. Interestingly, the functional divergence of TMEM16 proteins resembles similar diversities in the family of TM ATPases, the members of which can act as either electrogenic ion-pumping transporters (such as the sodium-potassium exchanger) or "nonelectrogenic" transporters to actively pump nonionic substrates, including phospholipids (such as P4-ATPase as lipid flippase).[69]

STRUCTURE AND FUNCTION OF THE CALCIUM-ACTIVATED TMEM16 PROTEINS

Oligomerization and Membrane Topology

Multiple lines of functional and structural evidence, including biochemistry, imaging, electrophysiology, and X-ray crystallography, have demonstrated that functional TMEM16 proteins are dimers.[11,70–72] According to the initial hydrophobicity analysis and the membrane accessibility of the human influenza hemagglutinin (HA) epitopes at various locations in TMEM16G, a TMEM16 protein was proposed to have eight TM segments with a "re-entry loop" between TM5 and TM6 and both the N- and C-terminus in the cytosol (Fig. 7.2C, Model A).[73] A thorough investigation conducted by the Hartzell laboratory using mutagenesis, epitope accessibility, and cysteine-scanning accessibility strongly suggests a revised topology model (Fig. 7.2C, Model B)[74] in which the N-terminal part of the re-entry loop transverses the membrane and forms an additional TM α-helix (TM6′). According to this revised model, E698 and E701, two acidic residues at the C-terminus of the re-entry loop, are located at the intracellular side and exposed to the cytosol, thereby nicely explaining the authors' observation that these two residues are critical for the calcium sensing of the TMEM16A-CaCC. As discussed in Section "Calcium-Binding Site and Activation Mechanism" two additional calcium-binding residues, E730 and D734, were recently discovered.[75] To allow the two pairs of acidic residues (E698/E701 and E730/D734) to coordinate calcium, the TM6 (in Model A) flanked by these four residues has to have both ends on the cytosolic side (Fig. 7.2C, Model C). Models B and C based on functional findings can now be further revised based on a crystal structure of the fungal homologue nhTMEM16.[11] In this structure, part of the re-entry loop indeed transverses the membrane and creates a new TM6, and the original TM6 in Model A corresponds to two TM α-helices (the newly minted TM7 and TM8) flanked by residues exposed to the cytosol (Fig. 7.2C, Model D and Fig. 7.2D). Preceding TM1, a short peptide forms an α-helical hairpin that is partially embedded into the inner leaflet of the membrane. This amphiphilic hairpin interacts with the cytosolic ends of TM1 and TM8 on its hydrophobic side. Therefore, according to the nhTMEM16 structure, the TM domain of TMEM16 proteins contains 10 TM α-helices, an arrangement that does not resemble any known membrane proteins.

The cytosolic N-terminal domain of nhTMEM16 consists of α-helices and β-strands that are organized in a ferredoxin-like fold, whereas the three long α-helices in the cytosolic C-terminus are wrapped around the N-terminal domain of the adjacent subunit, thereby constituting a large part of the dimer interface (Figs. 7.2D and 7.3A). In the TM region, the dimer interface is formed by residues in the central TM10 α-helices from both subunits at the extracellular side as well as residues of the neighboring TM3 and TM10 at their cytoplasmic ends (Fig. 7.3A). This arrangement thus creates the "dimer cavity," a large pore-like structure that contains two separate 15-Å wide V-shaped extracellular vestibules, which join a big vestibule that is approximately 30 Å wide within the cytosolic half of the membrane. The extracellular vestibule is predominantly composed of hydrophobic and aromatic residues presumably packed along with hydrophobic tails of lipids whereas the intracellular vestibule is hydrophilic with strong positive electrostatic potential, which could potentially interact with permeation ions during channel activation or the headgroups from lipid molecules during lipid scrambling (Fig. 7.3B). The exact role of this dimer cavity in TMEM16 protein functions requires further study.

On either side of the dimer cavity, a twisted 8- to 11-Å wide "subunit cavity" spanning almost the entire membrane is found in each subunit (Fig. 7.3A). The "spiral-staircase"-shaped subunit cavity has the TM3-TM7 TM α-helices on one side and presumably membrane lipids on the other side. The protein side of the subunit cavity, especially the intracellular half, is hydrophilic with strong negative electrostatic potentials (Fig. 7.3B). As discussed in Section "Calcium-Binding Site and Activation Mechanism", this cavity is in the vicinity of two highly conserved-calcium binding sites[11,74,75] and borders residues critical for ion permeation.[74–76] Because of its hydrophilicity, accessibility to the membrane lipids, and sufficient dimensions to accommodate a phospholipid headgroup, it has been proposed that the subunit cavity is a primary candidate for the translocation of lipid molecules, conceivably with cations accompanying the polar headgroups of lipids.[11] On the other hand, it is still puzzling how an aqueous pathway could be created in this "half membrane, half protein" subunit cavity to transport chloride ions in the TMEM16 ion channels. As Brunner et al. postulated,[11] it is possible that the two TMEM16 subunits might turn 180 degree so that two subunit cavities could form an enclosed, common aqueous pore to allow ions to go through. Thus structural and functional evidence are required to fully understand the mechanism of ion and lipid transportation in TMEM16 proteins.

Calcium-Binding Site and Activation Mechanism

Calcium-dependent activation is the central theme of the well-characterized TMEM16 proteins.[12] Nevertheless, the mechanism of calcium-dependent activation has been debated since the discovery of the TMEM16A-CaCC. Some laboratories believed that the calcium sensitivity of the TMEM16 channels is derived from a tightly associated calcium-sensing protein, calmodulin (CaM),[77,78] whereas others thought that CaM is dispensable for calcium activation.[11,72,74,79,80] Although a study strongly suggests that CaM might be important for the anion selectivity of TMEM16A-CaCC, recent functional studies indicate that CaM is not required for TMEM16 calcium sensing[75,80,81]: (1) barium, a divalent cation that cannot activate CaM, is capable of activating TMEM16A-CaCC; (2) manipulating CaM function with CaM antagonist and CaM antibody does not influence the calcium sensitivity of TMEM16A; and (3) disrupting the predicted CaM binding motifs in TMEM16A does not reduce its calcium sensitivity. Whereas CaM is not essential for calcium activation of CaCC, preassociation of calcium-free apoCaM with TMEM16A/16B channel complexes could infer calcium-dependent sensitization of activation or calcium-dependent inactivation of channels containing certain splice isoforms because of calcium binding to the N or the C lobes of CaM, respectively.[82] This CaM modulation may also depend on the cell type given that CaM is not found in the TMEM16A interactome in a recent study.[83] Recently, the Hartzell group elegantly demonstrated that two acidic residues, E698 and E701, are critical to the calcium sensing of TMEM16A.[74] In addition, the Jan laboratory recently systematically mutated all evolutionarily conserved potential intracellular acidic residues and identified three more (E650, E730, and D734 in addition to E698 and E701) as putative calcium-binding residues.[75] Further cysteine crosslinking experiments and functional analyses support the physical proximity of these acidic residues, potentially forming a calcium-binding pocket. This conclusion was further supported by the nhTMEM16 structure and additional functional characterization from the Dutzler laboratory.[11] In this structure, the carboxylate groups from these five highly conserved acidic residues, together with the side chain carboxyl group from a less conserved Asn residue (Fig. 7.2F) that corresponds to N646 in TMEM16A, form a calcium-binding pocket in the immediate vicinity of the subunit cavity where two closely packed calcium ions (4.9 Å apart) are coordinated (Fig. 7.2E) in the middle of the membrane. Mutations of these calcium-binding residues not only diminishes the calcium-dependent lipid scrambling activity of nhTMEM16 and afTMEM16[8,11] but also abolishes or significantly reduces the calcium sensitivity of ion channel activity in TMEM16A, TMEM16F, afTMEM16, and Subdued channels.[7,8,39,74,75] It is unclear if both calcium-binding sites are simultaneously occupied by calcium under physiological conditions. Nevertheless, mutating the side chains of these calcium coordinates indeed causes differential reduction in the calcium sensitivity of TMEM16A-CaCC,[75] implying that the calcium-binding residues may play slightly different roles in binding calcium. It is also conceivable that they could dynamically adjust their relative positions within the binding pocket to coordinate calcium in the wild-type and mutant channels. Interestingly, the calcium-binding sites are located in the middle of the membrane electrical field and only have accessibility to the intracellular aqueous solution (Fig. 7.2D). This novel location of the calcium-binding site is different from the cytosolic location of the calcium sensors in the large cytosolic "gating ring" domain of the large conductance BK-type potassium channels that also harbor voltage sensors similar to those of voltage-gated potassium channels.[84] Because all characterized TMEM16 channels exhibit voltage dependence (Fig. 7.2B), it remains to be established whether their voltage dependence is derived from the voltage-dependent binding of calcium ions inside of the membrane electrical field. Future studies are also needed to demonstrate how the bound calcium ions trigger the opening of the ion channel and lipid transport pathway. The crystal structure of the nhTMEM16 has laid a solid foundation for future exploration of the activation mechanism of TMEM16 proteins.

Substrate Selectivity and Transportation

The mechanism underlying the ability of TMEM16 proteins to recognize and transport different ionic and lipid substrates is still an open question. As discussed in Section "The Discoveries concerning the TMEM16 family", TMEM16 proteins are fairly nonselective (Fig. 7.1). The TMEM16 scramblases usually do not discriminate different lipid species.[8,9,11] Moreover, both TMEM16F and fungal afTMEM16 act as both lipid scramblases and nonselective ion channels.[7–9] Even for TMEM16A and TMEM16B that could be defined as "pure" CaCCs without lipid scrambling activity,[9] their ion permeation pores allow anions with various properties and even a small amount of sodium ions to go through.[4–6] Mutagenesis studies of the selectivity of the TMEM16 proteins have identified some key residues, which are concentrated in the subunit cavity according to the nhTMEM16 structure,[7,74] and additional basic residues in TM3-9 that surround the subunit cavity and are important for the anion selectivity of TMEM16A-CaCC.[76] For instance, a residue in the middle of TM5 pointing to the subunit cavity (Fig. 7.3B) is important in determining the anion and cation selectivity of the TMEM16 CaCC and SCAN channel selectivity. Swapping the Lys residue (K584) in TMEM16A-CaCC and the corresponding Gln residue (Q559) in TMEM16F-SCAN makes the TMEM16A-CaCC more permeable to sodium and the TMEM16F-SCAN more permeable to chloride. Because the ion selectivity was only partially switched by these single mutations, other residues must also contribute to their difference in ion selectivity.

Thus far, the key region for lipid selectivity has not been identified. Therefore it is unknown if the ion permeation shares the same pathway with the lipid transport for those TMEM16 proteins that have both functions. One possibility is that ions might co-transport with lipid headgroups, which may explain the small conductance (0.5 pS) of the TMEM16F-SCAN channel. However, it may be difficult to rely on this model to explain the fast ion conduction rate in afTMEM16 (~300 pS), the lipid scrambling activity of which seems to operate at a much slower rate (on the scale of tens to hundreds seconds). Thus identification of substrate pathway(s) and understanding the mechanism of substrate selectivity and ion permeation remain as intriguing open questions for future studies.

TMEM16 PROTEINS IN HEALTH AND DISEASE

Recent studies have begun to shed some light on the physiological and pathological roles of the 10 mammalian TMEM16 family members. This section will summarize the current understanding of the TMEM16 proteins on human health and disease (Table 7.1).

TMEM16A CaCCs in Health and Diseases

TMEM16A is by far the most well-studied TMEM16 protein in terms of its function and its roles in physiological and pathological conditions. TMEM16A has been found to be widely expressed in various cell types, including epithelial cells, smooth muscle cells (SMCs), interstitial cells of Cajal (ICC), and dorsal root ganglion (DRG) neurons.[12] The function of TMEM16A-CaCC is mainly attributed to its permeability to chloride ions in response to the elevation of intracellular calcium and membrane depolarization, as well as the driving force for chloride flux. Calcium elevation is a prerequisite for TMEM16A activation under physiological conditions. This requires the spatial proximity between the CaCC and various calcium sources, which include calcium entry through calcium-permeable channels or calcium release from internal calcium stores (Fig. 7.1). Depending on the electrochemical gradient of chloride in specific cell types, chloride could either enter or leave the cell, thereby altering intracellular chloride homeostasis and leading to changes of calcium signaling, chloride secretion, membrane potential, cell volume, and cell migration and proliferation.

It is well known that CaCCs provide one of the major chloride conductance responsible for chloride secretion in epithelial cells.[24] Owing to the coordinated activity of various channels and transporters, particularly the combination of the sodium-potassium ATPase and the sodium-potassium-chloride cotransporter (NKCC), epithelial cells accumulate high concentrations of chloride ions in the cytosol. Opening of CaCC in response to intracellular calcium elevation will result in chloride ion efflux/secretion. It has been proposed that TMEM16A is the major CaCC in various polarized epithelia. In 2008 the Galietta laboratory

TABLE 7.1 TransMEMbrane Protein 16 Proteins in Health and Disease

TMEM	Cytogenetic Location	Subcellular Localization	Molecular Function	Major Expression Tissue	Physiological Role	Phenotype in Knockout Mouse	Implication in Human Disease
16A	11q13.3	Plasma membrane	CaCC	Epithelial cells, ICC cells, smooth muscle cells, DRG neurons,	Chloride transport, mucus secretion, cell volume regulation, migration, membrane excitability, proliferation?	Tracheal malformation, impaired mucociliary clearance, block of gastrointestinal peristalsis	Upregulated/ amplified in head and neck squamous cell carcinomas, gastrointestinal stromal tumors, breast cancer; upregulated in asthma; partially compensates for defects of cystic fibrosis?
16B	12p13.31	Plasma membrane	CaCC	Hippocampal neurons, photoreceptors, olfactory receptors	Neuronal excitability, olfactory transduction, phototransduction	Absence of CaCC in olfactory sensory epithelium, no impairment in olfaction	Associated with panic disorder
16C	11p14.2	Intracellular organelles, plasma membrane?	Unclear, regulator of sodium-activated Slack channel, scramblase?	Central neurons, DRG neurons	Neuronal excitability	Increased mechano- and heat-induced pain sensation	Mutated in cervical dystonia; GWAS association with febrile seizures; associated with late-onset Alzheimer's disease
16D	12q23.1	Intracellular organelles	Unclear, scramblase?	Nervous system	Unclear	Unknown	Alzheimer's disease, schizophrenia, neuroticism?

Continued

TABLE 7.1   TransMEMbrane Protein 16 Proteins in Health and Disease—cont'd

TMEM	Cytogenetic Location	Subcellular Localization	Molecular Function	Major Expression Tissue	Physiological Role	Phenotype in Knockout Mouse	Implication in Human Disease
16E	11p14.3	ER	Unclear	Skeletal muscle, cardiac muscle, chondrocytes, osteoblasts	Muscle membrane repair? ER homeostasis?	Unknown	Mutated in gnathodyaphyseal dysplasia, limb-girdle muscular dystrophy, and distal Miyoshi myopathy
16F	12q12	Plasma membrane	Ca-activated scramblase, SCAN, CaCC, and ORCC?	Blood cell lineage, osteoblasts, muscles?	Lipid scrambling, calcium homeostasis? membrane excitability?	Impaired blood coagulation; skeletal abnormalities	Mutated in Scott syndrome with bleeding phenotype; ankylosing spondylitis
16G	2q37.3	Intracellular organelles, plasma membrane?	Unclear, scramblase?	Prostate epithelial cells	Unclear	Unknown	Prostate cancer?
16H	19p13.11	Intracellular organelles	Unknown	Unclear	Unclear	Unknown	Unknown
16J	11p15.5	Intracellular organelles	Unclear, scramblase? Inhibits TMEM16A?	Unclear	Unclear	Unknown	P53-induced gene 5 protein; susceptibility to tuberculosis
16K	3p22.1	Intracellular organelles	Unclear	Central neurons?	Unclear	Unknown	Mutated in cerebellar ataxia

CaCC, calcium-activated chloride channel; DRG, dorsal root ganglion cells; ER, endoplasmic reticulum; GWAS, genome-wide association studies; ICC, interstitial cells of Cajal; ORCC, outward rectifying chloride channel; SCAN, small conductance, calcium-activated nonselective cation channel; TMEM16, transMEMbrane protein 16.

discovered TMEM16A-CaCC through a global gene expression analysis of IL4-induced CaCC upregulation in bronchial epithelial cells.[6] Recent studies indicate that TMEM16A is specifically expressed in mucus-producing goblet cells with little expression in ciliated epithelia.[85,86] In fact, TMEM16A was found to have a low expression level even in the resting goblet cells, suggesting that the low level of TMEM16A conductance is sufficient to support calcium-dependent chloride secretion under resting conditions.[86] Only upon allergic and asthmatic inflammatory stimulation did the airway goblet cells exhibit hyperplasia/metaplasia and the TMEM16A-CaCC conductance upregulate, resulting in enhanced chloride and mucin secretion. Consistent with this model, potent TMEM16A-CaCC blockers significantly impair mucus secretion in IL-13-treated human airway epithelial cells.[85] In airway epithelial cells, CaCCs have been proposed to serve as an alternative route for chloride secretion to that provided by the cystic fibrosis transmembrane conductance regulator (CFTR) channel.[87] It appears that CFTR and the TMEM16A-CaCC have different expression patterns in the airway, with the former to be primarily expressed in the ciliated epithelial cells and the later in the goblet cells,[85,86] although it remains possible that one chloride channel could be upregulated in compensation for the loss-of-function of the other. Outside of the lung, TMEM16A is also expressed in the acinar cells of the pancreatic, mammary, and salivary glands[4,48,88,89]; the epithelial cells in the intestine, colon, biliary duct,[90,91] and kidney collecting duct; as well as the epithelium lining the cysts of autosomal polycystic kidney disease patients.[92]

TMEM16A is also highly expressed in the SMCs in the airway and reproductive tracts[89] as well as some blood vessels.[93] Because chloride ions prefer to exit SMCs because of the NKCC1 activity, opening of TMEM16A-CaCCs by either calcium entry through the voltage-gated calcium (Ca_V) channels or calcium release from ryanodine receptors will depolarize the membrane, which in turn induces more Ca_V channels to open. Through this positive feedback loop, TMEM16A can be actively involved in the regulation of SMC contraction; thus TMEM16A may represent a potential therapeutic target for treating asthma and hypertension.[94,95]

In the gastrointestinal tract, TMEM16A is absent from SMCs.[89] Instead, it is highly expressed in the ICCs, the pacemaker cells that generate rhythmic electrical slow waves in the SMCs via gap junction coupling, to control synchronized smooth muscle contractions that give rise to peristalsis. The physiological importance of TMEM16A in ICCs is underscored by the diminished slow waves and rhythmic contraction of gastric smooth muscle from TMEM16A knockout mice.[16,89,96,97] TMEM16A also plays an important role in regulating the electrical activity of the ICCs in the myosalpinx, the contractions of which are critical for oocyte transport along the oviduct.[98]

Endogenous CaCCs were observed in rat DRG neurons, where they contribute to the generation of afterdepolarization and may play important roles in somatosensation.[99] TMEM16A is the molecular basis for this CaCC conductance.[100,101] In small nociceptive DRGs, activation of phospholipase C by bradykinin induces intracellular calcium release to activate the TMEM16A-CaCC. An alternative activation mechanism of the DRG CaCC was also proposed.[100] According to this model, CaCC is not only downstream of a calcium-permeable TRPV1 channel, which is a capsaicin-sensitive heat sensor in the nociceptive DRGs, but also sensitive to heat. When TMEM16A is knocked down by RNA silencing or removed via conditional gene knockout in DRG neurons, the mice showed a marked decreased response to noxious heat. The localization of TMEM16A in these nociceptive DRGs and whether the Ca_V channels play any role acting on the TMEM16A remain to be explored, as well as the question about TMEM16A expression in non-nociceptive DRG neurons.

Whereas no inherited disease has been linked to TMEM16A thus far, there is upregulation or amplification of this gene in multiple cancers. The *TMEM16A* gene is located at the CCND1-EMS1 locus on human chromosome 11q13[26,102] that is frequently amplified in head and neck, parathyroid, bladder, and ovarian tumors; GISTs; esophageal squamous cell carcinoma; and breast cancers. Hence, TMEM16A has been named DOG1 (discovered on GIST 1) and TAOS2 (tumor amplified and overexpressed sequence 2) because of its remarkably high expression level in GISTs and oral squamous cell carcinomas, respectively. It remains an open question why and how TMEM16A is upregulated in tumors, although it seems likely to involve its effects on controlling cell proliferation and/or cell migration.

Interestingly, TMEM16A was recently discovered to be important in regulating and the maintaining primary cilium,[103] a sensory organelle that responds to mechanical and chemical stimuli in the environment and transduces the external signals to the cell's interior.[104] In this study, TMEM16A was found to be located in or near the primary cilia in certain types of epithelial cell lines. Pharmacological of CaCC function or knockdown of TMEM16A expression greatly shortens cilium length. Because the primary cilium is critical for the Wnt and Hedgehog signaling pathways, it is conceivable that TMEM16A-CaCC may regulate cell growth and differentiation during development and tumorigenesis by controlling the primary ciliogenesis. Future studies are needed to test this hypothesis, especially for the tumors with upregulated TMEM16 expression.

TMEM16B CaCC, Neuronal Excitability, and Sensory Transduction

Closely relative to TMEM16A, TMEM16B is another CaCC in the TMEM16 family. TMEM16B-CaCC exhibits characteristics similar to those of TMEM16A-CaCC, such as calcium and voltage sensitivity, halide ion permeability, and small permeability to sodium ($P_{Na}/P_{Cl} = 0.23$).[5,50,105] Compared with TMEM16A, TMEM16B is less sensitive to calcium, with an EC_{50} of approximately 3–5 µM, and it has a smaller single-channel conductance of 1.2 pS.

In vertebrate main olfactory epithelium (MOE), CaCC conductance has long been believed to serve as a major electrical amplifier for the initial olfactory stimulus in the cilia, accounting for up to 80–90% of the odorant-induced depolarizing current.[106] This depolarizing CaCC is activated by the opening of the sodium- and calcium-permeable cyclic nucleotide-gated (CNG) channels in response to the activation of odorant receptors in the cilia and the resultant elevation of intracellular cyclic adenosine monophosphate. Because of its high expression level in olfactory sensory neurons (OSNs) and its functional similarities to the endogenous CaCC, TMEM16B was proposed to be the main CaCC-forming protein in the MOE.[50,105] The CaCC current was completely absent in the MOE of a TMEM16B-deficient mouse model, strongly supporting this hypothesis.[107] Nevertheless, the electro-olfactograms showed that TMEM16B knockout mice only had an approximately 40% reduction in the electrical responses to odorants. In contrast to the CNG2 knockout animal, the TMEM16B knockout mice show no impairment in their ability to sense odorants, suggesting that CaCCs are dispensable for olfaction.[107] It remains to be seen whether there is any upregulation of some compensatory conductance in the MOE of TMEM16B knockout mice to maintain the normal olfactory sensation.

TMEM16B is also highly expressed in the vomeronasal sensory neurons (VSNs)[107,108] and the photoreceptors in the retina.[109] Activated by the mobilizing calcium and diacylglycerol signals downstream of phospholipase C activation, TMEM16B in the VSNs helps rodents to detect pheromones.[108] In photoreceptors, TMEM16B CaCC is associated with adaptor proteins PSD95, VELI3, and MPP4 at the ribbon synapses because of its interaction with the PDZ domains of PSD95 through its C-terminal PDZ binding motif.[109] Thus TMEM16B is the long sought-after CaCC in the photoreceptor synapses, and TMEM16B activation regulate synaptic transmission from photoreceptors to second-order neurons.

In addition to its expression in these sensory neurons, TMEM16B CaCC is also functionally expressed in hippocampal neurons. Huang and colleagues demonstrated that TMEM16B can control action potential firing in the somatodendritic region of hippocampal neurons and modulate the synaptic transmission between CA1 and CA3 neurons.[110] Small hairpin RNA knockdown and pharmacological blockade of TMEM16B significantly inhibits the endogenous CaCC current. In these neurons, TMEM16B is in close physical proximity to *N*-methyl D-aspartate receptors and voltage-gated calcium channels and can be directly activated by the calcium influx through these calcium-permeable channels. Because of the low intracellular chloride concentration in mature hippocampal neurons, influx of chloride through TMEM16B-CaCC hyperpolarizes the neuron and exerts an inhibitory effect on neuronal excitability.

Interestingly, TMEM16B is associated with a panic disorder in a recent genome-wide association study (GWAS) in the Japanese population.[111] Panic disorder is an anxiety disorder characterized by panic attacks and anticipatory anxiety.[112] A recent survey of putative anxiety-associated genes in panic disorder patients has identified a single-nucleotide polymorphism of TMEM16B that is strongly associated with bladder syndrome.[113] These patients with panic disorder are characterized by urological and bladder pain and possibly interstitial cystitis.[114] The exact pathophysiological function of TMEM16B in this disorder requires further investigation.

TMEM16F, Lipid Scrambling, and the Bleeding Disorder of Scott Syndrome

TMEM16F, a potential dual-function protein with ion channel and lipid scramblase activity, is widely expressed in different cell types, such as blood cells of hematopoietic linage (including platelets, erythrocytes, and immune cells), epithelial cells, muscles, neurons, osteoblasts, and chondrocytes.[48,115,116] In these cells, especially the blood cells, TMEM16F is mainly responsible for the lipid scrambling activity[7,9] that is activated by the elevation of intracellular calcium. Lipid scrambling results in the exposure of PS on the platelet cell surface. The exposed PS will create a platform on the platelet surface to recruit blood clotting factors factor Va, factor Xa, and prothrombin.[117,118] This protease cascade will convert prothrombin to thrombin, which will trigger the formation of blood clots. Consistent with its importance for blood coagulation, the loss-of-function mutations of TMEM16F[9,45] have been identified from patients with the bleeding disorder Scott syndrome,[119,120] whose blood cells exhibit diminished PS exposure. The TMEM16F knockout mice recapitulate the defective scramblase activity and the bleeding phenotype observed in the Scott syndrome patients.[7] In addition to these defects in hemostasis, knockout mice also exhibit significant protection from thrombosis, suggesting that TMEM16F could serve as a novel anticoagulant target to treat or prevent thrombotic disorders, such as stroke and heart attack.

The TMEM16F knockout mice also showed skeletal phenotype with reduced skeleton size and skeletal deformities,[115] a phenotype that was not reported in the Scott patients. These skeletal defects are mainly due to the diminished PS exposure in osteoblasts, which leads to decreased mineral deposition in skeletal tissues. Interestingly, a GWAS in a Chinese population identified TMEM16F as a new susceptibility gene for ankylosing spondylitis.[121] It is not known whether TMEM16F is directly involved in the defects of the skeletal cells or indirectly through its function in immune cells. The role of TMEM16F in bone development, immune cells, and pathogenesis of ankylosing spondylitis requires further investigation.

TMEM16F has also been implicated in cell migration, cell volume regulation, and apoptosis.[47,49] It is unknown whether its lipid scramblase activity, channel activity, or both is involved in these cellular processes. Thus one of the urgent tasks is to fully characterize the channel function and its relationship with the lipid scramblase activity of TMEM16F. Studies of the fungal TMEM16 proteins with channel/scramblase activity, aided by structural studies, will greatly facilitate the understanding of this mysterious protein.

TMEM16C, Sodium-Activated Slack Potassium Channel, and Neuronal Excitability

TMEM16C is mainly expressed in neuronal tissues from both the central and peripheral nervous system of rodents[68] and is highly expressed in human striatum, hippocampus, and cortex.[122] A microarray analysis of human and chimpanzee brains positions TMEM16C at a hub in the modules of coexpressed genes in the caudate nucleus,[123] and analyses of a high-density genomic variant suggest its association with late-onset Alzheimer's disease.[124] A recent study reveals that TMEM16C forms a complex with the sodium-activated Slack potassium (K_{Na}) channels in DRG neurons to enhance its sodium sensitivity.[68] DRG neurons from the TMEM16C-deficient rats have diminished Slack expression and display broadened action potentials and increased excitability, leading to increased thermal and mechanical pain sensitivity. TMEM16C also bears some lipid scrambling activity in an immortalized fetal thymocyte cell line derived from TMEM16F knockout mice.[10] Moreover, loss of TMEM16C function causes hyperexcitability of hippocampal neurons, likely because of deficient potassium channel activity.[125]

By linkage analysis combined with exome sequencing and confirmed by Sanger sequencing, a heterozygous mutation in the *TMEM16C* gene (R494W) was recently identified in affected members of a British family with autosomal-dominant dystonia-24 (DYT24).[122] A different heterozygous mutation (W490C) was subsequently identified in affected members of another family with autosomal-dominant cervical dystonia. High-throughput sequencing of this gene in 188 samples yielded four additional putative pathogenic variants. Most of the patients in these studies have adult-onset cervical dystonia, often with laryngeal involvement and tremor of the upper limbs. Patient fibroblasts with the W490C mutation show a defect in ER-related calcium handling. On the basis of this observation, Charlesworth et al. postulated that mutations in the TMEM16C gene may lead to abnormal striatal-neuron excitability, manifested as dystonia. Interestingly, a recent well-designed GWAS discovered that TMEM16C is one of the four common variants associated with febrile seizures.[125] Consistent with this finding, TMEM16C knockout rats not only have hyperexcitable hippocampal neurons but also a reduced proportion of temperature-sensitive neurons in slices from the thermoregulatory anterior hypothalamic nucleus. Together with the finding that the TMEM16C knockout rats exhibit hyperexcitability of nociceptive neurons and a decreased threshold for pain,[68] it is apparent that TMEM16C is important for neuronal excitability. Whether TMEM16C itself acts as an ion channel, lipid scramblase and/or regulates potassium channels to regulate neuronal excitability in different brain regions requires further investigation.

TMEM16E and Musculoskeletal Disorders

Among the TMEM16 family members, TMEM16E was the first to be associated with human inherent disease. In 2004 the missense mutations (C356R and C356G) of TMEM16E were identified at chromosome 11p14.3–15.1 in patients with.[35,126] GDD is a rare, familial, and autosomal-dominant inherited disorder characterized by florid osseous dysplasia in the jaw bones, bone fragility, and bowing associated with diaphyseal sclerosis in the limb bones.[127,128] Consistent with its important role in bone formation, TMEM16E is highly expressed in growth-plate chondrocytes and osteoblasts at sites of active bone turnover.[129]

TMEM16E is also expressed in somites and in developing skeletal and cardiac muscle, and it is upregulated during the skeletal muscle cell line C2Cl2 myogenic differentiation.[129] The physiological importance of TMEM16E in skeletal muscle is underscored by the discoveries of the loss-of-function recessive alleles in individuals with limb girdle muscular dystrophy (LGMD2L) and distal Miyoshi myopathy (MMD3).[130,131] These mutations include a splice site and base pair duplication, which result in premature stop codons, and two missense mutations, R758C and G231V. The LGMD2L phenotype is characterized by late-onset proximal scapular and pelvic girdle muscle weakness and asymmetrical muscle atrophy, whereas the MMD3 patients exhibit distal muscle weakness, of calf muscles in particular. Multifocal sarcolemmal lesions are associated with both phenotypes via electron microscopy studies.[130] The phenotypes resulting from the TMEM16E mutations are reminiscent of dysferlinopathies, in which a deficiency in dysferlin can cause both LGMD2B and Miyoshi myopathy (MMD1). Dysferlin is a key molecule of the sarcolemmal repair machinery and has been proposed to act as a fusogen in the formation of the patch membrane required for membrane resealing.[132] It is likely that TMEM16E also plays an important role in muscle membrane repair because TMEM16E expression was increased in the skeletal muscle of the dystrophin-deficient *mdx* mice,[129] a mouse model of Duchenne muscular dystrophy. In contrast to the lethal human pathology, the *mdx* mice are viable and can effectively regenerate affected muscle tissue.[133] The over-expression of TMEM16E might be one of the compensatory factors that rescue the defects induced by knocking out dysferlin.

Biochemical studies indicated that TMEM16E is an integral membrane glycoprotein that resides predominantly in the ER.[35,129] Overexpression of the GDD1 Cys356 mutations in COS-7 cells decreases cell adhesion and alters the cell morphology to a round shape without sign of apoptosis. It remains unknown whether TMEM16E is a CaCC[134] and its role on membrane repair. Within the TMEM16 family, TMEM16E is the closest relative of TMEM16F (Fig. 7.2A). Nonetheless, no lipid scrambling activity was observed in a TMEM16E-expressing cell line,[10] which was immortalized from the TMEM16F-deficient mice. Thus molecular characterization of TMEM16E is required to further understand its physiology and pathogenesis in musculoskeletal disorders.

TMEM16K and Autosomal-Recessive Cerebellar Ataxia

TMEM16K is another TMEM16 family member that directly links to human genetic disease. By whole-exome sequencing, a homozygous p.Leu510Arg mutation was first discovered in two siblings of a Dutch family who had autosomal-recessive spinocerebellar ataxia-10 (SCAR10).[135] This neurodegenerative disorder is characterized by onset in the teenage or young adult years of gait and limb ataxia, dysarthria, hyperreflexia, normal plantar reflex, lower motor neuron involvement, and downbeat nystagmus associated with marked cerebellar atrophy on brain imaging.[136] Subsequent genetic analysis identified more missense, in-frame exon skipping; in-frame exon deletion; and truncation TMEM16K mutations from different patient families with SCAR10[135,137–139] or autosomal-recessive cerebellar ataxia type 3.[140] TMEM16K is highly expressed in the cerebellum and the cerebral cortex in adult brains, consistent with the late onset of ataxia.[135] Considering the facts that alteration of calcium signaling in the cerebellar Purkinje cells (PCs) is one of the most important pathological mechanisms for cerebellar ataxia[141] and TMEM16K belongs to a calcium regulated protein family, it is reasonable to hypothesize that these potentially loss-of-function TMEM16K mutations in the ataxia patients may alter the neuronal excitability and calcium signaling of PCs. This hypothesis, as well as the exact molecular function of TMEM16K, needs to be further investigated.

Other TMEM16 Proteins in Health and Disease

TMEM16D, TMEM16G, TMEM16H, and TMEM16J are the least understood family members (Table 7.1). Sharing 39% sequence identity with

TMEM16A, TMEM16D was first identified by bioinformatic analysis using TMEM16A as a query.[26] TMEM16D transcripts are restricted to the murine nervous system,[48] and its function and subcellular localization are controversial. One study claimed that heterologously expressed TMEM16D cannot target to the PM of HEK-293 cells, thereby no CaCC activity was detected.[142] In contrast, another study found its surface expression and CaCC current using the same cell line.[54] More intriguingly, no CaCC activity was recorded when TMEM16D was heterologously expressed in an immortalized murine lymphocyte cell line devoid of TMEM16F expression.[10] Instead, expression of TMEM16D induced calcium-activated lipid scramblase activity. To date, no clear human genetic disorder has been identified in correlation with the *TMEM16D* gene, although it has some correlation with Alzheimer's disease, schizophrenia, and neuroticism in a few GWAS.[143–146]

TMEM16G, also known as NGEP, has been found in normal prostate and prostate cancer.[34,73] RNA analysis revealed two splice variants of TMEM16G mRNA: a short form encoding a soluble protein and a long form encoding a polytopic membrane protein. Overexpression of TMEM16G in the prostate cancer cell line LNCaP promotes the formation of large cell aggregates.[73] TMEM16G has been proposed as a CaCC[54] or a calcium-activated lipid scramblase[10] on the PM, whereas another study showed that TMEM16G is trapped in the cytosolic compartments with no cell surface expression.[142] The exact function of TMEM16G and its role in prostate need further investigation.

TMEM16H was first cloned from a size-fractionated human adult brain cDNA library.[147] Designated as KIAA1623, TMEM16H was found to be widely expressed in specific brain regions examined, with highest expression in the cerebellum and substantia nigra. By in silico analysis of expressed sequence tag (EST) sequences, Katoh and Katoh detected TMEM16H expression in embryonic stem cells, fetal brain, and neural tissues.[30] Nevertheless, subsequent heterologous expression studies identified neither CaCC activity[48] nor lipid scramblase activity.[10] With the lowest homology to TMEM16A, TMEM16H may reside in intracellular compartments[54] where its molecular function and physiological roles need to be identified in the future.

TMEM16J was first identified as the p53-induced gene five protein (TP53I5).[28] One copy of a p53-binding sequence was identified within intron 1 of the *TMEM16J* gene. TMEM16J mRNA is expressed in human colorectal, lung, and breast cancer, and any connection between the *TMEM16J* gene and the p53 transcription factor still remains to be established. Heterologous expression of TMEM16J in HEK-293 cells does not generate CaCC activity. Instead, its coexpression with TMEM16A inhibits TMEM16A CaCC activity.[48] Immunolocalization of transfected cells showed that TMEM16J is mainly expressed in intracellular compartments[54] whereas heterologous expression of TMEM16J in an immortalized murine lymphocyte cell line devoid of TMEM16F expression showed that TMEM16J may also behave similar to a calcium-dependent scramblase.[10] It is intriguing how the intracellular localized TMEM16J proteins might control the lipid scrambling activity on the PM. Interestingly, a recent genomic association study[148] discovered that a single-nucleotide polymorphism of *TMEM16J* (rs7111432) was found to associate with the susceptibility to both pulmonary tuberculosis and tuberculous meningitis in a Vietnamese Kinh population, implying its potential role in host immunity and infectious disease pathogenesis.

CONCLUSIONS

The functions of TMEM16 proteins are rather diverse based on the current knowledge of vertebrate TMEM16s and their homologues in lower organisms. A TMEM16 protein could be an intracellular membrane protein that regulates organelle interactions, or serves as an ion channel, a lipid scramblase, or both, on the PM. Because of the collective effort of many laboratories, the molecular and cellular characterization of the TMEM16 proteins has been emerging. The recent characterization of the evolutionarily conserved calcium-binding sites[11,74,75] has laid a solid foundation on understanding how TMEM16 proteins are activated under physiological and pathological conditions; the X-ray crystal structure of the fungal nhTMEM16[11] will surely accelerate the studies of TMEM16-mediated ion/lipid transport and the identification of novel chemical or biological modulators for the TMEM16 proteins.

TMEM16 proteins are important in human health and disease. The cellular and physiological functions of individual TMEM16 members are being dissected. The ion channel functions of TMEM16A and TMEM16B CaCCs have been shown to be important in regulating chloride secretion, cell volume, calcium signaling, neuronal excitability, smooth muscle contraction, primary ciliogenesis, and cell proliferation and migration. Although the channel function of TMEM16F in physiology is still unclear, its critical role on membrane lipid scrambling has been firmly established. The loss-of-function mutations identified in Scott syndrome patients cause reduced PS exposure on platelet surface,[9,45] resulting in a prolonged bleeding phenotype.[119] In addition to TMEM16F, mutations in TMEM16C, TMEM16E, and TMEM16K have also been associated with human inherited diseases ranging from skeletomuscular disorders to neurological disorders such as dystonia and ataxia, although the molecular and cellular functions of these TMEM16 proteins remain to be established. In addition, the expression level and properties of TMEM16 proteins can be altered under pathological conditions, such as during inflammation and cancer development. Future studies are needed to understand the underlying pathological mechanism of the TMEM16 proteins and their mutations.

The 10 TMEM16 family members are widely expressed throughout the human body. Their properties are likely to be further fine-tuned by heterodimerization between different family members, RNA splicing, various post-translational modifications, and possibly some yet unidentified auxiliary subunits. These modulatory mechanisms may help to explain the diverse CaCC properties observed in different native tissues as well as the lipid scrambling activity in different cell types. Future studies on the tissue-specific modulatory mechanisms will facilitate the understanding of the diverse roles of the TMEM16 proteins in health and disease, as well as the rational design of TMEM16- and cell-specific pharmacological interventions.

Acknowledgment

This work was supported by National Institutes of Health grants R01NS069229 (to Lily Jan) and K99NS086916 (to Huanghe Yang). Lily Jan is an investigator at Howard Hughes Medical Institute.

References

1. van Meer G. Dynamic transbilayer lipid asymmetry. *Cold Spring Harb Perspect Biol* 2011;**3**(5):a004671.
2. Almen MS, Nordstrom KJ, Fredriksson R, Schioth HB. Mapping the human membrane proteome: a majority of the human membrane proteins can be classified according to function and evolutionary origin. *BMC Biol* 2009;**7**:50.
3. Babcock JJ, Li M. Deorphanizing the human transmembrane genome: a landscape of uncharacterized membrane proteins. *Acta Pharmacol Sin* 2014;**35**(1):11–23.
4. Yang YD, Cho H, Koo JY, Tak MH, Cho Y, Shim WS, et al. TMEM16A confers receptor-activated calcium-dependent chloride conductance. *Nature* 2008;**455**(7217):1210–5.
5. Schroeder BC, Cheng T, Jan YN, Jan LY. Expression cloning of TMEM16A as a calcium-activated chloride channel subunit. *Cell* 2008;**134**(6):1019–29.
6. Caputo A, Caci E, Ferrera L, Pedemonte N, Barsanti C, Sondo E, et al. TMEM16A, a membrane protein associated with calcium-dependent chloride channel activity. *Science* 2008;**322**(5901):590–4.
7. Yang H, Kim A, David T, Palmer D, Jin T, Tien J, et al. TMEM16F forms a Ca2+-activated cation channel required for lipid scrambling in platelets during blood coagulation. *Cell* 2012;**151**(1):111–22.
8. Malvezzi M, Chalat M, Janjusevic R, Picollo A, Terashima H, Menon AK, et al. Ca2+-dependent phospholipid scrambling by a reconstituted TMEM16 ion channel. *Nat Commun* 2013;**4**:2367.
9. Suzuki J, Umeda M, Sims PJ, Nagata S. Calcium-dependent phospholipid scrambling by TMEM16F. *Nature* 2010;**468**(7325):834–8.
10. Suzuki J, Fujii T, Imao T, Ishihara K, Kuba H, Nagata S. Calcium-dependent phospholipid scramblase activity of TMEM16 protein family members. *J Biol Chem* 2013;**288**(19):13305–16.
11. Brunner JD, Lim NK, Schenck S, Duerst A, Dutzler R. X-ray structure of a calcium-activated TMEM16 lipid scramblase. *Nature* 2014;**516**(7530):207–12.
12. Pedemonte N, Galietta LJ. Structure and function of TMEM16 proteins (anoctamins). *Physiol Rev* 2014;**94**(2):419–59.

13. Duran C, Hartzell HC. Physiological roles and diseases of Tmem16/Anoctamin proteins: are they all chloride channels? *Acta Pharmacol Sin* 2011;**32**(6):685–92.
14. Ferrera L, Caputo A, Galietta LJ. TMEM16A protein: a new identity for Ca(2+)-dependent Cl(−) channels. *Physiology (Bethesda)* 2010;**25**(6):357–63.
15. Picollo A, Malvezzi M, Accardi A. TMEM16 proteins: unknown structure and confusing functions. *J Mol Biol* 2015;**427**(1):94–105.
16. Sanders KM, Zhu MH, Britton F, Koh SD, Ward SM. Anoctamins and gastrointestinal smooth muscle excitability. *Exp Physiol* 2012;**97**(2):200–6.
17. Kunzelmann K, Tian Y, Martins JR, Faria D, Kongsuphol P, Ousingsawat J, et al. Anoctamins. *Pflugers Arch* 2011;**462**(2):195–208.
18. Duran C, Thompson CH, Xiao Q, Hartzell HC. Chloride channels: often enigmatic, rarely predictable. *Annu Rev Physiol* 2010;**72**:95–121.
19. Kunzelmann K, Kongsuphol P, Chootip K, Toledo C, Martins JR, Almaca J, et al. Role of the Ca^{2+}-activated Cl$^-$ channels bestrophin and anoctamin in epithelial cells. *Biol Chem* 2011;**392**(1–2):125–34.
20. Jang Y, Oh U. Anoctamin 1 in secretory epithelia. *Cell Calcium* 2014;**55**(6):355–61.
21. Huang F, Wong X, Jan LY. International union of basic and clinical pharmacology. LXXXV: calcium-activated chloride channels. *Pharmacol Rev* 2012;**64**(1):1–15.
22. Berg J, Yang H, Jan LY. Ca2+-activated Cl− channels at a glance. *J Cell Sci* 2012;**125**(Pt 6):1367–71.
23. Wong X, Jan L. Ca-activated chloride channels. In: Zheng J, Trudeau MC, editors. *Handbook of ion channels*. CRC Press; 2015. p. 477–88.
24. Hartzell C, Putzier I, Arreola J. Calcium-activated chloride channels. *Annu Rev Physiol* 2005;**67**:719–58.
25. Scudieri P, Sondo E, Ferrera L, Galietta LJ. The anoctamin family: TMEM16A and TMEM16B as calcium-activated chloride channels. *Exp Physiol* 2012;**97**(2):177–83.
26. Katoh M. FLJ10261 gene, located within the CCND1-EMS1 locus on human chromosome 11q13, encodes the eight-transmembrane protein homologous to C12orf3, C11orf25 and FLJ34272 gene products. *Int J Oncol* 2003;**22**(6):1375–81.
27. Katoh M. GDD1 is identical to TMEM16E, a member of the TMEM16 family. *Am J Hum Genet* 2004;**75**(5):927–8. author reply 8–9.
28. Katoh M. Identification and characterization of human TP53I5 and mouse Tp53i5 genes in silico. *Int J Oncol* 2004;**25**(1):225–30.
29. Katoh M. Identification and characterization of TMEM16E and TMEM16F genes in silico. *Int J Oncol* 2004;**24**(5):1345–9.
30. Katoh M. Identification and characterization of TMEM16H gene in silico. *Int J Mol Med* 2005;**15**(2):353–8.
31. West RB, Corless CL, Chen X, Rubin BP, Subramanian S, Montgomery K, et al. The novel marker, DOG1, is expressed ubiquitously in gastrointestinal stromal tumors irrespective of KIT or PDGFRA mutation status. *Am J Pathol* 2004;**165**(1):107–13.
32. Huang X, Godfrey TE, Gooding WE, McCarty Jr KS, Gollin SM. Comprehensive genome and transcriptome analysis of the 11q13 amplicon in human oral cancer and synteny to the 7F5 amplicon in murine oral carcinoma. *Genes Chromosom Cancer* 2006;**45**(11):1058–69.
33. Kiessling A, Weigle B, Fuessel S, Ebner R, Meye A, Rieger MA, et al. D-TMPP: a novel androgen-regulated gene preferentially expressed in prostate and prostate cancer that is the first characterized member of an eukaryotic gene family. *Prostate* 2005;**64**(4):387–400.
34. Bera TK, Das S, Maeda H, Beers R, Wolfgang CD, Kumar V, et al. NGEP, a gene encoding a membrane protein detected only in prostate cancer and normal prostate. *Proc Natl Acad Sci USA* 2004;**101**(9):3059–64.
35. Tsutsumi S, Kamata N, Vokes TJ, Maruoka Y, Nakakuki K, Enomoto S, et al. The novel gene encoding a putative transmembrane protein is mutated in gnathodiaphyseal dysplasia (GDD). *Am J Hum Genet* 2004;**74**(6):1255–61.

36. Bader CR, Bertrand D, Schwartz EA. Voltage-activated and calcium-activated currents studied in solitary rod inner segments from the salamander retina. *J Physiol* 1982;**331**:253–84.

37. Barish ME. A transient calcium-dependent chloride current in the immature Xenopus oocyte. *J Physiol* 1983;**342**:309–25.

38. Miledi RA. calcium-dependent transient outward current in *Xenopus laevis* oocytes. *Proc R Soc Lond B Biol Sci* 1982;**215**(1201):491–7.

39. Wong XM, Younger S, Peters CJ, Jan YN, Jan LY. Subdued, a TMEM16 family Ca(2) (+)-activated Cl(−)channel in *Drosophila melanogaster* with an unexpected role in host defense. *eLife* 2013;**2**:e00862.

40. Cross NL. Initiation of the activation potential by an increase in intracellular calcium in eggs of the frog, *Rana pipiens. Dev Biol* 1981;**85**(2):380–4.

41. Pomorski T, Menon AK. Lipid flippases and their biological functions. *Cell Mol Life Sci* 2006;**63**(24):2908–21.

42. Contreras FX, Sanchez-Magraner L, Alonso A, Goni FM. Transbilayer (flip-flop) lipid motion and lipid scrambling in membranes. *FEBS Lett* 2010;**584**(9):1779–86.

43. Bevers EM, Williamson PL. Phospholipid scramblase: an update. *FEBS Lett* 2010;**584**(13):2724–30.

44. Sanyal S, Menon AK. Flipping lipids: why an' what's the reason for? *ACS Chem Biol* 2009;**4**(11):895–909.

45. Castoldi E, Collins PW, Williamson PL, Bevers EM. Compound heterozygosity for 2 novel TMEM16F mutations in a patient with Scott syndrome. *Blood* 2011;**117**(16):4399–400.

46. Satta N, Toti F, Fressinaud E, Meyer D, Freyssinet JM. Scott syndrome: an inherited defect of the procoagulant activity of platelets. *Platelets* 1997;**8**(2–3):117–24.

47. Almaca J, Tian Y, Aldehni F, Ousingsawat J, Kongsuphol P, Rock JR, et al. TMEM16 proteins produce volume-regulated chloride currents that are reduced in mice lacking TMEM16A. *J Biol Chem* 2009;**284**(42):28571–8.

48. Schreiber R, Uliyakina I, Kongsuphol P, Warth R, Mirza M, Martins JR, et al. Expression and function of epithelial anoctamins. *J Biol Chem* 2010;**285**(10):7838–45.

49. Martins JR, Faria D, Kongsuphol P, Reisch B, Schreiber R, Kunzelmann K. Anoctamin 6 is an essential component of the outwardly rectifying chloride channel. *Proc Natl Acad Sci USA* 2011;**108**(44):18168–72.

50. Pifferi S, Dibattista M, Menini A. TMEM16B induces chloride currents activated by calcium in mammalian cells. *Pflugers Arch* 2009;**458**(6):1023–38.

51. Piper AS, Large WA. Multiple conductance states of single Ca2+-activated Cl− channels in rabbit pulmonary artery smooth muscle cells. *J Physiol* 2003;**547**(Pt 1):181–96.

52. Grubb S, Poulsen KA, Juul CA, Kyed T, Klausen TK, Larsen EH, et al. TMEM16F (Anoctamin6), an anion channel of delayed Ca(2+) activation. *J Gen Physiol* 2013;**141**(5): 585–600.

53. Juul CA, Grubb S, Poulsen KA, Kyed T, Hashem N, Lambert IH, et al. Anoctamin 6 differs from VRAC and VSOAC but is involved in apoptosis and supports volume regulation in the presence of Ca2+. *Pflugers Arch* 2014;**466**(10):1899–910.

54. Tian Y, Schreiber R, Kunzelmann K. Anoctamins are a family of Ca2+-activated Cl− channels. *J Cell Sci* 2012;**125**(Pt 21):4991–8.

55. Kunzelmann K, Nilius B, Owsianik G, Schreiber R, Ousingsawat J, Sirianant L, et al. Molecular functions of anoctamin 6 (TMEM16F): a chloride channel, cation channel, or phospholipid scramblase? *Pflugers Arch* 2014;**466**(3):407–14.

56. Launay P, Fleig A, Perraud AL, Scharenberg AM, Penner R, Kinet JP. TRPM4 is a Ca^{2+}-activated nonselective cation channel mediating cell membrane depolarization. *Cell* 2002;**109**(3):397–407.

57. Milenkovic VM, Brockmann M, Stohr H, Weber BH, Strauss O. Evolution and functional divergence of the anoctamin family of membrane proteins. *BMC Evol Biol* 2010;**10**:319.

58. Galindo BE, Vacquier VD. Phylogeny of the TMEM16 protein family: some members are overexpressed in cancer. *Int J Mol Med* 2005;**16**(5):919–24.

59. Wang Y, Alam T, Hill-Harfe K, Lopez AJ, Leung CK, Iribarne D, et al. Phylogenetic, expression, and functional analyses of anoctamin homologs in *Caenorhabditis elegans. Am J Physiol Regul Integr Comp Physiol* 2013;**305**(11):R1376–89.

60. Entian KD, Schuster T, Hegemann JH, Becher D, Feldmann H, Guldener U, et al. Functional analysis of 150 deletion mutants in *Saccharomyces cerevisiae* by a systematic approach. *Mol Gen Genet* 1999;**262**(4–5):683–702.

61. Fischer MA, Temmerman K, Ercan E, Nickel W, Seedorf M. Binding of plasma membrane lipids recruits the yeast integral membrane protein Ist2 to the cortical ER. *Traffic* 2009;**10**(8):1084–97.

62. Ercan E, Momburg F, Engel U, Temmerman K, Nickel W, Seedorf M. A conserved, lipid-mediated sorting mechanism of yeast Ist2 and mammalian STIM proteins to the peripheral ER. *Traffic* 2009;**10**(12):1802–18.

63. Maass K, Fischer MA, Seiler M, Temmerman K, Nickel W, Seedorf M. A signal comprising a basic cluster and an amphipathic alpha-helix interacts with lipids and is required for the transport of Ist2 to the yeast cortical ER. *J Cell Sci* 2009;**122**(Pt 5):625–35.

64. Wolf W, Kilic A, Schrul B, Lorenz H, Schwappach B, Seedorf M. Yeast Ist2 recruits the endoplasmic reticulum to the plasma membrane and creates a ribosome-free membrane microcompartment. *PLoS One* 2012;**7**(7):e39703.

65. Kramer J, Hawley RS. The spindle-associated transmembrane protein Axs identifies a new family of transmembrane proteins in eukaryotes. *Cell Cycle* 2003;**2**(3):174–6.

66. Zitron AE, Hawley RS. The genetic analysis of distributive segregation in *Drosophila melanogaster*. I. Isolation and characterization of Aberrant X segregation (Axs), a mutation defective in chromosome partner choice. *Genetics* 1989;**122**(4):801–21.

67. Jang W, Kim JY, Cui S, Jo J, Lee BC, Lee Y, et al. The anoctamin family channel subdued mediates thermal nociception in Drosophila. *J Biol Chem* 2015;**290**(4):2521–8.

68. Huang F, Wang X, Ostertag EM, Nuwal T, Huang B, Jan YN, et al. TMEM16C facilitates Na(+)-activated K+ currents in rat sensory neurons and regulates pain processing. *Nat Neurosci* 2013;**16**(9):1284–90.

69. Lopez-Marques RL, Poulsen LR, Bailly A, Geisler M, Pomorski TG, Palmgren MG. Structure and mechanism of ATP-dependent phospholipid transporters. *Biochim Biophys Acta* 2015;**1850**(3):461–75.

70. Fallah G, Romer T, Detro-Dassen S, Braam U, Markwardt F, Schmalzing G. TMEM16A(a)/anoctamin-1 shares a homodimeric architecture with CLC chloride channels. *Mol Cell Proteomics* 2011;**10**(2). M110 004697.

71. Sheridan JT, Worthington EN, Yu K, Gabriel SE, Hartzell HC, Tarran R. Characterization of the oligomeric structure of the Ca(2+)-activated Cl– channel Ano1/TMEM16A. *J Biol Chem* 2011;**286**(2):1381–8.

72. Tien J, Lee HY, Minor Jr DL, Jan YN, Jan LY. Identification of a dimerization domain in the TMEM16A calcium-activated chloride channel (CaCC). *Proc Natl Acad Sci USA* 2013;**110**(16):6352–7.

73. Das S, Hahn Y, Nagata S, Willingham MC, Bera TK, Lee B, et al. NGEP, a prostate-specific plasma membrane protein that promotes the association of LNCaP cells. *Cancer Res* 2007;**67**(4):1594–601.

74. Yu K, Duran C, Qu Z, Cui YY, Hartzell HC. Explaining calcium-dependent gating of anoctamin-1 chloride channels requires a revised topology. *Circ Res* 2012;**110**(7):990–9.

75. Tien J, Peters CJ, Wong XM, Cheng T, Jan YN, Jan LY, et al. A comprehensive search for calcium binding sites critical for TMEM16A calcium-activated chloride channel activity. *eLife* 2014;**3**.

76. Peters CJ, Yu H, Tien J, Jan YN, Li M, Jan LY. Four basic residues critical for the ion selectivity and pore blocker sensitivity of TMEM16A calcium-activated chloride channels. *Proc Natl Acad Sci USA* 2015;**112**(11):3547–52.

77. Tian Y, Kongsuphol P, Hug M, Ousingsawat J, Witzgall R, Schreiber R, et al. Calmodulin-dependent activation of the epithelial calcium-dependent chloride channel TMEM16A. *FASEB J* 2011;**25**(3):1058–68.

78. Vocke K, Dauner K, Hahn A, Ulbrich A, Broecker J, Keller S, et al. Calmodulin-dependent activation and inactivation of anoctamin calcium-gated chloride channels. *J Gen Physiol* 2013;**142**(4):381–404.

79. Terashima H, Picollo A, Accardi A. Purified TMEM16A is sufficient to form Ca2+-activated Cl– channels. *Proc Natl Acad Sci USA* 2013;**110**(48):19354–9.

80. Yu K, Zhu J, Qu Z, Cui YY, Hartzell HC. Activation of the Ano1 (TMEM16A) chloride channel by calcium is not mediated by calmodulin. *J Gen Physiol* 2014;**143**(2):253–67.

81. Xiao Q, Yu K, Perez-Cornejo P, Cui Y, Arreola J, Hartzell HC. Voltage- and calcium-dependent gating of TMEM16A/Ano1 chloride channels are physically coupled by the first intracellular loop. *Proc Natl Acad Sci USA* 2011;**108**(21):8891–6.

82. Yang T, Hendrickson WA, Colecraft HM. Preassociated apocalmodulin mediates Ca2+-dependent sensitization of activation and inactivation of TMEM16A/16B Ca2+-gated Cl– channels. *Proc Natl Acad Sci USA* 2014;**111**(51):18213–8.

83. Bill A, Gutierrez A, Kulkarni S, Kemp C, Bonenfant D, Voshol H, et al. ANO1 interacts with EGFR and correlates with sensitivity to EGFR-targeting therapy in head and neck cancer. *Oncotarget* 2015;**6**.

84. Yang H, Zhang G, Cui J. BK channels: multiple sensors, one activation gate. *Front Physiol* 2015;**6**:29.

85. Huang F, Zhang H, Wu M, Yang H, Kudo M, Peters CJ, et al. Calcium-activated chloride channel TMEM16A modulates mucin secretion and airway smooth muscle contraction. *Proc Natl Acad Sci USA* 2012;**109**(40):16354–9.

86. Scudieri P, Caci E, Bruno S, Ferrera L, Schiavon M, Sondo E, et al. Association of TMEM16A chloride channel overexpression with airway goblet cell metaplasia. *J Physiol* 2012;**590**(Pt 23):6141–55.

87. Tarran R, Loewen ME, Paradiso AM, Olsen JC, Gray MA, Argent BE, et al. Regulation of murine airway surface liquid volume by CFTR and Ca2+-activated Cl– conductances. *J Gen Physiol* 2002;**120**(3):407–18.

88. Romanenko VG, Catalan MA, Brown DA, Putzier I, Hartzell HC, Marmorstein AD, et al. Tmem16A encodes the Ca^{2+}-activated Cl$^-$ channel in mouse submandibular salivary gland acinar cells. *J Biol Chem* 2010;**285**(17):12990–3001.

89. Huang F, Rock JR, Harfe BD, Cheng T, Huang X, Jan YN, et al. Studies on expression and function of the TMEM16A calcium-activated chloride channel. *Proc Natl Acad Sci USA* 2009;**106**(50):21413–8.

90. Dutta AK, Khimji AK, Kresge C, Bugde A, Dougherty M, Esser V, et al. Identification and functional characterization of TMEM16A, a Ca2+-activated Cl– channel activated by extracellular nucleotides, in biliary epithelium. *J Biol Chem* 2011;**286**(1):766–76.

91. Ousingsawat J, Martins JR, Schreiber R, Rock JR, Harfe BD, Kunzelmann K. Loss of TMEM16A causes a defect in epithelial Ca^{2+}-dependent chloride transport. *J Biol Chem* 2009;**284**(42):28698–703.

92. Buchholz B, Faria D, Schley G, Schreiber R, Eckardt KU, Kunzelmann K. Anoctamin 1 induces calcium-activated chloride secretion and proliferation of renal cyst-forming epithelial cells. *Kidney Int* 2014;**85**(5):1058–67.

93. Davis AJ, Forrest AS, Jepps TA, Valencik ML, Wiwchar M, Singer CA, et al. Expression profile and protein translation of TMEM16A in murine smooth muscle. *Am J Physiol Cell Physiol* 2010;**299**(5):C948–59.

94. Zhang CH, Li Y, Zhao W, Lifshitz LM, Li H, Harfe BD, et al. The transmembrane protein 16A Ca(2+)-activated Cl– channel in airway smooth muscle contributes to airway hyperresponsiveness. *Am J Respir Crit Care Med* 2013;**187**(4):374–81.

95. Wang M, Yang H, Zheng LY, Zhang Z, Tang YB, Wang GL, et al. Downregulation of TMEM16A calcium-activated chloride channel contributes to cerebrovascular remodeling during hypertension by promoting basilar smooth muscle cell proliferation. *Circulation* 2012;**125**(5):697–707.

96. Hwang SJ, Blair PJ, Britton FC, O'Driscoll KE, Hennig G, Bayguinov YR, et al. Expression of anoctamin 1/TMEM16A by interstitial cells of Cajal is fundamental for slow wave activity in gastrointestinal muscles. *J Physiol* 2009;**587**(Pt 20):4887–904.
97. Gomez-Pinilla PJ, Gibbons SJ, Bardsley MR, Lorincz A, Pozo MJ, Pasricha PJ, et al. Ano1 is a selective marker of interstitial cells of Cajal in the human and mouse gastrointestinal tract. *Am J Physiol Gastrointest Liver Physiol* 2009;**296**(6):G1370–81.
98. Dixon RE, Hennig GW, Baker SA, Britton FC, Harfe BD, Rock JR, et al. Electrical slow waves in the mouse oviduct are dependent upon a calcium activated chloride conductance encoded by Tmem16a. *Biol Reprod* 2012;**86**(1):1–7.
99. Mayer ML. A calcium-activated chloride current generates the after-depolarization of rat sensory neurones in culture. *J Physiol* 1985;**364**:217–39.
100. Cho H, Yang YD, Lee J, Lee B, Kim T, Jang Y, et al. The calcium-activated chloride channel anoctamin 1 acts as a heat sensor in nociceptive neurons. *Nat Neurosci* 2012;**15**(7):1015–21.
101. Liu B, Linley JE, Du X, Zhang X, Ooi L, Zhang H, et al. The acute nociceptive signals induced by bradykinin in rat sensory neurons are mediated by inhibition of M-type K+ channels and activation of Ca2+-activated Cl– channels. *J Clin Invest* 2010;**120**(4):1240–52.
102. Wilkerson PM, Reis-Filho JS. The 11q13-q14 amplicon: clinicopathological correlations and potential drivers. *Genes Chromosom Cancer* 2013;**52**(4):333–55.
103. Ruppersburg CC, Hartzell HC. The Ca2+-activated Cl– channel ANO1/TMEM16A regulates primary ciliogenesis. *Mol Biol Cell* 2014;**25**(11):1793–807.
104. Gerdes JM, Davis EE, Katsanis N. The vertebrate primary cilium in development, homeostasis, and disease. *Cell* 2009;**137**(1):32–45.
105. Stephan AB, Shum EY, Hirsh S, Cygnar KD, Reisert J, Zhao H. ANO2 is the cilial calcium-activated chloride channel that may mediate olfactory amplification. *Proc Natl Acad Sci USA* 2009;**106**(28):11776–81.
106. Kleene SJ. The electrochemical basis of odor transduction in vertebrate olfactory cilia. *Chem Senses* 2008;**33**(9):839–59.
107. Billig GM, Pal B, Fidzinski P, Jentsch TJ. Ca^{2+}-activated Cl^- currents are dispensable for olfaction. *Nat Neurosci* 2011;**14**(6):763–9.
108. Yang C, Delay RJ. Calcium-activated chloride current amplifies the response to urine in mouse vomeronasal sensory neurons. *J Gen Physiol* 2010;**135**(1):3–13.
109. Stohr H, Heisig JB, Benz PM, Schoberl S, Milenkovic VM, Strauss O, et al. TMEM16B, a novel protein with calcium-dependent chloride channel activity, associates with a presynaptic protein complex in photoreceptor terminals. *J Neurosci* 2009;**29**(21): 6809–18.
110. Huang WC, Xiao S, Huang F, Harfe BD, Jan YN, Jan LY. Calcium-activated chloride channels (CaCCs) regulate action potential and synaptic response in hippocampal neurons. *Neuron* 2012;**74**(1):179–92.
111. Otowa T, Yoshida E, Sugaya N, Yasuda S, Nishimura Y, Inoue K, et al. Genome-wide association study of panic disorder in the Japanese population. *J Hum Genet* 2009;**54**(2):122–6.
112. Katon WJ. Clinical practice. Panic disorder. *N Engl J Med* 2006;**354**(22):2360–7.
113. Subaran RL, Talati A, Hamilton SP, Adams P, Weissman MM, Fyer AJ, et al. A survey of putative anxiety-associated genes in panic disorder patients with and without bladder symptoms. *Psychiatr Genet* 2012;**22**(6):271–8.
114. Chung KH, Liu SP, Lin HC, Chung SD. Bladder pain syndrome/interstitial cystitis is associated with anxiety disorder. *Neurourol Urodyn* 2014;**33**(1):101–5.
115. Ehlen HW, Chinenkova M, Moser M, Munter HM, Krause Y, Gross S, et al. Inactivation of anoctamin-6/Tmem16f, a regulator of phosphatidylserine scrambling in osteoblasts, leads to decreased mineral deposition in skeletal tissues. *J Bone Min Res* 2013;**28**(2):246–59.

116. Zhao P, Torcaso A, Mariano A, Xu L, Mohsin S, Zhao L, et al. Anoctamin 6 regulates C2C12 myoblast proliferation. *PLoS One* 2014;**9**(3):e92749.

117. Bevers EM, Comfurius P, Zwaal RF. Changes in membrane phospholipid distribution during platelet activation. *Biochim Biophys Acta* 1983;**736**(1):57–66.

118. Rosing J, van Rijn JL, Bevers EM, van Dieijen G, Comfurius P, Zwaal RF. The role of activated human platelets in prothrombin and factor X activation. *Blood* 1985;**65**(2):319–32.

119. Zwaal RF, Comfurius P, Bevers EM. Scott syndrome, a bleeding disorder caused by defective scrambling of membrane phospholipids. *Biochim Biophys Acta* 2004;**1636**(2–3):119–28.

120. Weiss HJ, Vicic WJ, Lages BA, Rogers J. Isolated deficiency of platelet procoagulant activity. *Am J Med* 1979;**67**(2):206–13.

121. Lin Z, Bei JX, Shen M, Li Q, Liao Z, Zhang Y, et al. A genome-wide association study in Han Chinese identifies new susceptibility loci for ankylosing spondylitis. *Nat Genet* 2012;**44**(1):73–7.

122. Charlesworth G, Plagnol V, Holmstrom KM, Bras J, Sheerin UM, Preza E, et al. Mutations in ANO3 cause dominant craniocervical dystonia: ion channel implicated in pathogenesis. *Am J Hum Genet* 2012;**91**(6):1041–50.

123. Oldham MC, Horvath S, Geschwind DH. Conservation and evolution of gene coexpression networks in human and chimpanzee brains. *Proc Natl Acad Sci USA* 2006;**103**(47):17973–8.

124. Briones N, Dinu V. Data mining of high density genomic variant data for prediction of Alzheimer's disease risk. *BMC Med Genet* 2012;**13**:7.

125. Feenstra B, Pasternak B, Geller F, Carstensen L, Wang T, Huang F, et al. Common variants associated with general and MMR vaccine-related febrile seizures. *Nat Genet* 2014;**46**(12):1274–82.

126. Tsutsumi S, Kamata N, Maruoka Y, Ando M, Tezuka O, Enomoto S, et al. Autosomal dominant gnathodiaphyseal dysplasia maps to chromosome 11p14.3-15.1. *J Bone Min Res* 2003;**18**(3):413–8.

127. Akasaka Y, Nakajima T, Koyama K, Furuya K, Mitsuka Y. Familial cases of a new systemic bone disease, hereditary gnatho-diaphyseal sclerosis. *Nihon Seikeigeka Gakkai Zasshi* 1969;**43**(5):381–94.

128. Riminucci M, Collins MT, Corsi A, Boyde A, Murphey MD, Wientroub S, et al. Gnathodiaphyseal dysplasia: a syndrome of fibro-osseous lesions of jawbones, bone fragility, and long bone bowing. *J Bone Min Res* 2001;**16**(9):1710–8.

129. Mizuta K, Tsutsumi S, Inoue H, Sakamoto Y, Miyatake K, Miyawaki K, et al. Molecular characterization of GDD1/TMEM16E, the gene product responsible for autosomal dominant gnathodiaphyseal dysplasia. *Biochem Biophys Res Commun* 2007;**357**(1):126–32.

130. Bolduc V, Marlow G, Boycott KM, Saleki K, Inoue H, Kroon J, et al. Recessive mutations in the putative calcium-activated chloride channel Anoctamin 5 cause proximal LGMD2L and distal MMD3 muscular dystrophies. *Am J Hum Genet* 2010;**86**(2):213–21.

131. Hicks D, Sarkozy A, Muelas N, Koehler K, Huebner A, Hudson G, et al. A founder mutation in Anoctamin 5 is a major cause of limb-girdle muscular dystrophy. *Brain* 2011;**134**(Pt 1):171–82.

132. Han R, Campbell KP. Dysferlin and muscle membrane repair. *Curr Opin Cell Biol* 2007;**19**(4):409–16.

133. Turk R, Sterrenburg E, de Meijer EJ, van Ommen GJ, den Dunnen JT, Hoen PA. Muscle regeneration in dystrophin-deficient mdx mice studied by gene expression profiling. *BMC Genomics* 2005;**6**:98.

134. Tran TT, Tobiume K, Hirono C, Fujimoto S, Mizuta K, Kubozono K, et al. TMEM16E (GDD1) exhibits protein instability and distinct characteristics in chloride channel/pore forming ability. *J Cell Physiol* 2014;**229**(2):181–90.

135. Vermeer S, Hoischen A, Meijer RP, Gilissen C, Neveling K, Wieskamp N, et al. Targeted next-generation sequencing of a 12.5 Mb homozygous region reveals ANO10 mutations in patients with autosomal-recessive cerebellar ataxia. *Am J Hum Genet* 2010;**87**(6):813–9.
136. Anheim M, Tranchant C, Koenig M. The autosomal recessive cerebellar ataxias. *N Engl J Med* 2012;**366**(7):636–46.
137. Balreira A, Boczonadi V, Barca E, Pyle A, Bansagi B, Appleton M, et al. ANO10 mutations cause ataxia and coenzyme Q(1)(0) deficiency. *J Neurol* 2014;**261**(11):2192–8.
138. Chamova T, Florez L, Guergueltcheva V, Raycheva M, Kaneva R, Lochmuller H, et al. ANO10 c.1150_1151del is a founder mutation causing autosomal recessive cerebellar ataxia in Roma/Gypsies. *J Neurol* 2012;**259**(5):906–11.
139. Maruyama H, Morino H, Miyamoto R, Murakami N, Hamano T, Kawakami H. Exome sequencing reveals a novel ANO10 mutation in a Japanese patient with autosomal recessive spinocerebellar ataxia. *Clin Genet* 2014;**85**(3):296–7.
140. Renaud M, Anheim M, Kamsteeg EJ, Mallaret M, Mochel F, Vermeer S, et al. Autosomal recessive cerebellar ataxia type 3 due to ANO10 mutations: delineation and genotype-phenotype correlation study. *JAMA Neurol* 2014;**71**(10):1305–10.
141. Kasumu A, Bezprozvanny I. Deranged calcium signaling in Purkinje cells and pathogenesis in spinocerebellar ataxia 2 (SCA2) and other ataxias. *Cerebellum* 2012;**11**(3):630–9.
142. Duran C, Qu Z, Osunkoya AO, Cui Y, Hartzell HC. ANOs 3-7 in the anoctamin/Tmem16 Cl– channel family are intracellular proteins. *Am J Physiol Cell Physiol* 2012;**302**(3):C482–93.
143. Sherva R, Tripodis Y, Bennett DA, Chibnik LB, Crane PK, de Jager PL, et al. Genome-wide association study of the rate of cognitive decline in Alzheimer's disease. *Alzheimers Dement* 2014;**10**(1):45–52.
144. Athanasiu L, Mattingsdal M, Kahler AK, Brown A, Gustafsson O, Agartz I, et al. Gene variants associated with schizophrenia in a Norwegian genome-wide study are replicated in a large European cohort. *J Psychiatr Res* 2010;**44**(12):748–53.
145. Webb BT, Guo AY, Maher BS, Zhao Z, van den Oord EJ, Kendler KS, et al. Meta-analyses of genome-wide linkage scans of anxiety-related phenotypes. *Eur J Hum Genet* 2012;**20**(10):1078–84.
146. Terracciano A, Sanna S, Uda M, Deiana B, Usala G, Busonero F, et al. Genome-wide association scan for five major dimensions of personality. *Mol Psychiatry* 2010;**15**(6):647–56.
147. Nagase T, Kikuno R, Nakayama M, Hirosawa M, Ohara O. Prediction of the coding sequences of unidentified human genes. XVIII. The complete sequences of 100 new cDNA clones from brain which code for large proteins in vitro. *DNA Res* 2000;**7**(4):273–81.
148. Horne DJ, Randhawa AK, Chau TT, Bang ND, Yen NT, Farrar JJ, et al. Common polymorphisms in the PKP3-SIGIRR-TMEM16J gene region are associated with susceptibility to tuberculosis. *J Infect Dis* 2012;**205**(4):586–94.

K_{ATP} Channels in the Pancreas: Hyperinsulinism and Diabetes

M.S. Remedi, C.G. Nichols

Washington University School of Medicine, St. Louis, MO, United States

ATP-sensitive potassium (K_{ATP}) channels are present in the surface membranes of many organs and cell types. Significant advances have been made in understanding the molecular basis of K_{ATP} channel activity, as well as their role in physiology and pathophysiology. Pancreatic K_{ATP} channels are heterooctameric complexes of four pore-forming Kir6.2 subunits and four regulatory SUR1 subunits (encoded by KCNJ11 and ABCC8 genes, respectively). K_{ATP} channels are inhibited by intracellular ATP, and activated by ADP. In the pancreatic β-cell, K_{ATP} channels play a critical role in coupling glucose metabolism to insulin secretion. Normally, glucose oxidation leads to a rise in [ATP]:[ADP] ratio, which reduces K_{ATP} channel activity and causes membrane depolarization. This leads to subsequent opening of voltage-dependent Ca^{2+} channels, and increase in intracellular [Ca^{2+}], which in turn promotes insulin vesicles to fuse to the plasma membrane and release insulin. Sulfonylureas (SUs), oral hypoglycemic agents widely used in the treatment of diabetes, act by binding to the SUR1 subunit thus inhibiting K_{ATP} current and inducing insulin secretion, independently of the metabolic state of the cell. Conversely, K_{ATP}-specific channel openers (ie, diazoxide) suppress insulin release by activating K_{ATP} and preventing depolarization-dependent rise in intracellular [Ca^{2+}]. Loss-of-function mutations in the KCNJ11 and ABCC8 genes underlie hyperinsulinism that can in some cases be treated by potassium channel openers. Conversely, gain-of-function mutations are the main cause of human neonatal diabetes mellitus which can now be treated by SUs. In addition, a common Kir6.2variant (E23K) is highly associated as a risk factor for development of type-2 diabetes. This chapter will focus on advances in K$_{ATP}$ channel biochemistry and physiology, as well as on the mechanistic basis by which K$_{ATP}$ mutations underlie these insulin secretory disorders.$_{ATPATP}$

MOLECULAR BASIS AND STRUCTURE OF THE K$_{ATP}$ CHANNEL

ATP-sensitive K$^+$ (K$_{ATP}$) channels are present in the membranes of many organs and cell types and play a critical role in linking metabolism to electrical activity of the cell.[1] These K$_{ATP}$ channels are heterooctameric complexes of two distinct proteins: sulfonylurea receptor (SUR) subunits and pore-forming (Kir6.x) subunits (Fig. 8.1A)[1]. Kir6.1 and Kir6.2, encoded by the *KCNJ8* and *KCNJ11* genes, respectively, are members of the inwardly rectifying potassium channel (Kir) subfamily of pore-loop cation channels. The regulatory subunits, SUR1 and SUR2, encoded by *ABBC8* and *ABCC9* genes, belong to the ATP-binding cassette (ABC) superfamily of membrane proteins.[1] In recombinant expression, functional channels can be generated by C-terminally truncated Kir6.x subunits without SUR subunits[2] but, in this case, C-terminal truncation removes a sequence (Arg-Lys-Arg) that normally causes retention in the endoplasmic reticulum (ER), in the absence of SUR. [3] SUR, which contains a similar ER-retention sequence, is also not efficiently trafficked to the surface membrane in the absence of Kir6.x. Thus, tight association of one subunit with the other is normally required for efficient trafficking of each to the plasma membrane.[3]

Our focus in this chapter is on the K$_{ATP}$ channels in the insulin-secreting β-cells of the islets of Langerhans, comprised of Kir6.2 and SUR1 subunits, for which the physiological and pathological roles of the channel are now understood in some detail. Crystallographic studies of bacterial and eukaryotic Kir channels[4] demonstrate a conserved architecture for all Kir channels consisting of two transmembrane helices bridged by an extracellular loop that generates the narrow portion of the pore and controls ion selectivity. No crystal structures are available for SUR subunits, which possess 17 transmembrane spanning helices divided into three membrane domains (TMD0, TMD1, and TMD2), and two intracellular nucleotide-binding folds (NBF1, NBF1), the first of which is located between TMD1 and TMD2, and the second after TMD2.[1]

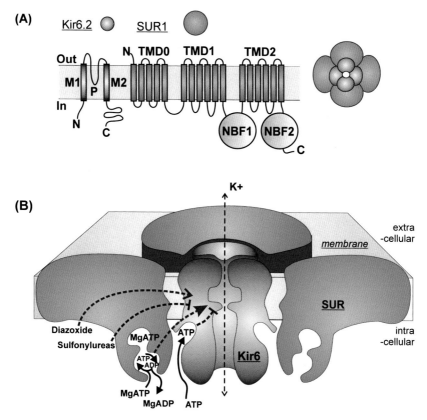

FIGURE 8.1   **Molecular basis of K$_{ATP}$ channel activity.** (A) The pancreatic β-cell channel is generated as an octamer of four pore-forming Kir6.2 and four regulatory SUR1 subunits. (B) Cytoplasmic ATP binding to the Kir6.2 subunits stabilizes channel closure. MgATP binds to the two ATP-binding sites which are formed at the NBF1–NBF2 interface on the SUR1 subunits. MgATP hydrolysis results in a conformational "activated" state that can open the channel. The "activated state" persists through ADP dissociation and can be maintained by MgADP rebinding. Sulfonylureas or diazoxide, interacting with the SUR1 subunit, causes channel closure or opening respectively.

MODULATION OF THE PANCREATIC K$_{ATP}$ ACTIVITY: NUCLEOTIDE BINDING TO THE KIR6.2 SUBUNIT

Nucleotide regulation of K$_{ATP}$ channels is unique among K channels: the channel is rapidly and reversibly inhibited by binding of cytoplasmic adenosine nucleotides and exhibits complex activation by nucleotide tri- and diphosphates[1,5] (Fig. 8.1B). The putative ATP-binding pocket in Kir6.2, as predicted by homology modeling,[6] automated ligand docking,[7] and mutagenesis studies,[8–11] is formed by multiple key residues in the cytoplasmic N- and C-terminus (I182, R50, R201, K185, and G334) that group together in the tertiary structure to collectively coordinate the docked ATP at the subunit interface, thus controlling ATP affinity. Paradoxically, the half-inhibitory ATP concentration ($K_{1/2,ATP}$) of the native K$_{ATP}$ in isolated membrane patches is only ~10 µM[12;] yet, cellular ATP concentrations in the energized cell is in the low millimolar range.[13,14] As a result, greater than 99% of K$_{ATP}$ channels in the pancreatic β-cell are predicted to be closed at any one time.[15] However, measured K$_{ATP}$ activity "on-cell" is significantly higher, and this likely reflects the net stimulatory inputs from MgADP,[16–19] phospholipids (eg, phosphatidylinositol (4,5)-bisphosphate (PIP$_2$)),[20,21] adenylate kinase,[22] and long-chain acyl CoAs[23,24] which collectively reduce the ATP inhibition in the cellular milieu.

MODULATION OF THE PANCREATIC K$_{ATP}$ ACTIVITY: NUCLEOTIDE AND DRUG INTERACTION WITH THE SUR1 SUBUNITS

Channel activation arises from the activating effects of Mg-nucleotides on the SUR subunit, countering the inhibitory effects of ATP.[16] Thus a decrease in sensitivity to inhibitory ATP or an increase in sensitivity to activation by MgADP will enhance absolute currents at any physiological condition and thereby decrease excitability of the cell. Electrophysiological and biochemical studies suggested that both NBFs are involved in channel stimulation by MgADP, and there is biochemical evidence for nucleotide hydrolysis at NBF2.[25–27] NBFs from bacterial ABC proteins crystallize as "head-to-tail" dimers, and this is likely the functional arrangement between NBF1 and NBF2 in SUR.[1] It has also been shown that while ATP binds to NBF1 in an Mg^{2+}-independent manner, ATP and ADP binding to NBF2 requires Mg^{2+}.[28] The conformational changes associated with MgATP binding/hydrolysis at the NBFs are presumably transduced to the activation gates of Kir6.2, stabilizing the opening state of the channel pore. In the case of MgADP, direct binding to NBF2 is sufficient to activate the channel in isolated membrane patches, presumably by stabilizing the post-hydrolytic state of the protein complex.[25,29]

Physical interaction between the cytoplasmic N-terminus of Kir6.2 and the TMD0 domain of SUR, including the cytoplasmic L0 linker, is demonstrated to be important in regulation of Kir6.2 gating by SUR.[30,31] The first clue to the role of the SUR subunit in K$_{ATP}$ channel gating was provided by a mutation (G1479R) in NBF2 that caused human hyperinsulinism[16] (Fig. 8.3). This mutation and other disease-causing NBF2 mutations selectively abolish MgADP-mediated stimulation of channel activity, with no effect on ATP inhibition.[1] Mutation of the conserved lysine residues in the Walker A motifs of either NBF1 (mutation K719A) or NBF2 (mutation K1384M), which are predicted to reduce ATP hydrolytic activity, also blocks the stimulatory effect of MgADP[32] (Fig. 8.3). Furthermore, nucleotide binding to NBF2 stabilizes ATP binding to NBF1, either as a direct result of binding to NBF2 or by hydrolysis of NBF2-bound MgATP,[33] and mutations in the nucleotide-binding motifs of NBF2 (K13845M and D1506N) abolish this cooperative binding of nucleotides to the two NBFs.

LIPID MODULATION OF K_{ATP} CHANNEL ACTIVITY

K_{ATP} channels are also modulated by cytoplasmic factors and by the lipid composition of the membrane, especially by negatively charged phospholipids. The application of PIP_2 to inside-out patches causes an increase in channel open probability and a decrease in sensitivity to ATP. PIP_2 interacting residues (ie, R54, R176, R177, R206) are also on the top surface of the domain. Thus the overlapping nature of the ATP-binding sites and the PIP_2 interacting sites is very consistent with the proposed negative interaction of these two groups in channel regulation. There is also evidence that protein kinase A (PKA) and protein kinase C (PKC) mediate phosphorylation of Kir6.2, increasing channel open probability and decreasing ATP sensitivity.[34] Neurotransmitters that couple to adenylate cyclase and therefore PKA may also affect Kir6.2 phosphorylation and hence provide further complexity to the regulation of channel activity.

THE ROLE OF β-CELL K_{ATP} CHANNEL IN INSULIN SECRETION

The unique and complex dependence of K_{ATP} activity on nucleotides provides the pancreatic β-cell with a mechanism to couple cellular metabolism to electrical activity and thus to insulin release.[35] The pancreatic K_{ATP} (Kir6.2 and SUR1[17,36]) provides the primary K+-conductance of the resting β-cell and sets the membrane potential, V_m, of the cell close to equilibrium potential for potassium (approximately−70~mV) under conditions of low glucose.

When glucose metabolism is increased, as following a meal, β-cell K_{ATP} channels become inhibited due to elevated intracellular [ATP]/[ADP] ratio. This will cause membrane depolarization, leading to Ca^{2+} entry through voltage-dependent Ca^{2+} channels and calcium influx. The resultant rise in intracellular calcium concentration ($[Ca^{2+}]_i$) triggers insulin secretion (Fig. 8.2A) in two phases, an early transient first phase and a sustained secondary phase.[1] A rise in circulating insulin then leads to increased glucose uptake in the peripheral tissues and a compensatory drop in blood glucose.[35] Conversely, a fall in intracellular [ATP]/[ADP] ratio in the fasted state relieves inhibition of K_{ATP} channels, resulting in membrane hyperpolarization and cessation of insulin release (Fig. 8.2A).[35]

However, this straightforward scenario does not account for the entire picture of insulin secretion. K_{ATP} channel opening drugs (KCOs) such as diazoxide bind directly to the SUR1 subunit and activate K_{ATP} channels, thereby inhibiting insulin secretion.[1,37–39] Under conditions in which K_{ATP} channels are maximally activated by diazoxide, glucose-dependent insulin secretion (GSIS) can still be detected in isolated islets. Thus, the

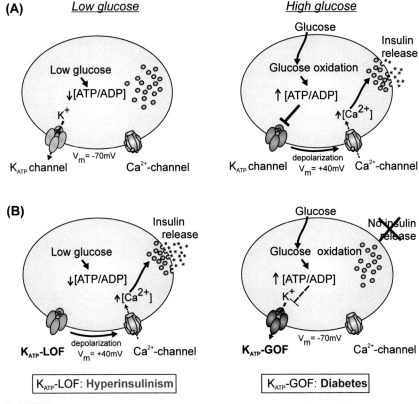

FIGURE 8.2 **Physiologic and pathologic role of K_{ATP} in islet β-cells.** (A) In normal β-cells, metabolism is inhibited in low glucose, lowering the [ATP]/[ADP] ratio, promoting K_{ATP} channel opening, which hyperpolarizes the cell, inhibiting voltage-dependent Ca^{2+} channels, and inhibiting insulin secretion. When glucose rises, metabolism is stimulated, raising the [ATP]/[ADP] ratio. This promotes K_{ATP} channel closure, which depolarizes the cell, opening Ca^{2+} channels, causing Ca^{2+} to rise, and triggering insulin secretion. (B) Loss-of-function (LOF) mutations in K_{ATP} channels reduce channel activity, causing β-cells to become hyperexcitable and to persistently hypersecrete insulin. Gain-of-function (GOF) K_{ATP} mutants that increase channel activity cause underexcitability of the β-cell and failure to secrete insulin.

K_{ATP}-dependent (triggering) pathway is modulated by K_{ATP}-independent amplifying mechanisms which do not alter electrical activity per se, but enhance GSIS once the triggering pathway is initiated (ie, when β-cell $[Ca^{2+}]$ is elevated).[40,41] In addition, glucose metabolites and incretins (eg, glucagon-like-peptide-1, GLP1) can affect secretion at various stages downstream of K_{ATP}.[42–45] Nevertheless, the ability of highly selective K_{ATP} drugs (potassium channel openers and inhibitory sulfonylurea drugs, see later discussion) to modulate GSIS both in vivo and in vitro underscores the central role of K_{ATP}-dependent regulation.[46,47]

CONGENITAL HYPERINSULINISM

The SUR1 subunit also confers sensitivity of the β-cell K_{ATP} to sulfonylureas (SUs), such as tolbutamide and glibenclamide (glyburide), which bind directly to the SUR1 subunit (Fig. 8.1B) leading to inhibition of K_{ATP} channel activity, membrane depolarization, and triggering of insulin secretion.[1] These drugs are widely used to treat diabetes because of their ability to secrete insulin independently of the metabolic state of the β-cell. Reduced or absent K_{ATP} channel activity will result in enhanced or constitutive membrane depolarization, elevated $[Ca^{2+}]_i$, and insulin hypersecretion. In support of this prediction, genetic studies identified reduced or absent K_{ATP} in the β-cell as causal in congenital hyperinsulinism of infancy (HI), also called persistent hyperinsulinemia hypoglycemia of infancy, a rare disorder with an incidence of 1 in 50,000 live births in the general population.[16,48–52] HI is characterized by excessive insulin secretion despite low blood glucose levels and, in the absence of treatment, severe mental retardation and epilepsy can occur.[53]

Such loss-of-function (inactivating) mutations in Kir6.2 (KCNJ11) and SUR1 (ABCC8) K_{ATP} subunits are the most common cause of HI, accounting for ~70% of all HI cases with >200 mutations reported to date.[54–57] In most cases, the mutations are recessively inherited, although there are increasing reports of dominantly inherited inactivating mutations that are generally associated with milder, drug-responsive forms of HI[58,59] (Fig. 8.3).

Implicit in the paradigm of GSIS is the notion that alterations in the metabolic signal, in the sensitivity of K_{ATP} to metabolites, or in the number of active K_{ATP}, should all impair electrical signaling in the β-cell and alter insulin release.[60,61] HI-associated K_{ATP} mutations can be classified into two functional groups: (1) mutations that reduce net channel expression at the cell surface by affecting either channel biosynthesis, trafficking, or assembly, or (2) mutations that decrease intrinsic channel activity, but without affecting surface expression.[61] In either scenario, decreased K_{ATP} activity results in constitutive membrane depolarization, a persistently elevated $[Ca^{2+}]_i$ stimulus, and unregulated

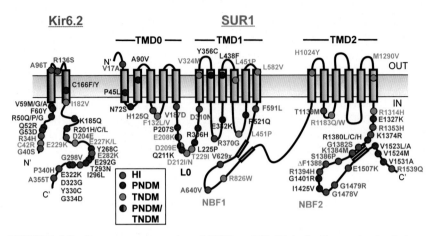

FIGURE 8.3   Cartoon representation of SUR1 and Kir6.2, indicating approximate locations of mutations associated with hyperinsulinism (HI), permanent (PNDM) or transient neonatal diabetes (TNDM).

insulin secretion (Figs. 8.2B and 8.3), irrespective of the blood glucose concentration.

A majority of HI-associated mutations map to the SUR1 subunit with fewer being associated with the Kir6.2 subunit.[62] Of the SUR1 mutations, many localize to the NBFs and are shown in heterologous expression to inhibit the stimulatory effect of MgADP.[63] In a subset of NBF mutations, however, residual channel activity is still present due to a partial response to MgADP,[59] and, as a consequence, the HI phenotype associated with these NBF mutations is generally milder than those associated with complete loss-of-function.[58]

Morphologically, HI can be either diffuse or focal: in the diffuse form all β-cells in the pancreas are affected, whereas in the focal form the affected cell population is present as a localized region of the pancreas.[18] In patients with diffuse HI, potassium channel openers (KCO drugs, eg, diazoxide, chlorothiazide) can often be administered as an initial treatment since opening of quiescent K_{ATP} at the cell surface will hyperpolarize the β-cell and suppress insulin hypersecretion. However, in the majority of cases of K_{ATP}-induced hyperinsulinism, KCO therapy is ineffective as K_{ATP} are either absent at the cells' surface or drug resistant and consequently a partial or total pancreatectomy is frequently required.[64] For patients with the focal form of HI, surgical resection is usually effective if the lesion has been preoperatively located.[65] Both sulfonylurea and K+-channel opener drugs, in addition to their potent secretory effects, have been shown to rescue surface expression of mistrafficked mutant channels in vitro, raising the possibility of drug-specific molecular chaperones as a novel therapy to treat trafficking defects in HI.[66,67]

ANIMAL MODELS OF REDUCED K_{ATP} CHANNEL ACTIVITY: INSIGHTS TO HYPERINSULINEMIA

Genetic manipulation of K_{ATP} channel subunits in mice has been carried out to probe the role of the channels in insulin secretion and in the etiology of disease. Genetically modified mice with reduced or absent K_{ATP} provide further insight into disease progression and collectively recapitulate primary features of HI, including elevated $[Ca^{2+}]_i$ and neonatal hyperinsulinemia. However, the HI phenotype in transgenic mouse models is complex and not entirely consistent with human findings. Mice completely lacking SUR1 or Kir6.2 subunits do not reiterate the HI disease in any simple way. In these animals, hypersecretion reportedly occurs immediately after birth, but rapidly progresses to a relative undersecretion with mild to severe glucose intolerance in adults and an overall loss of secretory capacity.[68–70] A full explanation for the complexity of the results is yet to be achieved, but in terms of modeling HI, it is clear that postneonatal mice with complete absence of β-cell K_{ATP} channels are glucose intolerant and neither persistently hyperinsulinemic nor hypoglycemic.[71] One important potential caveat to consider is whether patients really have a complete loss of K_{ATP} channels and therefore whether complete knockout mice are the appropriate models for the disease. There is a very nonlinear dependence of electrical activity and secretion on K_{ATP} conductance, and significant reduction of K_{ATP} conductance can occur without big shifts in glucose sensitivity of secretion: mice expressing a dominant-negative Kir6.2[AAA] transgene in β-cells, as well as mice heterozygous for either Kir6.2 or SUR1, show marked (~60–70%) reduction of K_{ATP} channel activity but only a small shift in glucose sensitivity (~1–2 mM).[72,73]

In various mouse models in which K_{ATP} is specifically suppressed in the pancreatic β-cell utilizing dominant-negative Kir6.2 transgenes, hyperinsulinism is present and persists into adulthood,[72,74] although, in one transgenic model the mice eventually progress to a hyperglycemic phenotype due to a loss of β-cell mass.[74] Normally, heterozygous Kir6.2 knockout (Kir6.2$^{+/-}$) or Kir6.2[AAA] mice do not reach the necessary threshold, and only when excitability is further enhanced, for instance by dietary stress,[75] a rapid progression to undersecretion can ensue. The phenotype of many HI mutations would actually suggest that a reduction, but not complete absence, of K_{ATP} channels is likely to be the case in human patients, and persistent hyperinsulinism in humans might indeed typically reflect an incomplete loss of K_{ATP} conductance. Some HI patients with K_{ATP} channel mutations must maintain some K_{ATP} channel activity, since the patients are responsive to the K_{ATP} channel drugs tolbutamide and diazoxide,[76] and carriers of loss-of-function K_{ATP} mutations may have enhanced glucose tolerance and subclinical hyperinsulinism (comparable to Kir6.2$^{+/-}$ or SUR1$^{+/-}$ mice).[71,73,75]

We have therefore proposed an "inverse-U" model to explain the β-cell response to hyperexcitability, resulting from reduced K_{ATP} conductance.[71,75] Initially, any reduction of K_{ATP} will cause hyperexcitability that will lead to enhanced insulin secretion, but above some threshold of enhanced excitability, progression to the "descending limb" of the relationship, and a consequent failure of glucose sensing and secretion is triggered. K_{ATP} knockout animals may reach this threshold spontaneously, such that adult mice exhibit only basal secretion and are glucose intolerant.[68–70] There are also increasing reports of nonpancreatectomized HI patients who become diabetic later in life.[77,78,79] Thus, one implication is that in humans with more complete loss of K_{ATP} channels, a progression from hyperinsulinemia to glucose intolerance will be expected, and there are indeed reports that some nonsurgically treated HI patients also spontaneously progress to type-2 diabetes,[80] ie, onto the descending limb of the "inverse-U" progression. Interestingly, the "inverse-U" progression occurs in wild-type animals with chronically suppressed K_{ATP} activity: within a short period of treatment with high dose glibenclamide, mice progress from a hyperinsulinemic state to undersecretion, just like K_{ATP} knockout animals.[77] However, complete reversibility of this undersecretion following removal of glibenclamide[77] suggests the possibility that suppression of excitability with diazoxide or dietary restriction may prove beneficial in controlling diabetes in such patients. The fact that the hyperinsulinemia in certain animal models does not necessarily persist into adulthood implicates compensatory mechanisms that counterbalance the expected hyperinsulinism, and can cross over to a diabetic phenotype, of currently unexplained mechanism.

NEONATAL DIABETES MELLITUS

In contrast to loss-of-function mutations associated with HI, gain-of-function (GOF) (activating) mutations in K_{ATP} are predicted to give rise to the corollary disorder of hyperglycemia and diabetes due to insulin undersecretion. In this paradigm, overactive K_{ATP} remain open despite the metabolic signal to close and, as a result, intracellular $[Ca^{2+}]$ does not rise and insulin secretion is suppressed, giving rise to a hyperglycemic phenotype (Fig. 8.2B). This prediction was originally confirmed by the striking neonatal diabetic phenotype of transgenic mice expressing β-cell K_{ATP} channels with decreased sensitivity to inhibitory ATP, which thereby renders channels more active at higher ATP concentrations.[81] In this mouse model, acute neonatal hyperglycemia with severe ketoacidosis, followed by neonatal lethality, was observed.

Extrapolation from this diabetic animal predicted that overactive β-cell K_{ATP} may underlie impaired insulin release and neonatal diabetes mellitus (NDM) in humans, a rare disorder with an incidence of ~1:300,000 live births that is diagnosed within the first 6 months of life and, until recently, typically required insulin therapy.[82,83] This prediction was borne out by the discovery of activating mutations in Kir6.2 as causal in NDM.[84] Since then, activating mutations in both Kir6.2 and SUR1 subunits have been identified as the most common cause of NDM.[85–88] NDM can present as a permanent (PNDM) form which requires antidiabetic treatment for life or as a milder transient (TNDM) form in which hyperglycemia usually resolves within 18 months of life, but with a risk of relapse typically during pubescence. The exact mechanism(s) underlying the remission phase of TNDM is not known, but may reflect a secondary increase in insulin sensitivity, a compensatory increase in β-cell function, or both (see later discussion).

MECHANISMS OF K_{ATP}-INDUCED DIABETES AND IMPLICATIONS FOR DISEASE SEVERITY

Many NDM-causing K_{ATP} mutations have been found to arise de novo during embryogenesis, with relatively few examples of familial transmission. All result in enhanced channel activity at any given [ATP]:[ADP] ratio.[89] Detailed electrophysiological studies of reconstituted K_{ATP} demonstrate that NDM-associated mutations in both Kir6.2 and SUR1, enhance channel currents by either: (1) reducing the ability of inhibitory ATP to block channel activity through the Kir6.2 subunit,[90–92] or in rarer cases, (2) by increasing sensitivity of channels to stimulatory Mg^{2+}-adenosine nucleotides through the SUR1 subunit.[85,91,93,94] With respect to the Kir6.2 subunit, missense mutations of putative ATP-binding residues are a common cause of PNDM (Fig. 8.3) and predictably reduce sensitivity of the channel to inhibitory ATP, both in the presence and absence of Mg^{2+}.[90,92,95]

Several Kir6.2 residues (ie, I182, R201, R50, K185) previously been identified in biochemical and mutagenesis studies as important coordinates in the ATP-binding pocket and, therefore predicted to be key determinants in physiological regulation of the channel.[8,9,96] Each of these residues, and others (Y330 and F333) predicted to lie close to the phosphate tail of the ATP-binding pocket,[97] have been found to be mutated in patients with NDM (Fig. 8.3). Other NDM mutations located in Kir6.2 but outside the ATP-binding site, or particularly in the TM0 domain of the SUR1 subunit, are involved in altered channel gating in the absence of ATP and thereby allosterically result in reduction of ATP sensitivity without changes in ATP binding affinity itself (eg, I296L, V59G, F132L, Q52R, C42R) (Fig. 8.3).[90,95] Specifically, such gating mutations increase the steady-state open probability of the channel in the absence of ATP (Po,zero). Since ATP interacts predominantly with the closed state of the channel,[98,99] access to the binding site is reduced when the open state is favored, and, as a result, the apparent ATP sensitivity ($K_{1/2,ATP}$) is decreased.$_{ATP,1/2}$

Notably, in approximately one-third of PNDM cases, K_{ATP} mutations are associated with developmental delay (both in motor and intellectual skills), epilepsy, and neonatal diabetes, DEND syndrome, and these extrapancreatic symptoms likely reflect overactive K_{ATP} in skeletal muscle, peripheral nerves, and brain.[100] In the absence of epilepsy, an intermediate DEND (i-DEND) phenotype is also reported.[101–103] It is important to note that the disease severity is generally correlated with the magnitude of channel overactivity such that mutations which demonstrate the largest increase in basal K_{ATP} currents in vitro are more likely to be associated with severe forms of NDM,[104,105] regardless of the mechanism of overactivity, ie, gating mutations versus ATP-binding mutations (Fig. 8.3). For example, the mildly overactive I82V mutation (ATP-binding) underlies TNDM, the R201H (ATP-binding), and F35V (gating) mutations, which are moderately overactive underlie an isolated PNDM phenotype, and the V59M mutation (gating) which is a more severe activating mutation, is the most common cause of i-DEND.[106] In the most extreme cases of overactivity, eg, the G334D (ATP-binding) and I296L (gating) mutations, a full DEND phenotype is observed.[91,106] As only the more severe activating mutations are correlated with DEND symptoms, this suggests that extrapancreatic tissues (muscle, nerve, and brain) are less sensitive to changes in K_{ATP} activity than the pancreatic β-cell and that a threshold of channel overactivity must be reached before neuronal firing rates are suppressed and neurological features become evident. In contrast to the pancreatic β-cell (where K_{ATP} provides the primary K^+-conductance), non-ATP-sensitive K^+-currents are present at proportionally higher levels in neurons and striated muscle and are likely to contribute more to the control of membrane potential. Consequently, a greater degree of K_{ATP} overactivity may be necessary to significantly change V_m, and, therefore, electrical activity in these extrapancreatic tissues.[105]

At present the exact mechanisms by which overactive K_{ATP} predisposes to neurological features is unknown, and the interpretation is confounded by the fact that K_{ATP} is expressed in both stimulatory and inhibitory (GABAergic) neurons throughout the brain where they presumably contribute to neuronal excitability, and, hence, neurotransmitter release.[107,108] Generation of appropriate transgenic mice expressing either neuronal GOF K_{ATP} (Kir6.2 + SUR1) in specific brain regions (ie, substantia nigra, forebrain) or GOF skeletal muscle K_{ATP} (Kir6.2 + SUR2A) should provide insight into the origins of neurological features in DEND and whether the extrapancreatic features are secondary to hyperglycemic or hypoglycemic episodes in NDM.

SECONDARY MECHANISMS IN NEONATAL DIABETES MELLITUS

Finally, phenotypic heterogeneity in NDM has been reported in multiple carriers with the same disease-causing K_{ATP} mutation. In a pedigree with the activating C42R mutation in Kir6.2, the cross-generational phenotypes of affected family members ranged from PNDM, to gestational diabetes, to a late-onset, type-2 diabetes.[109] Similarly, in several pedigrees with SUR1 mutations, carriers are variably reported to be unaffected, type-2 diabetic, or neonatal diabetic.[85,110,111] Taken together, these data indicate that the severity of the β-cell defect in K_{ATP}-induced NDM can be modulated by additional, unknown genetic, and/or environmental factors that integrate to determine the clinical presentation.

TREATABILITY OF NEONATAL DIABETES MELLITUS

The realization that NDM results from mutations in the K_{ATP} channel rapidly shifted the therapy from insulin injections to oral sulfonylurea drugs: by directly inhibiting the overactive K_{ATP} channels, these drugs can provide successful control of blood glucose levels, and insulin requirements can be avoided (or reduced) in the majority of K_{ATP}-induced NDM,[89] although the SU-dose requirement is generally higher than that used to treat type-2 diabetics.[102] Unlike exogenous insulin which acts rather indiscriminately to lower blood glucose by promoting peripheral glucose uptake, SU drugs circumvent the metabolic signal in the β-cell and directly target K_{ATP} overactivity to restore endogenous insulin secretion. The stimulation of endogenous insulin secretion by SUs may be augmented by the presence of potentiating hormones (glucagon and GLP-1) and by other K_{ATP}-independent mechanisms, such that insulin secretion becomes quasiphysiological in such treated patients and blood glucose control may be better than that afforded by insulin injections.

The majority of NDM patients with K_{ATP} mutations have successfully switched from insulin injections to oral SUs with significant improvements in both glycemic control and overall lifestyle, although the success rate is lower in patients with the more severe DEND symptoms.[91,102,112–115] Nevertheless, resolution of diabetes is reported in several DEND patients following SU treatment, concurrent with improved neurological function.[101,116–118] This implies that SUs, in addition to targeting the β-cell, are capable of crossing the blood–brain barrier and acting on overactive K_{ATP} in central neurons and improving motor signal output. Interestingly, despite a comparatively high level of K_{ATP} expression in sarcolemma of the heart (Kir6.2 + SUR2) there are no reports of cardiac dysfunction associated with activating K$_{ATP}$ mutations in NDM[84,112] which likely reflects either the differential regulation of the cardiac K$_{ATP}$ by adenosine nucleotides and/or a different metabolic state in the cardiac myocyte.[27,119]

VARIANTS IN THE β-CELL K$_{ATP}$ AS A RISK FACTOR IN TYPE-2 DIABETES

Multiple genetic studies over the past 20 years have examined the association of K$_{ATP}$ variants (defined as a nucleotide change that has an allelic incidence of ≥1%), in both Kir6.2 and SUR1 with the development of type-2 diabetes.[120,121] By analogy with NDM, variants that increase K$_{ATP}$ activity are predicted, in combination with other genetic and environmental factors, to underlie impaired β-cell response and insufficient serum insulin level for a given blood glucose level. On the background of diet-induced insulin resistance, for example, this may be expected to exacerbate the hyperglycemic phenotype.

Control-based genetic studies,[122,123] meta-analysis,[124–126] and genome wide association studies[127,128] consistently demonstrate association of the E23K variant in Kir6.2 with reduced insulin secretion in both glucose-tolerant and intolerant cohorts (as reflected by reduced plasma insulin levels during hyperglycemic clamp or glucose challenge),[122,125,129,130] and as a risk allele in the development of type-2 diabetes.[123,128,131]

Although the individual risk is low (odds ratio of 1.2 or ~20% increased risk for carriers of the E23K variant), given the high allelic frequency of E23K in the general population (frequency of heterozygous EK genotype ~47%; homozygous KK genotype ~12%), the variant equates to a significant population attributable risk.[121] However, the mechanistic basis of the Kir6.2[E23K] association remains controversial despite examination by multiple investigators. While one early study observed no effects of the E23K variant on ATP-sensitive K$^+$ currents in Xenopus oocytes,[132] more recent electrophysiological studies of reconstituted K$_{ATP}$ in mammalian cells have consistently reported overactivity of E23K channels which is variably attributed to (1) an increase in intrinsic channel open probability (Po,zero) concomitant with a decrease in channel sensitivity to inhibitory ATP[124,130]; (2) unaltered ATP sensitivity, but with an enhanced activation by long-chain acyl CoAs[133]; or (3) a decrease in ATP sensitivity that is attributed not to E23K, but to a SUR1 variant (S1369A) that occurs in linkage disequilibrium with E23K.[134]

In the most extensive electrophysiological studies of reconstituted E23K channels,[124,130] enhanced open probability and a concomitant decrease in ATP sensitivity has been observed, but this is not as severe as for Kir6.2

mutations that underlie monogenic NDM (Fig. 8.3). As a consequence, the E23K variant is predicted to underlie a mild β-cell dysfunction, which may therefore predispose to type-2 diabetes, whereas the more severe activating K$_{ATP}$ mutations significantly impair GSIS and are causal in monogenic NDM.

ANIMAL MODELS OF INCREASED K$_{ATP}$ CHANNEL ACTIVITY: INSIGHTS TO DIABETES

A striking neonatal diabetic phenotype was first demonstrated in transgenic mice expressing GOF mutations in Kir6.2, specifically in β-cells. Acute neonatal diabetes was accompanied by severe ketoacidosis, followed by neonatal lethality.[81] In subsequent transgenic models of β-cell K$_{ATP}$-GOF (overactivity), including inducible animals in which the transgene could be induced at a specific time, a spectrum of diabetic phenotypes has been reported, ranging from severe neonatal diabetes to mild glucose intolerance in adulthood.[135–137] The variable diabetic phenotypes are likely to reflect differential expression of the overactive K$_{ATP}$ transgene in the β-cell with increasing penetrance correlated with earlier and more severe diabetes. In the most severe diabetic models reduced insulin secretion, lower insulin content, and abnormal morphology with loss of β-cell mass are consistently observed.[136,137] Importantly, maintaining normoglycemia in mouse models of K$_{ATP}$-induced diabetes, through either SU therapy or syngenic islet transplantation, prevents the progressive loss of β-cell mass.[136] Moreover, acute aggressive SU treatment at the onset of transgene induction in inducible K$_{ATP}$-GOF resulted in a striking divergent phenotype; some mice (70%) developed severe diabetes as naively expected, as soon as SU treatment was terminated, but the remaining 30% subsequently demonstrated persistently near-normal glucose levels.[138] This finding predicts that early intervention with SUs in NDM patients with K$_{ATP}$ mutations may have the added benefit of preserving β-cell function and might conceivably account for a mechanism of TNDM. Interestingly, there appears to be a correlation between the age at which SU transfer is attempted in NDM, and overall β-cell responsiveness, such that the younger the patient at the time of SU transfer, the greater the chance for a complete transition from insulin to SU therapy.[102,139] It is interesting to speculate that age-dependent differences in apparent SU responsivity reflect a difference in β-cell mass, although this is currently not known.

Finally, the long-term consequences of SU therapy, including the influences of aging and obesity, should be considered. The possibility exists that by analogy to the progressive loss of SU-responsiveness in type-2 diabetes,[140] NDM patients may eventually fail on SUs during their lifetime—perhaps because of β-cell degeneration of some kind, in which case insulin therapy may need to be reinitiated. Parenthetically, however, one 46-year-old NDM patient with a K$_{ATP}$ mutation was directly treated with SUs in childhood without prior insulin and thereafter maintained good glycemic control, suggesting that in certain cases long-term SU treatment may be safe and effective.[102]

Secondary consequences of chronic diabetes on β-cell function are evident in K$_{ATP}$-GOF mice. Apoptotic β-cell death is frequently assumed in long-term and poorly controlled diabetics, but dedifferentiation in common forms of β-cell failure has also been inferred from partial pancreatectomy studies.[141] As previously reported for mice with hyperglycemia after multiparity or stress,[142] β-cells in severely diabetic K$_{ATP}$-GOF islets dedifferentiate to progenitor cells.[143] Importantly, the same dedifferentiated cells can redifferentiate to mature insulin-secreting β-cells after normalization of blood glucose by intensive insulin therapy.[143] These results may not only inform gradual decrease in β-cell mass in other forms of longstanding diabetes, including recovery of β-cell function and drug responsivity in type-2 diabetic patients following intensive glucose control,[144–146] but may also suggest an approach to rescuing "exhausted" β-cells in NDM and other forms of diabetes.

References

1. Nichols CG. KATP channels as molecular sensors of cellular metabolism. *Nature* 2006;**440**(7083):470–6.
2. Tucker SJ, Ashcroft FM. Mapping of the physical interaction between the intracellular domains of an inwardly rectifying potassium channel, Kir6.2. *J Biol Chem* 1999;**274**(47):33393–7.
3. Zerangue N, et al. A new ER trafficking signal regulates the subunit stoichiometry of plasma membrane K(ATP) channels. *Neuron* 1999;**22**(3):537–48.
4. Kuo A, et al. Crystal structure of the potassium channel KirBac1.1 in the closed state. *Science* 2003;**300**(5627):1922–6.
5. Ashcroft SJ, Ashcroft FM. Properties and functions of ATP-sensitive K-channels. [Review] *Cell Signal* 1990;**2**(3):197–214.
6. Antcliff JF, et al. Functional analysis of a structural model of the ATP-binding site of the KATP channel Kir6.2 subunit. *EMBO J* 2005;**24**(2):229–39.
7. Trapp S, et al. Identification of residues contributing to the ATP binding site of Kir6.2. *EMBO J* 2003;**22**(12):2903–12.
8. Tucker SJ, et al. Molecular determinants of KATP channel inhibition by ATP. *EMBO J* 1998;**17**(12):3290–6.
9. Drain P, Li L, Wang J. KATP channel inhibition by ATP requires distinct functional domains of the cytoplasmic C terminus of the pore-forming subunit. *Proc Natl Acad Sci USA* 1998;**95**(23):13953–8.
10. John SA, et al. Molecular mechanism for ATP-dependent closure of the K$^+$ channel Kir6.2. *J Physiol* 2003;**552**(Pt 1):23–34.
11. Enkvetchakul D, Nichols CG. Gating mechanism of KATP channels: function fits form. *J Gen Physiol* 2003;**122**(5):471–80.
12. Inagaki N, et al. Reconstitution of IKATP: an inward rectifier subunit plus the sulfonylurea receptor. *Science* 1995;**270**(5239):1166–70.

13. Gribble FM, et al. A novel method for measurement of submembrane ATP concentration. *J Biol Chem* 2000;**275**(39):30046–9.
14. Kennedy HJ, et al. Glucose generates sub-plasma membrane ATP microdomains in single islet beta-cells. Potential role for located mitochondria. *J Biol Chem* 1999;**274**(19):13281–91.
15. Cook DL, et al. ATP-sensitive K⁺ channels in pancreatic beta-cells. Spare-channel hypothesis. *Diabetes* 1988;**37**(5):495–8.
16. Nichols CG, et al. Adenosine diphosphate as an intracellular regulator of insulin secretion. *Science* 1996;**272**(5269):1785–7.
17. Aguilar-Bryan L, et al. Cloning of the beta cell high-affinity sulfonylurea receptor: a regulator of insulin secretion. *Science* 1995;**268**(5209):423–6.
18. Dunne MJ, et al. The gating of nucleotide-sensitive K⁺ channels in insulin-secreting cells can be modulated by changes in the ratio ATP4-/ADP3- and by nonhydrolyzable derivatives of both ATP and ADP. *J Membr Biol* 1988;**104**(2):165–77.
19. Gribble FM, et al. MgATP activates the beta cell KATP channel by interaction with its SUR1 subunit. *Proc Natl Acad Sci USA* 1998;**95**(12):7185–90.
20. Shyng S, Nichols CG. Octameric stoichiometry of the KATP channel complex. *J Gen Physiol* 1997;**110**(6):655–64.
21. Baukrowitz T, et al. PIP2 and PIP as determinants for ATP inhibition of KATP channels. 282:1059–60. *Science* 1998;**282**(5391):1141–4.
22. Schulze DU, et al. An adenylate kinase is involved in KATP channel regulation of mouse pancreatic beta cells. *Diabetologia* 2007;**50**(10):2126–34.
23. Gribble FM, et al. Mechanism of cloned ATP-sensitive potassium channel activation by oleoyl-CoA. *J Biol Chem* 1998;**273**(41):26383–7.
24. Larsson O, et al. Activation of the ATP-sensitive K⁺ channel by long chain acyl-CoA. A role in modulation of pancreatic beta-cell glucose sensitivity. *J Biol Chem* 1996;**271**(18):10623–6.
25. Zingman LV, et al. Signaling in channel/enzyme multimers: ATPase transitions in SUR module gate ATP-sensitive K⁺ conductance. *Neuron* 2001;**31**(2):233–45.
26. Bienengraeber M, et al. ATPase activity of the sulfonylurea receptor: a catalytic function for the KATP channel complex. [In Process Citation] *FASEB J* 2000;**14**(13):1943–52.
27. Masia R, Enkvetchakul D, Nichols CG. Differential nucleotide regulation of KATP channels by SUR1 and SUR2A. *J Mol Cell Cardiol* 2005;**39**(3):491–501.
28. Ueda K, Inagaki N, Seino S. MgADP antagonism to Mg²⁺-independent ATP binding of the sulfonylurea receptor SUR1. *J Biol Chem* 1997;**272**(37):22983–6.
29. Zingman LV, et al. Tandem function of nucleotide binding domains confers competence to SUR in gating KATP channels. *J Biol Chem* 2002;**1**:1.
30. Babenko AP, Bryan J. Sur domains that associate with and gate KATP pores define a novel gatekeeper. *J Biol Chem* 2003;**278**(43):41577–80.
31. Chan KW, Zhang H, Logothetis DE. N-terminal transmembrane domain of the SUR controls trafficking and gating of Kir6 channel subunits. *EMBO J* 2003;**22**(15):3833–43.
32. Gribble FM, Tucker SJ, Ashcroft FM. The essential role of the Walker A motifs of SUR1 in K-ATP channel activation by Mg-ADP and diazoxide. *EMBO J* 1997;**16**(6): 1145–52.
33. Ueda K, et al. Cooperative binding of ATP and MgADP in the sulfonylurea receptor is modulated by glibenclamide. *Proc Natl Acad Sci USA* 1999;**96**(4):1268–72.
34. Beguin P, et al. PKA-mediated phosphorylation of the human K(ATP) channel: separate roles of Kir6.2 and SUR1 subunit phosphorylation. *EMBO J* 1999;**18**(17):4722–32.
35. Ashcroft FM, Gribble FM. ATP-sensitive K⁺ channels and insulin secretion: their role in health and disease. [comment] *Diabetologia* 1999;**42**(8):903–19.
36. Inagaki N, et al. Cloning and functional characterization of a novel ATP-sensitive potassium channel ubiquitously expressed in rat tissues, including pancreatic islets, pituitary, skeletal muscle, and heart. *J Biol Chem* 1995;**270**(11):5691–4.

37. Bryan J, et al. Toward linking structure with function in ATP-sensitive K$^+$ channels. *Diabetes* 2004;**53**(Suppl. 3):S104–12.
38. Bryan J, et al. Insulin secretagogues, sulfonylurea receptors and K(ATP) channels. *Curr Pharm Des* 2005;**11**(21):2699–716.
39. Moreau C, et al. SUR, ABC proteins targeted by KATP channel openers. *J Mol Cell Cardiol* 2005;**38**(6):951–63.
40. Aizawa T, et al. Glucose action 'beyond ionic events' in the pancreatic beta cell. *Trends Pharmacol Sci* 1998;**19**(12):496–9.
41. Henquin JC. Triggering and amplifying pathways of regulation of insulin secretion by glucose. *Diabetes* 2000;**49**(11):1751–60.
42. Gromada J, et al. Glucagon-like peptide 1 (7-36) amide stimulates exocytosis in human pancreatic beta-cells by both proximal and distal regulatory steps in stimulus-secretion coupling. *Diabetes* 1998;**47**(1):57–65.
43. Britsch S, et al. Glucagon-like peptide-1 modulates Ca^{2+} current but not K$^+$ ATP current in intact mouse pancreatic B-cells. *Biochem Biophys Res Commun* 1995;**207**(1):33–9.
44. Gembal M, et al. Mechanisms by which glucose can control insulin release independently from its action on adenosine triphosphate-sensitive K$^+$ channels in mouse B cells. *J Clin Invest* 1993;**91**(3):871–80.
45. Detimary P, Van den Berghe G, Henquin JC. Concentration dependence and time course of the effects of glucose on adenine and guanine nucleotides in mouse pancreatic islets. *J Biol Chem* 1996;**271**(34):20559–65.
46. Trube G, Rorsman P, Ohno ST. Opposite effects of tolbutamide and diazoxide on the ATP-dependent K$^+$ channel in mouse pancreatic beta-cells. *Pflugers Archiv Eur J Physiol* 1986;**407**(5):493–9.
47. Simonson DC, et al. Mechanism of improvement in glucose metabolism after chronic glyburide therapy. *Diabetes* 1984;**33**(9):838–45.
48. Thomas PM, et al. Homozygosity mapping, to chromosome 11p, of the gene for familial persistent hyperinsulinemic hypoglycemia of infancy. *Am J Hum Genet* 1995;**56**(2):416–21.
49. Thomas PM, et al. Mutations in the sulfonylurea receptor gene in familial persistent hyperinsulinemic hypoglycemia of infancy. [see comments] *Science* 1995;**268**(5209):426–9.
50. Nestorowicz A, et al. Mutations in the sulfonylurea receptor gene are associated with familial hyperinsulinism in Ashkenazi Jews. *Hum Mol Genet* 1996;**5**(11):1813–22.
51. Cartier EA, et al. Defective trafficking and function of KATP channels caused by a sulfonylurea receptor 1 mutation associated with persistent hyperinsulinemic hypoglycemia of infancy. *Proc Natl Acad Sci USA* 2001;**98**(5):2882–7.
52. de Lonlay P, et al. Hyperinsulinemic hypoglycemia in children. *Ann D Endocrinol* 2004;**65**(1):96–8.
53. Dunne MJ, et al. Hyperinsulinism in infancy: from basic science to clinical disease. *Physiol Rev* 2004;**84**(1):239–75.
54. Giurgea I, et al. Congenital hyperinsulinism and mosaic abnormalities of the ploidy. *J Med Genet* 2006;**43**(3):248–54.
55. Fournet JC, Junien C. Genetics of congenital hyperinsulinism. *Endocr Pathol* 2004;**15**(3):233–40.
56. Cosgrove KP, Carroll ME, Cosgrove KE. Genetics and pathophysiology of hyperinsulinism in infancy. *Neurosci Biobehav Rev* 2004;**28**(6):533–46.
57. Sharma N, et al. Familial hyperinsulinism and pancreatic beta-cell ATP-sensitive potassium channels. *Kidney Int* 2000;**57**(3):803–8.
58. Pinney SE, et al. Clinical characteristics and biochemical mechanisms of congenital hyperinsulinism associated with dominant KATP channel mutations. *J Clin Invest* 2008;**118**(8):2877–86.
59. Huopio H, et al. Dominantly inherited hyperinsulinism caused by a mutation in the sulfonylurea receptor type 1. *J Clin Invest* 2000;**106**(7):897–906.

60. Koster JC, Permutt MA, Nichols CG. Diabetes and insulin secretion: the ATP-sensitive K+ channel (K ATP) connection. *Diabetes* 2005;**54**(11):3065–72.

61. Ashcroft FM. ATP-sensitive potassium channelopathies: focus on insulin secretion. *Diabetes* 2005;**54**(9):2503–13.

62. Flanagan SE, et al. Update of mutations in the genes encoding the pancreatic beta-cell K(ATP) channel subunits Kir6.2 (KCNJ11) and sulfonylurea receptor 1 (ABCC8) in diabetes mellitus and hyperinsulinism. *Hum Mutat* 2009;**30**(2):170–80.

63. Shyng SL, et al. Functional analyses of novel mutations in the sulfonylurea receptor 1 associated with persistent hyperinsulinemic hypoglycemia of infancy. *Diabetes* 1998;**47**(7):1145–51.

64. De Leon DD, Stanley CA. Mechanisms of disease: advances in diagnosis and treatment of hyperinsulinism in neonates. *Nat Clin Pract Endocrinol Metab* 2007;**3**(1):57–68.

65. Hardy OT, Litman RS. Congenital hyperinsulinism - a review of the disorder and a discussion of the anesthesia management. *Paediatr Anaesth* 2007;**17**(7):616–21.

66. Partridge CJ, Beech DJ, Sivaprasadarao A. Identification and pharmacological correction of a membrane trafficking defect associated with a mutation in the sulfonylurea receptor causing familial hyperinsulinism. *J Biol Chem* 2001;**276**(38):35947–52.

67. Yan FF, Casey J, Shyng SL. Sulfonylureas correct trafficking defects of disease-causing ATP-sensitive potassium channels by binding to the channel complex. *J Biol Chem* 2006;**281**(44):33403–13.

68. Miki T, et al. Defective insulin secretion and enhanced insulin action in KATP channel-deficient mice. *Proc Natl Acad Sci USA* 1998;**95**(18):10402–6.

69. Shiota C, et al. Sulfonylurea receptor type 1 knock-out mice have intact feeding-stimulated insulin secretion despite marked impairment in their response to glucose. *J Biol Chem* 2002;**277**(40):37176–83.

70. Seghers V, et al. Sur1 knockout mice. A model for K(ATP) channel-independent regulation of insulin secretion. *J Biol Chem* 2000;**275**(13):9270–7.

71. Nichols CG, Koster JC, Remedi MS. Beta-cell hyperexcitability: from hyperinsulinism to diabetes. *Diabetes Obes Metab* 2007;**9**(Suppl. 2):81–8.

72. Koster JC, et al. Hyperinsulinism induced by targeted suppression of beta cell KATP channels. *Proc Natl Acad Sci USA* 2002;**99**(26):16992–7.

73. Remedi MS, et al. Hyperinsulinism in mice with heterozygous loss of K(ATP) channels. *Diabetologia* 2006;**49**(10):2368–78.

74. Miki T, et al. Abnormalities of pancreatic islets by targeted expression of a dominant-negative KATP channel. *Proc Natl Acad Sci USA* 1997;**94**(22):11969–73.

75. Remedi MS, et al. Diet-induced glucose intolerance in mice with decreased {beta}-cell ATP-sensitive K+ channels. *Diabetes* 2004;**53**(12):3159–67.

76. Henwood MJ, et al. Genotype-phenotype correlations in children with congenital hyperinsulinism due to recessive mutations of the adenosine triphosphate-sensitive potassium channel genes. *J Clin Endocrinol Metab* 2005;**90**(2):789–94.

77. Remedi MS, Nichols CG. Chronic antidiabetic sulfonylureas in vivo: reversible effects on mouse pancreatic beta-cells. *PLoS Med* 2008;**5**(10):e206.

78. Grimberg A, et al. Dysregulation of insulin secretion in children with congenital hyperinsulinism due to sulfonylurea receptor mutations. *Diabetes* 2001;**50**(2):322–8.

79. Huopio H, et al. A new subtype of autosomal dominant diabetes attributable to a mutation in the gene for sulfonylurea receptor 1. *Lancet* 2003;**361**(9354):301–7.

80. Huopio H, et al. K(ATP) channels and insulin secretion disorders. *Am J Physiol Endocrinol Metab* 2002;**283**(2):E207–16.

81. Koster JC, et al. Targeted overactivity of beta cell K(ATP) channels induces profound neonatal diabetes. *Cell* 2000;**100**(6):645–54.

82. Polak M, Cave H. Neonatal diabetes mellitus: a disease linked to multiple mechanisms. *Orphanet J Rare Dis* 2007;**2**:12.

83. Slingerland AS, Hattersley AT. Activating mutations in the gene encoding Kir6.2 alter fetal and postnatal growth and also cause neonatal diabetes. *J Clin Endocrinol Metab* 2006;**91**(7):2782–8.

84. Gloyn AL, et al. Activating mutations in the gene encoding the ATP-sensitive Potassium-channel subunit Kir6.2 and permanent neonatal diabetes. *N Engl J Med* 2004;**350**(18):1838–49.

85. Babenko AP, et al. Activating mutations in the ABCC8 gene in neonatal diabetes mellitus. *N Engl J Med* 2006;**355**(5):456–66.

86. Gloyn AL, Siddiqui J, Ellard S. Mutations in the genes encoding the pancreatic beta-cell KATP channel subunits Kir6.2 (KCNJ11) and SUR1 (ABCC8) in diabetes mellitus and hyperinsulinism. *Hum Mutat* 2006;**27**(3):220–31.

87. Vaxillaire M, et al. New ABCC8 mutations in relapsing neonatal diabetes and clinical features. *Diabetes* 2007;**56**(6):1737–41.

88. Tornovsky S, et al. Hyperinsulinism of infancy: novel ABCC8 and KCNJ11 mutations and evidence for additional locus heterogeneity. *J Clin Endocrinol Metab* 2004;**89**(12):6224–34.

89. Remedi MS, Koster JC. KATP channelopathies in the pancreas. *Pflugers Arch* 2010;**460**(2):307–20.

90. Koster JC, et al. ATP and sulfonylurea sensitivity of mutant ATP-sensitive K$^+$ channels in neonatal diabetes: implications for pharmacogenomic therapy. *Diabetes* 2005;**54**(9):2645–54.

91. Masia R, et al. A mutation in the TMD0-L0 region of sulfonylurea receptor-1 (L225P) causes permanent neonatal diabetes mellitus (PNDM). *Diabetes* 2007;**56**(5):1357–62.

92. Proks P, et al. Molecular basis of Kir6.2 mutations associated with neonatal diabetes or neonatal diabetes plus neurological features. *Proc Natl Acad Sci USA* 2004;**101**(50):17539–44.

93. de Wet H, et al. Increased ATPase activity produced by mutations at arginine-1380 in nucleotide-binding domain 2 of ABCC8 causes neonatal diabetes. *Proc Natl Acad Sci USA* 2007;**104**(48):18988–92.

94. de Wet H, et al. A mutation (R826W) in nucleotide-binding domain 1 of ABCC8 reduces ATPase activity and causes transient neonatal diabetes. *EMBO Rep* 2008;**9**(7):648–54.

95. Proks P, Girard C, Ashcroft FM. Functional effects of KCNJ11 mutations causing neonatal diabetes: enhanced activation by MgATP. *Hum Mol Genet* 2005;**14**(18):2717–26.

96. Shyng SL, et al. Modulation of nucleotide sensitivity of ATP-sensitive potassium channels by phosphatidylinositol-4-phosphate 5-kinase. *Proc Natl Acad Sci USA* 2000;**97**(2):937–41.

97. Haider S, et al. Focus on Kir6.2: a key component of the ATP-sensitive potassium channel. *J Mol Cell Cardiol* 2005;**38**(6):927–36.

98. Enkvetchakul D, et al. The kinetic and physical basis of K(ATP) channel gating: toward a unified molecular understanding. *Biophys J* 2000;**78**(5):2334–48.

99. Enkvetchakul D, et al. ATP interaction with the open state of the K(ATP) channel. *Biophys J* 2001;**80**(2):719–28.

100. Gloyn AL, et al. KCNJ11 activating mutations are associated with developmental delay, epilepsy and neonatal diabetes syndrome and other neurological features. *Eur J Hum Genet* 2006;**14**(7):824–30.

101. Mlynarski W, et al. Sulfonylurea improves CNS function in a case of intermediate DEND syndrome caused by a mutation in KCNJ11. *Nat Clin Pract Neurol* 2007;**3**(11):640–5.

102. Pearson ER, et al. Switching from insulin to oral sulfonylureas in patients with diabetes due to Kir6.2 mutations. *N Engl J Med* 2006;**355**(5):467–77.

103. Hattersley AT, Pearson ER. Minireview: pharmacogenetics and beyond: the interaction of therapeutic response, beta-cell physiology, and genetics in diabetes. *Endocrinology* 2006;**147**(6):2657–63.

104. Proks P, et al. A gating mutation at the internal mouth of the Kir6.2 pore is associated with DEND syndrome. *EMBO Rep* 2005;**6**(5):470–5.

105. Aguilar-Bryan L, Bryan J. Neonatal diabetes mellitus. *Endocr Rev* 2008;**29**(3):265–91.
106. Tammaro P, Proks P, Ashcroft FM. Functional effects of naturally occurring KCNJ11 mutations causing neonatal diabetes on cloned cardiac KATP channels. *J Physiol* 2006;**571**(Pt 1):3–14.
107. Zawar C, et al. Cell-type specific expression of ATP-sensitive potassium channels in the rat hippocampus. *J Physiol* 1999;**514**(Pt 2):327–41.
108. Griesemer D, Zawar C, Neumcke B. Cell-type specific depression of neuronal excitability in rat hippocampus by activation of ATP-sensitive potassium channels. *Eur Biophys J* 2002;**31**(6):467–77.
109. Yorifuji T, et al. The C42R mutation in the Kir6.2 (KCNJ11) gene as a cause of transient neonatal diabetes, childhood diabetes, or later-onset, apparently type 2 diabetes mellitus. *J Clin Endocrinol Metab* 2005;**90**(6):3174–8.
110. Flanagan SE, et al. Mutations in ATP-sensitive K+ channel genes cause transient neonatal diabetes and permanent diabetes in childhood or adulthood. *Diabetes* 2007;**56**(7):1930–7.
111. Patch AM, et al. Mutations in the ABCC8 gene encoding the SUR1 subunit of the KATP channel cause transient neonatal diabetes, permanent neonatal diabetes or permanent diabetes diagnosed outside the neonatal period. *Diabetes Obes Metab* 2007;**9**(Suppl. 2):28–39.
112. Sagen JV, et al. Permanent neonatal diabetes due to mutations in KCNJ11 encoding Kir6.2: patient characteristics and initial response to sulfonylurea therapy. *Diabetes* 2004;**53**(10):2713–8.
113. Sumnik Z, et al. Sulphonylurea treatment does not improve psychomotor development in children with KCNJ11 mutations causing permanent neonatal diabetes mellitus accompanied by developmental delay and epilepsy (DEND syndrome). *Diabetes Med* 2007;**24**(10):1176–8.
114. Della Manna T, et al. Glibenclamide unresponsiveness in a Brazilian child with permanent neonatal diabetes mellitus and DEND syndrome due to a C166Y mutation in KCNJ11 (Kir6.2) gene. *Arq Bras Endocrinol Metabol* 2008;**52**(8):1350–5.
115. Tonini G, et al. Sulfonylurea treatment outweighs insulin therapy in short-term metabolic control of patients with permanent neonatal diabetes mellitus due to activating mutations of the KCNJ11 (KIR6.2) gene. *Diabetologia* 2006;**49**(9):2210–3.
116. Shimomura K, et al. A novel mutation causing DEND syndrome: a treatable channelopathy of pancreas and brain. *Neurology* 2007;**69**(13):1342–9.
117. Slingerland AS, et al. Improved motor development and good long-term glycaemic control with sulfonylurea treatment in a patient with the syndrome of intermediate developmental delay, early-onset generalised epilepsy and neonatal diabetes associated with the V59M mutation in the KCNJ11 gene. *Diabetologia* 2006;**49**(11):2559–63.
118. Koster JC, et al. The G53D mutation in Kir6.2 (KCNJ11) is associated with neonatal diabetes and motor dysfunction in adulthood that is improved with sulfonylurea therapy. *J Clin Endocrinol Metab* 2008;**93**(3):1054–61.
119. Koster JC, et al. Tolerance for ATP-insensitive K_{ATP} channels in transgenic mice. *Circulation Res* 2001;**89**:1022–9.
120. Tarasov AI, et al. A rare mutation in ABCC8/SUR1 leading to altered ATP-sensitive K+ channel activity and beta-cell glucose sensing is associated with type 2 diabetes in adults. *Diabetes* 2008;**57**(6):1595–604.
121. Riedel MJ, Steckley DC, Light PE. Current status of the E23K Kir6.2 polymorphism: implications for type-2 diabetes. *Hum Genet* 2005;**116**(3):133–45.
122. Florez JC, et al. Haplotype structure and genotype-phenotype correlations of the sulfonylurea receptor and the islet ATP-sensitive potassium channel gene region. *Diabetes* 2004;**53**(5):1360–8.
123. Gloyn AL, et al. Large-scale association studies of variants in genes encoding the pancreatic beta-cell KATP channel subunits Kir6.2 (KCNJ11) and SUR1 (ABCC8) confirm that the KCNJ11 E23K variant is associated with type 2 diabetes. *Diabetes* 2003;**52**(2):568–72.

124. Schwanstecher C, Meyer U, Schwanstecher M. K(IR)6.2 polymorphism predisposes to type 2 diabetes by inducing overactivity of pancreatic beta-cell ATP-sensitive K(+) channels. *Diabetes* 2002;**51**(3):875–9.

125. Nielsen EM, et al. The E23K variant of Kir6.2 associates with impaired post-OGTT serum insulin response and increased risk of type 2 diabetes. *Diabetes* 2003;**52**(2):573–7.

126. Hani EH, et al. Missense mutations in the pancreatic islet beta cell inwardly rectifying K$^+$ channel gene (KIR6.2/BIR): a meta-analysis suggests a role in the polygenic basis of Type II diabetes mellitus in Caucasians. *Diabetologia* 1998;**41**(12):1511–5.

127. Saxena R, et al. Genome-wide association analysis identifies loci for type 2 diabetes and triglyceride levels. *Science* 2007;**316**(5829):1331–6.

128. Zeggini E, et al. Replication of genome-wide association signals in UK samples reveals risk loci for type 2 diabetes. *Science* 2007;**316**(5829):1336–41.

129. Chistiakov DA, et al. Genetic variations in the pancreatic ATP-sensitive potassium channel, beta-cell dysfunction, and susceptibility to type 2 diabetes. *Acta Diabetol* 2009;**46**(1):43–9.

130. Villareal DT, et al. Kir6.2 variant E23K increases ATP-sensitive K$^+$ channel activity and is associated with impaired insulin release and enhanced insulin sensitivity in adults with normal glucose tolerance. *Diabetes* 2009;**58**(8):1869–78.

131. Barroso I, et al. Candidate gene association study in type 2 diabetes indicates a role for genes involved in beta-cell function as well as insulin action. *PLoS Biol* 2003;**1**(1):E20.

132. Sakura H, et al. Sequence variations in the human Kir6.2 gene, a subunit of the beta-cell ATP-sensitive K-channel: no association with NIDDM in while Caucasian subjects or evidence of abnormal function when expressed in vitro. *Diabetologia* 1996;**39**(10):1233–6.

133. Riedel MJ, et al. Kir6.2 polymorphisms sensitize beta-cell ATP-sensitive potassium channels to activation by acyl CoAs: a possible cellular mechanism for increased susceptibility to type 2 diabetes? *Diabetes* 2003;**52**(10):2630–5.

134. Hamming KS, et al. Coexpression of the type 2 diabetes susceptibility gene variants KCNJ11 E23K and ABCC8 S1369A alter the ATP and sulfonylurea sensitivities of the ATP-sensitive K(+) channel. *Diabetes* 2009;**58**(10):2419–24.

135. Koster JC, et al. Expression of ATP-insensitive KATP channels in pancreatic beta-cells underlies a spectrum of diabetic phenotypes. *Diabetes* 2006;**55**(11):2957–64.

136. Remedi MS, et al. Secondary consequences of beta cell inexcitability: identification and prevention in a murine model of K(ATP)-induced neonatal diabetes mellitus. *Cell Metab* 2009;**9**(2):140–51.

137. Girard CA, et al. Expression of an activating mutation in the gene encoding the KATP channel subunit Kir6.2 in mouse pancreatic beta cells recapitulates neonatal diabetes. *J Clin Invest* 2009;**119**(1):80–90.

138. Remedi MS, et al. Acute sulfonylurea therapy at disease onset can cause permanent remission of KATP-induced diabetes. *Diabetes* 2011;**60**(10).

139. Wambach JA, et al. Successful sulfonylurea treatment of an insulin-naive neonate with diabetes mellitus due to a KCNJ11 mutation. *Pediatr Diabetes* 2009;**11**(4).

140. Matthews DR, et al. UKPDS 26: sulphonylurea failure in non-insulin-dependent diabetic patients over six years. UK Prospective Diabetes Study (UKPDS) Group. *Diabet Med* 1998;**15**(4):297–303.

141. Jonas JC, et al. Chronic hyperglycemia triggers loss of pancreatic beta cell differentiation in an animal model of diabetes. *J Biol Chem* 1999;**274**(20):14112–21.

142. Talchai C, et al. Pancreatic beta cell dedifferentiation as a mechanism of diabetic beta cell failure. *Cell* 2012;**150**(6):1223–34.

143. Wang Z, et al. Pancreatic β-cell dedifferentiation in diabetes and re-differentiation following insulin therapy. *Cell Metab* 2014;**19**:872–82.

144. UKPDS-group. Intensive blood-glucose control with sulphonylureas or insulin compared with conventional treatment and risk of complications in patients with type 2 diabetes (UKPDS 33).UK Prospective Diabetes Study Group. *Lancet* 1998;**352**(9131):837–53.

145. Alvarsson M, et al. Effects of insulin vs glibenclamide in recently diagnosed patients with type 2 diabetes: a 4-year follow-up. *Diabetes Obes Metab* 2008;**10**(5):421–9.
146. Alvarsson M, et al. Beneficial effects of insulin versus sulphonylurea on insulin secretion and metabolic control in recently diagnosed type 2 diabetic patients. *Diabetes Care* 2003;**26**(8):2231–7.

The Role of Sperm Ion Channels in Reproduction

P.V. Lishko, M.R. Miller, S.A. Mansell

University of California, Berkeley, CA, United States

SPERM CELL: MORPHOLOGY AND ORGANIZATION

Sperm cells are terminally differentiated and are thought to be transcriptionally and translationally silent, meaning that spermatozoa are largely unable to synthesize new mRNA or translate it into new polypeptides.[2,3] They are divided into two easily recognized parts: the head and the tail (flagellum). The sperm head comprises a condensed nucleus, a redundant nuclear envelope, and an acrosomal vesicle. The shape and size of the sperm head differs dramatically among species: from the spatula-like head in primates and ruminants to the hook-like pointed head that usually defines rodent sperm. Motility originates from the flagellum and is powered by adenosine triphosphate (ATP) hydrolysis within the sperm tail. The flagellum has a specialized cytoskeleton called an axoneme surrounded by specialized structural components and is subdivided into three functional parts: a mitochondria-containing midpiece, a principal piece, and the endpiece (Fig. 9.1). The flagellum is in essence a motile cilium[4,5] with a flagellar plasma membrane tightly attached to all underlying structures along the sperm body. This arrangement provides spermatozoa with a ridged structure to which membrane proteins, including ion channels, can be tethered to the fibrous sheath[6] to ensure their strict compartmentalization.[7] Sperm basal motility depends on three crucial factors: the presence of ATP, a low concentration of intracellular calcium $[Ca^{2+}]_i$, and normal to alkaline intracellular pH. Intraflagellar ATP is generated during glycolysis and oxidative phosphorylation whereas the intraflagellar concentration of protons and calcium is controlled by ion channels and transporters. Because of its high motility, powered by glycolysis and oxidative phosphorylation, the sperm flagellum quickly acidifies. The removal of protons from the sperm

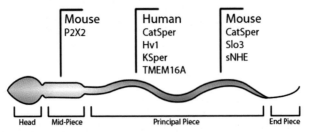

FIGURE 9.1 **Mammalian spermatozoon.** Schematic representation of spermatozoon with cellular compartments labeled and distribution of species-specific ion channels found within each section. *CatSper,* cation channels of sperm; *sNHE,* sperm-specific Na$^+$/H$^+$; *TMEM16,* Transmembrane protein 16. *Reproduced from Miller MR, Mansell SA, Meyers SA, Lishko PV. Cell Calcium. 2015 Jul;58(1):105-13, Elsevier.*

flagellum is an essential because the axonemal motility is inhibited by acidic pH. To expel protons and support motility, spermatozoa possess specialized proton channels and/or transporters. With the help of bioinformatic approaches,[8] the whole-cell sperm patch clamp method,[9,10] and several genetic models,[8–35] sperm ion channels have been comprehensively characterized, among which calcium, potassium, proton, and nonselective and various ligand-gated channels have been identified.[21,27,30,36]

CATSPER CHANNEL: A PRINCIPAL CALCIUM CHANNEL OF SPERM

In spermatozoa, swimming behavior is controlled by rises in flagellar $[Ca^{2+}]_i$ that changes the basal flagellar beat pattern through Ca^{2+}-sensing proteins called calaxins.[37,38] Calcium-bound calaxins inhibit the activity of the dynein motors within the axoneme of the cell, resulting in asymmetrical, whip-like bending of the flagellum, commonly referred to as hyperactivation.[39] This high-amplitude asymmetric flagellar bending is essential for sperm fertility because it enables them to overcome the protective vestments of the oocyte. The propagation of a Ca^{2+}-induced wave produced from the opening of Ca^{2+} channels along the flagella is an important step in sperm maturation that triggers hyperactivated motility.[40–42] Therefore the identification of the molecule comprising Ca^{2+} permeation pathways in mammalian spermatozoa represented a significant milestone in understanding the basic molecular principles of sperm activation.[8]

Discovery of the CatSper Channel and CatSper Complex Organization

The founding member of the family of cation channels of sperm (CatSper) was cloned in 2001 as a result of the bioinformatics search for the DNA sequences with similarities to other calcium channels.[8] Named

Catsper1, this gene showed selective expression only in developing sperma-tozoa,[29] and its product was likely to be a pore-forming α subunit of an ion channel related to the two-pore channels and distantly related to transient receptor potential (TRP) channels. CatSper1 contained six transmembrane (TM) segments, making it more similar to the voltage-gated potassium channels, but the ion selectivity pore resembled that of voltage-gated cal-cium channels. Male mice deficient in *Catsper1* exhibited infertility, but no other phenotypical abnormalities were noted. Female *Catsper1*-deficient mice were healthy and fertile.[8] Later studies revealed that CatSper1 func-tions as a calcium channel that controls calcium entry to mediate hyperac-tivated motility, a key flagellar function.[14] Three other homologous genes were later identified based on their similarity to *Catsper1*.[11,18,23,43] Similar to CatSper1 they also encode pore-forming subunits that contain six TM seg-ments. In addition, all four are required for fertility, which indicates that CatSper1, CatSper2, CatSper3, and CatSper4 form a functional calcium channel complex that is likely assembled as a tetramer.[28] The availability of sperm electrophysiology opened the door to comprehensive characteriza-tion of sperm channels.[9] CatSper activity was eventually directly recorded and CatSper was established as the principal Ca^{2+} channel of mouse sperm that is activated by intracellular alkalinization.[9] In addition, three other CatSper auxiliary subunits, β, γ, and δ, have been shown to colocalize and associate with the CatSper complex, suggesting that CatSper is a hetero-meric ion channel complex of at least seven different subunits.[15,22,32] Loss of any single member of the complex is detrimental to male fertility because of their interdependent protein expression in spermatozoa, although CatSperβ- and CatSperγ-null mice have not been generated.[8,21,28,30,43,44] CatSper localization is restricted to the principal piece of the sperm tail, and recent work by Chung et al. suggests that CatSper is localized to lin-early arranged Ca^{2+} signaling domains along the length of the flagellum.[44] Loss of CatSper expression results in aberrant regulation of these Ca^{2+} signaling domains, indicating that CatSper not only acts as a mechanism of Ca^{2+} entry into the cell but also as an organizing, possibly anchoring, unit of various intracellular signaling.

The Role of CatSper in Male Fertility

The critical role of the CatSper channel in human fertility was con-firmed when patients with mutations within the *CatSper1* or *CatSper2* genes were also shown to be infertile.[11,17,45–49] Human CatSper current was finally recorded in 2010 with the adaptation of the patch-clamp technique for human ejaculated sperm.[20,36] Interestingly, comparison of the CatSper currents originated from human and mouse sperm revealed signifi-cant differences in CatSper regulation. Although both currents were pH dependent, as human CatSper relies on intracellular alkalinity the same way as murine CatSper does, the voltage–current relationship of human

CatSper was different: $V_{1/2\text{human}} = +85\,\text{mV}$, whereas $V_{1/2\text{mouse}} = +11\,\text{mV}$ under similar conditions. This indicated that a smaller fraction of human CatSper channels would be open under normal physiological conditions in comparison to murine CatSper.[9,50] This suggested that to open, human CatSper requires not only a combination of intracellular alkalinization and membrane voltage but also an additional activator. This activator was later revealed to be the steroid hormone progesterone,[50,51] which is produced and released by the cumulus cells surrounding the oocyte. Progesterone has long been known as a stimulator of an immediate increase in intracellular Ca^{2+} within human sperm cells[52,53] that coincides with onset of hyperactivated motility and initiation of the acrosome reaction.[52–59] Indeed, extracellular application of progesterone resulted in fast activation of the CatSper channel, and taking into account that sperm cells are transcriptionally silent, it was suggested that the nuclear progesterone receptor was not involved in CatSper activation.[50,51] Human CatSper is acutely sensitive to progesterone, with an EC_{50} of $7\,\text{nM}$.[50] Progesterone activates the CatSper by shifting its $V_{1/2}$ toward more physiological membrane potential.[50] Further evidence for CatSper as the sole source of progesterone-induced Ca^{2+} influx into sperm flagellum came from studies on a CatSper-deficient infertile patient with a homozygous microdeletion in the *Catsper2* gene.[11,48] These CatSper-less spermatozoa did not produce any progesterone-activated current and even lacked basal CatSper activity.[48] Interestingly, rodent CatSper is insensitive to progesterone,[50] which indicates the evolutionary diversity of these channels.

CatSper activation by progesterone happens in the absence of all intracellular soluble secondary messengers, such as Ca^{2+}, ATP, and guanosine triphosphate, suggesting that the progesterone effect observed acts through a receptor directly associated with the CatSper channel and not through G-proteins or protein kinases. Thus sperm cells represent an example of obscure nongenomic progesterone signaling that also takes place in other tissues.[60] Steroid hormones control fundamental organism functions via two pathways: by modulating gene expression in the nucleus and by signaling at the plasma membrane (Fig. 9.2).[60–62]

The membrane or nongenomic pathway plays a vital role in human sperm cell activation, modulation of pain perception by dorsal root ganglion (DRG) neurons, and oocyte maturation, but the molecular determinants of this pathway are poorly understood. It is possible that DRG neurons and spermatozoa are activated similarly by progesterone and share common molecular features of this cascade. Interestingly, pain thresholds in women tend to be higher during the follicular phase of the menstrual cycle, when levels of estradiol and progesterone are increased. Transcriptionally and translationally silent spermatozoa lack conventional genomic progesterone signaling; therefore they represent an ideal

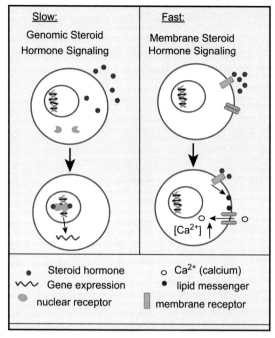

FIGURE 9.2 **Models of genomic and membrane steroid hormone actions.** According to the classical model, steroid hormones bind to a genomic (nuclear) receptor that resides in the cytoplasm. The receptor/hormone complex then migrates to the nucleus and initiates and/or alters gene expression. The timescale of such an event is slow and may take several days. According to the nongenomic (membrane) model, steroids bind to their corresponding, and in many cases yet-to-be-identified, membrane receptors, which often are associated with various ion channels or G-protein-coupled receptors. These receptors then trigger ion channel opening or potentially initiate a lipid signaling cascade, resulting in an immediate cellular response (within seconds to minutes).

model to identify the nongenomic progesterone receptor. Although it is still possible that progesterone may bind to CatSper channel directly and hence activate the channel, it may also initiate a lipid cascade through its nongenomic progesterone receptor. The latter would produce a bioactive signaling lipid, thereby altering CatSper activity. This bioactive lipid could be a true CatSper modulator, which also can alter the activity of other ion channels. Although the molecular identity of sperm progesterone receptor is still unknown, the binding site for a cell impermeant analog of this steroid is accessible from the extracellular space.[50] Furthermore, evidence for nongenomic progesterone signaling in cells lacking CatSper channel expression[13] suggests that the initiation of membrane progesterone signaling is through a yet unidentified protein separate from the CatSper complex.

Together, voltage, intracellular alkalinization, and progesterone work collectively to regulate Ca^{2+} influx into human spermatozoa, although other regulatory mechanisms may also exist. Further modulation of CatSper activity has been observed through direct recording of CatSper in the presence of prostaglandins.[21,50] Environmental toxins including those known to disrupt the endocrine system have also been shown to induce intracellular Ca^{2+} elevation via the CatSper mechanism.[63,64] The regulation of $[Ca^{2+}]_i$ is critical for proper sperm function; thus compounds that directly modulate CatSper or affect $[Ca^{2+}]_i$ pose a genuine threat to sperm fertilization potential. Alternatively, exclusive CatSper expression in sperm cells makes this channel an excellent target for novel contraceptives for both women and men.

Evolutionary Diversity of CatSper Channels

The CatSper channel is evolutionarily conserved in the genome of species from mammals to invertebrates such as sea urchins and sea squirts. However, the CatSper genes are lost in teleosts, amphibians, and birds.[13] So far, the only species with an electrophysiologically confirmed CatSper current were human and mouse. Characterization of Ca^{2+} influx within sperm of different species has been attempted using optical methods including fluorescent Ca^{2+} dyes and motility studies. For instance, bovine[65] and equine[66] sperm are both sensitive to intracellular alkalization, resulting in an influx of Ca^{2+} into the cell. However, the presence of a functional CatSper channel does not guarantee that it is regulated similarly in different species. Murine CatSper is insensitive to the classical activators of human CatSper, progesterone and prostaglandins, which may indicate that the spermatozoa of different species evolved their CatSpers to adjust to specific activators of the corresponding female reproductive tract. Additional work is needed to confirm the molecular arrangement of the sperm Ca^{2+} channel complexes within these species and characterize the role of these channels in sperm function.

VOLTAGE-GATED PROTON CHANNEL OF SPERM

Intracellular pH is a key regulator of many sperm physiological processes, including initiation of motility, capacitation, hyperactivation, chemotaxis, and acrosome reaction. Even the basal sperm motility is pH sensitive because dynein's ability to hydrolyze ATP and provide axonemal bending greatly increases with the rise of intracellular pH (pH_i). The motile sperm flagellum constantly generates intracellular protons via glycolysis, ATP hydrolysis, and proton/calcium exchange.[21] The faster a flagellum moves, the more acidic it becomes. In 1983 Babcock et al.[67,68]

suggested that the mechanism for proton efflux from bovine sperm was via a voltage-gated proton channel based on the fact that the sperm cytosol becomes alkaline upon membrane depolarization.[67,68] Later studies focused on the role of sperm-specific Na^+/H^+ (sNHE) and Cl^-/HCO_3^- exchangers as potential mechanisms for intracellular alkalization in rodent spermatozoa.[69–71] An intriguing model was recently proposed[72] suggesting that sNHE may function as a hyperpolarization activated proton extrusion mechanism because it possesses a putative voltage sensor.[69,73–75]

In 2010 direct electrophysiological recordings of human sperm revealed a large voltage-activated H^+ current[20] that was sensitive to Zn^{2+}.[20] It was shown that this sperm proton channel closely resembles the voltage-gated channel Hv1.[76,77] Interestingly, Hv1 was highly expressed in the flagellum, indicating a potential role in the regulation of CatSper.[20] Although Hv1 is ubiquitously expressed and plays an important role in several physiological processes such as the innate immune system, apoptosis, and cancer metastasis, the presence of Hv1 in human spermatozoa was intriguing. Hv1 is not a true ion channel but rather a hybrid between a transporter and an ion channel without the pore that provides rapid movement of protons across a lipid bilayer via a voltage-gated mechanism.[78–80]

Hv1: Principles of Proton Transport

The first electrophysiological recording of voltage-dependent proton current in snail neurons was performed by Roger Thomas and Robert Meech in 1982.[82] Later studies, primarily done by the Thomas DeCoursey group,[83–85] found similar currents in other cell types.[86] It was not until 2006 when two publications—one from the David Clapham group[76] and another from the Yasushi Okamura group[77]—reported the molecular identity of this channel to be voltage-gated proton channel Hv1. Molecular and electrophysiological studies of Hv1 have indicated four distinct features: (1) the channel is activated by membrane depolarization and intracellular acidification, (2) is inhibited by zinc, (3) is H^+ selective, and (4) it lacks pore domain. Hv1 forms four TM spanning segments homologous in structure to voltage-sensor of voltage-gated cation channels.[76,77] Unlike the traditional voltage-gated ion channel, Hv1 does not possess a classical selectivity pore region. Several models for the proton conductance pathway have been proposed.[78,79,87] Among them the "water wire" model[79] suggests the movement of protons via the Grotthus mechanism from the intracellular water vestibules to the extracellular vestibule formed by the TM portions of the channel. Two vestibules are connected by a narrow bottleneck[79] where proton selectivity occurs via a highly conserved and unique aspartate moiety (Asp112 in human Hv1[88]). Another model[78] suggests that the Hv1 closed state favors electrostatic interactions between hydrophobic residues of TM segments and pulls them together to form a hydrophobic plug. This hydrophobic layer is formed at the center of the voltage-sensing domain and prevents proton permeation. The latter occurs once the plug is removed by a voltage-dependent rearrangement of TMs and is followed by insertion of the protonatable residues in the center of the channel.[78] The recent crystal structure of the Hv1 in the resting state sheds[80] light on the possible mechanism of H^+ permeation, suggesting that a combination of two hydrophobic plugs and a protonatable aspartate is required for proton movement. Hv1 forms a functional dimer in the plasma membrane through a coil–coil interaction of the C-termini[89–91]; however, each Hv1 subunit can function independently as a channel.

Human Sperm Hv1

At the time of ejaculation, spermatozoa are combined with the seminal plasma, which contains high concentrations of zinc (~2 mM)[36] and as such directly inhibits Hv1 function by binding to two histidine residues that stabilize the channel in the closed state.[92] As sperm progress through the female reproductive tract, divalent zinc is chelated by proteins in the oviductal fluid, resulting in gradual activation of the Hv1 channel.[36,93] To fertilize an oocyte, sperm cells have to undergo a process known as capacitation, which comprises several physiological changes including intracellular alkalization. In humans this could be triggered via proton extrusionthrough Hv1[20], especially because Hv1 is activated by capacitation.[20] In contrast to human sperm, mouse spermatozoa do not possess Hv1, and, not surprisingly, $Hv1^{-/-}$ mice are fertile. Natural mutations in Hv1 appear to be extremely rare, with only one case of a single substitution mutation being reported in humans,[94] which was assessed only in airway epithelial cells. The role of this channel in human fertility is still unclear and will remain so until an Hv1-deficient patient can be identified. In species that lack sperm Hv1, it is still unclear how intracellular pH is regulated in these cells, although sNHE remainsas possible mechanism.

OTHER ION CHANNELS OF SPERM

Sperm membrane potential is important for fertility because both Hv1 and CatSper channels are voltage dependent (Fig. 9.3). As in other cells, sperm membrane potential is defined by the gradients of K^+, Na^+, and Cl^-, with potassium channels playing a crucial role in its regulation. Epididymal murine spermatozoa keep their membrane potential slightly depolarized at approximately −40 mV; however, they tend to hyperpolarize up to −60 mV upon in vitro capacitation.[95] This effect is attributed to

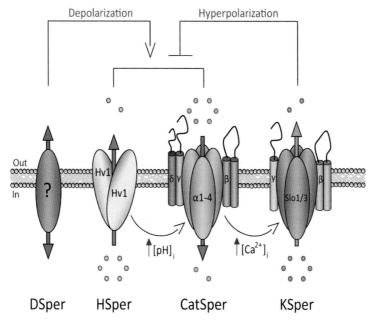

FIGURE 9.3   **Flagellar ion channels of sperm.** Murine potassium channel (KSper) is composed of the Slo3 channel and associated subunit LRRC52 (gamma) as well as possible beta subunits. The identity of human KSper is still under debate but seems to share characteristics of the Slo family of ion channels. Cation channels of sperm (CatSper) in mouse and human share the same molecular composition consisting of at least seven different subunits to form calcium-permeable channel. Proton conductance via the Hv1 channel (HSper) is only detected in human sperm cells. Both CatSper and Hv1 require membrane depolarization to function. This could be achieved by inhibition of KSper. In addition, CatSper is activated by intracellular alkalinization that could be provided by the Hv1. By moving protons out of the flagellum, Hv1 further hyperpolarizes the cell and provides negative feedback inhibiting both Hv1 and CatSper. This model misses the fourth member, yet-to-be-identified "depolarizing channel of sperm" or "DSper." This hypothetical DSper could be activated by either membrane hyperpolarization as proposed for sperm-specific Na^+/H^+ (sNHE)[87] or other mechanisms. DSper would provide positive net charge influx resulting in depolarization, consequent activation of Hv1/CatSper, and sperm hyperactivation. *Reproduced from Miller MR, Mansell SA, Meyers SA, Lishko PV. Cell Calcium. 2015 Jul;58(1):105-13, Elsevier.*

specific potassium permeability, or KSper, and two members of the Slo family of potassium channels have been recently proposed to play a role in this process.[27,31,96–103] Slo3 (*Kcnu1*), a pH-sensitive, calcium-independent, and weakly voltage-sensitive channel, has been identified as the principal potassium channel in murine sperm.[31,98,99,101–105] Although Slo3 was also expected to form the potassium channel of human sperm, recent recording from ejaculated human spermatozoa indicates that in contrast to murine sperm, human sperm potassium current (KSper) is less pH dependent and sensitive to $[Ca^{2+}]_i$.[24] Furthermore, it could also be inhibited by progesterone.[12,24] From these biophysical properties as well as the pharmacological profile, human KSper resembled the calcium-activated big conductance potassium channel Slo1 (KCNMA1).[24] Recently other models were proposed, suggesting that human capacitated spermatozoa possess a different type of Slo channel[25] or even a modified version of Slo3 that is calcium-sensitive and weakly pH-dependent,[12] an unusual set of properties for this type of ion channel. Slo channels exist as tetrameric complexes of alpha subunits (Slo1 or Slo3) and auxiliary gamma subunits (LRCC26 or LRCC52); additional beta subunits can change their biophysical and pharmacological properties.[100,106,107] Uncovering the precise molecular identity of the human potassium channel remains an essential task for understanding the regulation of potassium homeostasis in human spermatozoa.

Navarro et al. recently reported the presence of an ATP-gated P2X2 ion channel (also known as purinergic receptor P2X, ligand gated ion channel, 2) that is cation nonselective and originates from the midpiece of murine sperm cells.[108] P2rx2-deficient male mice are fertile and have normal sperm morphology and other sperm parameters; however, their sperm lack ATP-evoked currents and fertility of $P2rx2^{-/-}$ males declines with frequent mating.[108] Calcium-activated chloride channels (anoctamins) have been recently found in human sperm,[109] but not in mouse,[34] as well as aquaporins, which are water channels that are required for sperm osmoadaptation.[110] Several members of the TRP ion channel family[111–113] were thought to function in mammalian sperm cells.[16,114] These include TRPM8, TRPV1, TRPA1, and others. However, mice deficient in TRPV1–4, TRPA1, and TRPM8 have no obvious defects in sperm morphology or male fertility.[111,112] Moreover, CatSper promiscuity toward high concentrations of exogenous activators, such as menthols, may account for the mistaken identity of TRPM8 in human spermatozoa.[115] However, it is possible that other species that do not have CatSper activity rely on different flagellar channels; therefore the role of TRP channels in sperm is yet to be determined.

CONCLUSIONS

On their route to the egg, mammalian spermatozoa encounter multiple obstacles: viscous mucus, the narrow lumen of the uterotubal junction, the complex maze formed by the epithelial folds of the fallopian tubes, and finally the protective layers of the egg. To overcome these barriers, the sperm cell must sense the cues released by the egg and adapt its swimming behavior. Sperm can achieve this by increasing the amplitude and driving force of their tail bending, changing their direction of movement, and releasing special enzymes to dissolve the egg's protective vestments. Such sperm responses depend upon activity of the sperm ion channels that open in response to environmental cues within the female reproductive tract. This in turn changes conductance of the sperm plasma membrane and sperm behavior. Influx of calcium through the flagellar calcium channel CatSper results in the flagellar bending and hyperactivation of motility. In turn, CatSper is activated by intracellular alkalinization, provided by perhaps the sNHE or Hv1 channel (in human sperm) and membrane depolarization, which results from KSper inhibition by progesterone. The same compartmental flagellar localization of sperm ion channels provides fine-tuned regulation of sperm motility. It is possible that many cases of idiopathic male infertility can be attributed to malfunctioning of sperm ion channels and their regulators.

References

1. Deleted in review.
2. Kierszenbaum AL, Tres LL. Structural and transcriptional features of the mouse spermatid genome. *J Cell Biol* 1975;**65**(2):258–70.
3. Kramer JA, Krawetz SA. RNA in spermatozoa: implications for the alternative haploid genome. *Mol Hum Reprod* 1997;**3**(6):473–8.
4. Afzelius B. Electron microscopy of the sperm tail; results obtained with a new fixative. *J Biophys Biochem Cytol* 1959;**5**(2):269–78.
5. Baccetti B, Afzelius BA. The biology of the sperm cell. *Monogr Dev Biol* 1976;**10**:1–254.
6. Eddy EM, Toshimori K, O'Brien DA. Fibrous sheath of mammalian spermatozoa. *Microsc Res Tech* 2003;**61**(1):103–15.
7. Inaba K. Sperm flagella: comparative and phylogenetic perspectives of protein components. *Mol Hum Reprod* 2011;**17**(8):524–38.
8. Ren D, Navarro B, Perez G, et al. A sperm ion channel required for sperm motility and male fertility. *Nature* 2001;**413**(6856):603–9.
9. Kirichok Y, Navarro B, Clapham DE. Whole-cell patch-clamp measurements of spermatozoa reveal an alkaline-activated Ca^{2+} channel. *Nature* 2006;**439**(7077):737–40.
10. Lishko P, Clapham DE, Navarro B, Kirichok Y. Sperm patch-clamp. *Methods Enzymol* 2013;**525**:59–83.
11. Avidan N, Tamary H, Dgany O, et al. CATSPER2, a human autosomal nonsyndromic male infertility gene. *Eur J Hum Genet* 2003;**11**(7):497–502.
12. Brenker C, Zhou Y, Muller A, et al. The Ca^{2+}-activated K^+ current of human sperm is mediated by Slo3. *eLife* 2014;**3**:e01438.
13. Cai X, Clapham DE. Evolutionary genomics reveals lineage-specific gene loss and rapid evolution of a sperm-specific ion channel complex: CatSpers and CatSperbeta. *PLoS One* 2008;**3**(10):e3569.
14. Carlson AE, Westenbroek RE, Quill T, et al. CatSper1 required for evoked Ca^{2+} entry and control of flagellar function in sperm. *Proc Natl Acad Sci USA* 2003;**100**(25):14864–8.
15. Chung JJ, Navarro B, Krapivinsky G, Krapivinsky L, Clapham DE. A novel gene required for male fertility and functional CATSPER channel formation in spermatozoa. *Nat Commun* 2011;**2**:153.
16. Darszon A, Acevedo JJ, Galindo BE, et al. Sperm channel diversity and functional multiplicity. *Reproduction* 2006;**131**(6):977–88.
17. Hildebrand MS, Avenarius MR, Fellous M, et al. Genetic male infertility and mutation of CATSPER ion channels. *Eur J Hum Genet* 2010;**18**(11):1178–84.

18. Jin J, Jin N, Zheng H, et al. Catsper3 and Catsper4 are essential for sperm hyperactivated motility and male fertility in the mouse. *Biol Reprod* 2007;**77**(1):37–44.
19. Kirichok Y, Lishko PV. Rediscovering sperm ion channels with the patch-clamp technique. *Mol Hum Reprod* 2011;**17**(8):478–99.
20. Lishko PV, Botchkina IL, Fedorenko A, Kirichok Y. Acid extrusion from human spermatozoa is mediated by flagellar voltage-gated proton channel. *Cell* 2010;**140**(3):327–37.
21. Lishko PV, Kirichok Y, Ren D, Navarro B, Chung JJ, Clapham DE. The control of male fertility by spermatozoan ion channels. *Annu Rev Physiol* 2012;**74**:453–75.
22. Liu J, Xia J, Cho KH, Clapham DE, Ren D. CatSperbeta, a novel transmembrane protein in the CatSper channel complex. *J Biol Chem* 2007;**282**(26):18945–52.
23. Lobley A, Pierron V, Reynolds L, Allen L, Michalovich D. Identification of human and mouse CatSper3 and CatSper4 genes: characterisation of a common interaction domain and evidence for expression in testis. *Reprod Biol Endocrinol* 2003;**1**:53.
24. Mannowetz N, Naidoo NM, Choo SA, Smith JF, Lishko PV. Slo1 is the principal potassium channel of human spermatozoa. *eLife* 2013;**2**:e01009.
25. Mansell SA, Publicover SJ, Barratt CL, Wilson SM. Patch clamp studies of human sperm under physiological ionic conditions reveal three functionally and pharmacologically distinct cation channels. *Mol Hum Reprod* 2014;**20**(5):392–408.
26. Martinez-Lopez P, Trevino CL, de la Vega-Beltran JL, et al. TRPM8 in mouse sperm detects temperature changes and may influence the acrosome reaction. *J Cell Physiol* 2011;**226**(6):1620–31.
27. Navarro B, Kirichok Y, Chung JJ, Clapham DE. Ion channels that control fertility in mammalian spermatozoa. *Int J Dev Biol* 2008;**52**(5–6):607–13.
28. Qi H, Moran MM, Navarro B, et al. All four CatSper ion channel proteins are required for male fertility and sperm cell hyperactivated motility. *Proc Natl Acad Sci USA* 2007;**104**(4):1219–23.
29. Quill TA, Ren D, Clapham DE, Garbers DL. A voltage-gated ion channel expressed specifically in spermatozoa. *Proc Natl Acad Sci USA* 2001;**98**(22):12527–31.
30. Ren D, Xia J. Calcium signaling through CatSper channels in mammalian fertilization. *Physiology (Bethesda)* 2010;**25**(3):165–75.
31. Schreiber M, Wei A, Yuan A, Gaut J, Saito M, Salkoff L. Slo3, a novel pH-sensitive K⁺ channel from mammalian spermatocytes. *J Biol Chem* 1998;**273**(6):3509–16.
32. Wang H, Liu J, Cho KH, Ren D. A novel, single, transmembrane protein CATSPERG is associated with CATSPER1 channel protein. *Biol Reprod* 2009;**81**(3):539–44.
33. Yeung CH, Anapolski M, Depenbusch M, Zitzmann M, Cooper TG. Human sperm volume regulation. Response to physiological changes in osmolality, channel blockers and potential sperm osmolytes. *Hum Reprod* 2003;**18**(5):1029–36.
34. Zeng XH, Navarro B, Xia XM, Clapham DE, Lingle CJ. Simultaneous knockout of Slo3 and CatSper1 abolishes all alkalization- and voltage-activated current in mouse spermatozoa. *J general Physiol* 2013;**142**(3):305–13.
35. Zeng XH, Yang C, Xia XM, Liu M, Lingle CJ. SLO3 auxiliary subunit LRRC52 controls gating of sperm KSPER currents and is critical for normal fertility. *Proc Natl Acad Sci USA* 2015;**112**(8):2599–604.
36. Lishko PV, Kirichok Y. The role of Hv1 and CatSper channels in sperm activation. *J Physiol* 2010;**588**(Pt 23):4667–72.
37. Mizuno K, Padma P, Konno A, Satouh Y, Ogawa K, Inaba K. A novel neuronal calcium sensor family protein, calaxin, is a potential Ca(2+)-dependent regulator for the outer arm dynein of metazoan cilia and flagella. *Biol Cell* 2009;**101**(2):91–103.
38. Mizuno K, Shiba K, Okai M, et al. Calaxin drives sperm chemotaxis by Ca(2)(+)-mediated direct modulation of a dynein motor. *Proc Natl Acad Sci USA* 2012;**109**(50):20497–502.
39. Shiba K, Baba SA, Inoue T, Yoshida M. Ca²⁺ bursts occur around a local minimal concentration of attractant and trigger sperm chemotactic response. *Proc Natl Acad Sci USA* 2008;**105**(49):19312–7.

40. White DR, Aitken RJ. Relationship between calcium, cyclic AMP, ATP, and intracellular pH and the capacity of hamster spermatozoa to express hyperactivated motility. *Gamete Res* 1989;**22**(2):163–77.

41. Suarez SS, Varosi SM, Dai X. Intracellular calcium increases with hyperactivation in intact, moving hamster sperm and oscillates with the flagellar beat cycle. *Proc Natl Acad Sci USA* 1993;**90**(10):4660–4.

42. Ho HC, Granish KA, Suarez SS. Hyperactivated motility of bull sperm is triggered at the axoneme by Ca^{2+} and not cAMP. *Dev Biol* 2002;**250**(1):208–17.

43. Carlson AE, Quill TA, Westenbroek RE, Schuh SM, Hille B, Babcock DF. Identical phenotypes of CatSper1 and CatSper2 null sperm. *J Biol Chem* 2005;**280**(37):32238–44.

44. Chung JJ, Shim SH, Everley RA, Gygi SP, Zhuang X, Clapham DE. Structurally distinct Ca(2+) signaling domains of sperm flagella orchestrate tyrosine phosphorylation and motility. *Cell* 2014;**157**(4):808–22.

45. Zhang Y, Malekpour M, Al-Madani N, et al. Sensorineural deafness and male infertility: a contiguous gene deletion syndrome. *J Med Genet* 2007;**44**(4):233–40.

46. Avenarius MR, Hildebrand MS, Zhang Y, et al. Human male infertility caused by mutations in the CATSPER1 channel protein. *Am J Hum Genet* 2009;**84**(4):505–10.

47. Hildebrand MS, Avenarius MR, Smith RJH. *CATSPER-related male infertility*. 2009.

48. Smith JF, Syritsyna O, Fellous M, et al. Disruption of the principal, progesterone-activated sperm Ca^{2+} channel in a CatSper2-deficient infertile patient. *Proc Natl Acad Sci USA* 2013;**110**(17):6823–8.

49. Jaiswal D, Singh V, Dwivedi US, Trivedi S, Singh K. Chromosome microarray analysis: a case report of infertile brothers with CATSPER gene deletion. *Gene* 2014;**542**(2): 263–5.

50. Lishko PV, Botchkina IL, Kirichok Y. Progesterone activates the principal Ca^{2+} channel of human sperm. *Nature* 2011;**471**:387–91.

51. Strünker T, Goodwin N, Brenker C, et al. The CatSper channel mediates progesterone-induced Ca^{2+} influx in human sperm. *Nature* 2011;**471**:382–6.

52. Blackmore PF, Beebe SJ, Danforth DR, Alexander N. Progesterone and 17 alpha-hydroxyprogesterone. Novel stimulators of calcium influx in human sperm. *J Biol Chem* 1990;**265**(3):1376–80.

53. Baldi E, Casano R, Falsetti C, Krausz C, Maggi M, Forti G. Intracellular calcium accumulation and responsiveness to progesterone in capacitating human spermatozoa. *J Androl* 1991;**12**(5):323–30.

54. Osman RA, Andria ML, Jones AD, Meizel S. Steroid induced exocytosis: the human sperm acrosome reaction. *Biochem Biophys Res Commun* 1989;**160**(2):828–33.

55. Ralt D, Goldenberg M, Fetterolf P, et al. Sperm attraction to a follicular factor(s) correlates with human egg fertilizability. *Proc Natl Acad Sci USA* 1991;**88**(7):2840–4.

56. Uhler ML, Leung A, Chan SY, Wang C. Direct effects of progesterone and antiprogesterone on human sperm hyperactivated motility and acrosome reaction. *Fertil Steril* 1992;**58**(6):1191–8.

57. Roldan ER, Murase T, Shi QX. Exocytosis in spermatozoa in response to progesterone and zona pellucida. *Science* 1994;**266**(5190):1578–81.

58. Schaefer M, Hofmann T, Schultz G, Gudermann T. A new prostaglandin E receptor mediates calcium influx and acrosome reaction in human spermatozoa. *Proc Natl Acad Sci USA* 1998;**95**(6):3008–13.

59. Shimizu Y, Yorimitsu A, Maruyama Y, Kubota T, Aso T, Bronson RA. Prostaglandins induce calcium influx in human spermatozoa. *Mol Hum Reprod* 1998;**4**(6):555–61.

60. Luconi M, Francavilla F, Porazzi I, Macerola B, Forti G, Baldi E. Human spermatozoa as a model for studying membrane receptors mediating rapid nongenomic effects of progesterone and estrogens. *Steroids* 2004;**69**(8–9):553–9.

61. Revelli A, Massobrio M, Tesarik J. Nongenomic actions of steroid hormones in reproductive tissues. *Endocr Rev* 1998;**19**(1):3–17.

62. Wendler A, Albrecht C, Wehling M. Nongenomic actions of aldosterone and progesterone revisited. *Steroids* 2012;**77**(10):1002–6.
63. Schiffer C, Muller A, Egeberg DL, et al. Direct action of endocrine disrupting chemicals on human sperm. *EMBO Rep* 2014;**15**(7):758–65.
64. Tavares RS, Mansell S, Barratt CL, Wilson SM, Publicover SJ, Ramalho-Santos J. p,p′-DDE activates CatSper and compromises human sperm function at environmentally relevant concentrations. *Hum Reprod* 2013;**28**(12):3167–77.
65. Marquez B, Suarez SS. Bovine sperm hyperactivation is promoted by alkaline-stimulated Ca^{2+} influx. *Biol Reprod* 2007;**76**(4):660–5.
66. Loux SC, Crawford KR, Ing NH, et al. CatSper and the relationship of hyperactivated motility to intracellular calcium and pH kinetics in equine sperm. *Biol Reprod* 2013;**89**(5):123.
67. Babcock DF. Examination of the intracellular ionic environment and of ionophore action by null point measurements employing the fluorescein chromophore. *J Biol Chem* 1983;**258**(10):6380–9.
68. Babcock DF, Rufo Jr GA, Lardy HA. Potassium-dependent increases in cytosolic pH stimulate metabolism and motility of mammalian sperm. *Proc Natl Acad Sci USA* 1983;**80**(5):1327–31.
69. Garcia MA, Meizel S. Regulation of intracellular pH in capacitated human spermatozoa by a Na^+/H^+ exchanger. *Mol Reprod Dev* 1999;**52**(2):189–95.
70. Woo AL, James PF, Lingrel JB. Roles of the Na,K-ATPase alpha4 isoform and the Na^+/H^+ exchanger in sperm motility. *Mol Reprod Dev* 2002;**62**(3):348–56.
71. Zeng Y, Oberdorf JA, Florman HM. pH regulation in mouse sperm: identification of Na(+)-, Cl(−)-, and HCO_3(−)-dependent and arylaminobenzoate-dependent regulatory mechanisms and characterization of their roles in sperm capacitation. *Dev Biol* 1996;**173**(2):510–20.
72. Chavez JC, Ferreira Gregorio J, Butler A, et al. SLO3 K^+ Channels control calcium entry through CATSPER channels in sperm. *J Biol Chem* 2014;**289**(46).
73. Donowitz M, Ming Tse C, Fuster D. SLC9/NHE gene family, a plasma membrane and organellar family of Na(+)/H(+) exchangers. *Mol Aspects Med* 2013;**34**(2–3):236–51.
74. Wang D, Hu J, Bobulescu IA, et al. A sperm-specific Na^+/H^+ exchanger (sNHE) is critical for expression and in vivo bicarbonate regulation of the soluble adenylyl cyclase (sAC). *Proc Natl Acad Sci USA* 2007;**104**(22):9325–30.
75. Wang D, King SM, Quill TA, Doolittle LK, Garbers DL. A new sperm-specific Na^+/H^+ exchanger required for sperm motility and fertility. *Nat Cell Biol* 2003;**5**(12):1117–22.
76. Ramsey IS, Moran MM, Chong JA, Clapham DE. A voltage-gated proton-selective channel lacking the pore domain. *Nature* 2006;**440**(7088):1213–6.
77. Sasaki M, Takagi M, Okamura Y. A voltage sensor-domain protein is a voltage-gated proton channel. *Science* 2006;**312**(5773):589–92.
78. Chamberlin A, Qiu F, Rebolledo S, Wang Y, Noskov SY, Larsson HP. Hydrophobic plug functions as a gate in voltage-gated proton channels. *Proc Natl Acad Sci USA* 2014;**111**(2):E273–82.
79. Ramsey IS, Mokrab Y, Carvacho I, Sands ZA, Sansom MS, Clapham DE. An aqueous H^+ permeation pathway in the voltage-gated proton channel Hv1. *Nat Struct Mol Biol* 2010;**17**(7):869–75.
80. Takeshita K, Sakata S, Yamashita E, et al. X-ray crystal structure of voltage-gated proton channel. *Nat Struct Mol Biol* 2014;**21**(4):352–7.
81. Fogel M, Hastings JW. Bioluminescence: mechanism and mode of control of scintillon activity. *Proc Natl Acad Sci USA* 1972;**69**(3):690–3.
82. Thomas RC, Meech RW. Hydrogen ion currents and intracellular pH in depolarized voltage-clamped snail neurones. *Nature* 1982;**299**(5886):826–8.
83. DeCoursey TE. Hydrogen ion currents in rat alveolar epithelial cells. *Biophys J* 1991;**60**(5):1243–53.

84. DeCoursey TE, Cherny VV. Potential, pH, and arachidonate gate hydrogen ion currents in human neutrophils. *Biophys J* 1993;**65**(4):1590–8.
85. Cherny VV, Markin VS, DeCoursey TE. The voltage-activated hydrogen ion conductance in rat alveolar epithelial cells is determined by the pH gradient. *J Gen Physiol* 1995;**105**(6):861–96.
86. DeCoursey TE. Voltage-gated proton channels: molecular biology, physiology, and pathophysiology of the H(V) family. *Physiol Rev* 2013;**93**(2):599–652.
87. Hong L, Pathak MM, Kim IH, Ta D, Tombola F. Voltage-sensing domain of voltage-gated proton channel Hv1 shares mechanism of block with pore domains. *Neuron* 2013;**77**(2):274–87.
88. Musset B, Smith SM, Rajan S, Morgan D, Cherny VV, Decoursey TE. Aspartate 112 is the selectivity filter of the human voltage-gated proton channel. *Nature* 2011;**480**(7376):273–7.
89. Koch HP, Kurokawa T, Okochi Y, Sasaki M, Okamura Y, Larsson HP. Multimeric nature of voltage-gated proton channels. *Proc Natl Acad Sci USA* 2008;**105**(26):9111–6.
90. Lee SY, Letts JA, Mackinnon R. Dimeric subunit stoichiometry of the human voltage-dependent proton channel Hv1. *Proc Natl Acad Sci USA* 2008;**105**(22):7692–5.
91. Tombola F, Ulbrich MH, Isacoff EY. The voltage-gated proton channel Hv1 has two pores, each controlled by one voltage sensor. *Neuron* 2008;**58**(4):546–56.
92. Cherny VV, DeCoursey TE. pH-dependent inhibition of voltage-gated H(+) currents in rat alveolar epithelial cells by Zn(2+) and other divalent cations. *J Gen Physiol* 1999;**114**(6):819–38.
93. Lu J, Stewart AJ, Sadler PJ, Pinheiro TJ, Blindauer CA. Albumin as a zinc carrier: properties of its high-affinity zinc-binding site. *Biochem Soc Trans* 2008;**36**(Pt 6):1317–21.
94. Iovannisci D, Illek B, Fischer H. Function of the HVCN1 proton channel in airway epithelia and a naturally occurring mutation, M91T. *J Gen Physiol* 2010;**136**(1):35–46.
95. Munoz-Garay C, De la Vega-Beltran JL, Delgado R, Labarca P, Felix R, Darszon A. Inwardly rectifying K(+) channels in spermatogenic cells: functional expression and implication in sperm capacitation. *Dev Biol* 2001;**234**(1):261–74.
96. Navarro B, Kirichok Y, Clapham DE. KSper, a pH-sensitive K$^+$ current that controls sperm membrane potential. *Proc Natl Acad Sci USA* 2007;**104**(18):7688–92.
97. Nishigaki T, Jose O, Gonzalez-Cota AL, Romero F, Trevino CL, Darszon A. Intracellular pH in sperm physiology. *Biochem Biophys Res Commun* 2014;**450**(3).
98. Santi CM, Martinez-Lopez P, de la Vega-Beltran JL, et al. The SLO3 sperm-specific potassium channel plays a vital role in male fertility. *FEBS Lett* 2010;**584**(5):1041–6.
99. Tang QY, Zhang Z, Xia XM, Lingle CJ. Block of mouse Slo1 and Slo3 K$^+$ channels by CTX, IbTX, TEA, 4-AP and quinidine. *Channels (Austin, Tex)* 2010;**4**(1):22–41.
100. Yang C, Zeng XH, Zhou Y, Xia XM, Lingle CJ. LRRC52 (leucine-rich-repeat-containing protein 52), a testis-specific auxiliary subunit of the alkalization-activated Slo3 channel. *Proc Natl Acad Sci USA* 2011;**108**(48):19419–24.
101. Yang CT, Zeng XH, Xia XM, Lingle CJ. Interactions between beta subunits of the KCNMB family and Slo3: beta4 selectively modulates Slo3 expression and function. *PLoS One* 2009;**4**(7):e6135.
102. Zeng XH, Yang C, Kim ST, Lingle CJ, Xia XM. Deletion of the Slo3 gene abolishes alkalization-activated K$^+$ current in mouse spermatozoa. *Proc Natl Acad Sci USA* 2011;**108**(14):5879–84.
103. Zhang X, Zeng X, Lingle CJ. Slo3 K$^+$ channels: voltage and pH dependence of macroscopic currents. *J Gen Physiol* 2006;**128**(3):317–36.
104. Leonetti MD, Yuan P, Hsiung Y, Mackinnon R. Functional and structural analysis of the human SLO3 pH- and voltage-gated K$^+$ channel. *Proc Natl Acad Sci USA* 2012;**109**(47):19274–9.
105. Tang QY, Zhang Z, Xia J, Ren D, Logothetis DE. Phosphatidylinositol 4,5-bisphosphate activates Slo3 currents and its hydrolysis underlies the epidermal growth factor-induced current inhibition. *J Biol Chem* 2010;**285**(25):19259–66.

106. Behrens R, Nolting A, Reimann F, Schwarz M, Waldschutz R, Pongs O. hKCNMB3 and hKCNMB4, cloning and characterization of two members of the large-conductance calcium-activated potassium channel beta subunit family. *FEBS Lett* 2000;**474**(1):99–106.

107. Yan J, Aldrich RW. LRRC26 auxiliary protein allows BK channel activation at resting voltage without calcium. *Nature* 2010;**466**(7305):513–6.

108. Navarro B, Miki K, Clapham DE. ATP-activated P2X2 current in mouse spermatozoa. *Proc Natl Acad Sci USA* 2011;**103**(34).

109. Orta G, Ferreira G, Jose O, Trevino CL, Beltran C, Darszon A. Human spermatozoa possess a calcium-dependent chloride channel that may participate in the acrosomal reaction. *J Physiol* 2012;**590**(Pt 11):2659–75.

110. Chen Q, Peng H, Lei L, et al. Aquaporin3 is a sperm water channel essential for post-copulatory sperm osmoadaptation and migration. *Cell Res* 2011;**21**(6):922–33.

111. Clapham DE, Montell C, Schultz G, Julius D, International Union of Pharmacology. International Union of Pharmacology. XLIII. Compendium of voltage-gated ion channels: transient receptor potential channels. *Pharmacol Rev* 2003;**55**(4):591–6.

112. Julius D. TRP channels and pain. *Annu Rev Cell Dev Biol* 2013;**29**:355–84.

113. Montell C, Birnbaumer L, Flockerzi V, et al. A unified nomenclature for the superfamily of TRP cation channels. *Mol Cell* 2002;**9**(2):229–31.

114. Jungnickel MK, Marrero H, Birnbaumer L, Lemos JR, Florman HM. Trp2 regulates entry of Ca^{2+} into mouse sperm triggered by egg ZP3. *Nat Cell Biol* 2001;**3**(5):499–502.

115. Brenker C, Goodwin N, Weyand I, et al. The CatSper channel: a polymodal chemosensor in human sperm. *EMBO J* 2012;**31**(7):1654–65.

Mutations of Sodium Channel SCN8A (Na$_V$1.6) in Neurological Disease

J.L. Wagnon, R.K. Bunton-Stasyshyn,
M.H. Meisler

University of Michigan, Ann Arbor, MI, United States

EVOLUTION OF THE SODIUM CHANNEL GENE FAMILY

SCN8A is a member of the multigene family of voltage-gated sodium channels that includes *SCN1A* and *SCN2A*. These three major neuronal sodium channels share 70–85% amino acid sequence identity and are derived from the ancestral prokaryotic and invertebrate sodium channels through a series of duplication events. As a result, mammalian genomes contain four chromosomal loci with a total of nine functional voltage-gated sodium channel α subunits[1–4] (Fig. 10.1A). The large pore-forming α subunits (~260 kDa), in association with the small sodium channel β subunits (33–36 kDa), are localized in a protein complex that regulates cell membrane excitability. The tissue specificity of the mammalian sodium channel genes and their encoded proteins is summarized in Fig. 10.1B. Differential sensitivity to tetrodotoxin (TTX), which is used to distinguish among the channels, is also indicated. Mutations in the sodium channel α subunit genes have been linked to epilepsy, neuropsychiatric disease, cardiac disease, chronic pain, and motor disorders.[5–9]

Mutations that inactivate *Scn1a*, *Scn2a*, or *Scn8a* in the mouse cause early lethality, demonstrating that each of the major neuronal channels has unique and nonredundant functions. Mutations of human *SCN1A* are found in more than 80% of individuals with Dravet Syndrome, also known as EIEE6 (OMIM #607208). Mutations of *SCN2A* contribute to

(A)

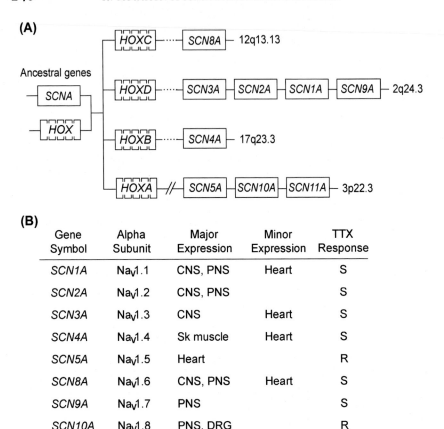

(B)

Gene Symbol	Alpha Subunit	Major Expression	Minor Expression	TTX Response
SCN1A	Na$_v$1.1	CNS, PNS	Heart	S
SCN2A	Na$_v$1.2	CNS, PNS		S
SCN3A	Na$_v$1.3	CNS	Heart	S
SCN4A	Na$_v$1.4	Sk muscle	Heart	S
SCN5A	Na$_v$1.5	Heart		R
SCN8A	Na$_v$1.6	CNS, PNS	Heart	S
SCN9A	Na$_v$1.7	PNS		S
SCN10A	Na$_v$1.8	PNS, DRG		R
SCN11A	Na$_v$1.9	PNS, DRG		R

FIGURE 10.1 **Voltage-gated sodium channel α subunit gene family.** (A) Chromosomal linkage. The sodium channel genes are located in four paralogous chromosome segments that were generated by genome duplications in a prevertebrate ancestor that also affected the linked HOX gene cluster.[3] (B) Gene characteristics. *CNS*, central nervous system; *PNS*, peripheral nervous system; *Sk*, skeletal muscle; *DRG*, dorsal root ganglion; *TTX-S*, tetrodotoxin-sensitive channels inhibited by nanomolar concentration of TTX; *TTX-R*, tetrodotoxin-resistant channels inhibited by micromolar concentration of TTX.

autism, intellectual disability, and epileptic encephalopathy (EIEE11). In this chapter, we focus on the unique biological features of Na$_v$1.6, the channel encoded by *SCN8A* (EIEE13) and its emerging role in human disease.

GENE STRUCTURE, TRANSCRIPTION, AND ALTERNATIVELY SPLICED EXONS OF SCN8A

The *SCN8A* gene is located on human chromosome 12q13.13 and the orthologous segment of mouse distal chromosome 15. The 26 coding

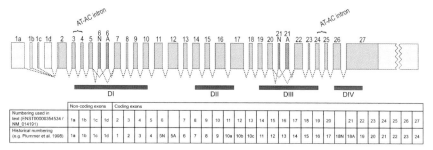

FIGURE 10.2   **Exon organization of *SCN8A*.** The upstream noncoding exons 1a–1d were identified by 5′ RACE.[15] Two minor class introns with AT-AC splice sites are indicated.[10] Two pairs of alternatively spliced exons are shown in *pink*. UTRs (untranslated regions) are shown in *light blue*.

exons, 2 pairs of alternatively spliced exons, and upstream noncoding exons of *SCN8A* (Fig. 10.2) span 170 kb and encode a protein of 1980 residues[10] (GenBank AH007414.1). The full-length *SCN8A* transcript is expressed throughout the brain and expression increases after birth, reaching adult level between 2 and 4 weeks of age in the mouse[11–14] and later than 9 months of age in human.[15]

Four alternative transcription start sites for *Scn8a* were mapped to a cluster of noncoding exons located 70 kb upstream of the translation start site[16] (Fig. 10.2). Exon 1c appears to contain the ancestral start site as it is evolutionarily conserved in vertebrate species including fish. Upstream of this noncoding exon there are predicted binding sites for the neuronal transcription factors Pou6f1/Brn5, YY1, and REST/NRSF.[17] A 470 bp promoter construct containing exon 1c and the splice acceptor from the first coding exon of *Scn8a* were sufficient to direct LacZ expression to neurons of transgenic mice.[17]

SCN8A contains two pairs of alternatively spliced, mutually exclusive coding exons whose regulated expression modulates channel function. Exons 6N and 6A (previously referred to as 5N and 5A, see Fig. 10.2) encode the S3–S4 transmembrane segments of domain I.[10] At the protein level, the neonatal (N) and adult (A) isoforms are distinguished by substitution of one neutral amino acid residue in the neonatal form by an aspartate residue in the adult form, close to the voltage sensor of the channel.[10] In newborn mice, the neonatal exon 6N accounts for approximately 40% of *Scn8a* transcripts in cortex and hippocampus, but by postnatal day 15 this is reduced to 20%.[12] There are functional differences between the corresponding isoforms of *SCN2A*, with the neonatal splice form being less excitable than the adult form.[18,19] The missense mutation p.Leu1563Val in a patient with benign familial neonatal-infantile seizures increases the excitability of the neonatal form but not the adult form, which may explain the resolution of seizures with age in this individual.[19]

The alternatively spliced exons 21N and 21A in *SCN8A* (previously referred to as exons 18N and 18A, see Fig. 10.2) encode the S3–S4 transmembrane segments of domain III.[10] Exon 21N contains an inframe stop codon and is expressed at a low level in many nonneuronal cells, including glia.[14,15] The low-level transcript expressed in nonneuronal cells encodes a truncated, two-domain protein that is predicted to lack channel activity.[14] Inclusion of exon 21A instead of 21N is promoted by the neuronal splice factors RBFOX1 and RBFOX2 and leads to neuron-specific expression of the full-length Na$_V$1.6.[15,20,21] Splice enhancers and silencers within in exons 21A and 21N are also reported to regulate temporal and spatial expression.[21] Alternative sites of polyadenylation are located 4 and 6.5kb downstream of the translation termination codon and generate two major transcripts of 9 and 12kb that are present at comparable abundance in brain RNA.[16] No specific functions have been associated with the alternatively polyadenylated transcripts.

Na$_V$1.6 PROTEIN STRUCTURE AND FUNCTION

The sodium channel α subunit is comprised of four homologous domains, DI–DIV, each containing six transmembrane segments, S1–S6[22] (Fig. 10.3). Two large cytoplasmic loops separate DI from DII, and DII from DIII (Fig. 10.3). DIII and DIV are separated by a short, highly conserved

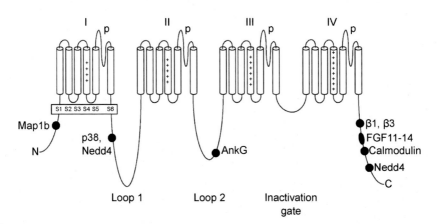

FIGURE 10.3 **Architecture of the Na$_V$1.6 channel.** Four homologous protein domains, I–IV, each contain six transmembrane segments, S1–S6, which are highly evolutionarily conserved. The large interdomain cytoplasmic loops 1 and 2 are less well conserved. The short loop 3 functions as the inactivation gate. *p*, pore loops; +, positively charged residues in S4 segments that function in voltage sensitivity. *Circles* mark sites of protein interaction: MAP1B (77–80), p38 (553), ankyrin G (1089–1122), calmodulin (1902–1912), and NEDD4 (551–554 and 1943–1945); β1, includes residue 1846; FGF, fibroblast growth factor.

inactivation gate. The sodium channels are among the most highly conserved proteins in the mammalian genome,[23] especially within the transmembrane segments, where *SCN8A* exhibits 50% amino acid sequence identity to invertebrate sodium channels.

In each domain, transmembrane segment S4 contains four to eight positively charged residues, most often arginine. These S4 segments function as voltage sensors, mediating channel opening in response to changes in resting membrane potential of the neuron. The sodium-selective pore is formed by the residues connecting segments S5 and S6 in the four domains.

The opening of Na$_V$1.6 permits the transient flow of sodium ions down a concentration gradient into the neuron to initiate or propagate action potentials. Rapid inactivation of the channel follows within milliseconds.[24] By 100 ms after initiation of the action potential, the persisting inward sodium current is reduced to between 1% and 10% of peak current, depending on cell type.[25-27] This persistent current (I_{NaP}) is higher for Na$_V$1.6 than the other major sodium channels[25] and exhibits slow inactivation.[28] I_{NaP} operates in the subthreshold voltage range where it modulates near-threshold membrane potentials, amplifies synaptic currents, and facilitates repetitive firing.[29-31] I_{NaP} can amplify the neuronal response to excitation, and an increase of only a few percent can dramatically alter cell firing by reducing the threshold for action potentials and enhancing repetitive firing. Mutations that increase Na$_V$1.6 persistent sodium current can lead to seizure activity.[27,32,33]

SUBCELLULAR LOCALIZATION OF Na$_V$1.6 IN NEURONS

The axon initial segment (AIS) is the cellular membrane domain near the proximal end of the axon that contains a concentrated cluster of sodium and potassium channels that are estimated to be much higher than in other membrane domains, based on intensity of immunostaining.[34] As a result, the threshold for initiation of action potentials is lowest at the AIS.[35-38] Electrical signals from dendrites and soma are summed at the AIS, and the channel composition of the AIS appears to determine whether an action potential is initiated.[34] Differences in firing threshold between neurons are largely determined by the composition of the AIS.[34,39]

Na$_V$1.6 is the major sodium channel at the AIS of both excitatory and inhibitory neurons. Na$_V$1.6 is also the major sodium channel in mature nodes of Ranvier in myelinated axons[40-45] (Fig. 10.4). At lower concentration, Na$_V$1.6 is detected by immunostaining in neuronal soma and dendrites and in nonmyelinated axons.[43,46]

During the postnatal development of excitatory neurons in the mouse, the AIS is first occupied by Na$_V$1.2, which is replaced by Na$_V$1.6 between

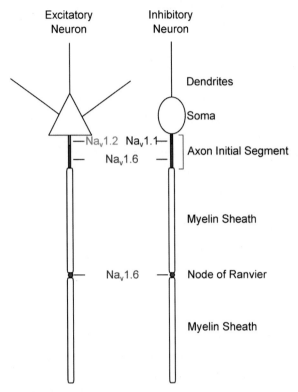

FIGURE 10.4   **Subcellular localization of sodium channels in neurons.** $Na_v1.6$ is localized to the distal axon initial segment (AIS) and to the nodes of Ranvier in excitatory and inhibitory neurons. The α subunits $Na_v1.2$ and $Na_v1.1$ are localized to the proximal AIS in excitatory and inhibitory neurons, respectively.

weeks 2 and 4 of postnatal life.[13,41] There is a similar developmental switch in sodium channel composition of the nodes of Ranvier.[40] In mature excitatory and inhibitory neurons, $Na_v1.6$ is concentrated at the distal end of the AIS[34,45] (Fig. 10.4), where it regulates the initiation of action potentials.[35] $Na_v1.1$ and $Na_v1.2$ are concentrated at the proximal end of the AIS[34,38,44] (Fig. 10.4), where they contribute to backpropagation of action potentials.[35] As a consequence of its localization at the distal AIS, mutations affecting the properties of $Na_v1.6$ can have a large impact on neuronal firing.

SUBCELLULAR LOCALIZATION OF $Na_V1.6$ IN CARDIAC MYOCYTES

In addition to its major expression in neurons, full-length $Na_v1.6$ is expressed at a low level in the heart. The TTX-resistant α subunit $Na_v1.5$ is the major sodium channel in the cell membrane of cardiac myocytes

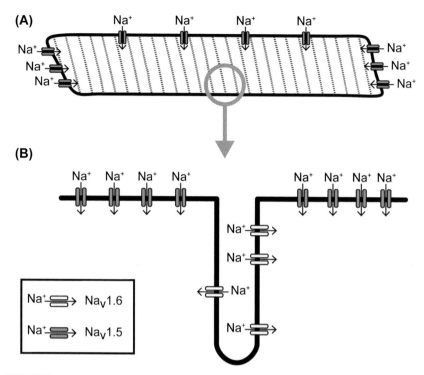

FIGURE 10.5 **Subcellular localization of sodium channels in cardiac myocytes.** (A) The TTX (tetrodotoxin)-resistant α subunit Na$_V$1.5 is expressed at a high level and is located on the cell surface and at intercalated disks. (B) TTX-sensitive α subunits, including Na$_V$1.6, are expressed at low levels at the transverse tubules.

and at the intercalated disk, where intercellular electrical signals are generated.[47,48] Mutations of *SCN5A* are a major cause of cardiac disease including long QT syndrome and Brugada syndrome (see Chapter 14: Complexity of Molecular Genetics in the Inherited Cardiac Arrhythmias). The TTX-sensitive channels Na$_V$1.6 and Na$_V$1.1 are the major channels in the rat sinoatrial node, the pacemaking tissue where sinus rhythm is generated.[49,50] In ventricular myocytes, TTX-sensitive channels appear to be localized to the transverse tubules (Fig. 10.5), where they coordinate excitation–contraction coupling.[47,48,51–57] Low concentrations of TTX that inhibit Na$_V$1.1 and Na$_V$1.6 desynchronize excitation–contraction coupling and reduce cardiac contractility.[48] In Na$_V$1.6 null mice studied at 3 weeks of age, TTX-sensitive current in isolated ventricular myocytes is reduced by 50–60%.[55] Concurrently, PR and QRS intervals and action potential duration are prolonged, and calcium transient duration is increased.[55] Conversely, expression of a hyperactive Na$_V$1.6 mutation caused prolongation of the early repolarization phases of the action potential and generated delayed afterdepolarizations (unpublished observations).

BIOPHYSICAL PROPERTIES OF Na$_V$1.6 AND THEIR INFLUENCE ON NEURONAL FIRING PATTERNS

The activation threshold for Na$_V$1.6 is lower than that of the other neuronal sodium channels. In transfected neurons, the voltage dependence of rapid activation of sodium currents for Na$_V$1.6 is shifted by 15mV leftward compared with transfected Na$_V$1.2, indicating that Na$_V$1.6 would be activated earlier during depolarization.[58] Na$_V$1.6 is also less likely to inactivate at high stimulation frequencies (20–100Hz)[58] and exhibits a more positive voltage dependence of slow inactivation, passing ~10% more current in the −35 to −25mV range than Na$_V$1.2.[25]

In null neurons lacking expression of *Scn8a*, Na$_V$1.6 is replaced in the distal AIS by Na$_V$1.1 and Na$_V$1.2.[59,60] In hippocampal CA1 pyramidal neurons, this change results in a 5mV depolarizing (rightward) shift in the voltage dependence of activation, a 60% reduction in persistent current, and a 75% reduction in resurgent current.[61] *Scn8a* null neurons are thus less excitable than wild-type cells, with an 8mV depolarizing shift in the threshold for spike initiation.[61]

Another unique property of Na$_V$1.6 is its high proportion of persistent current. In transfected human embryonic kidney cells, Na$_V$1.6 generates fivefold more persistent current than Na$_V$1.2.[25] Persistent current facilitates the generation of rapid bursts of action potentials, and in the absence of *Scn8a* expression, repetitive firing is greatly reduced.[59,62–65] In cerebellar Purkinje cells from *Scn8a* null mice, persistent current was reduced to 30% wild-type levels, and repetitive firing was eliminated.[65]

Resurgent current is a small, transient current that is initiated during repolarization after an action potential is fired, facilitating repetitive firing of additional action potentials.[66] In cerebellar Purkinje cells lacking Na$_V$1.6, the reduction in resurgent current contributes to loss of spontaneous firing and multipeaked action potentials.[62,65,67]

These unique biophysical characteristics of Na$_V$1.6 contribute to its important role in neuronal excitability. Perturbation of Na$_V$1.6 kinetics by mutation alters neuronal excitability and causes disease.

PROTEIN INTERACTIONS WITH Na$_V$1.6

At the AIS and nodes of Ranvier, Na$_V$1.6 is localized within large macromolecular complexes. Binding sites for some colocalized proteins are shown in Fig. 10.3. The light chain of microtubule-associated protein MAP1B binds at the cytoplasmic N-terminus of Na$_V$1.6 and enhances its trafficking to the cell surface.[68] Binding of ankyrin G to the second intracellular loop of Na$_V$1.6 is necessary and sufficient for localization to the AIS and nodes of Ranvier.[69–72] The E3 ubiquitin ligase NEDD4 interacts with Na$_V$1.6 at the PXpS/pTP motif at residues 551–554 of the DI/DII loop, and the PXY motif at residues 1943–1945 of the cytoplasmic C-terminus.[73–75] Phosphorylation of Na$_V$1.6 by the stress-activated MAP kinase p38 facilitates binding of NEDD4 and reduces peak current amplitude.[73,76] Ubiquitination of Na$_V$1.6 by NEDD4 may target Na$_V$1.6 for internalization and degradation as part of the neuronal stress response.[73]

Intracellular fibroblast growth factors FGF11, FGF12, FGF13, and FGF-14 bind to the C-terminus of Na$_V$1.6 (Fig. 10.3) and to the corresponding positions in Na$_V$1.2 and Na$_V$1.5.[77–80] Binding of FGF13-1a and 1b increases Na$_V$1.6 current density.[81,82] FGF13-1a enhances the accumulation of inactivated channels, while FGF13-1b inhibits accumulation of inactivated channels.[81] FGF14-1b suppresses Na$_V$1.6 current density.[78] Knockout of *Fgf14* in the mouse reduces Na$_V$1.6 concentration at the AIS of cerebellar Purkinje cells, resulting in reduced repetitive firing and development of ataxia, phenotypes that resemble the *Scn8a* null mice.[60,83] The missense mutation p.Phe150Ser in *FGF14* was identified in a Dutch family with autosomal dominant spinocerebellar ataxia. This mutation disrupts interaction of FGF14 with neuronal sodium channels, including Na$_V$1.6 and reduces neuronal excitability.[77,78,84]

The sodium channel subunits β1, β1B, β2, β3, and β4 are single-transmembrane cell-adhesion molecules that modulate the cell surface expression and kinetics of the α subunit.[85–87] Beta subunits are important regulators of central nervous system development, including cell proliferation and migration, axon outgrowth, and neuronal pathfinding.[88] Stimulation of neurite outgrowth in cerebellar granule neurons by β1 is dependent on sodium influx mediated by Na$_V$1.6.[89] Knockout of β1 in the mouse reduces Na$_V$1.6 concentration at the AIS of cerebellar granule neurons and causes impaired excitability and reduced resurgent current.[89] These mice exhibit seizures, ataxia, and cardiac arrhythmias.[90–92] Mutations of human β1 are responsible for some cases of inherited epileptic encephalopathy.[93,94]

The calcium sensor calmodulin binds the IQ motif at residues 1902–1912[95] in the C-terminal domain of Na$_V$1.6. Apo-calmodulin (calcium-free) accelerates inactivation of Na$_V$1.6.[96] Mutations of Na$_V$1.6 residues Arg1902, Tyr1904, and Arg1905 in the IQ motif reduce peak sodium current in the absence of calcium.[97] Residues upstream of the IQ motif may also have an effect upon sodium channel–calmodulin interaction. The mutation p.Arg1902Cys in a family with autism[98] is located 10 residues upstream of the IQ motif of *SCN2A*[99] and reduces binding affinity for apo-calmodulin.[80]

Mutations in proteins that bind and regulate Na$_V$1.6 may also generate disorders caused by mutation of *SCN8A*, as demonstrated by the *FGF14* mutation described earlier. Common variants of these binding partners may act as genetic modifiers of the clinical severity resulting from mutation of *SCN8A*.

MUTATIONS OF *SCN8A* IN MICE WITH MOVEMENT DISORDERS

More than a dozen mutations of *Scn8a* have been identified in mutant mice with visible movement disorders that include tremor, ataxia, dystonia, and hindlimb paralysis[100–114] (Table 10.1). Most of these mutations cause partial or complete loss of $Na_v1.6$ function and exhibit recessive inheritance. The most severe effects are seen in null mice, which have onset of motor defects within 2 weeks after birth, when $Na_v1.6$ replaces $Na_v1.1$ and $Na_v1.2$ at the AIS and nodes of Ranvier. Ataxia and tremor progress to paralysis and death by 3 weeks of age in *Scn8a* null mice. Protein truncation mutations and the missense mutation p.Ser21Pro generate the null phenotype.[108]

Partial loss-of-function mutations of mouse *Scn8a* result in chronic, progressive movement abnormalities including ataxia, tremor, muscle weakness, and dystonia (Table 10.1). In *Scn8a^medjo* mice, the mutation p.Ala1319Thr in DIII S4–S5 causes a depolarizing (rightward) shift in the voltage dependence of channel activation. This reduction in channel excitability results in impaired Purkinje cell firing and cerebellar ataxia.[106] In *Scn8a^8J* mice, the mutation p.Val929Phe in the pore region of DII causes a similar depolarizing shift in voltage dependence of activation and a hyperpolarizing shift in the voltage dependence of inactivation.[115] Both of these cause a reduction in channel activity leading to ataxia and juvenile lethality.[115,116]

Heterozygous carriers of loss-of-function mutations exhibit mild abnormalities including disrupted sleep architecture and elevated anxiety that may be models for related human disorders.[117–119] On certain strain backgrounds, heterozygous null mice exhibit spike-wave discharges on electroencephalogram recordings, the hallmark of absence (nonconvulsive) epilepsy.[116] Interestingly, the *Scn8a* heterozygous null genotype has the beneficial effect of reducing the severity of epilepsy resulting from mutations of *Scn1a*, the β1 subunit, and genetically complex epilepsy.[120,121] The protective effect of haploinsufficiency of $Na_v1.6$ is probably mediated by reduction of neuronal excitability. These observations in mice have led to the suggestion that downregulation of $Na_v1.6$ could be a therapeutic strategy for patients with epileptic encephalopathy, and this approach is currently under investigation. It would be important to titrate the extent of $Na_v1.6$ inhibition to avoid the severe dystonia and other movement disorders seen in mice with less than 50% of the normal level of $Na_v1.6$.

TABLE 10.1 Phenotypic mutants of mouse *Scn8a*

Allele	Mutation	Domain	Effect on Na$_v$1.6 function	Phenotype	Refs.
med	LINE insertion, exon 2	N/A	No functional channel	Null[a]	104
med-J	4 bp deletion in splice site, exon 3	N/A	5 to 10% functional channel, strain dependent	Dystonia; splicing modified by *Scnm1*	104, 111
med-tg	Transgene insertion, 20 kb deletion	N/A	No functional channel	Null	105
ataxia3	p.Ser21Pro	N-term	Approximately 5% functional channel	Null; retention in Golgi	108
dmu	c.1538delA	Loop 1	No functional channel	Null	102
m10J	p.Arg914Ser	DIIS5-6	Predict no activity	Null	114
8J	p.Val929Phe	DIIS5-6	Reduced activity	SWD in heterozygotes; altered voltage dependence of activation/inactivation	115, 116
tremorD	p.Trp935Leu	DIIS5-6	Predict reduced activity	Tremor and gait disturbance	112
clth	p.Asp981Val	Near DIIS6	Predict reduced activity	Hearing loss	107
med-jo	p.Ala1319Thr	DIIIS4-5	Altered activity	Ataxia, jolting gate; altered voltage dependence of activation	106, 109
nmf2	p.Asn1370Thr	DIIIS5-6	Predict no activity	Null	100
nmf5	p.Ile1392Phe	DIIIS5-6	Predict no activity	Null	100
nmf58	p.Leu1404His	DIIIS5-6	Predict reduced activity	Ataxia	100
dan	p.Lys1570X	DIVS2	Predict no activity	Null	113
nymph	p.Tyr1577Cys	DIVS2-3	Predict reduced activity	Tremor, ataxia, small size	110
9J	p.Ile1750del	DIVS6	Partial loss of function	Movement disorder; impaired glycosylation	103

[a] *The null phenotype includes tremor, progressive hindlimb paralysis, and juvenile lethality at 3 weeks of age. SWD, spike wave discharge.*

DISCOVERY OF MUTATIONS OF HUMAN *SCN8A*

The role of *SCN8A* in human disease was first examined by a candidate gene approach in families segregating inherited movement disorders such as ataxia, dystonia, and tremor. One family with a mutation of *SCN8A* was identified by this approach.[122] The proband, with ataxia and intellectual disability, was heterozygous for a protein truncation mutation of *SCN8A*. Three other heterozygous family members had intellectual disabilities but did not exhibit ataxia. A screen of 80 patients with familial essential tremor did not detect any *SCN8A* variants,[123] and analysis of 80 ataxia families was also negative (unpublished observations).

Large-scale sequencing of patients with nonfamilial epilepsy has revealed a large number of pathogenic mutations of *SCN8A*. The first de novo mutation was found in a patient with early infantile epileptic encephalopathy (EIEE) in 2012.[27] Since then, more than 60 de novo mutations of *SCN8A* have been identified by genome and exome sequencing and by the inclusion of *SCN8A* in commercial epilepsy gene panels.[124–135]

MUTATIONS OF SCN8A IN EARLY INFANTILE EPILEPTIC ENCEPHALOPATHY (EIEE13)

More than 100 mutations of *SCN8A* have been identified to date in patients with epileptic encephalopathy, a severe, early onset seizure disorder with developmental delay. All of the causal mutations are missense mutations that arose de novo in the patient or were inherited from a mosaic parent in two cases. The locations of these mutations in $Na_v1.6$ are shown in Fig. 10.6 and described in Table 10.2. All of the missense mutations result in substitution of an evolutionarily conserved amino acid residue.

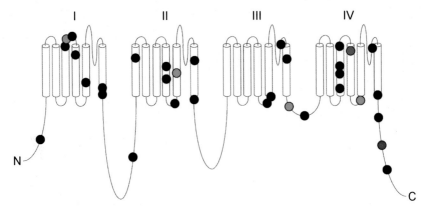

FIGURE 10.6 **Positions of de novo missense mutations of *SCN8A* in patients with EIEE13 or intellectual disability.** Number of individual cases: *Black*, 1; *green*, 2; *blue*, 5; *red*, 6.

TABLE 10.2 *De novo* mutations of SCN8A in epileptic encephalopathy

Amino acid substitution	Nucleotide	Exon	Domain	CpG	#	Premature death	Refs.
p.Gly214Asp	c.641 G>A	6A	DIS3	no	1		126
p.Asn215Asp[a]	c.643 A>G	6A	DIS3	no	2		131, 134
p.Val216Asp	c.647 T>A	6A	DIS3-4	no	1		132
p.Arg223Gly	c.667 A>G	6A	DIS4	no	1		127
p.Phe260Ser	c.779 T>C	7	DIS5	no	1		131
p.Leu407Phe	c.1221 G>C	10	DIS6	no	1	yes	130
p.Val410Leu	c.1228 G>C	10	DIS6	no	1		131
p.Arg662Cys	c.1984 C>T	12	Loop 1	yes	1		125
p.Thr767Ile	c.2003 C>T	14	DIIS1	no	1	yes	129
p.Phe846Ser	c.2537 T>C	15	DIIS4	no	1		132
p.Arg850Gln	c.2549 G>A	16	DIIS4	yes	1		130
p.Leu875Gln	c.2624 T>A	16	DIIS4-5	no	1		126
p.Ala890Thr	c.2668 G>A	16	DIIS5	no	2		130, 131
p.Val960Asp	c.2879 T>A	16	DIIS6	no	1		131
p.Asn984Lys	c.2952 C>G	17	Loop 2	no	1		124
p.Ile1327Val	c.3979 A>G	22	DIIIS4–5	no	1		135
p.Leu1331Val	c.3991 C>G	22	DIIIS5	no	1		125

Continued

TABLE 10.2 *De novo* mutations of SCN8A in epileptic encephalopathy—cont'd

Amino acid substitution	Nucleotide	Exon	Domain	CpG	#	Premature death	Refs.
p.Pro1428_Lys1473del	c.4419+1_+4del	24	DIIIS5–6	n/a	1		131
p.Gly1451Ser	c.4351 G>A	24	DIIIS6	yes	1		124
p.Asn1466Lys	c.4398 C>A	24	Inact. gate	no	1		132
p.Asn1466Thr	c.4397 A>C	24	Inact. gate	no	1		132
p.Ile1479Val	c.4435 A>G	25	Inact. gate	no	1		131
p.Val1592Leu	c.4774 C>G	26	DIVS3	no	1		131
p.Ser1596Cys	c.4787 C>G	26	DIVS3	no	1		130
p.Ile1605Arg	c.4813 A>G	27	DIVS3	no	1		131
p.Arg1617Gln	c.4850 G>A	27	DIVS4	yes	5	yes	128, 130–133
p.Gly1625Arg[b]	c.4873 G>A	27	DIVS4	no	1		134
p.Ala1650Thr	c.4948 G>A	27	DIVS4–5	no	2		131, 132
p.Asn1768Asp	c.5302 A>G	27	DIVS6	no	1	yes	27
p.Gln1801Glu	c.5401 C>G	27	C-term	no	1		131
p.Arg1872Gln	c.5614 C>T	27	C-term	yes	2		125, 131
p.Arg1872Trp	c.5615G>A	27	C-term	yes	3	yes	131, 132

[a] *One allele originally reported as N215R.*
[b] *Originally reported as G1584R. Mutations listed were published prior to Spring 2015. Exons are numbered according to transcript NM_0141k91.*

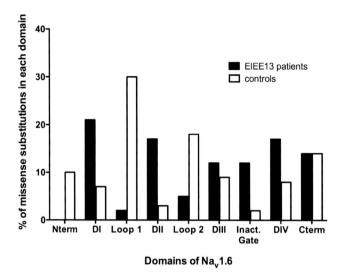

FIGURE 10.7 **Differential distribution of pathogenic and nonpathogenic variants in the protein domains of SCN8A.** Mutations of patients with EIEE13 are compared with variation in the ExAC database (exac.broadinstitute.org), which excludes severe pediatric disease. Each amino acid substitution (missense variant) was counted once, regardless of the number of individuals with the variant. Total number of EIEE13 mutations was 42, and total number of ExAC variants was 257. Two polymorphic variants in the normal population are p.Arg1026Cys and p.Ile700Leu.

Most of the mutated residues are located in the transmembrane segments, inactivation gate, and proximal half of the C-terminus of $Na_V1.6$. This distribution differs from the locations of nonpathogenic variants in the ExAC database of >60,000 exomes from individuals without severe pediatric disease (Exome Aggregation Consortium (ExAC), Cambridge, MA (URL: http://exac.broadinstitute.org)). Nonpathogenic mutations are concentrated in the nonconserved cytoplasmic loops of the channel (Fig. 10.7).

Interestingly, approximately one-third of the de novo mutations identified in EIEE13 have occurred more than once in unrelated individuals (Table 10.2). The recurrent mutations include five occurrences of the substitution p.Arg1617Gln[128,130–133] and six reports of mutation of p.Arg1872.[125,131,132] The codons for Arg1617 and Arg1872 contain CpG dinucleotides that are mutation hotspots due to cytosine methylation followed by deamination that can explain the amino acid substitutions (Fig. 10.8). The data suggest that only a subset of $Na_V1.6$ residues can generate the types of gain-of-function properties that lead to epileptic encephalopathy.

The clinical features of patients with EIEE13 include onset of seizures between birth and 18 months of age and presentation with multiple seizure types including tonic, tonic/clonic, generalized, and focal and absence seizures. There is also developmental delay and moderate to

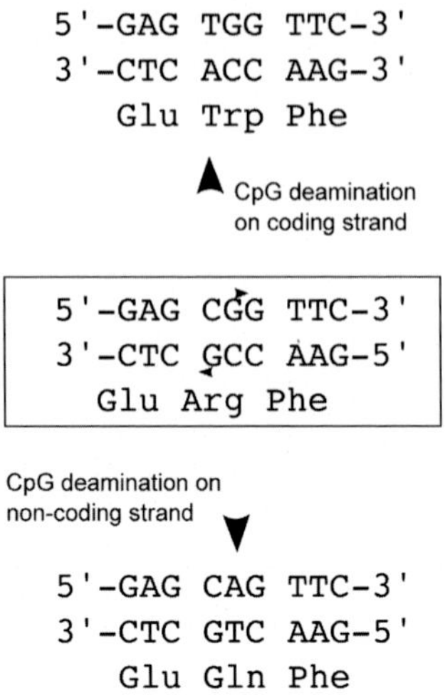

```
5'-GAG TGG TTC-3'
3'-CTC ACC AAG-3'
   Glu Trp Phe
```

▲ CpG deamination
on coding strand

```
5'-GAG CGG TTC-3'
3'-CTC GCC AAG-5'
   Glu Arg Phe
```

CpG deamination on
non-coding strand   ▼

```
5'-GAG CAG TTC-3'
3'-CTC GTC AAG-5'
   Glu Gln Phe
```

FIGURE 10.8   **Mechanism for recurrent mutation of Arg1872.** Arginine 1872 is encoded by a CGG codon, which yields a hypermutable CpG site on the coding and noncoding strands (indicated by *small arrows*). Methylation of cytosine followed by deamination on the coding or noncoding strand leads to substitution of tryptophan (Trp) or glutamine (Gln), respectively.

severe intellectual disability. Nearly half of reported patients have severely impaired motor control resulting in immobility. The frequency of SUDEP (sudden unexpected death in epilepsy) is also high in these patients, with 5/40 deaths in the published cases (Table 10.2). The low frequency of febrile seizures distinguishes EIEE13 from Dravet Syndrome (EIEE6).

TREATMENT OF EPILEPSY ASSOCIATED WITH *SCN8A* MUTATIONS

Seizures in EIEE13 patients are typically refractory to treatment, but approximately 1/3 of patients have had limited periods of seizure freedom. Approximately 20% of patients for whom we have detailed clinical histories have responded favorably to anticonvulsants that modulate sodium channel activity including valproate, phenytoin, carbamazepine, oxcarbazepine, lamotrigine, and topiramate. Carbamazepine and its derivative oxcarbazepine were successful in the largest number of patients. Patients with the same *SCN8A* mutation do not always respond to the same drug. The available clinical evidence suggests that sodium channel "blockers" may be the best first line treatments, rather than modulators of GABA signaling or other synaptic function. This is an important difference from Dravet Syndrome, where the use of sodium channel blockers appears to exacerbate seizures. Early differential diagnosis of EIEE13 and Dravet Syndrome is therefore important for the selection of appropriate treatment. The ketogenic diet is an alternative therapy that is effective for as many as 1/3 of patients with Dravet Syndrome,[136,137] but efficacy in EIEE13 has not been evaluated to date.

MODELING *SCN8A* EPILEPTIC ENCEPHALOPATHY IN THE MOUSE

The EIEE13 mutation p.Asn1768Asp (N1768D) results in gain-of-function changes that include delayed channel inactivation and increased persistent sodium current.[27] The N1768D mutation was introduced into the mouse genome using TALEN technology with a targeting vector.[138] Heterozygous $Scn8a^{N1768D/+}$ mice recapitulate many of the features of EIEE13 patients, including spontaneous convulsive seizures, ataxia, and SUDEP.[139] In homozygous mutants, seizure onset at 3 weeks of age is followed by SUDEP within 24 h. In heterozygous mice, seizure onset occurs at 2–4 months, with 0–3 generalized tonic-clonic seizures per day and progression to SUDEP within one month. There is no evidence of intellectual disability or behavioral dysfunction.

In intercrosses between $Scn8a^{N1768D/+}$ heterozygotes, heterozygous, and homozygous mutant mice are born at the expected Mendelian frequencies. The mutant $Na_v1.6$ protein is stable in vivo, and gross brain morphology and histology are unaffected. This novel mouse model is being used to investigate *Scn8a* pathogenesis in vivo, including investigation of SUDEP mechanisms.

MUTATIONS OF *SCN8A* IN INTELLECTUAL DISABILITY WITHOUT SEIZURES

A heterozygous loss-of-function mutation was identified in a proband with intellectual disability, pancerebellar atrophy, and ataxia.[122] The proband had a 2 bp deletion in the last coding exon of *SCN8A*. The resulting p.Pro1719ArgfsX1724 mutation introduces a premature termination codon into the pore loop of DIV, resulting in truncation of the C-terminal cytoplasmic domain. This mutation cosegregated with cognitive deficits in family members.[122] Three additional heterozygotes in the family exhibited cognitive and behavioral defects, including attention-deficit hyperactivity disorder.

The de novo mutation p.Asp58Asn was identified in a patient with intellectual disability without epilepsy.[124] The clinical picture includes developmental delay since birth, autism spectrum disorder, anxiety, and cerebral atrophy. However, the mutant $Na_v1.6$ exhibited normal activity in transfected cells and may not be pathogenic. The finding of another de novo missense mutation in a ubiquitin transferase in this patient makes the role of the *SCN8A* mutation unclear.

EFFECTS OF PATIENT MUTATIONS OF SCN8A ON CHANNEL FUNCTION

Pathogenic mutations of *SCN8A* in epileptic encephalopathy cause single amino acid substitutions whose functional impact is difficult to predict. Experimental analysis is therefore essential for evaluation of pathogenicity of novel mutations. The results from functional analyses of eight patient mutations are summarized in Table 10.3. Five mutations in patients with epileptic encephalopathy result in a clear gain-of-function phenotype and lead to hyperactivity of $Na_v1.6$. Two distinct mechanisms were observed, affecting channel inactivation or activation. p.Asn1768Asp and p.Arg1872Trp cause hyperexcitability by elevation of persistent current due to delayed channel inactivation.[27,140] p.Thr767Ile and p.Asn984Lys exhibit a hyperpolarized shift in voltage dependence of activation leading to premature channel opening.[124,129] Both classes of mutations resulted in elevated $Na_v1.6$ activity leading to seizures.

Two de novo mutations causing apparent loss of $Na_v1.6$ have also been identified in patients with epileptic encephalopathy. p.Arg223Gly results in protein instability at 37°C and loss of channel activity in transfected cells.[127] p.Gly1451Ser generates a stable protein that lacks channel activity in transfected cells[124] (Table 10.3). In a human pedigree segregating a protein truncation mutation of *SCN8A*, there was intellectual disability but no seizures.[122]

The very low frequency of nonsense, frameshift and splice site mutations in the ExAC database of human exomes predicts that loss-of-function mutations in this gene are not tolerated. In the mouse, heterozygous loss-of-function mutations do not cause tonic/clonic seizures but they are associated with absence seizures in certain strain backgrounds.[116] Variants at additional modifier loci may be necessary to cause seizures in mice or patients with haploinsufficiency of *SCN8A*. Analysis of additional patients with isolated intellectual disability will be essential to clarify the causal role of haploinsufficiency of *SCN8A*.

CONCLUSIONS

The important role of the sodium channel $Na_v1.6$ in neuronal function was discovered in the 1990s through analysis of spontaneous mouse mutants with movement disorders. Large-scale sequencing of patient exomes, beginning in the 2010s, has revealed the role of *SCN8A* mutations in sporadic human epilepsy and intellectual disability. The predominance of de novo mutations in children with epileptic encephalopathy reflects the severity of the disorder, which prevents transmission to the next generation. Accurate genetic diagnosis ("genotype first") is important for the

TABLE 10.3   Functional effects of SCN8A mutations in human disease

Amino acid substitution	Channel domain	ID	EE	Cell assay	Effect on function	Refs.
p.Asp58Asn	N-term	+	-	HEK[a]	No effect; probably not causal	124
p.Arg223Gly	DIS4	+	+	HEK	Partial LOF, protein unstable at 37C	127
p.Thr767Ile	DIIS1	+	+	ND7/23[b]	GOF, hyperpolarizing shift in voltage dependence of activation	129
p.Asn984Lys	near DIIS6	+	+	HEK	GOF, hyperpolarizing shift in voltage dependence of activation	124
p.Gly1451Ser	DIIIS6	+	+	HEK	LOF, no activity at 37C	124
p.Pro1719ArgfsX6	DIVS5-6	+	-	none	LOF, protein truncation	122
p.Asn1768Asp	DIVS6	+	+	ND7/23	GOF, increased persistent current	27
p.Arg1872Trp	C-term	+	+	ND7/23	GOF, increased persistent current	140

[a] HEK 293, human embryonic kidney.
[b] ND7/23, mouse neuroblastoma and rat dorsal root ganglia chimera.

choice of treatment in patients with epileptic encephalopathy. Isolated intellectual disability has been reported in one family with an inherited *SCN8A* mutation. The broad distribution of $Na_v1.6$ throughout the central and peripheral nervous system, and its unique role in the regulation of neuronal signaling, suggest that milder alleles may play a yet-undiscovered role in common neurological disorders.

References

1. Bailey WJ, Kim J, Wagner GP, Ruddle FH. Phylogenetic reconstruction of vertebrate Hox cluster duplications. *Mol Biol Evol* 1997;**14**(8):843–53.
2. Holland PW, Garcia-Fernandez J, Williams NA, Sidow A. Gene duplications and the origins of vertebrate development. *Dev Suppl* 1994:125–33.
3. Plummer NW, Meisler MH. Evolution and diversity of mammalian sodium channel genes. *Genomics* 1999;**57**(2):323–31.

4. Ruddle FH, Bartels JL, Bentley KL, Kappen C, Murtha MT, Pendleton JW. Evolution of Hox genes. *Annu Rev Genet* 1994;**28**:423–42.

5. Catterall WA, Dib-Hajj S, Meisler MH, Pietrobon D. Inherited neuronal ion channelopathies: new windows on complex neurological diseases. *J Neurosci* 2008;**28**(46):11768–77.

6. Meisler MH, Kearney JA. Sodium channel mutations in epilepsy and other neurological disorders. *J Clin Invest* 2005;**115**(8):2010–7.

7. O'Brien JE, Meisler MH. Sodium channel SCN8A (Nav1.6): properties and de novo mutations in epileptic encephalopathy and intellectual disability. *Front Genet* 2013;**4**:213.

8. O'Malley HA, Isom LL. sodium channel beta subunits: emerging targets in channelopathies. *Annu Rev Physiol* 2015;**77**:481–504.

9. Priori SG, Napolitano C. Genetics of cardiac arrhythmias and sudden cardiac death. *Ann NY Acad Sci* 2004;**1015**:96–110.

10. Plummer NW, Galt J, Jones JM, et al. Exon organization, coding sequence, physical mapping, and polymorphic intragenic markers for the human neuronal sodium channel gene SCN8A. *Genomics* 1998;**54**(2):287–96.

11. Garcia KD, Sprunger LK, Meisler MH, Beam KG. The sodium channel *Scn8a* is the major contributor to the postnatal developmental increase of sodium current density in spinal motoneurons. *J Neurosci* 1998;**18**(14):5234–9.

12. Gazina EV, Richards KL, Mokhtar MB, Thomas EA, Reid CA, Petrou S. Differential expression of exon 5 splice variants of sodium channel alpha subunit mRNAs in the developing mouse brain. *Neuroscience* 2010;**166**(1):195–200.

13. Liao Y, Deprez L, Maljevic S, et al. Molecular correlates of age-dependent seizures in an inherited neonatal-infantile epilepsy. *Brain* 2010;**133**(Pt 5):1403–14.

14. Plummer NW, McBurney MW, Meisler MH. Alternative splicing of the sodium channel SCN8A predicts a truncated two-domain protein in fetal brain and non-neuronal cells. *J Biol Chem* 1997;**272**(38):24008–15.

15. O'Brien JE, Drews VL, Jones JM, Dugas JC, Barres BA, Meisler MH. Rbfox proteins regulate alternative splicing of neuronal sodium channel SCN8A. *Mol Cell Neurosci* 2012;**49**(2):120–6.

16. Drews VL, Lieberman AP, Meisler MH. Multiple transcripts of sodium channel SCN8A (Na(V)1.6) with alternative 5′- and 3′-untranslated regions and initial characterization of the SCN8A promoter. *Genomics* 2005;**85**(2):245–57.

17. Drews VL, Shi K, de Haan G, Meisler MH. Identification of evolutionarily conserved, functional noncoding elements in the promoter region of the sodium channel gene SCN8A. *Mamm Genome* 2007;**18**(10):723–31.

18. Gazina EV, Leaw BT, Richards KL, et al. 'Neonatal' Nav1.2 reduces neuronal excitability and affects seizure susceptibility and behaviour. *Hum Mol Genet* 2014;**24**(5).

19. Xu R, Thomas EA, Jenkins M, et al. A childhood epilepsy mutation reveals a role for developmentally regulated splicing of a sodium channel. *Mol Cell Neurosci* 2007;**35**(2): 292–301.

20. Gehman LT, Meera P, Stoilov P, et al. The splicing regulator Rbfox2 is required for both cerebellar development and mature motor function. *Genes Dev* 2012;**26**(5):445–60.

21. Zubovic L, Baralle M, Baralle FE. Mutually exclusive splicing regulates the Nav 1.6 sodium channel function through a combinatorial mechanism that involves three distinct splicing regulatory elements and their ligands. *Nucleic Acids Res* 2012;**40**(13): 6255–69.

22. Marban E, Yamagishi T, Tomaselli GF. Structure and function of voltage-gated sodium channels. *J Physiol* 1998;**508**(Pt 3):647–57.

23. Petrovski S, Wang Q, Heinzen EL, Allen AS, Goldstein DB. Genic intolerance to functional variation and the interpretation of personal genomes. *PLoS Genet* 2013;**9**(8):e1003709.

24. Hodgkin AL, Huxley AF. The dual effect of membrane potential on sodium conductance in the giant axon of Loligo. *J Physiol* 1952;**116**(4):497–506.

25. Chen Y, Yu FH, Sharp EM, Beacham D, Scheuer T, Catterall WA. Functional properties and differential neuromodulation of Na(v)1.6 channels. *Mol Cell Neurosci* 2008;**38**(4):607–15.

26. Cummins TR, Xia Y, Haddad GG. Functional properties of rat and human neocortical voltage-sensitive sodium currents. *J Neurophysiol* 1994;**71**(3):1052–64.

27. Veeramah KR, O'Brien JE, Meisler MH, et al. De novo pathogenic SCN8A mutation identified by whole-genome sequencing of a family quartet affected by infantile epileptic encephalopathy and SUDEP. *Am J Hum Genet* 2012;**90**(3):502–10.

28. Fleidervish IA, Gutnick MJ. Kinetics of slow inactivation of persistent sodium current in layer V neurons of mouse neocortical slices. *J Neurophysiol* 1996;**76**(3):2125–30.

29. French CR, Sah P, Buckett KJ, Gage PW. A voltage-dependent persistent sodium current in mammalian hippocampal neurons. *J Gen Physiol* 1990;**95**(6):1139–57.

30. Stafstrom CE, Schwindt PC, Crill WE. Repetitive firing in layer V neurons from cat neocortex in vitro. *J Neurophysiol* 1984;**52**(2):264–77.

31. Yue C, Remy S, Su H, Beck H, Yaari Y. Proximal persistent Na$^+$ channels drive spike afterdepolarizations and associated bursting in adult CA1 pyramidal cells. *J Neurosci* 2005;**25**(42):9704–20.

32. Chen S, Su H, Yue C, et al. An increase in persistent sodium current contributes to intrinsic neuronal bursting after status epilepticus. *J Neurophysiol* 2011;**105**(1): 117–29.

33. Stafstrom CE. Persistent sodium current and its role in epilepsy. *Epilepsy Curr* 2007;**7**(1): 15–22.

34. Lorincz A, Nusser Z. Cell-type-dependent molecular composition of the axon initial segment. *J Neurosci* 2008;**28**(53):14329–40.

35. Hu W, Tian C, Li T, Yang M, Hou H, Shu Y. Distinct contributions of Na(v)1.6 and Na(v)1.2 in action potential initiation and backpropagation. *Nat Neurosci* 2009;**12**(8): 996–1002.

36. Kole MH, Ilschner SU, Kampa BM, Williams SR, Ruben PC, Stuart GJ. Action potential generation requires a high sodium channel density in the axon initial segment. *Nat Neurosci* 2008;**11**(2):178–86.

37. Kole MH, Stuart GJ. Is action potential threshold lowest in the axon? *Nat Neurosci* 2008;**11**(11):1253–5.

38. Van Wart A, Trimmer JS, Matthews G. Polarized distribution of ion channels within microdomains of the axon initial segment. *J Comp Neurol* 2007;**500**(2):339–52.

39. Bean BP. The action potential in mammalian central neurons. *Nat Rev Neurosci* 2007;**8**(6):451–65.

40. Boiko T, Rasband MN, Levinson SR, et al. Compact myelin dictates the differential targeting of two sodium channel isoforms in the same axon. *Neuron* 2001;**30**(1):91–104.

41. Boiko T, Van Wart A, Caldwell JH, Levinson SR, Trimmer JS, Matthews G. Functional specialization of the axon initial segment by isoform-specific sodium channel targeting. *J Neurosci* 2003;**23**(6):2306–13.

42. Caldwell JH, Schaller KL, Lasher RS, Peles E, Levinson SR. Sodium channel Na(v)1.6 is localized at nodes of ranvier, dendrites, and synapses. *Proc Natl Acad Sci USA* 2000;**97**(10):5616–20.

43. Krzemien DM, Schaller KL, Levinson SR, Caldwell JH. Immunolocalization of sodium channel isoform NaCh6 in the nervous system. *J Comp Neurol* 2000;**420**(1):70–83.

44. Li T, Tian C, Scalmani P, et al. Action potential initiation in neocortical inhibitory interneurons. *PLoS Biol* 2014;**12**(9):e1001944.

45. Tian C, Wang K, Ke W, Guo H, Shu Y. Molecular identity of axonal sodium channels in human cortical pyramidal cells. *Front Cell Neurosci* 2014;**8**:297.

46. Lorincz A, Nusser Z. Molecular identity of dendritic voltage-gated sodium channels. *Science* 2010;**328**(5980):906–9.

47. Maier SK, Westenbroek RE, McCormick KA, Curtis R, Scheuer T, Catterall WA. Distinct subcellular localization of different sodium channel alpha and beta subunits in single ventricular myocytes from mouse heart. *Circulation* 2004;**109**(11):1421–7.
48. Maier SK, Westenbroek RE, Schenkman KA, Feigl EO, Scheuer T, Catterall WA. An unexpected role for brain-type sodium channels in coupling of cell surface depolarization to contraction in the heart. *Proc Natl Acad Sci USA* 2002;**99**(6):4073–8.
49. Du Y, Huang X, Wang T, et al. Downregulation of neuronal sodium channel subunits Nav1.1 and Nav1.6 in the sinoatrial node from volume-overloaded heart failure rat. *Pflugers Arch* 2007;**454**(3):451–9.
50. Maier SK, Westenbroek RE, Yamanushi TT, et al. An unexpected requirement for brain-type sodium channels for control of heart rate in the mouse sinoatrial node. *Proc Natl Acad Sci USA* 2003;**100**(6):3507–12.
51. Dhar Malhotra J, Chen C, Rivolta I, et al. Characterization of sodium channel alpha- and beta-subunits in rat and mouse cardiac myocytes. *Circulation* 2001;**103**(9):1303–10.
52. Duclohier H. Neuronal sodium channels in ventricular heart cells are localized near T-tubules openings. *Biochem Biophys Res Commun* 2005;**334**(4):1135–40.
53. Haufe V, Camacho JA, Dumaine R, et al. Expression pattern of neuronal and skeletal muscle voltage-gated Na$^+$ channels in the developing mouse heart. *J Physiol* 2005;**564**(Pt 3):683–96.
54. Kaufmann SG, Westenbroek RE, Maass AH, et al. Distribution and function of sodium channel subtypes in human atrial myocardium. *J Mol Cell Cardiol* 2013;**61**:133–41.
55. Noujaim SF, Kaur K, Milstein M, et al. A null mutation of the neuronal sodium channel NaV1.6 disrupts action potential propagation and excitation-contraction coupling in the mouse heart. *FASEB J* 2012;**26**(1):63–72.
56. Torres NS, Larbig R, Rock A, Goldhaber JI, Bridge JH. Na$^+$ currents are required for efficient excitation-contraction coupling in rabbit ventricular myocytes: a possible contribution of neuronal Na$^+$ channels. *J Physiol* 2010;**588**(Pt 21):4249–60.
57. Westenbroek RE, Bischoff S, Fu Y, Maier SK, Catterall WA, Scheuer T. Localization of sodium channel subtypes in mouse ventricular myocytes using quantitative immunocytochemistry. *J Mol Cell Cardiol* 2013;**64**:69–78.
58. Rush AM, Dib-Hajj SD, Waxman SG. Electrophysiological properties of two axonal sodium channels, Nav1.2 and Nav1.6, expressed in mouse spinal sensory neurones. *J Physiol* 2005;**564**(Pt 3):803–15.
59. Van Wart A, Matthews G. Impaired firing and cell-specific compensation in neurons lacking Nav1.6 sodium channels. *J Neurosci* 2006;**26**(27):7172–80.
60. Xiao M, Bosch MK, Nerbonne JM, Ornitz DM. FGF14 localization and organization of the axon initial segment. *Mol Cell Neurosci* 2013;**56**:393–403.
61. Royeck M, Horstmann MT, Remy S, Reitze M, Yaari Y, Beck H. Role of axonal NaV1.6 sodium channels in action potential initiation of CA1 pyramidal neurons. *J Neurophysiol* 2008;**100**(4):2361–80.
62. Aman TK, Raman IM. Subunit dependence of Na channel slow inactivation and open channel block in cerebellar neurons. *Biophys J* 2007;**92**(6):1938–51.
63. Enomoto A, Han JM, Hsiao CF, Chandler SH. Sodium currents in mesencephalic trigeminal neurons from Nav1.6 null mice. *J Neurophysiol* 2007;**98**(2):710–9.
64. Osorio N, Cathala L, Meisler MH, Crest M, Magistretti J, Delmas P. Persistent Nav1.6 current at axon initial segments tunes spike timing of cerebellar granule cells. *J Physiol* 2010;**588**(Pt 4):651–70.
65. Raman IM, Sprunger LK, Meisler MH, Bean BP. Altered subthreshold sodium currents and disrupted firing patterns in Purkinje neurons of Scn8a mutant mice. *Neuron* 1997;**19**(4):881–91.
66. Lewis AH, Raman IM. Resurgent current of voltage-gated Na(+) channels. *J Physiol* 2014;**592**(Pt 22):4825–38.
67. Raman IM, Bean BP. Resurgent sodium current and action potential formation in dissociated cerebellar Purkinje neurons. *J Neurosci* 1997;**17**(12):4517–26.

68. O'Brien JE, Sharkey LM, Vallianatos CN, et al. Interaction of voltage-gated sodium channel Nav1.6 (SCN8A) with microtubule-associated protein Map1b. *J Biol Chem* 2012;**287**(22):18459–66.

69. Davis JQ, Lambert S, Bennett V. Molecular composition of the node of Ranvier: identification of ankyrin-binding cell adhesion molecules neurofascin (mucin+/third FNIII domain-) and NrCAM at nodal axon segments. *J Cell Biol* 1996;**135**(5):1355–67.

70. Gasser A, Ho TS, Cheng X, et al. An ankyrinG-binding motif is necessary and sufficient for targeting Nav1.6 sodium channels to axon initial segments and nodes of Ranvier. *J Neurosci* 2012;**32**(21):7232–43.

71. Hill AS, Nishino A, Nakajo K, et al. Ion channel clustering at the axon initial segment and node of Ranvier evolved sequentially in early chordates. *PLoS Genet* 2008;**4**(12):e1000317.

72. Srinivasan Y, Elmer L, Davis J, Bennett V, Angelides K. Ankyrin and spectrin associate with voltage-dependent sodium channels in brain. *Nature* 1988;**333**(6169):177–80.

73. Gasser A, Cheng X, Gilmore ES, Tyrrell L, Waxman SG, Dib-Hajj SD. Two Nedd4-binding motifs underlie modulation of sodium channel Nav1.6 by p38 MAPK. *J Biol Chem* 2010;**285**(34):26149–61.

74. Sudol M, Hunter T. New wrinkles for an old domain. *Cell* 2000;**103**(7):1001–4.

75. Zarrinpar A, Lim WA. Converging on proline: the mechanism of WW domain peptide recognition. *Nat Struct Biol* 2000;**7**(8):611–3.

76. Wittmack EK, Rush AM, Hudmon A, Waxman SG, Dib-Hajj SD. Voltage-gated sodium channel Nav1.6 is modulated by p38 mitogen-activated protein kinase. *J Neurosci* 2005;**25**(28):6621–30.

77. Laezza F, Gerber BR, Lou JY, et al. The FGF14(F145S) mutation disrupts the interaction of FGF14 with voltage-gated Na+ channels and impairs neuronal excitability. *J Neurosci* 2007;**27**(44):12033–44.

78. Laezza F, Lampert A, Kozel MA, et al. FGF14 N-terminal splice variants differentially modulate Nav1.2 and Nav1.6-encoded sodium channels. *Mol Cell Neurosci* 2009;**42**(2):90–101.

79. Lou JY, Laezza F, Gerber BR, et al. Fibroblast growth factor 14 is an intracellular modulator of voltage-gated sodium channels. *J Physiol* 2005;**569**(Pt 1):179–93.

80. Wang C, Chung BC, Yan H, Wang HG, Lee SY, Pitt GS. Structural analyses of Ca(2)(+)/CaM interaction with NaV channel C-termini reveal mechanisms of calcium-dependent regulation. *Nat Commun* 2014;**5**:4896.

81. Rush AM, Wittmack EK, Tyrrell L, Black JA, Dib-Hajj SD, Waxman SG. Differential modulation of sodium channel Na(v)1.6 by two members of the fibroblast growth factor homologous factor 2 subfamily. *Eur J Neurosci* 2006;**23**(10):2551–62.

82. Wittmack EK, Rush AM, Craner MJ, Goldfarb M, Waxman SG, Dib-Hajj SD. Fibroblast growth factor homologous factor 2B: association with Nav1.6 and selective colocalization at nodes of Ranvier of dorsal root axons. *J Neurosci* 2004;**24**(30):6765–75.

83. Shakkottai VG, Xiao M, Xu L, et al. FGF14 regulates the intrinsic excitability of cerebellar Purkinje neurons. *Neurobiol Dis* 2009;**33**(1):81–8.

84. Brusse E, de Koning I, Maat-Kievit A, Oostra BA, Heutink P, van Swieten JC. Spinocerebellar ataxia associated with a mutation in the fibroblast growth factor 14 gene (SCA27): a new phenotype. *Mov Disord* 2006;**21**(3):396–401.

85. Isom LL, De Jongh KS, Catterall WA. Auxiliary subunits of voltage-gated ion channels. *Neuron* 1994;**12**(6):1183–94.

86. Isom LL, De Jongh KS, Patton DE, et al. Primary structure and functional expression of the beta 1 subunit of the rat brain sodium channel. *Science* 1992;**256**(5058):839–42.

87. Patton DE, Isom LL, Catterall WA, Goldin AL. The adult rat brain beta 1 subunit modifies activation and inactivation gating of multiple sodium channel alpha subunits. *J Biol Chem* 1994;**269**(26):17649–55.

88. Calhoun JD, Isom LL. The role of non-pore-forming beta subunits in physiology and pathophysiology of voltage-gated sodium channels. *Handb Exp Pharmacol* 2014;**221**:51–89.

89. Brackenbury WJ, Calhoun JD, Chen C, et al. Functional reciprocity between Na$^+$ channel Nav1.6 and beta1 subunits in the coordinated regulation of excitability and neurite outgrowth. *Proc Natl Acad Sci USA* 2010;**107**(5):2283–8.

90. Chen C, Westenbroek RE, Xu X, et al. Mice lacking sodium channel beta1 subunits display defects in neuronal excitability, sodium channel expression, and nodal architecture. *J Neurosci* 2004;**24**(16):4030–42.

91. Lin X, O'Malley H, Chen C, et al. Scn1b deletion leads to increased tetrodotoxin-sensitive sodium current, altered intracellular calcium homeostasis and arrhythmias in murine hearts. *J Physiol* 2014;**593**(Pt 6).

92. Lopez-Santiago LF, Meadows LS, Ernst SJ, et al. Sodium channel Scn1b null mice exhibit prolonged QT and RR intervals. *J Mol Cell Cardiol* 2007;**43**(5):636–47.

93. Ogiwara I, Nakayama T, Yamagata T, et al. A homozygous mutation of voltage-gated sodium channel beta(I) gene SCN1B in a patient with Dravet syndrome. *Epilepsia* 2012;**53**(12):e200–3.

94. Patino GA, Claes LR, Lopez-Santiago LF, et al. A functional null mutation of SCN1B in a patient with Dravet syndrome. *J Neurosci* 2009;**29**(34):10764–78.

95. Bahler M, Rhoads A. Calmodulin signaling via the IQ motif. *FEBS Lett* 2002;**513**(1):107–13.

96. Herzog RI, Liu C, Waxman SG, Cummins TR. Calmodulin binds to the C terminus of sodium channels Nav1.4 and Nav1.6 and differentially modulates their functional properties. *J Neurosci* 2003;**23**(23):8261–70.

97. Reddy Chichili VP, Xiao Y, Seetharaman J, Cummins TR, Sivaraman J. Structural basis for the modulation of the neuronal voltage-gated sodium channel NaV1.6 by calmodulin. *Sci Rep* 2013;**3**:2435.

98. Weiss LA, Escayg A, Kearney JA, et al. Sodium channels SCN1A, SCN2A and SCN3A in familial autism. *Mol Psychiatry* 2003;**8**(2):186–94.

99. Feldkamp MD, Yu L, Shea MA. Structural and energetic determinants of apo calmodulin binding to the IQ motif of the Na(V)1.2 voltage-dependent sodium channel. *Structure* 2011;**19**(5):733–47.

100. Buchner DA, Seburn KL, Frankel WN, Meisler MH. Three ENU-induced neurological mutations in the pore loop of sodium channel Scn8a (Na(v)1.6) and a genetically linked retinal mutation, rd13. *Mamm Genome* 2004;**15**(5):344–51.

101. Burgess DL, Kohrman DC, Galt J, et al. Mutation of a new sodium channel gene, Scn8a, in the mouse mutant 'motor endplate disease'. *Nat Genet* 1995;**10**(4):461–5.

102. De Repentigny Y, Cote PD, Pool M, et al. Pathological and genetic analysis of the degenerating muscle (dmu) mouse: a new allele of Scn8a. *Hum Mol Genet* 2001;**10**(17):1819–27.

103. Jones JM, Dionne L, Dell'Orco J, et al. A single amino acid deletion in mouse Scn8a alters the sodium channel pore and causes a chronic movement disorder. *Neurobiol Dis* 2016;**89**(5):36–45.

104. Kohrman DC, Harris JB, Meisler MH. Mutation detection in the med and medJ alleles of the sodium channel Scn8a. Unusual splicing due to a minor class AT-AC intron *J Biol Chem* 1996;**271**(29):17576–81.

105. Kohrman DC, Plummer NW, Schuster T, et al. Insertional mutation of the motor endplate disease (med) locus on mouse chromosome 15. *Genomics* 1995;**26**(2):171–7.

106. Kohrman DC, Smith MR, Goldin AL, Harris J, Meisler MH. A missense mutation in the sodium channel Scn8a is responsible for cerebellar ataxia in the mouse mutant jolting. *J Neurosci* 1996;**16**(19):5993–9.

107. Mackenzie FE, Parker A, Parkinson NJ, et al. Analysis of the mouse mutant Cloth-ears shows a role for the voltage-gated sodium channel Scn8a in peripheral neural hearing loss. *Genes Brain Behav* 2009;**8**(7):699–713.

108. Sharkey LM, Cheng X, Drews V, et al. The ataxia3 mutation in the N-terminal cytoplasmic domain of sodium channel Na(v)1.6 disrupts intracellular trafficking. *J Neurosci* 2009;**29**(9):2733–41.

109. Smith MR, Goldin AL. A mutation that causes ataxia shifts the voltage-dependence of the Scn8a sodium channel. *Neuroreport* 1999;**10**(14):3027–31.

110. SoRelle J, Chen Z, McAlpine W, Hutchins N, Murray AR, Beutler B. *Mutagenix entry for nymph*. MGI Direct Data Submission. 2014.

111. Sprunger LK, Escayg A, Tallaksen-Greene S, Albin RL, Meisler MH. Dystonia associated with mutation of the neuronal sodium channel Scn8a and identification of the modifier locus Scnm1 on mouse chromosome 3. *Hum Mol Genet* 1999;**8**(3):471–9.

112. Timms H, Smart NG, Beutler B. *Record for "TremorD"*. MGI Direct Data Submission. 2008.

113. Weatherly J, Purrington T, Murray AR, Beutler B. *Direct data submission for dan*. MGI Direct Data Submission. 2014.

114. Schroeder DG, Dziedzic J, Cox GA. *The Scn8a spontaneous point mutation*. MGI Direct Data Submission. 2013.

115. Oliva MK, McGarr TC, Beyer BJ, et al. Physiological and genetic analysis of multiple sodium channel variants in a model of genetic absence epilepsy. *Neurobiol Dis* 2014;**67**:180–90.

116. Papale LA, Beyer B, Jones JM, et al. Heterozygous mutations of the voltage-gated sodium channel SCN8A are associated with spike-wave discharges and absence epilepsy in mice. *Hum Mol Genet* 2009;**18**(9):1633–41.

117. McKinney BC, Chow CY, Meisler MH, Murphy GG. Exaggerated emotional behavior in mice heterozygous null for the sodium channel Scn8a (Nav1.6). *Genes Brain Behav* 2008;**7**(6):629–38.

118. Papale LA, Paul KN, Sawyer NT, Manns JR, Tufik S, Escayg A. Dysfunction of the Scn8a voltage-gated sodium channel alters sleep architecture, reduces diurnal corticosterone levels, and enhances spatial memory. *J Biol Chem* 2010;**285**(22):16553–61.

119. Sawyer NT, Papale LA, Eliason J, Neigh GN, Escayg A. Scn8a voltage-gated sodium channel mutation alters seizure and anxiety responses to acute stress. *Psychoneuroendocrinology* 2014;**39**:225–36.

120. Martin MS, Tang B, Papale LA, Yu FH, Catterall WA, Escayg A. The voltage-gated sodium channel Scn8a is a genetic modifier of severe myoclonic epilepsy of infancy. *Hum Mol Genet* 2007;**16**(23):2892–9.

121. Sun W, Wagnon JL, Mahaffey CL, Briese M, Ule J, Frankel WN. Aberrant sodium channel activity in the complex seizure disorder of Celf4 mutant mice. *J Physiol* 2013;**591**(Pt 1):241–55.

122. Trudeau MM, Dalton JC, Day JW, Ranum LP, Meisler MH. Heterozygosity for a protein truncation mutation of sodium channel SCN8A in a patient with cerebellar atrophy, ataxia, and mental retardation. *J Med Genet* 2006;**43**(6):527–30.

123. Sharkey LM, Jones JM, Hedera P, Meisler MH. Evaluation of SCN8A as a candidate gene for autosomal dominant essential tremor. *Park Relat Disord* 2009;**15**(4):321–3.

124. Blanchard MG, Willemsen MH, Walker JB, et al. De novo gain-of-function and loss-of-function mutations of SCN8A in patients with intellectual disabilities and epilepsy. *J Med Genet* 2015;**52**(5).

125. Carvill GL, Heavin SB, Yendle SC, et al. Targeted resequencing in epileptic encephalopathies identifies de novo mutations in CHD2 and SYNGAP1. *Nat Genet* 2013;**45**(7):825–30.

126. Epi4K Consortium, Epilepsy Phenome/Genome Project, Allen AS, et al. De novo mutations in epileptic encephalopathies. *Nature* 2013;**501**(7466):217–21.

127. de Kovel CG, Meisler MH, Brilstra EH, et al. Characterization of a de novo SCN8A mutation in a patient with epileptic encephalopathy. *Epilepsy Res* 2014;**108**(9):1511–8.

128. Dyment DA, Tetreault M, Beaulieu CL, et al. Whole-exome sequencing broadens the phenotypic spectrum of rare pediatric epilepsy: a retrospective study. *Clin Genet* 2014;**88**(1).

129. Estacion M, O'Brien JE, Conravey A, et al. A novel de novo mutation of SCN8A (Nav1.6) with enhanced channel activation in a child with epileptic encephalopathy. *Neurobiol Dis* 2014;**69**:117–23.

130. Kong W, Zhang Y, Gao Y, et al. SCN8A mutations in Chinese children with early onset epilepsy and intellectual disability. *Epilepsia* 2015;**56**(3).

131. Larsen J, Carvill GL, Gardella E, et al. The phenotypic spectrum of SCN8A encephalopathy. *Neurology* 2015;**84**(5):480–9.
132. Ohba C, Kato M, Takahashi S, et al. Early onset epileptic encephalopathy caused by de novo SCN8A mutations. *Epilepsia* 2014;**55**(7):994–1000.
133. Rauch A, Wieczorek D, Graf E, et al. Range of genetic mutations associated with severe non-syndromic sporadic intellectual disability: an exome sequencing study. *Lancet* 2012;**380**(9854):1674–82.
134. The Deciphering Developmental Disorders Study. Large-scale discovery of novel genetic causes of developmental disorders. *Nature* 2014;**519**(7542).
135. Vaher U, Noukas M, Nikopensius T, et al. De novo SCN8A mutation identified by whole-exome sequencing in a boy with neonatal epileptic encephalopathy, multiple congenital anomalies, and movement disorders. *J Child Neurol* 2013;**29**. http://dx.doi.org/10.1177/0883073813511300.
136. Caraballo RH, Cersosimo RO, Sakr D, Cresta A, Escobal N, Fejerman N. Ketogenic diet in patients with Dravet syndrome. *Epilepsia* 2005;**46**(9):1539–44.
137. Laux L, Blackford R. The ketogenic diet in Dravet syndrome. *J Child Neurol* 2013;**28**(8):1041–4.
138. Jones JM, Meisler MH. Modeling human epilepsy by TALEN targeting of mouse sodium channel Scn8a. *Genesis* 2014;**52**(2):141–8.
139. Wagnon JL, Korn MJ, Parent R, et al. Convulsive seizures and SUDEP in a mouse model of SCN8A epileptic encephalopathy. *Hum Mol Genet* 2015;**24**(2):506–15.
140. Wagnon JL, Barker BS, Hounsell JA, et al. Mechanisms of recurrent epileptogenic mutations in sodium channel SCN8A. *Ann Clin Transl Neurol* 2015;**3**(2):114–23.

Alternative Splicing and RNA Editing of Voltage-Gated Ion Channels: Implications in Health and Disease

J. Zhai[1,a], Q.-S. Lin[1,a], Z. Hu[1], R. Wong[1,2], T.W. Soong[1,2,3]

[1]National University of Singapore, Singapore; [2]NUS Graduate School for Integrative Sciences and Engineering, Singapore; [3]National Neuroscience Institute, Singapore

POSTTRANSCRIPTIONAL MODIFICATIONS OF VOLTAGE-GATED ION CHANNELS

Transcripts of ion channels undergo posttranscriptional modifications like alternative splicing and RNA editing. These modifications generate diversity of channel protein structures resulting in altered electrophysiological or pharmacological properties. Besides altered channel properties, some channel splice variants are expressed in a tissue-specific manner or are developmentally regulated. In this chapter, the implications of posttranscriptional modifications of voltage-gated ion channels on health and disease are discussed.

Alternative Splicing

Alternative splicing is an exquisite process to fine-tune ion channel function through inclusion or exclusion of between one to hundreds of amino acids in the protein structure. This process can be done via five

[a]These authors contribute equally to the chapter.

basic mechanisms: (1) cassette exon skipping, (2) mutually exclusive splicing of a pair of exons of same size, (3) use of alternative donor or acceptor sites, (4) intron retention, or (5) use of alternative promoter or polyadenylation sites. An RNA and protein complex known as spliceosome is required for splicing, and this complex contains small nuclear ribonucleic proteins that bind to pre-mRNA to mediate the removal of introns.

Alternative splicing of ion channels is widely found in mammalian cells, and the expressions of certain splice variants have been shown to change under physiological or pathological conditions. The functional consequences for these altered splicing patterns of ion channels in different cellular states have been investigated, and their predicted roles have been proposed. However, assessment of functional changes arising from alternative splicing of an ion channel in a single site should be considered in context of the combinatorial assortment of all possible alternatively spliced sites. As such, the need to know the major tissue or cellular selective splice pattern or combination of an ion channel is critical for proper evaluation of the effects ion channel mutations have on a particular cell, tissue, or organ dysfunction. As such, in some ion channels, changes in function because of missense mutations can be modulated in a splice variant–dependent manner. However, in other channels, channel mutations can cause altered splicing pattern and result in channel dysfunction. In another word, the same mutation present in context of different splice combinatorial assortments would result in a different functional change on the channel. The net result may have cell-type or organ-type selectivity in expressions of pathological effects.

A-to-I RNA Editing

A-to-I RNA editing is a type of posttranscriptional modification that results in the pinpoint recoding of genomic information. This mechanism contributes to the diversity of proteins and their functions. A-to-I RNA editing is catalyzed by a family of enzymes known as adenosine deaminase acting on RNA (ADARs) that bind to double-stranded RNA. This family has three mammalian members: ADAR1, ADAR2, and ADAR3. During RNA editing, single adenosine of an ion channel pre-mRNA is deaminated to inosine and in the mature edited mRNA the inosine is recognized as guanosine during translation resulting in recoding of a codon within the primary RNA transcript. Functional editing of ion channels has been shown to lead to alteration in their expression and/or functional properties.[1]

For RNA editing to proceed, the formation of a hairpin structure formed between the editing site and editing site complimentary sequence of the ion channel has to occur first. This hairpin structure will then be recognized and be bound by ADARs to initiate RNA editing. As for the regulation of RNA editing level, though the underlying mechanisms are still largely unknown, *cis*-acting elements have been reported to be involved in modulating editing level.[2] Various evidences have shown that the levels of RNA editing could be modulated under various physiological conditions and that changes in editing levels have been associated with neurological disorders.

ALTERNATIVE SPLICING OF VOLTAGE-GATED CALCIUM CHANNELS AND DISEASES

Opening of voltage-gated calcium (Ca_V) channels (VGCCs) initiates various physiological processes via rapid Ca^{2+} influx through them. These processes include release of neurotransmitters at the synapse, gene transcription, and muscle contraction.[3] According to their biophysical properties, VGCCs could be categorized into two broad subclasses: (1) high voltage–activated (HVA) calcium channels that are activated upon sensing large membrane depolarization, such as L-type ($Ca_V1.1$, 1.2, 1.3, and 1.4), P/Q-type ($Ca_V2.1$), N-type ($Ca_V2.2$), and R-type ($Ca_V2.3$) channels and (2) low voltage–activated (LVA) calcium channels that open in response to small voltage changes close to neuronal resting membrane potentials, such as T-type calcium channels ($Ca_V3.1$, 3.2, and 3.3) (Table 11.1).[4] On the basis of biochemical and molecular analysis, it has been well known that HVA channels are heterooligomeric transmembrane protein complexes featuring the coassembly of a pore-forming $Ca_V\alpha_1$ subunit with auxiliary $Ca_V\beta$ and $Ca_V\alpha_2\delta$ subunits; whereas LVA channels usually lack these auxiliary subunits (Table 11.1).[5]

The calcium channel subtypes are mainly determined by $Ca_V\alpha_1$ subunit which may undergo extensive alternative splicing in a tissue- and age-dependent manner, thereby resulting in functional diversity of calcium channels.[6] Recently, an increasing number of functional splice variants of the pore-forming $Ca_V\alpha_1$ subunit have been identified. These phenotypic variants affect the electrophysiological and pharmacological properties of the channels. More importantly, studies have shown exon-selective effects of mutations on severity or localization of disorders like in Timothy syndrome (TS) and pain.[3] Therefore, disease states could be affected by alternative splicing in two ways: (1) tissue-selective expression of splicing variants containing mutations related to a particular disease could affect the severity of the disease and (2) developmental stage of disease in which the expression level of alternatively spliced exons may be altered due to changed stimuli.

TABLE 11.1 Summary of VGCC Classification and Disorders Related to Alternative Splicing of $Ca_V\alpha_1$ Subunit[3,6]

Electrophysiological Nomenclature		Molecular Nomenclature	Main distribution	Alternative splicing-related disorders
High voltage–activated calcium channels (co-assembled with $Ca_V\beta$ and $Ca_V\alpha_2\delta$ subunit)	L	$Ca_V1.1$ (α_{1S})	Skeletal muscle	
		$Ca_V1.2$ (α_{1C})	Cardiac Smooth muscle Endocrine cells Neuronal	Timothy syndrome Hypertension Heart failure Chronic myocardial infarction
		$Ca_V1.3$ (α_{1D})	Neuronal (dentritic) Sinoatrial node Cochlea hair cells	Sinoatrial node dysfunction and deafness
		$Ca_V1.4$ (α_{1F})	Retina	Congenital stationary night blindness X-linked cone–rod dystrophy
	P/Q	$Ca_V2.1$ (α_{1A})	Neuronal (presynaptic)	Episodic ataxia type 2 Familial hemiplegic migraine Spinocerebellar ataxia type 6
	N	$Ca_V2.2$ (α_{1B})	Neuronal (presynaptic)	Pain
	R	$Ca_V2.3$ (α_{1E})	Neuronal	
Low voltage–activated calcium channels	T	$Ca_V3.1$ (α_{1G})	Cardiac Neuronal	
		$Ca_V3.2$ (α_{1H})	Heart Neuronal	Childhood absent epilepsy
		$Ca_V3.3$ (α_{1I})	Neuronal	

To date, a large number of alternatively spliced sites of VGCCs have been identified in the 10 genes encoding $Ca_V\alpha_1$ subunits. Potentially, alternative splicing could generate thousands of distinct splice variants.[6] Moreover, an increasing number of mutations of $Ca_V\alpha_1$ subunits have been reported that are associated with disorders affecting multiple organs and the phenotypes depend on where the channels are predominantly expressed. In this section, we highlight calcium channelopathies of two L-type channels, namely $Ca_V1.2$ and $Ca_V1.3$ channels, of P/Q-type $Ca_V2.1$ channel and of T-type $Ca_V3.2$ channel (Fig. 11.1A).

$Ca_V1.2$ and Timothy Syndrome

Human $Ca_V1.2$ gene (*CACNA1C*) contains at least 56 exons, among which 20 exons undergo alternative splicing.[6] Examinations of the combinatorial assortments of $Ca_V1.2$ alternatively spliced exons showed

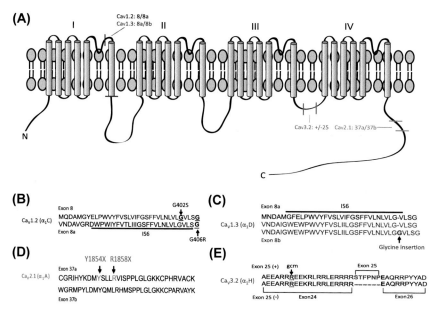

FIGURE 11.1 **Schematic representation of the α_1 subunits of the voltage-gated calcium channels.** (A) The alternatively spliced exons related to channelopathies are marked in different colors to reveal various subtypes of voltage-gated calcium channels. (B) Mutually exclusive exon 8/8a in *CACNA1C* is involved in the pathogenesis of Timothy syndrome. Both G402S and G406R mutations occurred in exon 8 of *CACNA1C*, while G406R mutation was found in exon 8a. (C) Mutually exclusive exon 8a/8b in *CACNA1D* with a glycine insertion (p.403_404insGly) is related to SANDD syndrome. (D) Mutually exclusive exon 37a/37b in EF-hand domain of *CACNA1A* with two premature stop codons (Y1854X and R1858X) is involved in the pathogenesis of EA2. (E) Inclusion or exclusion of exon 25 of *CACNA1H* and gcm mutation (R1584P) are related to childhood absence epilepsy.

tissue-selective splice patterns such as segregating cardiac muscle from smooth muscle splice patterns. Importantly these tissue-selective splice combinations produce distinct electrophysiological and pharmacological properties.[7] To illustrate this point, we consider a pair of mutually exclusive exons, exon 8a and 8, and this exon encodes the protein sequence of the IS6 segment of the $Ca_V1.2$ channels.[8] Even though the naming of exons 8/8a is different in the Splawski's article compared to other publications, nonetheless, the issue is that the expression levels of exons 8/8a govern the severity of cardiac disorder suffered by the patients.[9] According to Splawski's group, exon 8 is highly expressed in heart and brain, while exon 8a is found in approximately 20% of $Ca_V1.2$ transcripts.[10] This example reveals that interpretation of differences in disease phenotype or severity could be done through the lens of tissue-selective expression of alternative exons of any Ca_V channels. Two widely known de novo missense mutations, G402S and G406R in exons 8/8a of $Ca_V1.2$ channels, produce gain-of-function through impairing $Ca_V1.2$ channel inactivation. The clinical manifestations of these patients suffering from TS, which is life threatening, are lethal cardiac arrhythmias, congenital heart disease, immune deficiency, intermittent hypoglycemia, cognitive abnormalities and autism, and webbing of the fingers and toes.[11,12] However, the occurrence of the two mutations is different in two types of TS. Both G402S and G406R mutations in exon 8 of *CACNA1C* gene were identified in patients with atypical TS-2 phenotypes, while G406R mutation was found in exon 8a in patients with classical TS-1.[10] A novel *CACNA1C* gene mutation (p.Ala1473Gly) was detected in exon 38 in a patient with TS, and this discovery expanded the molecular basis of TS.[11]

Compared with WT channels, the G406R mutant $Ca_V1.2$ channel exhibited an impaired voltage-dependent inactivation (VDI) as shown by whole-cell patch clamp recordings conducted in both Chinese hamster ovary cells and *Xenopus* oocytes.[11] The results showed that time-dependent inactivation was largely decreased when the channel activity was recorded in *Xenopus* oocytes using Ba^{2+} as a charge carrier.[10] The reduction of VDI would lead to continuous influx of Ca^{2+}, a key reason for the prolongation of cardiac action potentials and the lengthening of the QT interval. The mutant channels were still sensitive to inhibition by the L-type channel blocker nisoldipine,[9] suggesting that calcium channel blockers may be useful for treatment of cardiac symptoms in TS patients. In contrast to slowed VDI in mutant channels, G406R mutation did not affect Ca^{2+}-dependent inactivation (CDI) of the channel.[13] Furthermore, the G406R mutation reduced VDI by about 45% when coexpressed with β_{1c} (predominant in brain), while only by about 27% with β_{2a} (predominant in heart), which may explain the different severity of TS-associated pathology in various tissues.[13]

In addition, G436R mutation in the alternatively spliced rabbit cardiac $Ca_V1.2$ channel containing exon 8a, corresponding to G406 position in the human $Ca_V1.2$ channel, was reported to impair open-state voltage-dependent inactivation (OS*vd*I) but not close-state voltage-dependent inactivation (CS*vd*I), and the mutant channel also exhibited slowed activation and deactivation kinetics.[14] Moreover, L-type channel agonist Bay K8644, which could prolong cardiac action potentials, failed to slow deactivation of the mutant channel.[14] Thus, slowed deactivation together with impaired OS*vd*I are factors predicted to synergistically increase cardiac action potential duration, and this outcome may lead to the development of arrhythmias in TS (Fig. 11.1B).

With the groundbreaking advances in stem cell research, new possibilities have been opened to uncover cellular phenotypes associated with TS.[15,16] Skin cells were reprogrammed from TS patients to generate induced pluripotent stem cells (iPSCs) which were then differentiated into cardiomyocytes. Whole-cell patch clamp recordings in differentiated human cardiomyocytes showed significant reduction in VDI and prolonged action potentials compared to control cells. Moreover, Ca^{2+} elevations in TS cardiomyocytes were more irregular, which appeared to be much larger and more prolonged. These results indicated that the activation of $Ca_V1.2$ channels was essential for maintaining the timing and amplitude of the ventricular Ca^{2+} release. Importantly, a compound named roscovitine was reported to increase VDI, and as such it might be a potential drug to test for efficacy in the management of cardiac problems in TS patients.[14]

$Ca_V1.3$ and SANDD Syndrome

$Ca_V1.3$ channels are widely expressed in neurons, sinoatrial node (SAN), inner hair cells (IHC) and neuroendocrine cells.[17] An insertion mutation (c.1208_1209insGGG) in exon 8b of *CACNA1D* gene led to a human channelopathy termed SANDD (sinoatrial node dysfunction and deafness) syndrome with a cardiac and auditory phenotype[18] that closely resembles that of $Ca_V1.3$-knockout mice which are deaf and which also exhibit SA node arrhythmia causing bradycardia.[19] Similar to $Ca_V1.2$ channels, mutually exclusive exon 8a/8b also encodes the IS6 segment in the $Ca_V1.3$ channel (Fig. 11.1C). Whole-cell patch-clamp recordings in tsA-201 cells expressing wild-type or mutant $Ca_V1.3$ channels showed that the mutant channels did not conduct significant Ca^{2+} currents. This is consistent with another report that no intact protein and current were detected when expressing human $Ca_V1.3$ splice variant containing exon 8b in tsA-201 cells.[20] Further investigations indicated that the ON-gating currents of the mutant channels were significantly reduced compared to the wild-type, although the mutant channels were also expressed at the cell surface. Therefore, it is possible that movement of the mutant channel voltage sensor did not induce pore opening, or pore opening was normal, but ion conduction was occluded. Of note, the expression pattern of $Ca_V1.3$ with exon 8a or 8b is distinct in different tissues: $Ca_V1.3$ channel with exon 8b is the predominant variant in cochlear IHCs and the SAN, while the expression levels of exon 8a and exon 8b are almost equal in rat brain (Huang et al unpublished data). Through targeting one of the two mutually exclusive $Ca_V1.3$ exons, the SANDD-causing mutation induced tissue-selective symptoms in the disorders that corresponded with the expression patterns of the splice variants.

$Ca_V2.1$ and Episodic Ataxia Type 2

The P/Q-type $Ca_V2.1$ channels are expressed presynaptically throughout the nervous system and abundantly in the cerebellar Purkinje and granule cells.[21] Similar to other VGCCs, $Ca_V2.1$ channels are also subject to extensive alternative splicing to produce several variants with different functions and expression patterns. Mutations in human *CACNA1A* gene encoding the $Ca_V2.1$ channel in certain splicing variants are associated with several disorders, including episodic ataxia type 2 (EA2), familial hemiplegic migraine (FHM1) and spinocerebellar ataxia type 6 (SCA6).[22] Here we mainly focus on EA2 to introduce the relationship between alternative splicing of *CACNA1A* and disorders. The C-terminus of $Ca_V2.1$ α_1-subunit contains the EF-hand-like domain, which is encoded by exon 36 and exon 37. The exon 37 is mutually exclusively spliced with exon 37a and exon 37b, forming two variants of $Ca_V2.1$.[23–25] $Ca_V2.1$ channels containing exon 37a, but not exon 37b, could sustain robust Ca^{2+}-dependent facilitation and which would affect synaptic transmission.[26,27] Two novel stop codon mutations, R1858X and Y1854X, in exon 37a (EFa) but not exon 37b (EFb), have been identified in a family suffering from EA2 (Fig. 11.1D).[28] These mutations led to a loss-of-function only of the $Ca_V2.1$ channels containing EFa, which again points to exon-selective phenotypes in neurological disorders. As exon 37a is highly expressed in the cerebellar Purkinje cells, the selective effect of the two mutations on $Ca_V2.1$ channel function in Purkinje cells is consistent with the clinical presentations of EA2. Besides, it has been reported that there is a switch of the $Ca_V2.1$ transcripts containing 37a to 37b in the fourth decade in men but not in women, suggesting that the two mutually exclusive exons expressed differently in different developmental stages.[27] As a result, such age and gender differences in the expression of alternatively spliced exons may help to provide a better understanding of the age-dependent decrease in frequency and severity of attacks of EA2 commonly reported in men.

$Ca_V3.2$ and Childhood Absence Epilepsy

Idiopathic generalized epilepsies (IGEs) involve both brain hemispheres.[29] It is reported that IGEs are associated with the genetic basis.

CACNA1H has been thought to be an epilepsy susceptibility gene. Mutations in this gene were first detected in patients with CAE which is one of the most common forms of IGEs.*CACNA1H*[30] *CACNA1H* is alternatively spliced at 12–14 sites, generating a wide range of different isoforms with distinct biophysical, modulatory, and pharmacological properties. Several evidences supported that the context of the splice variant is important in affecting the effects of mutations, such as genetic absence epilepsy rats from Strasbourg (GAERS) $Ca_V3.2$ mutation (gcm), on channel properties. The gcm mutation was found in the *CACNA1H* gene in the GAERS model of IGE.[31] The gcm mutation is a G to C single nucleotide mutation producing an R1584P substitution in exon 24 of *CACNA1H*, and this exon encodes a region of the linker of domain III-IV in the $Ca_V3.2$ channel (Fig. 11.1E).[31] In this study, Powell et al. identified two major thalamic splice variants, including or excluding exon 25. Interestingly, electrophysiological recordings showed that in the presence of exon 25, the mutant channels had increased rate of recovery from inactivation, which indicated that more of these channels were available during subsequent depolarizations. Such an alteration in channel property resulted in significantly larger Ca^{2+} currents that potentially increased excitability and promoted epileptogenesis. It seemed that gcm mutation induced a splice variant–specific gain-of-function in $Ca_V3.2$ (with exon 25), which could be highly related to neuronal burst firing. They also found that the ratio of $Ca_V3.2$ with exon 25 to $Ca_V3.2$ without exon 25 was about 1.5-fold greater in GAERS model compared to the nonepileptic control strain rats. As the gcm mutation is situated in exon 24 just upstream of the start of exon 25, it is possible that the mutation has an effect on the expression levels of the splice variants.

ALTERNATIVE SPLICING OF VOLTAGE-GATED POTASSIUM CHANNELS AND DISEASES

Voltage-gated potassium channels (K_V) are a large group of channels supporting K^+ efflux when they open in response to membrane depolarization. It has been reported that 40 human K_V genes could be categorized into 12 subfamilies. The K_V channels play crucial roles in coupling action potential, cell proliferation, and immunoresponse.[32] All mammalian K_V channels are assembled as four α-subunits, and each subunit contains six transmembrane segments (S1–S6). The S1–S4 sense changes in membrane potential, while the reentrant loop between S5 and S6 forms the P-loop which contains the selectivity filter to gate K^+ ion flux.[26,33] Many of the K_V channels have been cloned in the past few years and several studies have focused on characterizing their biophysical and pharmacology properties. However, native K_V channel function in various tissues remained unclear. As K_V channels are composed of four α-subunits, the α-subunits may form

TABLE 11.2   Summary of Alternative Splicing Related Diseases in K_V Channels

Channels	Mutation Site	Mutation Type	Splicing Site	Related Disease	References
KCNQ1	c1032G>A/C	Silent mutation A344A	Skip exon 7 and exon 8	LQT1	45
KCNQ1	c1251+1G>A	Intronic mutation	Skip exon 8 and exon 9	LQT1	47
KCNQ1	IVS7+3A>G	Intronic mutation	Skip exon 7	LQT1	46
KCNQ1	c477+1G>A	Intronic mutation	Skip exon 2 produce truncated channel	LQT1 and Jervell and Lange-Nielsen Syndrome	48
KCNQ1	c387-5T>A	Intronic mutation	Skip exon 2 produce truncated channel	LQT1 and Jervell and Lange-Nielsen Syndrome	49
KCNQ2	Ins5bp		5 nt insertion	Benign familial neonatal convulsions	50
KCNQ2	IVS14-6 C>A	Intronic mutation	4 nt of GTAG and a stop codon has been inserted	Benign familial neonatal convulsions	52
KCNQ2	c1187+2 T>G	Intronic mutation		Benign familial neonatal convulsions	53

tetramers of multiple configurations. In addition, K_V channel properties could be modulated with the association of β-subunit, which will further modulate channel function. K_V channel properties could be further diversified by posttranscriptional modification in a spatial and temporal manner. In this section, we will focus on the implication of functional diversity of K_V channel due to alternative splicing under physiological or pathological conditions (Table 11.2).

K_V7.1 and Long QT Syndrome

The K_V7.1 channels are highly expressed in cardiac muscles, and they contribute to the slow delayed rectifier K^+ currents. These channels also play a crucial role in cardiac late phase action potential repolarization.[34,35] Besides, K_V7.1 channels are also expressed in the inner ear, lung, and several endocrine cells.[36] The $K_V7.1$ α-subunit (KCNQ1) is usually associated with the β-subunit (KCNE1).[37] The presence of the β-subunit delays channel activation and increases macroscopic currents as compared with functional characteristics of $K_V7.1$ α-subunit expressed alone.[35] Dysfunction of $K_V7.1$ is one of the major causes of long QT syndrome (LQT), which is identified by prolongation of action potential, arrhythmias, and sudden cardiac death.[38,39] Some rare mutations of $K_V7.1$ are associated with less common Jervell and Lange-Nielsen syndrome (JLNS). Patients with JLNS suffer from hearing loss in addition to heart disorder. It has been shown that mutations of KCNQ1 and KCNE1 cause loss-of-function of the channels and results in delayed cardiac action potential.[36] It has also been reported that mutations of $K_V7.1$ result in protein misfoldings, disruption of subunit assembly, defective trafficking, and interaction between $K_V7.1$ and PIP2.[40–44]

However, some mutations of $K_V7.1$ do not change the amino acid sequences, and others are located in the intron of the *KCNQ1* gene. How do these mutations affect channel function? It has been shown that the silent mutations or intronic mutations may alter the splicing pattern of the $K_V7.1$ transcripts and lead to loss-of-function of the channels.

LQT type1-associated mutations c1032G>A and c1032G>C are found in the last nucleotide of exon 7. These mutations do not affect the amino acid sequence of $K_V7.1$ (A344A), but they alter its splicing pattern. The mutant KCNQ1 would skip exon 7 or exon 8. The exon-skipping $K_V7.1$ channels exhibit smaller current compared to the wild-type channel. When they coassemble with the wild-type channels to form tetramer, the exon-skipping $K_V7.1$ channels dominant negatively reduce the I_K currents.[45] There are also other mutations found in the intronic sequences of KCNQ1 that affect splicing. It has been reported that IVS7+3A>G results in skipping of exon 7, and 1251+1G>A induces cryptic splicing by skipping exons 8 and 9.[46,47] The splicing mutations alter alternative splicing events to various extents. Transcript scanning of patient samples have shown that mutation c1032G>A asserted a great effect to alter splicing than mutation IVS7+3A>G. Consistent with this observation, the family with IVS7+3A>G mutation also showed relatively mild LQT clinical phenotype compared to 1032G>A.[46]

Furthermore, studies have also found that mutation affecting splicing of KCNQ1 could cause even more severe consequences of Jervell and Lange-Nielsen Syndrome (JLNS). The splice donor site mutation of KCNQ1 (c477+1G>A) leads to almost 100% skipping of exon 2. This altered splicing would generate frameshift of transcripts and a premature stop codon in exon 4. c477+1G>A mutation totally disrupts the $K_V7.1$ function.[48] Another similar mutation of c387-5T>A on KCNQ1 leads to 90% of aberrant splicing channel and 10% normal splicing channel. The patients of c387-5T>A mutation suffer from severe LQT but the residue wild-type channels partially rescue hearing loss.[49] These results suggest the correlation of aberrant splicing of KCNQ1 and clinical phenotypes (Fig. 11.2).

K_V7.2 and Neonatal Epilepsy

K_V7.2 and K_V7.3 channels are highly expressed in the central nervous system, and they usually assemble as heterotetramers. K_V7.2 and K_V7.3 channel function sensitive to activation of G-protein coupled receptor, notably muscarinic acetylcholine receptor, and as such the K^+ currents of K_V7.2 and K_V7.3 are called M-currents. The K_V7.2 and K_V7.3 channels are involved in subthreshold excitability and mutations of K_V7.2 and K_V7.3 have been associated with neonatal epilepsy or benign familial neonatal convulsions (BFNS).[50,51] The patients suffer from frequent mild seizures on the first day of birth and the symptoms are usually relieved after several weeks. But there is about 15% occurrence of seizure later in life. Most of the mutations are associated with K_V7.2 and splicing of K_V7.2 exons is also affected by some mutations. Early studies have found that a 5 bp insertion into the genomic DNA sequence cause a frameshift and premature stop codon.[50] Later on, a single-point mutation in the intron generates an aberrant splice motif that leads to altered splicing. Therefore, an additional 4 nt is inserted into the transcripts, including a stop codon. This mutation produces a truncated K_V7.2 lacking the c-terminus. The truncated K_V7.2 may not be able to be assembled and trafficked to the cell membrane.[52] Another BFNS mutation is found in c1187+2T>G of KCNQ2, which is located at

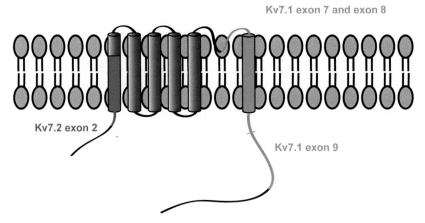

FIGURE 11.2   Schematic representation of the α_1 subunits of the voltage-gated potassium channels. The alternatively spliced exons related to channelopathies of K_V7.1 (exon 7, exon 8, and exon 9) and K_V7.2 (exon2) are shown.

the predicted splice motif. It is possible that aberrant splicing causes the loss-of-function of $K_V7.2$ (Fig. 11.2).[53]

$K_V7.4$ and Congenital Deafness

The KCNQ4 channels are expressed in the cochlear outer hair cells and neurons of the auditory system.[54] Dysfunction of KCNQ4 gives rise to deafness. The KCNQ4 channels produce a slow rectifying K^+ current, and they can be regulated by calmodulin. Alternative splicing of exons found in the C-terminus of KCNQ4 could selectively exclude exons 9, 10, or 11. The shorter version of KCNQ4 channels, without exon 9, 10, or 11, exhibits small currents and weak interaction with calmodulin. A dominant negative mutation on the KCNQ4 without exon 9, 10, or 11 cripples the entire KCNQ4 channels. These results indicate that alternative splicing of KCNQ4 pre-mRNA plays a crucial role in hearing.[55]

ALTERNATIVE SPLICING OF VOLTAGE-GATED SODIUM CHANNELS AND DISEASES

Voltage-gated sodium channels $Na_V1.1-Na_V1.9$ play crucial roles for generating and conducting electrical impulses in excitable cells. The predicted topology of the α-subunit of Na_V channels is similar to voltage-gated Ca^{2+} channels with four major domains (D1–D4) and six transmembrane segments for each domain (S1–S6). Na_V channels are widely expressed in multiple tissues; for example $Na_V1.1$ (SCN1A), $Na_V1.2$ (SCN2A), $Na_V1.3$ (SCN3A), and $Na_V1.6$ (SCN8A) channels are expressed in the peripheral tissues, and they are also expressed abundantly in the brain and CNS. By contrast, $Na_V1.4$ (SCN4A) and $Na_V1.5$ (SCN5A) channels are only selectively expressed in muscle tissues, with $Na_V1.4$ being expressed in adult striated skeletal muscle and $Na_V1.5$ in embryonic, denervated skeletal muscle and heart muscle respectively. More strikingly, $Na_V1.7$ (SCN9A), $Na_V1.8$ (SCN10A), and $Na_V1.9$ (SCN11A) channels are expressed selectively in the dorsal root ganglion in the peripheral nervous system. Three prominent features characterized the voltage-gated Na^+ channels namely rapid inactivation, high selectivity for Na^+ ion conductance, and inactivation by tetrodotoxins (Table 11.3).[56]

Alternative splicing has been described for most α-subunits of Na_V channels and their splice variations changes in development and disease. Much research has been devoted to the characterization of the effects single point mutations have on channel function. However, such effects on channel properties may be modified by the splicing combinations of alternative exons found in the Na_V channels expressed in a particular cellular

TABLE 11.3   Summary of Alternative Splicing Related Diseases in Na$_v$ Channels

Channels	Mutation Site	Splicing Site	Functional Alteration	Related Disease	References
SCN1A	IVS5N+5G>A	Exon 5	Expression level, pharmacological sensitivity	Epilepsy with febrile seizures	57
SCN2A	G211D and N212D (exon 5N), V213D (exon 5A)	Exon 5/5A	Neuronal excitability, seizure susceptibility	Epilepsy, autism	60
SCN4A	c.3912+6_3912+10delinsG	Exon 21–22	Disruption of fast inactivation	Skeletal myotonia	63
SCN5A		Exon 6/6A	Depolarized threshold of activation, slower kinetic of activation and inactivation, larger Na$^+$ influx, slower recovery rate, and bigger persistent Na$^+$ current	Neonatal and adult developmental stage, heart failure, Brugada syndrome	64,66
SCN5A		Exon 28	Premature termination, truncated protein	Heart failure	69
SCN5A		Exon 6	Upregulation of neonatal splice variant in metastatic	Breast cancer	65
SCN9A	I1461T	Adult isoform (exon 5A)	Increase of ramp current and shift activation to more hyperpolarized potential	Paroxysmal extreme pain disorder	70

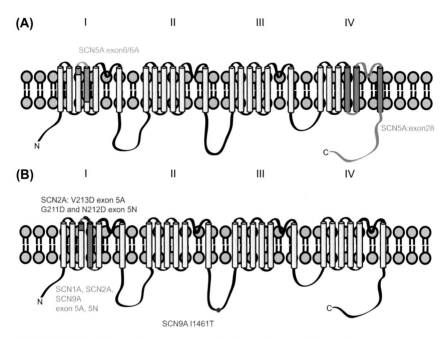

FIGURE 11.3 Schematic representation of the α_1 **subunits of the voltage-gated sodium channels.** (A) The alternatively spliced exons related to channelopathies of SCN5A (exon 6/6A and exon 28). (B) The alternatively spliced exon 5N/5A and point mutations related to channelopathies of SCN1A, SCN2A (G211D, N212D, and V213D), SCN4A, and SCN9A (I1461T).

milieu. Better understanding of the spatial and temporal expressions of alternative exons is therefore necessary to inform on disease pathogenesis.

Alternative splicing can be developmentally regulated to produce neonatal and adult isoforms of several Na_V channels. As such, mutations in either neonatal or adult Na_V channel isoforms could explain the onset of disorders such as epilepsy or pain disorders. In this section, we will focus on how functional diversity of Na_V channels generated by alternative splicing may contribute to adaptation to changing physiological or pathological conditions (Fig. 11.3).

$Na_V1.1$ (SCN1A) and Epilepsy With Febrile Seizure

SCN1A splice-site polymorphism rs3812718 (IVS5N+5 G>A) might be associated with susceptibility to epilepsy with febrile seizures (EFS) by contributing to the pathophysiology underlying genetic generalized epilepsies (Fig 11.3B).[57] It affected the ratio of adult and neonatal transcripts of *SCN1A*. The neonatal exon is preferentially expressed during early developmental periods. The 5′ splice donor site of exon 5N was disrupted by the major allele (A), which lead to drastically reduced expression of exon 5N as compared to the adult exon 5A. The mutation would increase relative risk of EFS. And it was responsible for almost half of patients suffering from febrile seizures.[58] The neonatal isoform has also been reported to have higher pharmacological sensitivity to the commonly used antiepileptic drugs with enhanced tonic block.[59]

Na$_V$1.2 (SCN2A) and Epilepsy as well as Autism

Na$_V$1.2 also has a neonatal and an adult isoform resulting from alternative splicing of exon 5N and 5A (Fig 11.3B). The neonatal channel splice variants were less excitable than the adult channels, which may result in reduced neuronal excitability and be seizure protective during early brain development. Dysregulation of the expression level of neonatal and adult isoforms during development has been shown to be associated with epilepsy and autism.[60,61]

Mutations in exon 5N (G211D and N212D) and 5A (V213D) were identified in patients with epileptic encephalopathies. The mutations generated a negative charge and formed an extracellular loop closer to the voltage sensor similar to the conformation of the adult isoform. The closer proximity seemed to affect the mutant channel's biophysical properties that then led to severe seizures.[60] The proportion of exon 5N-included channels to exon 5A-included channels is developmentally regulated. At an early developmental stage, the transcripts of SCN2A in the brain have more or equal amounts of neonatal isoforms than adult isoform, and the level of neonatal isoforms diminished during development. However, children with mutations did not show improved health condition during their development into adulthood.[61]

Similar developmentally regulated neonatal and adult isoforms of channels are also identified in Na$_V$1.5 in heart and Na$_V$1.7 in the peripheral nervous system. As for Na$_V$1.6 channel, inclusion or exclusion of splice variants exon 18N or 18A to produce the neonatal isoform that contains an in-frame stop-codon will result in a truncated two-domain protein. However, the function of this truncated channel is still unknown.[62]

Na$_V$1.4 (SCN4A) and Skeletal Myotonia

Many mutations have been identified in the skeletal muscle Na$_V$1.4 channel, and they are associated with myotonia. One of these is a deletion/insertion (c.3912+6_3912+10delinsG) mutation located in intron 21 (an AT-AC type II intron) of SCN4A, and the presence of these mutant channels cause myotonia.[63] Intron 21 is a rare kind of intron that can be spliced by the major or U2-type spliceosome even though it has the AT-AC boundaries.

In skeletal muscles from patients with myotonia, SCN4A mRNAs were spliced by activation of cryptic splice sites, and the result is the formation of aberrant isoforms. One of the isoforms contained a 35-amino acid insertion in the cytoplasmic loop between domains III and IV of $Na_V1.4$, and these splice variants exhibited disruption in fast inactivation, which is consistent with a gain-of-function in the channels observed in myotonia.[63]

$Na_V1.5$ (SCN5A) and Heart Failure as well as Breast Cancer

For $Na_V1.5$ channels, splice variation at exon 6 leads to generation of two variants: a 5′-exon adult variant and a 3′-exon neonatal variant. Although neonatal only differs from adult $Na_V1.5$ channels by seven amino acids, it is strongly linked to pathological heart failure.[64] Of particular interest, it is the replacement of a conserved negative aspartate residue in the "adult" with a positive lysine in the extracellular transmembrane domain preceding the voltage-sensing S4 segment, which alters the electrophysiological properties of the channel.[65] Characterization of the channels using patch-clamp electrophysiology showed a loss-of-function, including depolarized shift of activation, slower activation and inactivation kinetics, slower recovery, and reduced channel expression.[66] The neonatal channels shifted threshold of activation and peak current toward more depolarized voltage and exhibited slower activation and inactivation kinetics and recovery from inactivation. Therefore, both transient and persistent Na^+ influxes from neonatal channels are larger comparing to adult channels. Patients with Brugada syndrome have higher expression levels of neonatal channels containing exon 6A. Taking the modified kinetics and larger Na^+ current of neonatal channels into consideration, it was suggested that sudden cardiac arrest could result from dysregulated splicing at exon 6 of SCN5A.[66]

In addition to exon 6 and 6A splice variations in domain 1, c-terminal alternative splicing of the $Na_V1.5$ channels is also associated with heart failure. Three new exon 28 splice variants have been identified, and they lead to premature truncation of the Na_V channels that have missing in their structure the S3 and S4 segments of domain IV and the entire C-terminus.[67,68] The surface expression of truncated channels was strongly attenuated, and this resulted in reduction of the Na^+ currents. Chronic heart failure was related to the increased expression level of the truncated splicing variant and a decrease of full-length channels.[69] Besides heart failure, studies also showed that the neonatal-type of $Na_V1.5$ is crucial for invasive behavior and metastasis of the breast cancer cells (Fig. 11.3A).[65]

$Na_V1.7$ (SCN9A) and Pain Perception

$Na_V1.7$ channels are preferentially expressed in the nociceptive neurons. Gain-of-function mutation of $Na_V1.7$ channels lead to neuron hyperexcitability associated with pain. Two human pain perception disorders, inherited erythromelalgia (IEM) and paroxysmal extreme pain disorder (PEPD) are associated with $Na_V1.7$ mutations with age-dependent symptoms.[70] It has been reported that IEM is progressively more severe over age, while the frequency of pain episodes decreased in PEPD patients. It is known that the wild-type $Na_V1.7$ channels exhibit two exon 5 splice variants (5A and 5N). Exon 5A has two amino acid difference compared to exon 5N (L201V, N206D), and this channel splice variant displays hyperpolarizing activation thresholds. However, channels containing the PEPD mutation (I1461T) and exon 5A exhibited an additive effect in decreasing activation threshold. Such an effect is not observed in IEM-associated mutations I136V (Fig 11.3B).[70]

ADAR2-MEDIATED RNA EDITING

Ion channels and neurotransmitter receptors are the major targets of ADAR2 mediated A-to-I RNA editing in the mammalian nervous system. RNA editing of ion channels and receptors alters their gating properties or their ability to bind other interacting proteins (Table 11.4).

RNA Editing in Voltage-Gated Ion Channels

Functional A-to-I RNA editing of the first mammalian L-type $Ca_V1.3$ calcium channel was reported in which RNA editing of isoleucine (I) to methionine (M) or glutamine (Q) to arginine (R) at the IQ-domain was shown to reduce CDI in the edited channels (Fig. 11.4A). RNA editing of $Ca_V1.3$ channels was found to display a CNS-selective expression pattern.

TABLE 11.4   Summary of RNA Editing Related Diseases in Ion Channels and Receptors from Mammalian System

Channel/ Receptors	Editing Sites	Functional Changes	Relative Diseases	References
$Ca_V1.3$	I/M, Y/C, Q/R sites	Reduced calcium-dependent inactivation		71
$K_V1.1$	I/V site	Fast inactivation and increased recovery rate	Epileptic seizure	73
GluR-B	Q/R site	Enhanced rate of recovery from receptor desensitization	Amyotrophic lateral sclerosis	81
$GABA_A$ receptor	I/M site	Smaller amplitudes, slower activation, and faster deactivation	Synaptic formation and development of nervous system	88
$5\text{-}HT_{2C}$ receptor	A-E 5 sites	Reduction in G-protein coupling and lower level of constitutive activity	Mood disorders such as depression and anxiety	85

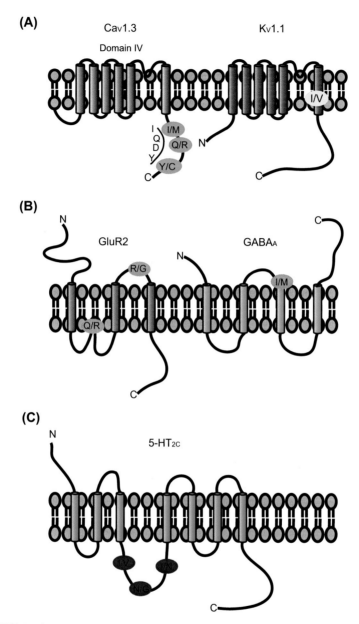

FIGURE 11.4 **Schematic representation of ADAR2-mediated RNA editing substrates.**
(A) I/M, Q/R, Y/C RNA editing sites in $Ca_V1.3$ channel IQ domain and I/V site in $K_V1.1$
channel. (B) Q/R and R/G editing sites in GluR2 AMPA receptor and I/M site in $GABA_A$
receptor. (C) A-E RNA editing sites (I/V, N/G, I/V) in $5\text{-}HT_{2C}$ receptor.

While the mechanism for such selectivity is unknown, it also begs the question of the different RNA editing levels found in different neurons in various brain regions. Although till now no disease relevance has been reported, the physiological significance of $Ca_V1.3$ IQ-domain RNA editing seemed to be associated with the role of $Ca_V1.3$ channel in pacemaking neurons. By using the ADAR2$^{-/-}$/GluRB$^{R/R}$ mouse model, it has been shown that deletion of RNA editing resulted in lower frequency of spontaneous action potential firing. This reduction of Na$^+$ spike frequency may be due to reduced underlying Ca^{2+} spikes, as there would be less Ca^{2+} entering through the unedited $Ca_V1.3$ channels.[71] However, the implication of IQ-domain RNA editing of $Ca_V1.3$ channels in disease is still unknown.

RNA editing at the I/V site of $K_V1.1$ channel affected interaction of β-subunits with different inactivation particles and reduced the binding affinity of channel blockers (Fig. 11.4A). The net result is enhanced recovery from inactivation.[72,73] The IC_{50} value of Psora-4 and 4-AP inhibition increased drastically to almost 70-fold when comparing to unedited channels, while the inhibition of arachidonic acid on $K_V1.1$ channel showed a negative correlation with editing level.[74] Elevated editing level at the I/V site of $K_V1.1$ channels has been reported in chronic epileptic rat model, which showed reduced 4-AP sensitivity.[73]

RNA editing in voltage-gated ion channels also play important roles in nonmammalian animals. RNA editing was shown to be widespread in *Drosophila melanogaster* both in exonic and intronic regions.[75] High level of RNA editing was found in the gene that encodes a voltage-gated Na$^+$ channel. The I/V site was detected in DmNa$_V$5.1 and DmNa$_V$4.2, while the other editing I/T site was detected in DmNa$_V$7.1. The edited DmNa$_V$5.1 channel produced a right-shift of activation potential toward the depolarization direction, suggesting that I/V RNA editing could be responsible for low voltage–dependent activation of these channels.[76]

Another interesting example is in octopus. RNA editing of the $K_V1.1$ channel at I/V site is important for their temperature adaptation, as the editing levels correlate with the water temperatures of their habitats. Octopuses from Antarctic and Arctic have higher editing levels at the I/V site in $K_V1.1$, while tropical octopuses have lower levels of RNA editing. The functional effect of I/V RNA editing of $K_V1.1$ is that edited channels close faster upon repolarization and together with reduced duration of afterhyperpolarization induce increased firing frequency of action potentials.[77]

RNA Editing in Neurotransmitter Receptors

Although the understanding of the physiological or pathophysiological significance of RNA editing of voltage-gated ion channels is still limited;

however, there is a greater body of published work on RNA editing of neurotransmitter receptors and the implications of editing in disease.

The level of RNA editing at the Q/R site of GluA2 subunit of AMPA receptors (GluRs) is approximately 99% (Fig 11.4B). AMPA receptors with edited GluA2 subunit are Ca^{2+} impermeable and editing will therefore modulate synaptic plasticity.[78] AMPA receptors containing edited GluA2 at the R/G site showed enhanced rate of recovery from desensitization.[79] Pharmacologically, RNA editing at the Q/R site resulted in reduced sensitivity of AMPA receptors to spider toxin (JSTX).[80] Reduced level of RNA editing at Q/R site in GluA2 was detected in the motor neurons from amyotrophic lateral sclerosis (ALS) patients, which contributed to increased Ca^{2+} permeability of AMPA receptors.[81]

While editing of serotonin 2C receptors occurred at five closely localized sites, the fully edited receptor exhibited the inactive conformation that had reduced coupling of receptors with G-protein and lower agonist affinity (Fig 11.4C).[82] RNA editing of 5-HT_{2C} receptors could also alter the agonist-directed trafficking of receptor stimulus, which means that because the edited receptor preferred to form the inactive conformation, the conformation promoted by receptor agonist that specifically couple to or activate certain downstream signaling cascades might be altered. Thus the edited receptor loses the ability to be directed by agonists to preferentially activate the PLC-PI or PLA2-AA pathways.[83] As different editing status results in activation of different downstream pathways, the effect of drug, besides the binding affinity, also strongly depends on the status of RNA editing. A study reported that higher level of expression of edited 5-HT_{2C} receptors was found in human suicide victims with high association of gene expression and DNA methylation.[84] Increased editing levels at sites A, E, and C in 5-HT_{2C} receptor were observed in patients suffering from schizophrenia or major depression, while the editing levels at sites B and D seemed to be decreased.[85]

Another RNA-editing substrate is $GABA_A$-$\alpha3$ receptor. Researchers reported that I/M site editing of $GABA_A$-$\alpha3$ receptor resulted in both reduction of current and alteration of gating kinetics (Fig 11.4B).[86,87] The reduced currents might be caused by the impaired trafficking of $GABA_A$-$\alpha3$ edited receptors to the surface. The editing level of I/M site of the $GABA_A$-$\alpha3$ receptor was shown to be developmentally regulated. The editing level in mouse brain was about 5% at E15, rising up to over 90% at P21. With this increased level of RNA editing, the protein level of $GABA_A$-$\alpha3$ receptor decreased after birth.[88] For regulation of RNA editing, beside the duplex hairpin structure, a *cis*-acting element 150 nt downstream of the editing site of $GABA_A$-$\alpha3$ receptor was reported to be able to induce site-specific editing.[2] It strongly supported the idea that there could be *cis*-acting elements responsible for the regulation of RNA editing. RNA editing of

GABRA3 and ADAR2 was also detected in regenerating tissues of red-spotted newt. Editing level of GABRA3 and ADAR2 increased early on during brain regeneration. While at late phase of brain injury, the editing level decreased. However, in the limb, increased editing level was shown at the late stage of injury.[89]

Besides the implication of these altered receptor properties in neurological diseases, such as in epileptic seizures, ALS, and mood disorders like schizophrenia, depression, and anxiety, RNA editing is also associated with other diseases. For example, increased level of RNA editing in thyroid hormone receptor associated protein 1 (THRAP1) was found in children suffering from cyanotic congenital heart disease.[90]

High Conservation of RNA Editing in Genome

RNA editing was demonstrated to be highly conserved among 15 mouse strains by using computational methods for analyzes. Over 7000 editing sites were identified with the majority being A-to-I RNA editing. The I/M and Y/C RNA editing sites in the IQ-domain of $Ca_V1.3$ channels were examples of those verified sites with consistently high editing level among all 15 mouse strains (Fig. 11.4).[91]

While in human, high-throughput sequencing had been conducted to identify differences between RNA and DNA coding sequences. It was reported that RNA–DNA differences (RDDs) were found in 10,210 exonic sites with all 12 possible categories including A-to-I and C-to-U known as RNA editing.[92] However, other groups seemed to have contradicting results. A deeper analysis seemed to show that only 1173 sites were real RDDs, while 89% of the rest of the identified sites may be false positive mismatches caused by gene duplications.[93]

Another study reported RNA editing sites in human Alu and non-Alu sequences identified also by computational approaches. Here, 147,029 sites were identified in Alu sequences with 95.8% A-to-G mismatch indicating potential A-to-I RNA editing, while in non-Alu and nonrepetitive sequences 2385 and 1451 editing sites were identified with respectively 97.4% and 86.6% A-to-G mismatches. Clustering of the identified editing sites seemed to suggest that the locally formed double-stranded RNA structure might be the explanation to the dependence of non-Alu editing sites on nearby edited Alu sites.[94]

CONCLUSION

As discussed earlier, voltage-gated ion channels and receptors undergo extensive posttranscriptional modifications. These modulations usually result in protein and functional diversities and further lead to increased complexity of ion channel functions. Voltage-gated ion channels serve as major routes for ion entry into cells in various tissues. Posttranscriptional modification of these channels provides a way for fine-tuning of channel function in response to various stimuli in different tissues and disease conditions. Ion channelopathies may be gradually developed due to chronic or drastic changes in the ion influx via channels following posttranscriptional modification. On the other hand, in disease state, those modifications may also contribute to the dynamic changes of channels within cells that may help to reset conditions to the physiological state. However, structural context of the channel proteins arising from alternative splicing should be evaluated, as exon selectivity in how the mutations may assert on phenotypic expressions should be carefully considered. Besides, alternative splicing and RNA editing affect pharmacology and therefore, it would be essential to investigate the patterns of posttranscriptional modification in diseases, as this knowledge may help us well understand the functional changes of these channels and receptors and provide valuable targets to screen for drugs tailored to manage the development of the disorders.

References

1. Ashani K, Bass BL. Mechanistic insights into editing-site specificity of ADARs. *PNAS* 2012;**109**(48).
2. Daniel C, Venø MT, Ekdahl Y, Kjems J, Ohman M. A distant cis acting intronic element induces site-selective RNA editing. *Nucleic Acids Res* 2012;**1**(11).
3. Lipscombe D, Allen SE, Toro CP. Control of neuronal voltage-gated calcium ion channels from RNA to protein. *Trends Neurosci* 2013;**36**(10):598–609.
4. Dolphin AC. Calcium channel diversity: multiple roles of calcium channel subunits. *Curr Opin Neurobiol* 2009;**19**(3):237–44.
5. Simms BA, Zamponi GW. Neuronal voltage-gated calcium channels: structure, function, and dysfunction. *Neuron* 2014;**82**(1):24–45.
6. Liao P, Zhang HY, Soong TW. Alternative splicing of voltage-gated calcium channels: from molecular biology to disease. *Pflugers Archiv Eur J Physiol* 2009;**458**(3):481–7.
7. Liao P, Yong TF, Liang MC, Yue DT, Soong TW. Splicing for alternative structures of $Ca_V 1.2$ Ca^{2+} channels in cardiac and smooth muscles. *Cardiovasc Res* 2005;**68**(2):197–203.
8. Tang ZZ, Sharma S, Zheng S, Chawla G, Nikolic J, Black DL. Regulation of the mutually exclusive exons 8a and 8 in the CaV1.2 calcium channel transcript by polypyrimidine tract-binding protein. *J Biol Chem* 2011;**286**(12):10007–16.
9. Splawski I, Timothy KW, Sharpe LM, Decher N, Kumar P, Bloise R, et al. Ca(V)1.2 calcium channel dysfunction causes a multisystem disorder including arrhythmia and autism. *Cell* 2004;**119**(1):19–31.
10. Splawski I, Timothy KW, Decher N, Kumar P, Sachse FB, Beggs AH, et al. Severe arrhythmia disorder caused by cardiac L-type calcium channel mutations. *Proc Natl Acad Sci USA* 2005;**102**(23):8089–96. discussion 6-8.
11. Gillis J, Burashnikov E, Antzelevitch C, Blaser S, Gross G, Turner L, et al. Long QT, syndactyly, joint contractures, stroke and novel CACNA1C mutation: expanding the spectrum of Timothy syndrome. *Am J Med Genet A* 2012;**158A**(1):182–7.

12. Dixon RE, Cheng EP, Mercado JL, Santana LF. L-type Ca^{2+} channel function during Timothy syndrome. *Trends Cardiovasc Med* 2012;**22**(3):72–6.

13. Barrett CF, Tsien RW. The Timothy syndrome mutation differentially affects voltage- and calcium-dependent inactivation of CaV1.2 L-type calcium channels. *Proc Natl Acad Sci USA* 2008;**105**(6):2157–62.

14. Yarotskyy V, Gao G, Peterson BZ, Elmslie KS. The Timothy syndrome mutation of cardiac CaV1.2 (L-type) channels: multiple altered gating mechanisms and pharmacological restoration of inactivation. *J Physiol* 2009;**587**(Pt 3):551–65.

15. Pasca SP, Portmann T, Voineagu I, Yazawa M, Shcheglovitov A, Pasca AM, et al. Using iPSC-derived neurons to uncover cellular phenotypes associated with Timothy syndrome. *Nat Med* 2011;**17**(12):1657–62.

16. Huttner A, Rakic P. Diagnosis in a dish: your skin can help your brain. *Nat Med* 2011;**17**(12):1558–9.

17. Vandael DH, Marcantoni A, Mahapatra S, Caro A, Ruth P, Zuccotti A, et al. Ca(v)1.3 and BK channels for timing and regulating cell firing. *Mol Neurobiol* 2010;**42**(3):185–98.

18. Baig SM, Koschak A, Lieb A, Gebhart M, Dafinger C, Nurnberg G, et al. Loss of Ca(v)1.3 (CACNA1D) function in a human channelopathy with bradycardia and congenital deafness. *Nat Neurosci* 2011;**14**(1):77–84.

19. Platzer J, Engel J, Schrott-Fischer A, Stephan K, Bova S, Chen H, et al. Congenital deafness and sinoatrial node dysfunction in mice lacking class D L-type Ca^{2+} channels. *Cell* 2000;**102**(1):89–97.

20. Koschak A, Reimer D, Huber I, Grabner M, Glossmann H, Engel J, et al. Alpha 1D (Cav1.3) subunits can form l-type Ca^{2+} channels activating at negative voltages. *J Biol Chem* 2001;**276**(25):22100–6.

21. Wheeler DB, Randall A, Tsien RW. Roles of N-type and Q-type Ca^{2+} channels in supporting hippocampal synaptic transmission. *Science* 1994;**264**(5155):107–11.

22. Pietrobon D. CaV2.1 channelopathies. *Pflugers Archiv Eur J Physiol* 2010;**460**(2):375–93.

23. Bourinet E, Soong TW, Sutton K, Slaymaker S, Mathews E, Monteil A, et al. Splicing of alpha 1A subunit gene generates phenotypic variants of P- and Q-type calcium channels. *Nat Neurosci* 1999;**2**(5):407–15.

24. Krovetz HS, Helton TD, Crews AL, Horne WA. C-Terminal alternative splicing changes the gating properties of a human spinal cord calcium channel alpha 1A subunit. *J Neurosci* 2000;**20**(20):7564–70.

25. Soong TW, DeMaria CD, Alvania RS, Zweifel LS, Liang MC, Mittman S, et al. Systematic identification of splice variants in human P/Q-type channel alpha1(2.1) subunits: implications for current density and Ca^{2+}-dependent inactivation. *J Neurosci* 2002;**22**(23): 10142–52.

26. Chaudhuri D, Chang SY, DeMaria CD, Alvania RS, Soong TW, Yue DT. Alternative splicing as a molecular switch for Ca^{2+}/calmodulin-dependent facilitation of P/Q-type Ca^{2+} channels (Chaudhuri, 2007 #114). *J Neurosci* 2004;**24**(28):6334–42.

27. Chang SY, Yong TF, Yu CY, Liang MC, Pletnikova O, Troncoso J, et al. Age and gender-dependent alternative splicing of P/Q-type calcium channel EF-hand. *Neuroscience* 2007;**145**(3):1026–36.

28. Graves TD, Imbrici P, Kors EE, Terwindt GM, Eunson LH, Frants RR, et al. Premature stop codons in a facilitating EF-hand splice variant of CaV2.1 cause episodic ataxia type 2. *Neurobiol Dis* 2008;**32**(1):10–5.

29. Nordli Jr DR. Idiopathic generalized epilepsies recognized by the International League against epilepsy. *Epilepsia* 2005;**46**(Suppl 9):48–56.

30. Khosravani H, Altier C, Simms B, Hamming KS, Snutch TP, Mezeyova J, et al. Gating effects of mutations in the Cav3.2 T-type calcium channel associated with childhood absence epilepsy. *J Biol Chem* 2004;**279**(11):9681–4.

31. Powell KL, Cain SM, Ng C, Sirdesai S, David LS, Kyi M, et al. A Cav3.2 T-type calcium channel point mutation has splice-variant-specific effects on function and segregates with seizure expression in a polygenic rat model of absence epilepsy. *J Neurosci* 2009;**29**(2):371–80.

32. Wulff H, Castle NA, Pardo LA. Voltage-gated potassium channels as therapeutic targets. *Nat Rev Drug Discov* 2009;**8**(12):982–1001.

33. Lujan R. Organisation of potassium channels on the neuronal surface. *J Chem Neuroanat* 2010;**40**(1):1–20.

34. Sanguinetti MC, Curran ME, Zou A, Shen J, Spector PS, Atkinson DL, et al. Coassembly of K(V)LQT1 and minK (IsK) proteins to form cardiac I(Ks) potassium channel. *Nature* 1996;**384**(6604):80–3.

35. Barhanin J, Lesage F, Guillemare E, Fink M, Lazdunski M, Romey G. K(V)LQT1 and lsK (minK) proteins associate to form the I(Ks) cardiac potassium current. *Nature* 1996;**384**(6604):78–80.

36. Jespersen T, Grunnet M, Olesen SP. The KCNQ1 potassium channel: from gene to physiological function. *Physiology* 2005;**20**:408–16.

37. Kang C, Tian C, Sonnichsen FD, Smith JA, Meiler J, George Jr AL, et al. Structure of KCNE1 and implications for how it modulates the KCNQ1 potassium channel. *Biochemistry* 2008;**47**(31):7999–8006.

38. Romano C. Congenital cardiac arrhythmia. *Lancet* 1965;**1**(7386):658–9.

39. Schwartz PJ, Stramba-Badiale M, Segantini A, Austoni P, Bosi G, Giorgetti R, et al. Prolongation of the QT interval and the sudden infant death syndrome. *N Engl J Med* 1998;**338**(24):1709–14.

40. Schmitt N, Schwarz M, Peretz A, Abitbol I, Attali B, Pongs O. A recessive C-terminal Jervell and Lange-Nielsen mutation of the KCNQ1 channel impairs subunit assembly. *EMBO J* 2000;**19**(3):332–40.

41. Krumerman A, Gao X, Bian JS, Melman YF, Kagan A, McDonald TV. An LQT mutant minK alters KvLQT1 trafficking. *Am J Physiol Cell Physiol* 2004;**286**(6):C1453–63.

42. Boulet IR, Raes AL, Ottschytsch N, Snyders DJ. Functional effects of a KCNQ1 mutation associated with the long QT syndrome. *Cardiovasc Res* 2006;**70**(3):466–74.

43. Dahimene S, Alcolea S, Naud P, Jourdan P, Escande D, Brasseur R, et al. The N-terminal juxtamembranous domain of KCNQ1 is critical for channel surface expression: implications in the Romano-Ward LQT1 syndrome. *Circulation Res* 2006;**99**(10):1076–83.

44. Seebohm G, Strutz-Seebohm N, Ureche ON, Henrion U, Baltaev R, Mack AF, et al. Long QT syndrome-associated mutations in KCNQ1 and KCNE1 subunits disrupt normal endosomal recycling of IKs channels. *Circulation Res* 2008;**103**(12):1451–7.

45. Murray A, Donger C, Fenske C, Spillman I, Richard P, Dong YB, et al. Splicing mutations in KCNQ1: a mutation hot spot at codon 344 that produces in frame transcripts. *Circulation* 1999;**100**(10):1077–84.

46. Tsuji-Wakisaka K, Akao M, Ishii TM, Ashihara T, Makiyama T, Ohno S, et al. Identification and functional characterization of KCNQ1 mutations around the exon 7-intron 7 junction affecting the splicing process. *Biochim Biophys acta* 2011;**1812**(11):1452–9.

47. Imai M, Nakajima T, Kaneko Y, Niwamae N, Irie T, Ota M, et al. A novel KCNQ1 splicing mutation in patients with forme fruste LQT1 aggravated by hypokalemia. *J Cardiol* 2014;**64**(2):121–6.

48. Zehelein J, Kathoefer S, Khalil M, Alter M, Thomas D, Brockmeier K, et al. Skipping of exon 1 in the KCNQ1 gene causes Jervell and Lange-Nielsen syndrome. *J Biol Chem* 2006;**281**(46):35397–403.

49. Bhuiyan ZA, Momenah TS, Amin AS, Al-Khadra AS, Alders M, Wilde AA, et al. An intronic mutation leading to incomplete skipping of exon-2 in KCNQ1 rescues hearing in Jervell and Lange-Nielsen syndrome. *Prog Biophys Mol Biol* 2008;**98**(2–3):319–27.

50. Biervert C, Schroeder BC, Kubisch C, Berkovic SF, Propping P, Jentsch TJ, et al. A potassium channel mutation in neonatal human epilepsy. *Science* 1998;**279**(5349):403–6.

51. Singh NA, Charlier C, Stauffer D, DuPont BR, Leach RJ, Melis R, et al. A novel potassium channel gene, KCNQ2, is mutated in an inherited epilepsy of newborns. *Nat Genet* 1998;**18**(1):25–9.

52. de Haan GJ, Pinto D, Carton D, Bader A, Witte J, Peters E, et al. A novel splicing mutation in KCNQ2 in a multigenerational family with BFNC followed for 25 years. *Epilepsia* 2006;**47**(5):851–9.

53. Lee WL, Biervert C, Hallmann K, Tay A, Dean JC, Steinlein OK. A KCNQ2 splice site mutation causing benign neonatal convulsions in a Scottish family. *Neuropediatrics* 2000;**31**(1): 9–12.

54. Maljevic S, Wuttke TV, Lerche H. Nervous system KV7 disorders: breakdown of a subthreshold brake. *J Physiol* 2008;**586**(7):1791–801.

55. Xu T, Nie L, Zhang Y, Mo J, Feng W, Wei D, et al. Roles of alternative splicing in the functional properties of inner ear-specific KCNQ4 channels. *J Biol Chem* 2007;**282**(33): 23899–909.

56. Raymond CK, Castle J, Garrett-Engele P, Armour CD, Kan Z, Tsinoremas N, et al. Expression of alternatively spliced sodium channel alpha-subunit genes. Unique splicing patterns are observed in dorsal root ganglia. *J Biol Chem* 2004;**279**(44):46234–41.

57. Tang L, Lu X, Tao Y, Zheng J, Zhao P, Li K, et al. SCN1A rs3812718 polymorphism and susceptibility to epilepsy with febrile seizures: a meta-analysis. *Gene* 2014;**533**(1): 26–31.

58. Schlachter K, Gruber-Sedlmayr U, Stogmann E, Lausecker M, Hotzy C, Balzar J, et al. A splice site variant in the sodium channel gene SCN1A confers risk of febrile seizures. *Neurology* 2009;**72**(11):974–8.

59. Thompson CH, Kahlig KM, George Jr AL. SCN1A splice variants exhibit divergent sensitivity to commonly used antiepileptic drugs. *Epilepsia* 2011;**52**(5):1000–9.

60. Gazina EV, Leaw BT, Richards KL, Wimmer VC, Kim TH, Aumann TD, et al. 'Neonatal' Nav1.2 reduces neuronal excitability and affects seizure susceptibility and behaviour. *Hum Mol Genet* 2015;**24**(5):1457–68.

61. Gazina EV, Richards KL, Mokhtar MB, Thomas EA, Reid CA, Petrou S. Differential expression of exon 5 splice variants of sodium channel alpha subunit mRNAs in the developing mouse brain. *Neuroscience* 2010;**166**(1):195–200.

62. Koopmann TT, Bezzina CR, Wilde AA. Voltage-gated sodium channels: action players with many faces. *Ann Med* 2006;**38**(7):472–82.

63. Kubota T, Roca X, Kimura T, Kokunai Y, Nishino I, Sakoda S, et al. A mutation in a rare type of intron in a sodium-channel gene results in aberrant splicing and causes myotonia. *Hum Mutat* 2011;**32**(7):773–82.

64. Onkal R, Mattis JH, Fraser SP, Diss JK, Shao D, Okuse K, et al. Alternative splicing of Nav1.5: an electrophysiological comparison of 'neonatal' and 'adult' isoforms and critical involvement of a lysine residue. *J Cell Physiol* 2008;**216**(3):716–26.

65. Brackenbury WJ, Chioni AM, Diss JK, Djamgoz MB. The neonatal splice variant of Nav1.5 potentiates in vitro invasive behaviour of MDA-MB-231 human breast cancer cells. *Breast Cancer Res Treat* 2007;**101**(2):149–60.

66. Wahbi K, Algalarrondo V, Becane HM, Fressart V, Beldjord C, Azibi K, et al. Brugada syndrome and abnormal splicing of SCN5A in myotonic dystrophy type 1. *Arch Cardiovasc Dis* 2013;**106**(12):635–43.

67. Brackenbury WJ, Isom LL. Na channel beta subunits: overachievers of the ion channel family. *Front Pharmacol* 2011;**2**:53.

68. Gao G, Dudley Jr SC. SCN5A splicing variants and the possibility of predicting heart failure-associated arrhythmia. *Expert Rev Cardiovasc Ther* 2013;**11**(2):117–9.

69. Shang LL, Pfahnl AE, Sanyal S, Jiao Z, Allen J, Banach K, et al. Human heart failure is associated with abnormal C-terminal splicing variants in the cardiac sodium channel. *Circulation Res* 2007;**101**(11):1146–54.

70. Jarecki BW, Sheets PL, Xiao Y, Jackson 2nd JO, Cummins TR. Alternative splicing of Na(V)1.7 exon 5 increases the impact of the painful PEPD mutant channel I1461T. *Channels* 2009;**3**(4):259–67.

71. Huang H, Tan Bao Z, Shen Y, Tao J, Jiang F, Sung Ying Y, et al. RNA editing of the IQ domain in Cav1.3 channels modulates their Ca^{2+}-dependent inactivation. *Neuron* 2012;**73**(2):304–16.

72. Bhalla T, Rosenthal JJ, Holmgren M, Reenan R. Control of human potassium channel inactivation by editing of a small mRNA hairpin. *Nat Struct Mol Biol* 2004;**11**(10):950–6.
73. Streit AK, Derst C, Wegner S, Heinemann U, Zahn RK, Decher N. RNA editing of Kv1.1 channels may account for reduced ictogenic potential of 4-aminopyridine in chronic epileptic rats. *Epilepsia* 2011;**52**(3):645–8.
74. Decher N, Streit AK, Rapedius M, Netter MF, Marzian S, Ehling P, et al. RNA editing modulates the binding of drugs and highly unsaturated fatty acids to the open pore of Kv potassium channels. *EMBO J* 2010;**29**(13):2101–13.
75. Hanrahan CJ, Palladino MJ, Ganetzky B, Reenan RA. RNA Editing of the Drosophila para Na(+) channel transcript - evolutionary conservation and developmental regulation. *Genetics* 2000;**155**.
76. Olson RO, Liu Z, Nomura Y, Song W, Dong K. Molecular and functional characterization of voltage-gated sodium channel variants from *Drosophila melanogaster. Insect Biochem Mol Biol* 2008;**38**(5):604–10.
77. Rosenthal JJ, Seeburg PH. A-to-I RNA editing: effects on proteins key to neural excitability. *Neuron* 2012;**74**(3):432–9.
78. Wright A, Vissel B. The essential role of AMPA receptor GluR2 subunit RNA editing in the normal and diseased brain. *Front Mol Neurosci* 2012;**5**:13.
79. Miyoko H, Single FN, Kohler M, Sommer B, Sprengel R, Seeburg PH. RNA editing of AMPA receptor subunit GluR-9-a base-paired intron-exon structure determines position and efficiency. *Cell* 1993;**75**.
80. Streit AK, Decher N. A-to-I RNA editing modulates the pharmacology of neuronal ion channels and receptors. *Biochemistry (Biokhimiia)* 2011;**76**(8):890–9.
81. Yukio K, Kyoko Ito HS, Aizawa H, Kanazawa I, Kwak S. RNA editing and death of motor neurons. *Nature* 2004;**427**.
82. Seeburg PH, Higuchi M, Sprengel R. RNA editing of brain glutamate receptor channels - mechanism and physiology. *Brain Res Rev* 1998;**26**:217–29.
83. Berg KA, Cropper JD, Niswender CM, Sanders, Bush E, Emeson RB, Clarke WP. RNA-editing of the 5-HT2C receptor alters agonist-receptor-effector coupling specificity. *Br J Pharmacol* 2001;**134**(2).
84. Di Narzo AF, Kozlenkov A, Roussos P, Hao K, Hurd Y, Lewis DA, et al. A unique gene expression signature associated with serotonin 2C receptor RNA editing in the prefrontal cortex and altered in suicide. *Hum Mol Genet* 2014;**23**(18).
85. Gurevich I, Tamir H, Arango V, Dwork AJ, John Mann J, Schmauss C. Altered editing of serotonin 2C receptor pre-mRNA in the prefrontal cortex of depressed suicide victims. *Neuron* 2002;**34**.
86. Rula EY, Lagrange AH, Jacobs MM, Hu N, Macdonald RL, Emeson RB. Developmental modulation of GABA(A) receptor function by RNA editing. *J Neurosci* 2008;**28**(24): 6196–201.
87. Ohlson J, Pedersen JS, Haussler D, Ohman M. Editing modifies the GABA(A) receptor subunit alpha3. *RNA* 2007;**13**(5):698–703.
88. Daniel C, Wahlstedt H, Ohlson J, Bjork P, Ohman M. Adenosine-to-inosine RNA editing affects trafficking of the gamma-aminobutyric acid type A (GABA(A)) receptor. *J Biol Chem* 2011;**286**(3):2031–40.
89. Witman NM, Behm M, Ohman M, Morrison JI. ADAR-related activation of adenosine-to-inosine RNA editing during regeneration. *Stem Cells Dev* 2013;**22**(16):2254–67.
90. Borik S, Simon AJ, Nevo-Caspi Y, Mishali D, Amariglio N, Rechavi G, et al. Increased RNA editing in children with cyanotic congenital heart disease. *Intensive Care Med* 2011;**37**(10):1664–71.
91. Danecek P, Nellåker C, McIntyre RE, Buendia-Buendia JE, Bumpstead S, Ponting CP, et al. High levels of RNA-editing site conservation amongst 15 laboratory mouse strains. *Genome Biol* 2012;**13**(26).

92. Li M, Wang IX, Li Y, Bruzel A, Richards AL, Toung JM, et al. Widespread RNA and DNA sequence differences in the human transcriptome. *Science* 2011;**333**.

93. Lin W, Piskol R, Tan MH, Li JB. Comment on "widespread RNA and DNA sequence differences in the human transcriptome". *Science* 2012;**335**.

94. Ramaswami G, Lin W, Piskol R, Tan MH, Davis C, Li JB. Accurate identification of human Alu and non-Alu RNA editing sites. *Nat Methods* 2012;**9**(6):579–81.

Computational Modeling of Cardiac K⁺ Channels and Channelopathies

L.R. Perez[1], S.Y. Noskov[2], C.E. Clancy[3]

[1]Polytechnic University of Valencia, Valencia, Spain; [2]University of Calgary, Calgary, AB, Canada; [3]University of California, Davis, CA, United States

INTRODUCTION

For the heart to perform its critical function as a mechanical pump, a regular and orderly electrical activation sequence must occur to drive the heartbeat. When drugs or diseases alter normal cardiac electrical activity, cardiac arrhythmias may occur, leading to compromised pump function and untimely death. One mechanism by which arrhythmias arise is from channelopathies, a term that describes a heterogeneous group of diseases that arise from the disruption of normal ion channel function. Because it is so difficult to predict how ion channel modification, studied in isolation, alters the functioning of the whole heart, computationally based modeling and simulation approaches have been developed and widely applied to link deviations in ion channel function to emergent electrical activity in cells, tissue, and even simulated whole hearts. Simulation studies have revealed plausible experimentally testable mechanisms for how perturbations to ion channels and associated processes alter emergent electrical behavior at higher system scales.

Mathematical models of cardiac ion channels have been used to clarify and expand experimental findings to improve understanding of cardiac electrical function in health and disease. The earliest mathematical models of cardiac ion channels were limited to descriptions of just the few key currents that had been identified in the cardiac myocyte.[1-4] These early channel models derived from modified Hodgkin–Huxley type equations were dependent on voltage and time.[5] In the last 20 years an explosion in development of sophisticated models has occurred, concomitant with improved experimental techniques that have allowed identification and characterization of numerous ion channel subtypes and their regulation in the heart from multiple species.[6–18]

Sophisticated ion channel models have even been developed that account for various genetic defects that alter the behavior of ion channels. These complex ion channel representations have been incorporated into numerous cardiac model "cells" from multiple species. The cellular level models have been widely replicated and coupled, creating mathematical representations of cardiac tissue in one, two, or three dimensions, with incorporation of complex anatomical heterogeneities including anisotropy, structural features, and distinct cells with specifically associated electrophysiological characteristics.[14,19–39]

In additional to the explosion in experimental data describing the function of numerous cardiac ion channels and alteration of channel function resulting from channelopathies, experimental data are increasingly available describing the fundamental structure of ion channels. For this reason, modeling and simulation studies at the molecular scale are now possible and are beginning to reveal fundamental biological principles and structural mechanisms of ion channel functional changes.

This chapter focuses on computational modeling and simulation approaches related to cardiac K⁺ channels. These approaches have been developed and applied from the atomic scale to the organ scale and have provided numerous physiological and biophysical insights on cardiac electrical activity in normal and disease states. Because multiple crystal structures for voltage-gated K⁺ channels have been solved, there have been substantial developments in homology modeling at the molecular scale for cardiac K⁺ channels. While K⁺ channels alone are covered here due to the full spectrum of model development, the approaches used for K⁺ channels are now being expanded to allow modeling of other key cardiac ion channels from channel to organ, including Na⁺ channels, Ca²⁺ channels, and the ryanodine receptor.

K⁺ Channelopathies: I_{Kr}, the Rapidly Activating Component of the Delayed Rectifier

While hERG (KCNH2, $K_V11.1$) is widely accepted as the α-subunit of I_{Kr}, other ancillary subunits are required to reconstitute native I_{Kr} currents. α- and β- subunits together modify gating,[40,41] regulation by external K⁺,[40,42,43] and sensitivity to antiarrhythmics.[44,45] A prime candidate is minK-related protein 1 (MiRP1 = KCNE2), which when coexpressed with hERG, results in currents similar to native I_{Kr}.[40,46] Coexpression with KCNE2 causes a measured positive depolarizing shift in steady state activation, accelerates the rate of deactivation, and causes a decrease in single channel conductance from 13 to 8 pS.[40] Several alternatively spliced ERG1 variants have also been demonstrated in the heart,[47,48] and there is evidence for posttranslational modification of hERG proteins.[49,50]

Numerous mutations in hERG have been linked to the congenital long QT syndrome (LQTS).[51–54] Defects in the hERG pore have been shown to have heterogeneous cellular phenotypes. Mutations in hERG resulting in LQTS reduce potassium current, either by dominant negative suppression, nonfunctional channels, or trafficking errors.[51,52] Pore mutations may result in a loss-of-function, sometimes due to a trafficking defect,[55] and may or may not coassemble with wild-type (WT) hERG subunits to exert dominant negative effects.[52,56] Other defects in the channel pore give rise to altered channel kinetics leading to decreased repolarizing current.[57,58]

Mutations in *KCNE2* alter the biophysical properties of I_{Kr} and act to reduce current.[40,52,59] *KCNE2* polymorphisms have also been associated with increased affinity of I_{Kr} to blockade by clarithromycin.[60]

ATOMIC SCALE MODELS TO REVEAL STRUCTURAL DETERMINANTS OF hERG CHANNELS AND CHANNELOPATHIES

Since the discovery of K^+ channels decades ago, great progress has been made in understanding their biophysical properties and structure–function relationships. A major breakthrough was the publication of the first high-resolution crystallographic structures of several K^+ channels by MacKinnon and coworkers.[61–65] These crystal structures have provided an unprecedented opportunity to finally begin to connect channel structure to its function, reveal locations of naturally occurring mutations, and highlight diverse mechanisms of channel–drug interactions. While the static structures of K^+ channels in different conformational states have provided a great deal of information, many issues regarding the functions and mechanism of drug block, gating, and fast C-type inactivation in K^+ channels remain unresolved.[66–68] Molecular scale approaches for structural modeling are beginning to yield important complementary insights in these areas that add breadth and depth to the experimentally obtained structural information.[69–71]

Most of the known families of K^+ channels can be broadly segregated into two groups: six transmembrane (TM) domain K^+ channels, and those with 2 TM domains (inwardly rectifying K-channels). 6TM and 2TM K^+ channels share similar topology for the pore domain that includes the S1-P-S2 structure. The 6TM channel has a voltage-sensing domain (VSD) consisting of four helices labeled S1-S2-S3-S4 resulting in the following topology: S1-S2-S3-S4-S5-P-S6. The 6TM superfamily includes several gene families that include the voltage-gated (K_V) channels, the ether-a-go-go-like (EAG) (also called ether-a-go-go related gene (ERG)) K-channels, and the KCNQ channels.[72,73] Cardiac specific isoforms of all these types of channels are found in the heart (hERG and KCNQ1), and mutations in the genes encoding these channels have been causally linked to inherited arrhythmia syndromes.[44,56,74] Much focus of molecular modeling efforts has been on understanding the specific and nonspecific block of K⁺ channels by drugs and blockers. These studies are critical, as revealing the structural determinants of the specific and nonspecific block of K⁺ channels by drugs and blockers is key to also reveal the structural bases of phenotypic manifestations of inherited channelopathies.

Structural Models of the Cardiac hERG Channel: Templates for Interpreting Functional Effects of Drugs and Channelopathies

The human ERG (hERG) channel is a primary delayed rectifier repolarizing K⁺ current in the heart derived from the EAG-like family that consists of four identical subunits containing six TM domains.[52,75,76] While the crystal structure of the hERG channel has not yet been solved, multiple in silico approaches including homology modeling, de novo protein design, and molecular dynamics (MD) simulations have been undertaken to reveal structural determinants of hERG function.[77] A reassuring fact is that the sequence conservation is fairly high between hERGs and K_V channels with known structures. However, hERG channels possess several unique soluble domains, extended loops with prominent helical elements, and notable differences in the chemical moieties present in the pore domain. These salient features of hERG channels result in the array of notable differences (compared to channels from Shaker family) in the voltage-dependent gating mechanism (very rapid C-type inactivation), channel conductance and selectivity (transient Na⁺ currents), promiscuity in drug binding, and specific toxin-binding properties.[78–81] Promising new studies suggest that computer-based homology models based on solved K_V structures can successfully be utilized to study human K⁺ channels.

The hERG models that have been developed are based on the available crystal structures of KcsA, MthK, K_V1.2, and K_VAP pore domains.[82–84] The predictions from molecular modeling studies–based bacterial KcsA and MthK[83,84] channels or the mammalian channel K_V1.2[77,85] concur with the experimental findings that have revealed two key residues responsible for drug stabilization in the hERG1 cavity, eg, Y652 and F656.[86,87] Model studies reproduced this feature in studies of hERG blockers such as dofetilide, KN-93, and other common high-affinity blockers (Fig. 12.1).[88,89]

Stary et al. reported a critical analysis of several homology models built on different alignments for the S5 domain.[77] It was found that the packing of the S5 helix depends critically on the original alignment. Accordingly,

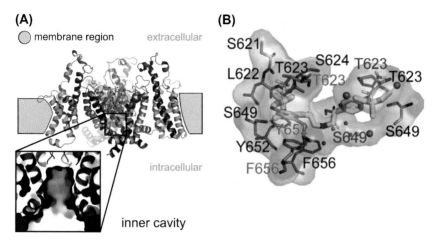

FIGURE 12.1   (A) Schematic illustration of the hERG channel in the lipid bilayer. (B) Organization of the binding site for dofetilide (shown in *green-and-yellow sticks*) binding in the inner cavity of hERG. The binding pocket is formed by residues from all four monomers shown in distinct colors (*blue, orange, magenta, and black*). The solvent molecules are shown as *red spheres* (Wang, Perisinotti, Noskov, unpublished data).

attempts have been made to refine the current methods for hERG modeling, which generally consists of mapping the hERG sequence to various existing K_V channels and refining the resultant structures, including the following: (1) using data-derived constraints in repacking side chains and relaxing the backbone of static models and (2) several applications of very long MD simulations (1–1.5 µs) to determine dynamics equilibrium states. The authors claimed that use of supercomputer resources allowed for relaxation of the pore domain structure and improved correlation between computed and experimentally obtained binding constants for a large panel of common blockers.[90]

While these results together strongly suggest resolution of the key hERG drug binding site in the intracellular cavity, conundrums remain: Experimental efforts to create a hERG-like cavity in the K_V channels to create an amenable template for studies of drug–channel interactions, have not been successful. Furthermore, the substitution of the presumed drug binding residues into the homologous positions in $K_V1.2$ has not led to a drug-sensitive channel. One potential reason may be the difference in the packing of the pore helices, where the I-F-G motif of the hERG channel (P-V-P in Shakers) results in a larger inner vestibule. It should be noted too, that a study of hERG1 gating suggests that mutations of residues adjacent to the I-F-G motif (G648) have only minor effects on C-type inactivation, another important component in drug targeting to hERG channel.[91] Other factors, related to the functional activities of the channel, will have to be included into a convincing atomistic model of the channel. Later, we will try to analyze additional structural and functional features of hERG that differentiate it from its "Shaker" cousins.

One of the most distinct structural features of the hERG structure is the presence of an extended S5-pore linker (turret). This structural element is formed by an approximately 40 amino acid sequence. It harbors a number of amino acids known to be critically important for a landmark feature of hERG gating—rapid C-type inactivation.[79,80,92–94] Several of the identified disease-causing mutations are localized to the S5-pore linker, namely G572, R582, N588, G601, and Y611.[52] To date, most of the computer-based structural studies of hERG have modeled this region with a very short connector loop. Yet, functionally the S5-pore linker is critical for the proper function of hERG, as indicated by the substantial number of mutations implicated into arrhythmogenicity localized to this region.

Durdagi et al. extended de novo ROSETTA modeling to assess the structure of the S5-pore linker in silico using close contacts and critical interactions to guide side-directed mutagenesis.[85] The de novo model of the turret exhibits a helical segment formed by residues 582–593 in accord with the NMR data.[81] Structural and biochemical data suggest that short amphipathic helices can also be formed by the I583-G590 and/or D591-I593 regions.[95]

The other common challenge to structural inference of hERG functional properties is modeling of the S5 helix. Lees-Miller et al. suggested that hydrogen-bonding and aromatic interactions between S5 and a pore helix (H562 region) are essential determinants of the rapid C-type inactivation.[96] J. Vandenberg and his team confirmed that residues in the region between A561 and W568 are involved in packing and stabilization of the pore helix.[97] Interestingly, there is now direct evidence that inherited mutations in the vicinity of H562 are implicated in the congenital form of long QT syndrome in humans.[98]

Many of the "hot spots" associated with QT prolongation are located in the pore domain of the channel.[52] The other clear indicator of the significance of the hERG1 pore in arrhythmogenicity is drug-induced QT prolongation. Drug sensitivity in patients has been linked to genetic alterations in hERG.[99] hERG1 channel blockade by most drugs requires opening of the activation gate, indicating that the blocker compounds bind to the inner cavity of the channel. The pore domain includes S5, pore, and S6 helices. As mentioned earlier, mutagenesis of the S6 helix has been used to characterize the binding sites of several drugs and suggests a critical role for the Y652 and F656 residues. However, many blockers contain not only aromatic rings or hydrophobic groups, but also polar functional moieties. Increasingly refined models to predict drug blockade in the inner cavity now include mapping of residues from the pore helix such as T623, S624, and V625 as part of the binding pocket for the blocker. In addition, there are several reports that G648 and V659 also reduce binding when mutated.

Studies such as these that implement structural models are examples of how drug interactions can be used to shed light on the thermodynamics of channel blockade and key determinants of the process.

It is important to note that hERG blockade is not a single entity: The process of drug interaction is nuanced and needs to consider both channel gating kinetics and the differential binding affinities of the blocker to distinct conformational states. For example, Perry et al. concluded that replacing the basicity of the nitrogen atom in blocker from quaternary to tertiary substantially accelerated the onset of block.[83] The IC_{50} and the blockade recovery kinetics were similar. Many drugs induce greater inhibition of the hERG ion channel at depolarizing membrane potentials. Lees-Miller et al. published a report that highlighted potential risks of state-dependent blockade by ivabradine.[100] Stork et al. studied the recovery of hERG from "use-dependent" inhibition by eight channel blockers.[101] They found that amiodarone, cisapride, haloperidol, and droperidol molecules exhibited frequency-dependent block of hERG channels. However, for the rest of the drugs (bepridil, terfenadine, E-4031, and domperidone) there is no enhancement of the channel inhibition at higher frequencies. These results suggest some blockers can be trapped by channel closures. The elucidation of exact molecular mechanism for such trapping is yet to come.

Conformation Dynamics of hERG1 Channel

It is well known that hERG1 displays a distinct multistate conformational dynamics depending on the membrane potential. The closed state of the channel is stabilized at negative membrane potentials (ie, −80 mV) while depolarizating voltages (to −60 mV) leads to channel opening (activation) enabling outward current during physiological ionic conditions. The channel opening is followed by a very rapid C-type inactivation where the activation gate is still open, but the permeation pathway is shut down. The exact mechanism of C-type inactivation in hERG remains elusive. It can be regulated by a multitude of factors including external cation concentrations and bath compositions, local mutations in the pore domain (S631A and S620T), and/or the presence of drugs. Molecular scale models may ultimately prove to be important to reveal the structure function relationship for C-type inactivation and its regulation. Importantly, the recovery from C-type inactivation generates a large repolarizing tail current, which opposes any spurious depolarizing forces. This may be one of the reasons that hERG is considered antiarrhythmic by contributing to repolarization reserve.[102–104]

Many of the mutations associated with the cardiac arrhythmias are localized in the VSD of hERG channels (S428X, T436X, R531Q, R534C), affecting voltage-dependent gating. The Noskov lab used a template-driven and ab initio ROSETTA modeling approach to refine of the full

S1–S6 model of hERG and reported channel models in open, closed, and open-inactivated states.[88] Colenso et al. also reported on the MD simulations of the TM portion (S1–S6) of the hERG1 in open state. The models were cross-validated against available experimental data,[105,106] thus providing initial glimpses into structural underpinnings of conformational dynamics of the TM domain.

The structural modeling suggests an activation mechanism for hERG channel based on a combination of MD simulations and gating charge computations.[88] The measured and computed total gating charge transfer between open and closed states suggests strong coupling between the channel conformation and the applied membrane potential. The voltage is coupled to the dynamics of the four positively charged residues K525, R528, R531, and R534 in the S4 helix in keeping with the experimental data.[107,108] Movement of S4 couples to the conformational rearrangement of the S6 helix leading to pore closure and channel sliding to the inactivated state. The proposed mechanism is reminiscent of that in Shaker channels.[109]

The conformational dynamics during a hERG gating cycle now allows the beginning revelation of the impact of arrhythmogenic mutations in terms of the structural stability and voltage coupling in the VSD. Mutations that augment hydrogen-bonding capacity at position 428 (S428L and S428STOP) and T436M have been directly associated with the occurrence of LQT2 in patients.[110] The structural model showed that stabilization of R531 is achieved by a combination of salt bridging to D460 and hydrogen bonding to S428, while T436 is essential to interactions with K525. Thus both of the amphipathic residues associated with LQT2 are essential for fine-tuning and spatial positioning of key gating positive charges. The shift of S4 upon inactivation induces a conformational change in the pore domain that is transmitted by the S4–S5 linker.

The charge reversal mutation D540K leads to a destabilization of the closed state of the channel leading to a U-shaped dependence of activation on voltage; opening occurs at both depolarized and hyperpolarized potentials.[86] The structure–function relation studies suggested that D540 interactions with S6 (R665) may be crucial stabilizing interaction in the closed state.[111]

The rapid conformational dynamics in the region of the outer mouth of the selectivity filter domain is thought to be one of the key drivers in the rapid C-type inactivation in potassium channels.[67,68] hERG channels display a very rapid conformational transition between the open and open-inactivated states on time scale of ~10–20 ms.[95] The C-type inactivation is very sensitive to ion concentrations, presence of blockers and mutations in the vicinity of the selectivity filter, including N629X mutations[87,112] implicated in the developmental defects in the right ventricle. Adenoviral overexpression of N629D resulted in depolarization of the resting potential to approximately $-45\,mV$ due to changes in the selectivity filter from a selective potassium conductance to a nonspecific cationic conductance.[113] It is yet to be established how rearrangements in this region are coupled to the dynamics of the voltage sensor.

Future Perspectives for the Structural Modeling of hERGs

While general mechanisms of the activation process in hERG channel are relatively well understood and are in a broad sense similar to many other voltage-gated K^+ channels,[95] there are several areas where structural mechanisms of hERG function have yet to be clarified. One of the grand challenges is on the understanding of the diverse roles played by cyclic nucleotide binding domain (CNBD) and Pert-Arnt-Sim (PAS) domains and, as a result, isoform-specific activity of hERG channels in heart tissue.[114–118] That is, the hERG1 channel has been shown to exist in two distinct isoforms, 1a and 1b, that are coexpressed.[117] $K_V11.1a$ (hERG) is the isoform originally described and is commonly referred to simply as hERG or hERG 1a.[44] It is the longest isoform, spanning 1159 residues and is considered to be the full-length transcript of the KCNH2 gene.

The $K_V11.1b$ (hERG1b) isoform is 340 amino acids shorter when compared to $K_V11.1a$ and does not contain the PAS domain.[115] However, the initial 36 residues of $K_V11.1b$ are unique to this isoform and are encoded by an additional exon located between the original exons 5 and 6 in the KCNH2 gene sequence. It has been suggested that the different isoforms play a key role in expression and trafficking of $K_V11.1$ (hERG) channels as well as in modulating the electrophysiological properties of the channels. Furthermore, changes in the relative expression levels of isoforms 1a and 1b have been described in the developing heart.[114] Expression of $K_V11.1b$ (mouse ERG 1b) protein was greater in neonatal than in adult mouse hearts, whereas the opposite is true for $K_V11.1a$. Thus, the relative expression of 1a and 1b isoforms appears to be strongly regulated and highly dynamic. Importantly, in terms of gating properties, the two isoforms differ in their gating profiles. Since the physiological role of I_{Kr} is to repolarize the late phase of cardiac action potentials (APs), hERG1 has a clear link to these arrhythmias.[119] In particular, a mutation in the unique N-terminus of hERG1b was discovered in a patient with LQTS,[120] highlighting the importance of this isoform in cardiac repolarization. Furthermore, many drugs have been shown to block $K_V11.1$ channels, resulting in an acquired form of long QT syndrome.[121] It was shown that EA4031, a selective blocker of hERG1 currents, differs in effectiveness on homomeric versus heteromeric channels.[122] Larsen et al.[123] also showed that activators such as NS1643 display differential effects on the homomeric isoforms.

Some of the PAS and CNBD mutations are thought to impact channel trafficking, yet many of them have kinetic effects on channel function. For instance, Doyle et al.[65] were able to show that the process of fast deactivation can be slowed by the addition of a recombinant PAS domain (residues 1–135). In addition to the deletion studies, several point mutations within the PAS domain were reported to have major impact on acceleration of the deactivation kinetics.[95]

The hERG1b isoform is characterized by faster kinetics of activation, recovery from inactivation, and most prominently, deactivation.[117,121] These differences in gating kinetics are mainly due to the differences in the N-terminal regions of the two isoforms. Specifically, steady state activation is altered by the absence of the proximal N-terminal region in hERG1b, and the activation rate is retarded by a short range of residues in the proximal portion of the hERG1a N-terminus.[124,125] As a result, activation rates are much faster in hERG1b channels where these residues are missing.[117] The mechanism by which fast inactivation occurs has been proposed to rely on voltage-induced changes in the structure of the outer mouth of the pore.[126] The slow deactivation of hERG1a channels has been shown to be facilitated by the first 16 residues of the N-terminus, among other factors.[127] Consequently, faster deactivation rates in hERG1b can be explained by the presence of a unique N-terminal. The rate of inactivation was shown to be similar between the two isoforms,[117] which is expected as the sequence spanning this region is identical in both isoforms.

Recovery from inactivation is significantly faster in hERG1b than hERG1a[117] potentially indicating that by some means, the N-terminus contributes to this process. Gustina and Trudeau[119,128] have shown that deletions in the CNBD resulted in rapid deactivation kinetics similar to the constructs missing PAS domains. If both of the soluble domains are deleted, then a recombinant EAG shows no slow deactivation. This implies that there should be structural interaction (binding) between the PAS domain and the CNBD domain. The structural data came from Haitin et al., who reported the first X-ray crystal structure for the PAS–CNBD complex for mEAG1.[129] The similarity of structure to hERG has been noted. The *working* hypothesis that may explain regulation of channel gating by soluble domains is that PAS-CNBD domains could interact through a complex allosteric mechanism to the S4-S5/distal S6 of the activation gate via a critical interaction between R665 and the C-linker distal to the S6, which couples the activation gate to the CNBD.

Ultimately structural modeling will pave the way for understanding of salient features responsible for voltage-dependent channel gating and inform development of better therapeutics. To achieve this ambitious goal, however, the molecular models have to be connected to cell- and tissue-level modeling allowing multiscale platform for predictive studies. Such studies are now beginning to be performed.

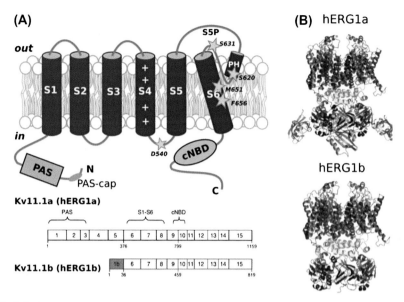

FIGURE 12.2   (B) Schematic illustration of the hERG1 organization for hERG1a and hERG1b isoforms. (A) The structural model illustrates packing of the linker (*yellow*), CNBD (*blue*) and PAS domains (*green*) developed with the published X-ray structure. *From Haitin Y, Carlson AE, Zagotta WN. The structural mechanism of KCNH-channel regulation by the eag domain. Nature 2013;501(7467):444–8.*

A study investigated the mechanism of action of the NS1643 hERG activator in WT and L5291 mutant channels (Fig. 12.2). This study is one of the first to link structure and function directly by combining electrophysiological studies with molecular and kinetic modeling.[130] Perissinotti et al. created Markov chain models with different drug-bound states representing different mechanisms of drug–channel interaction for both WT and L5291 mutant hERG. The kinetics of WT and L5291 mutant channels had been shown to be very similar, except that the L529I mutant shifts the activation curve to more negative potentials. However, the response to application of NS1643 is significantly different. Comparison between experimental and model generated currents using varied models for the NS1643 activator suggested that drug binds to closed and open states and that it influences early gating transitions. Molecular modeling suggested the structural mechanism underlying the functional effects—that the drug likely works by influencing movements of the voltage sensor that precede the opening of the activation gate.

In the next section of this chapter we will discuss progress made to date in the developing frameworks for functional models of cardiac K$^+$ channelopathies.

COMPUTATIONAL MODELS TO REVEAL FUNCTIONAL DETERMINANTS OF CARDIAC K⁺ CHANNELS AND CHANNELOPATHIES

Computational modeling and simulation approaches have been widely used as an approach to understand normal and abnormal cardiac ion channel behavior and to reveal mechanisms of mutation-induced channelopathies and resulting arrhythmia. In this section, we discuss the contributions that computational modeling of cardiac K⁺ channels has brought to bear on understanding the connection between channelopathies and resulting clinical syndromes. Such studies have provided critical information to reveal how disruption of normal cardiac ion channel function promotes and may ultimately provide a basis for the development of targeted genotype-specific therapeutic approaches.

Modeling and Simulation of Functional Effects of hERG-Linked Channelopathies

The first modeling and simulation studies of hERG-linked channelopathies and their effects on cardiac myocyte electrophysiology were published in 2001. Clayton et al. mimicked LQT1 and LQT2 syndromes by reducing the maximum conductances of I_{Kr} and I_{Ks}, respectively.[131] They also modeled LQT3 by increasing the Na channel open probability while simultaneously reducing the closing probability to produce a gain-of-function, the hallmark of Na channel–linked LQT3. Their findings suggested that reentry in the setting of LQT1 is more likely to self-terminate by moving to an unexcitable tissue boundary than in LQT2 and LQT3, which is consistent with clinical observations.[131]

Clancy and Rudy generated Markovian formulations to model the effects of specific mutations on I_{Kr} kinetics.[132] In doing so they shed light into the mechanisms by which severe LQT2 genotypes lead to prolongation of the AP duration (APD) and the QT interval. Markovian formulations of mutant models were developed to reproduce the gating alterations observed in measurements of cloned mutant channels in expression systems relative to WT. Simulations in the Luo–Rudy guinea pig ventricular myocyte model[133] showed that LQT2 mutations generally led to a loss of repolarizing current, which prolonged the APD, and in some cases, early afterdepolarizations (EADs) arose, which may trigger arrhythmias. Additionally, the severity of the phenotype and the disruption of the I_{Kr} time course depended on the specific kinetic changes produced by the genotype.

Clancy and Rudy modeled the T474I point mutation, which alters voltage dependence of activation and significantly reduces the macroscopic current density. A model was developed of the R56Q PAS domain mutation located in the amino-terminus (N-terminus) region of hERG, which normally interacts with the channel and reduces the rate of deactivation. The mutation altered deactivation kinetics of the channel and reduced the amount of time channels resided in the open state, thereby reducing current.[64] They also modeled the N626D a gain-of-function defect that causes loss of C-type inactivation together with loss of K^+ selectivity, which allowed the flow of Na^+ ions and therefore produced an inward current. hERG contains a pore selectivity sequence that is changed by the mutation from GFGN to GFGD, allowing for nonspecific passage of monovalent cations.[134]

Since the first studies, many interesting questions about I_{Kr} function and dysfunction have been answered with the help of computational modeling. For example, the mechanistic bases for LQT2 caused by hERG mutations that modify conserved arginine residues in the voltage sensor (R531Q, R531W, and R534L) have been studied.[135] Voltage clamp experiments revealed that R531Q and R531W profoundly altered activation, inactivation, and recovery from inactivation and deactivation. Surprisingly, coexpression of WT channels, to mimic the heterozygous genotype in most patients, mostly corrected the alterations produced by the mutations, except for deactivation. These were the first experiments that showed that coexpression with WT channels could correct activation and deactivation mutations produced by mutations in a critical voltage-sensing residue. Cellular simulations using the Hodgkin–Huxley formalism for I_{Kr} predicted that the APD prolongation resulting from mutations that hasten deactivation are less severe than those that reduced the maximal conductance. This work concluded that mutations that hasten deactivation may cause LQT2 but mutations that reduce the maximal conductance have a larger impact on the APD, which could be the reason why most mutations that cause LQT2 are trafficking deficient, resulting in a reduction in current.[135]

Computational models have been used to predict and quantify the effects of de novo hERG mutations from channel kinetic data not only on cellular but also on tissue electrophysiology, such as the risk of alternans and reentrant arrhythmias, and on the electrocardiogram. In a computational investigation by Benson et al., mutation-induced changes to hERG kinetics relative to WT channels that were observed were incorporated into a cellular model by varying the parameters in the Hodgkin–Huxley I_{Kr} formulation, with the aim of predicting the arrhythmogenic effects.[136] Mutations that led to reduced trafficking, faster activation kinetics, and positive shifts in the inactivation curve to different extents were modeled. At the cellular level, simulated reduction of I_{Kr} maximum conductance prolonged the APD and the QT interval, whereas a positive shift in the inactivation curve shortened them. Tissue vulnerability to reentrant arrhythmias, which was quantified by computing the mean vulnerable window in a transmural strand, was correlated with transmural dispersion of repolarization, QT interval and T-wave peak-to-end time. Accelerated I_{Kr} activation increased the range of pacing rates that produced alternans and the magnitude of the alternans. Moreover, spatial heterogeneity in these cellular alternans led to discordant alternans at the tissue level.

Computer-Based Investigation Into hERG Drugs and Mutations

Investigation of drug-induced or acquired LQTS (dLQTS) has also been the object of computational investigations, as approximately 40% of patients with dLQTS have been shown to exhibit allelic variants that disrupt the function of cardiac ionic channels.[99] These studies are so important because accurate identification of subjects who are susceptible to dLQTS is crucial for reducing the risk of cardiac arrhythmias. Experiments have shown that the functional changes of these mutations are mild, and most drug sensitivities for mutant channels are similar to that of the WT channels.[99] Computer simulations have revealed that the APD prolongation exerted by most mutant channels related (I_{Kr}, I_{Ks}, and I_{Na}) to dLQTS were shorter than those produced by mutations producing congenital LQTS.[99]

A combined clinical, experimental, and computational investigation was devoted to elucidate the role of the R1047L KCNH2 polymorphism in dofetilide-induced Torsades de pointes (TdP).[137] The R104L missense mutation was identified in two of seven individuals that developed TdP compared to 98 atrial fibrillation patients treated with dofetilide without experiencing TdP. The mutation caused a positive shift of the activation curve and slowed the activation and inactivation kinetics. Simulation of these abnormal activation properties by altering the parameters of a Hodgkin–Huxley I_{Kr} formulation resulted in a 15% prolongation of the APD, which suggests that 1047L may contribute to a higher incidence of TdP in the presence of I_{Kr} blockers.[137]

A systematic and comprehensive computational study has been conducted to reveal new insights into the impact of latent I_{Kr} channel kinetic dysfunction on I_{Kr} time course during the AP, susceptibility to dLQTS, and the potential for adjunctive therapy with I_{Kr} channel openers.[138] Specifically, this study predicted the most potentially lethal combinations of kinetic anomalies and drug properties and the ideal inverse therapeutic properties of I_{Kr} channel openers that would be expected to remedy a specific defect. This "in silico mutagenesis" was carried out by altering discrete kinetic transition rates in a Markov model of human I_{Kr}, corresponding to activation, inactivation, deactivation, and recovery from inactivation. The simulations predicted that drugs with disparate affinities to conformation states of the I_{Kr} channel markedly enhanced the susceptibility to dLQTS, especially at slow pacing rates. Specifically, drugs exhibiting high-affinity closed-state and low-affinity open-state block or high-affinity inactivated-state and low-affinity open-state block enhance dLQTS in the presence of I_{Kr} gene variants. Exposure to such drugs caused dramatic APD prolongation in cells with faster I_{Kr} deactivation. Virtual cells with defective activation developed the longest APDs after exposure to drugs exhibiting high-affinity closed-state binding and low-affinity open-state block while drugs with other characteristics resulted in shorter APD prolongation. I_{Kr} impaired inactivation and recovery from inactivation were unmasked in the presence of drugs with high-affinity inactivated-state binding and low-affinity open-state block whereas they were kept masked with other types of drugs.

In the same study, Romero et al. simulated the M54T MiRP1 latent mutation, which has been related to dLQTS and arrhythmias, in the presence of dofetilide, which drastically prolonged the QT interval duration in the M54T hMiRP1 mutation compared to wild type (Fig. 12.3). The study also predicted that application of a virtual potassium channel opener that only slows deactivation would be the ideal adjunctive therapy that could normalize the effect of dofetilide-induced AP prolongation in the presence of the M54T hMiRP1 mutation. Simulation of the addition of the I_{Kr} activator PRP260243, which slows the deactivation and increases the current magnitude by positively shifting the inactivation curve,[139] was predicted to correct the APD and QT interval prolongation, but it introduced the risk of developing short QT syndrome (SQTS).[138]

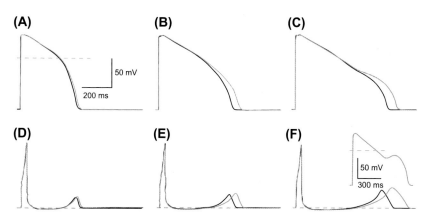

FIGURE 12.3 Simulated steady state action potential of isolated endocardial cells (top row, A–C) and pseudo-ECG (bottom row, D–F) for wild-type (*black*) and M54T hMiRP1 cells (*orange*) in the absence (left column) and in the presence of 16 nM (therapeutic dose) and 48 nM (supratherapeutic dose) of dofetilide (middle and right column, respectively). Inset of panel F shows the AP of midmyocardial cell #85 in the 1D model of the M54T hMiRP1 transmural wedge preparation. *From Romero L, et al. In silico screening of the impact of hERG channel kinetic abnormalities on channel block and susceptibility to acquired long QT syndrome. J Mol Cell Cardiol 2014;**72**:126–37.*

Finally, computational modeling has also been used to yield insight into the relationship between hERG 1a/1b channels, drug sensitivity, and arrhythmia proclivity.[122] These simulations showed that altered channel kinetics can explain reduced rectification and an increase in current during repolarization. The model also predicted that because the homomeric form of hERG possesses additional blocked states that the heteromeric form does not, E-4031 drug sensitivity is reduced for the heteromeric form. Finally, this study demonstrated that APD becomes longer with increasing doses of E-4031. Given that APD prolongation can be proarrhythmic, these simulations suggest that drugs that block hERG are more arrhythmogenic for the 1a homomer than the 1a/1b heteromer.

hERG Activators as Prospective Tools for Understanding Channel Structure and Function

A prime example of the need to apply computational modeling and simulation approaches is the case of the NS1643 hERG activator because it has many complex nonlinear effects.[140] In the computational modeling study, NS1643 functional effects were modeled by modifying the Hodgkin–Huxley equations of I_{Kr} in the 2004 ten Tusscher human ventricular model. As expected, NS1643 was predicted to decrease APD and triangulation, which is considered antiarrhythmic in the setting of LQTS. It also was predicted to increase postrepolarization refractoriness, but was shown to shorten the absolute refractory period, which could favor arrhythmia initiation. Simulations in one-dimensional tissue models predicted that NS1643 could increase the vulnerable window to reentry, which increases the risk of arrhythmias, but suppressed the development of premature AP and unidirectional block close to the APD, which is an antiarrhythmic property. An additional complication is that the effects of NS1643 are affected by the extracellular potassium concentration: While one of the primary drug effects is to increase I_{Kr} conductance during hypokalemia, the drug-induced positive shift of the inactivation curve is the effect that predominates during normokalemia. The modeling studies make prediction about situational dependence of drug efficacy. The NS1643 hERG channel activator are predicted to prove especially effective in preventing arrhythmic episodes related to EADs and long QT intervals or under hypokalemic conditions, but may be less effective in other proarrhythmic situations.

A computational modeling study combined with experimental approaches has also been used to reveal ideal properties of I_{Kr} activators to correct the APD prolonging effects of the R190Q-KCNQ1 LQT1 linked mutation.[141] LQT1 was modeled reducing I_{Ks} by 30% of control, which was the level of I_{Ks} registered in LQT1 iPSC–derived myocytes with the R190Q-KCNQ1 mutation. The authors set out to test the hypothesis that normalization of the APD could be achieved by shifting I_{Kr} inactivation to more positive voltages to slow the inactivation and increase channel availability during repolarization similarly to I_{Ks} channels. To test it, they undertook virtual experiments wherein the inactivation curve was progressively shifted in WT and LQT1 model cells, which was shown to lead to the progressive shortening of APD in both types of cells. They also tested the potential for increasing I_{Kr} via an alternative mechanism by slowing deactivation, which was simulated by increasing the time constant the activation (deactivation) gate. However the latter approach was not predicted to be effective in correcting the prolonged APDs in the LQT1 cardiomyocytes. The model predictions were validated by experiments using the ML-T531 I_{Kr} activator, whose main effect is a positive shift of the inactivation curve, which was shown experimentally to produce dose-dependent shortening of the APD and normalize the APD of cardiomyocytes in LQT1 patients.[141]

Gain-of-Function in hERG—Modeling the Short QT Syndrome

Insights into the mechanisms of the short QT syndrome type 1 produced by the N588K–hERG gain-of-function mutation have also been gained using computer simulations. Loss of I_{Kr} inactivation, a dysfunction produced by the N588K–hERG mutation, has been mimicked by eliminating the inactivation gate in an I_{Kr} Hodgkin–Huxley formulation.[142] Both single cell and one-dimensional simulations showed that loss of I_{Kr} inactivation produces a gain-of-function on I_{Kr} that leads to APD and QT shortening, without increasing transmural dispersion of AP repolarization, as AP shortening was increased in midmyocardial (M) cells versus endocardial and epicardial cells.[142]

In a different study, the parameters in a Hodgkin–Huxley I_{Kr} formulation were also modified to reproduce WT and N588K–hERG mutated currents in modified human embryonic kidney cells.[143] Simulation in both isolated cells and in a model of the human left ventricle showed that the N588K–hERG mutation increased I_{Kr} and reduced APD, dispersion of repolarization and the T-wave amplitude,[143] similarly to the previously described study.[142] The QT shortening reported in these works is consistent with clinical observations, but the reduction of the dispersion of repolarization and the T-wave amplitude are not.[144–147]

A very comprehensive investigation of the proarrhythmic effects of the N588K–hERG mutation was simulated using Markov models with the intent to clarify the apparent contradiction in the previous simulation studies and the clinical observations.[148] The investigation by Adeniran and coauthors predicted the arrhythmogenic consequences of the N588K mutation in multiscale models of the human ventricles (0D, 1D, 2D, and 3D anatomical model) and showed transmural heterogeneity of I_{Kr} density as a plausible mechanism to explain the discrepancies in earlier studies (Fig. 12.4).

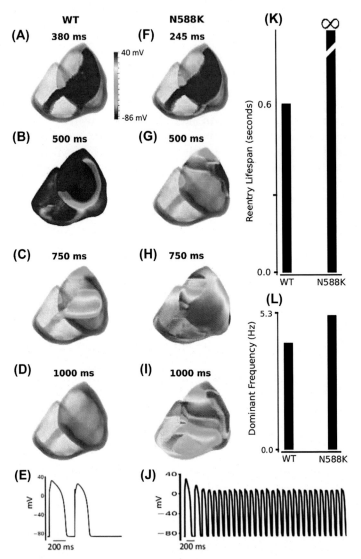

FIGURE 12.4 Snapshots of initiation and conduction of reentry in a 3D anatomical model of human ventricles. (A, F) Application of an S2 premature stimulus in a local region at refractory period of a previous conditioning excitation wave after a time delay of 380 ms for wild-type (WT) and 245 ms for N588K condition from the initial conditioning wave stimulus. (B, G) Developed scroll wave from the S2 stimulus. Snapshot at time = 500 ms. (C, H) Snapshot of scroll wave at time = 750 ms. The scroll wave self-terminated in WT, but persisted and broke up forming regenerative wavelets in N588K condition. (D and I) Snapshot of scroll wave at time = 1000 ms. Scroll wave self-terminated in WT before this recording point, but still persisted in N588K condition. (E and J) Evolution of the action potential of a cell in the left ventricle for WT and N588K conditions. (K) Measured lifespan of reentry scroll wave in WT and in N588K condition. (L) Computed-dominant frequency of electrical activity recorded from ventricle in WT and N588K conditions (2.7 Hz for WT and 5.3 Hz for N588K condition). *From Adeniran I, et al. Increased vulnerability of human ventricle to re-entrant excitation in hERG-linked variant 1 short QT syndrome. PLoS Comput Biol 2011;7(12):e1002313.*

To determine this mechanism, the authors fixed the I_{Kr} density in all cell types to the same value (as it was defined in the original Tusscher-Noble-Noble-Panfilov formulations of the different transmural cells types) and showed that in the absence of transmural heterogeneity in I_{Kr}, the N588K mutation reduced the dispersion of repolarization by decreasing the transmural heterogeneity of the APD and of the effective refractory period, like in the aforementioned studies.[142,143] These effects could be considered as antiarrhythmic rather than proarrhythmic. However, when the transmural heterogeneity of I_{Kr} density was included in the one-dimensional transmural ventricular model (ten Tusscher-Noble-Noble-Panfilov), the N588K mutation increased the T-wave amplitude, as observed in clinical observations, and the voltage gradient increased in some regions of ventricle tissue, which led to increased tissue vulnerability for unidirectional conduction block and predisposition to reentry. 2D and 3D simulations demonstrated that the N588K mutation facilitated and maintained reentries by reducing the minimum size necessary for reentry and increasing the dominant frequency of the electrical activity. This study concluded that the N588K-hERG mutation is a key factor in ventricular reentries as it prolongs the lifespan of reentrant spiral waves and stabilizes them in 3D tissue.

Another controversy regarding the N588K–hERG mutation is whether it also accelerates the deactivation process. Some experimental works have reported this feature,[144,149] while others have not.[150] An interesting study that used Markov models showed that not only alteration of the inactivation rate was needed to simulate this mutation, but also acceleration of the deactivation.[99] In this study, not only were the effects of the mutation simulated, but Itoh et al. also tested the sensitivity of the outcomes of the AP simulations to perturbations in each parameter. This study predicted that the parameter alteration that produced a gain-of-function (loss of inactivation) shortened the APD. The model simulations showed the accelerated deactivation of I_{Kr} also caused by the mutation trumped the gain-of-function effect and caused enough current reduction to promote arrhythmogenesis in SQTS.

Finally, anatomical 3D electromechanical models have been used to investigate consequences of the SQTS.[151] Two types of SQTS were considered, SQT1 and SQT3. To model the SQT1, a Markovian model of N588K mutated I_{Kr} was used and for SQT3 (I_{K1}), a Hodgkin–Huxley model formulation was modified. The 3D simulations were consistent with the evidence for dissociation between repolarization and the end of the mechanical systole.[151]

hERG Channel Mutations Linked to Brugada Syndrome

The role of sequence variations in the hERG gene that have been causally linked to the Brugada syndrome have also been studied using computer simulations.[152] The Brugada syndrome has been linked to mutations in *SCN5A* (the gene encoding the cardiac Na^+ channel) in a majority of patients but other gene defects have also been identified. Verkerk et al. identified hERG mutations (G873S and N985S) in two unrelated SCN5A mutation-negative patients. These mutations increased I_{Kr} density and produced a negative shift in the steady state inactivation curve, which causes increased rectification. In this study, AP clamp experiments showed augmented transient I_{Kr} peaks during phase-0 and phase-1 of the AP, especially at fast pacing. Computer simulations revealed that increased peak I_{Kr} facilitated the loss of the AP-dome typically in right ventricular subepicardial myocytes, which contributes to the BrS phenotype. Therefore, computer simulations helped to determine a role for I_{Kr} in BrS, which shed light on the alternative BrS mechanisms and may advance its diagnosis and therapy.[152]

Cell Type–Specific Computation Modeling of hERG Channelopathies

Several computational studies have also been conducted to assess the effects of mutations on distinct cardiac cell types. Indeed, the effect of the N588K on APD was found to be bigger in the atria compared with the ventricles in an aforementioned investigation.[153] However, another study suggested the opposite. Indeed, experiments and simulations of the N588K mutation with the Hodgkin–Huxley formalism showed that this mutation increased the peak current and it peaked earlier during the AP clamp. These effects were more severe for ventricular AP clamp, followed by Purkinje fiber AP clamp, and was shown to have smaller effects for atrial AP clamp. This study also demonstrated the influence of this mutation on the protection of I_{Kr} against premature stimulation.[154]

A theoretical study has evaluated the effects of different LQTS on ventricular myocytes and Purkinje fiber cells.[155] It is especially relevant because differences in cell type–specific responses may increase the risk for arrhythmogenesis by increasing the heterogeneity of repolarization, especially at the Purkinje-ventricular junctions. LQT1 and LQT2, which were modeled by reducing the maximal conductance of I_{Ks} and I_{Kr} respectively, produced much longer APD prolongation in ventricular than in Purkinje cells. The differential effects of the LQT3 mutations depended on the specific kinetic changes brought on by the mutations. The mutations were modeled using Markov type models. The Delta KPQ mutation markedly prolonged APD in both cell types. The F1475C mutation prominently affected Purkinje cells, even generating EADs. Finally, the S1904L also had a higher impact on Purkinje cells, especially at fast rates. This work suggested that the Purkinje fiber network could be a relevant therapeutic target in certain individuals with LQTS.

An interesting experimental and computational investigation has evaluated the effects of a mutation in two ionic channels. It evaluated the effects of the M54T hMiRP1 mutation on the sinus node.[156] This mutation affects the sinus node pacemaker current channels (I_f) (producing an 80% reduction of the current and slowing its activation) in addition to I_{Kr} channels. Computer simulations showed that the M54T hMiRP1 mutation slowed the pacing rate caused by both effects on I_f and I_{Kr}, which was consistent with the sinus bradycardia of the patient.

hERG Channelopathies in Atrial Arrhythmias

In addition to its ventricular associated phenotypes, the N588K and the L532P-hERG K$^+$ channel mutations have been associated with atrial fibrillation.[157] Computer simulations have also been conducted to assess and understand the arrhythmogenic mechanisms underlying the N588K and L532P I_{Kr} mutations in atrial fibrillation. Loewe et al. conducted simulations in virtual human atrial cells and tissues (2D)[153] by modifying the Hodgkin–Huxley I_{Kr} formulation to generate WT, N588K, and L532P models. Cellular simulations showed that both mutations augmented the peak amplitude and current integral of I_{Kr}, shortened the APD, and increased triangulation of the AP. Secondly, 1D simulations revealed that the reentry wavelength was reduced due to a shorter effective refractory period. Finally, at the 2D tissue level, model simulations predicted that both mutations enabled the initiation of spiral waves. The aforementioned effects were more severe for L532P than for N588K. Moreover, spiral waves were more stable and regular in the presence of L532P. These theoretical investigations may help to improve genotype-guided atrial fibrillation (AF) prevention and therapy strategies.

K$^+$ Channelopathies: I_{Ks}, the Slowly Activating Component of the Delayed Rectifier K$^+$ Current

The first LQTS locus (LQT1) was linked to mutations in *KCNQ1*, a gene coding for a voltage-gated potassium channel α-subunit of K$_V$7.1. I_{Ks}, the slowly activating component of the delayed rectifier K$^+$ current, results from coassembly of KCNQ1 (K$_V$LQT1) and KCNE1 (minK).[158,159] KCNQ1 was identified by positional cloning and mapped by linkage analysis to chromosome 11.[76] I_{Ks} is another primary contributor to repolarization of the cardiac AP.[160] KCNQ1 shares topological homology with other voltage-gated K$^+$ channels with 676 amino acids that form six TM domains and a pore-forming region. Four identical α-subunits form a tetramer with a central ion-conduction pore. KCNQ1 and its ancillary subunit, KCNE1, have been shown to recapitulate the gating properties of native I_{Ks}.[161]

The contribution of I_{Ks} to regulation of APD is augmented by the sympathetic branch of the autonomic nervous system, which increases I_{Ks} through primary and secondary effects on channel gating kinetics.[162–165] I_{Ks} is a key determinant of the physiological heart rate–dependent shortening of APD.[166–169] β-adrenergic receptor (β-AR) stimulation acts to increase the heart rate, which results in rate-dependent shortening of the APD resulting from the slow deactivation of I_{Ks}. Sympathetic stimulation and resulting fast pacing result in short diastolic (recovery) intervals that prevent complete deactivation of I_{Ks}, resulting in the buildup of instantaneous I_{Ks} repolarizing current at the AP onset.[166–169] At slower rates, less repolarizing current exists during each AP due to sufficient time between beats to allow for complete deactivation of I_{Ks}.[166–169]

I_{Ks} amplitude is also directly mediated by β-AR stimulation through protein kinase A (PKA) phosphorylation.[163–165] A leucine zipper motif in the C-terminus of KCNQ1 coordinates the binding of a targeting protein yotiao, which in turn binds to and recruits PKA and PP1 to the channel. The complex then regulates the phosphorylation of Ser[27] in the N-terminus of KCNQ1.[164] PKA phosphorylation of I_{Ks} considerably increases current amplitude, by increasing the rate of channel activation and reducing the rate of channel deactivation.[163–165] Each of these outcomes acts to increase the channel open probability leading to increased current amplitude and faster cardiac repolarization. Additional regulation of the I_{Ks} is via PIP2 and the serum- and glucocorticoid-inducible kinase-1 (SGK1).[170,171]

This section is focused on the contributions that computational modeling of the I_{Ks} channel has achieved in the understanding of the most common type of I_{Ks} channelopathy, the LQT syndrome type 1 (LQT1).[172] Then, we will focus on the insights revealed on the mechanisms of SQTS produced by I_{Ks} gain-of-function (SQT2) and its influence on atrial fibrillation.

Mutations in KCNQ1 or KCNE1 Can Underlie Long QT Syndrome

More than 30 mutations have been identified in KCNQ1 or KCNE1, and many act to reduce I_{Ks} through dominant negative effects,[173–175] reduced responsiveness to β-AR signaling,[165,176] or alterations in channel gating.[52,172,177–179] Mutations in KCNQ1 cause LQTS by a reduction in or loss of I_{Ks}, resulting in reduced repolarizing current during the AP.[76,180] These mutations often act by dominant negative suppression of normal subunits, which results in a significant reduction in current. In other instances, there is a complete loss of current, which has been demonstrated to result from the assembly of nonfunctional channels or the failure of channels to traffic to the plasma membrane.[174,175,181,182] Reduction or loss of I_{Ks} during the delicate plateau phase of the AP disrupts the balance of inward and outward current leading to delayed repolarization.

Prolongation of APD manifests clinically as forms of LQTS that are characterized by extended QT intervals on the electrocardiogram. Gene defects in KCNQ1 and KCNE1 are associated with distinct disease forms, LQT1 and LQT5, respectively.[174,175,183,184]

Mutations can also alter channel gating properties, which typically manifest as either reduction in the rate of channel activation, such as R539W KCNQ1,[185] R555C KCNQ1,[186] or an increased rate of channel deactivation including S74L,[75] V47F, W87R[174] KCNE1, and W248R KCNQ1.[187] An LQTS-associated KCNQ1 C-terminal mutation, G589D, disrupts the leucine zipper motif and prevents cAMP-dependent regulation of I_{Ks}.[164] The reduction of sensitivity to sympathetic activity likely prevents appropriate shortening of the APD in response to increases in heart rate.

As with KCNQ1, homozygous mutations in KCNE1 can cause the more severe Jervell and Lange-Nielsen phenotype,[75] while heterozygous mutations can manifest via dominant negative suppression or changes in channel kinetics resulting in reduced outward K^+ current.[188]

Even though KCNQ1 phosphorylation is independent of coassembly with KCNE1, transduction of the phosphorylated channel into the physiologically required increase in reserve channel activity requires the presence of KCNE1.[165] Several LQT-5 linked point mutations of KCNE1 can severely disrupt the important physiologically relevant functional consequences KCNQ1 phosphorylation including the D76N and W87R mutations.[165] Both the D76N and W87R mutations reduce basal current density when expressed with WT KCNQ1 subunits.[165] This effect would be expected to reduce repolarizing current, prolong cellular APs, and contribute to prolonged QT intervals in mutation carriers even in the absence of sympathetic nervous system (SNS) stimulation. However, in addition, the mutations would be expected to eliminate or reduce physiologically important reserve K^+ channel activity in the face of SNS stimulation. The W87R mutation speeds channel deactivation kinetics, which then are not slowed by cAMP. Thus, the data suggest that the W87R mutation eliminates the important cAMP-dependent accumulation of channel activity that is the normal response to the SNS. The D76N mutation ablates functional regulation of the channels by cAMP. The result is an expected delay in the onset of repolarization that is more pronounced in the face of SNS stimulation.

Despite their distinct origins, congenital and drug-induced forms of LQTS related to alterations in I_{Ks} are remarkably similar: Reduction in I_{Ks} results in prolongation of the QT interval on the ECG without an accompanying broadening of the T-wave, as observed in other forms of LQTS.[189] Reduced I_{Ks} leads to the loss of rate-dependent adaptation in APD, which is consistent with the clinical manifestation of arrhythmias associated with LQT1 and LQT5, which tend to occur due to sudden increases in heart rate. This strongly suggests that investigation of congenital forms of electrical abnormalities may act as a paradigm for drug-induced forms of clinical syndromes.

Combined clinical, functional, and computational modeling work revealed that mutant-specific I_{Ks} channel characteristics to improve risk stratification in LQT1 patients.[190] For this purpose, the cellular electrophysiological characteristics produced by 17 I_{Ks} mutations found in 387 LQT1 patients were simulated using the Hodgkin–Huxley formulism, and they were correlated with their clinical phenotype. Mutations that decreased the rate of activation were associated with increased risk of cardiac arrhythmias, independently of the common clinical parameters considered for risk stratification, such as the QT interval. In patients with subclinical LQT (QT$_c$ shorter than 500 ms) slower activation was an independent indicator of arrhythmogenicity.[190] Another combined clinical, functional, and computational modeling study has also been performed with the aim of improving risk stratification in LQT1 patients. In this case, computer-simulated electrocardiography parameters were analyzed. The cellular electrophysiology function of 34 I_{Ks} variants observed in 633 LQT1 patients was determined, and models of these mutations were formulated using the Hodgkin–Huxley formulism and used to simulate the transmural repolarization. It was shown that prolongation of transmural repolarization was linked to increased risk for arrhythmias.[191]

Computer simulation has also been a valuable tool to analyze the different mechanisms underlying arousal- and exercise-triggered arrhythmias in LQT1 patients.[192] Indeed, several mutations that affected both channel function and β-AR activation were modeled using the Hodgkin–Huxley formulism by taking into account cPKC-mediated effects (so the maximal conductance of I_{Ks} and voltage dependence of activation parameters were modified), the mutant-specific functional impairment and β-AR regulation on I_{Ks}. Two situations were simulated: β-AR stimulation alone and the coupling of α1- and β-AR stimulation. This study showed that impaired I_{Ks} activation by Ca^{2+}-dependent protein kinase C is correlated with emotion/arousal-triggered episodes in LQT1.[192]

Provocative tests have been shown to be useful in patients to unmask subclinical or latent LQT. One especially useful provocative test that can help to unmask LQT1 syndrome is exercise. Tsuji-Wakisaka and coauthors utilized an in silico provocative test in their study of the effects of several KCNQ1 splicing mutations/variant around the exon 7-intron 7 junction. They showed that slight prolongation of the QT that this variant produces is markedly enhanced under β-adrenergic stimulation.[193] In the presence of the mutation β-adrenergic stimulation provoked delayed after depolarizations and EADs in simulated epicardial cells and midmyocardial cells, which were reflected on the pseudo-ECG as monomorphic ventricular tachycardia and fibrillation-like episodes, respectively.[193] Another multidisciplinary study showed that I235N-K$_V$7.1 carriers have normal resting QT intervals but they exhibit long QT during the recovery of the treadmill stress test. Coexpression of WT and I235N K$_V$7.1 showed that mutant channels were insensitive to PKA activation and computer simulations showed that mutant AP was prolonged during β-adrenergic stimulation, especially at slow rates, due to limited upregulation of I_{Ks} by PKA activation.[194]

In another computational work, Markov models of WT and the Q357R KCNQ1 variant of the α-subunit of I_{Ks} were fitted to reproduce the characteristics of both channels as observed in CHO-K1 cells.[195] These models were included into a human ventricular AP model and AP simulations reproduced the minimal AP prolongation and the dominant negative loss of I_{Ks} characteristic of the Q357R mutation. Simulations showed that EADs were triggered in Q357R mutant epicardial cells only at fast rates and during simultaneous isoproterenol and I_{Kr} blocker application, while these conditions did not produce EADs in WT. Simulations also predicted that it was the reduced channel expression that caused EADs, not the I_{Ks} kinetic alterations. In addition, simulated exercise caused EADs even in the absence of I_{Kr} block in mutant midmyocardial cells. Also, isoproterenol together with partial I_{Kr} block resulted in drastic QT prolongation in the presence of Q357R variant.[195] This work concluded that arrhythmia generation in silent Q357R I_{Kr} mutants requires multiple insults and that reduced membrane I_{Ks} expression, not kinetic changes, are the basis of the arrhythmic phenotype.[195]

Although dLQT is usually related to I_{Kr} block, I_{Ks} block also contributes to dLQT, and cardiac modeling has contributed to explain the role of I_{Ks} blockade in dLQT.[196] Fluoxetine and its metabolite norfluoxetine, which predominantly block I_{Ks}, produced an excessive QT prolongation in a K422T KCNQ1 (α-subunit of I_{Ks} channel) mutation carrier. This mutation, which produces a positive shift in voltage dependence of activation and prolongs the deactivation rate, exerts a small influence on the QT interval. Mutation-induced changes in kinetics that were experimentally observed were modeled as relative changes in I_{Ks} gates. As norfluoxetine concentration dependence of blockade of WT and heterozygously expressed KCNQ1/KCNE1 currents were very similar, drug application to WT and mutant cells was mimicked by reducing the I_{Ks} maximal conductance to the same extent. Simulations showed that this mutation only produced a slight prolongation of the AP, and I_{Ks} block by norfluoxetine produced similar APD prolongation in WT and in mutant cells. However, when I_{Kr} blockade by norfluoxetine was also simulated, much longer prolongations were predicted in mutant cells. Therefore, this work suggests that I_{Kr} blockade plays a relevant role in dLQT, especially under conditions of reduced repolarization reserve.[196]

SQTS produced by I_{Ks} gain-of-function (SQT2) has also been an object of interesting mathematical modeling and simulation studies. A combined clinical, functional, and simulation study revealed for the first time the existence of SQT2.[197] Genetic screening was performed in an individual that presented ventricular fibrillation, whose QT interval was abnormally short after the arrhythmic episode. It revealed the presence of the KCNQ1 V307L mutant. Functional studies of the mutant revealed a severe shift of the activation curve and an acceleration of the activation which produced a gain-of-function in I_{Ks}. The parameters governing the steady state activation of I_{Ks} in a model following the Hodgkin–Huxley formulism were modified to simulate the effects of the mutation. Simulated mutant AP showed a significant shortening of the AP.[197]

The arrhythmogenic effects of KCNQ1 V307L mutation was also the subject of another computational work. Specifically, models based on the Hodgkin–Huxley formulism were fitted to reproduce the functional properties of the homozygous and heterozygous KCNQ1 V307L mutation in transfected cells. 1D and 2D simulations reproduced the typical shortening of the QT, the increase of the T-wave amplitude and of the prolongation of T_{peak}–T_{end} duration. Simulations of isolated V307L cells showed abbreviated AP duration and increased maximal transmural heterogeneity of the APD and the refractory period. 1D and 2D simulations predicted that the vulnerable window for the unidirectional conduction block was prolonged in the presence of the V307L mutation and that it favored the generation and maintenance of reentries.[198]

KCNQ1 Channelopathies in Atrial Arrhythmias

I_{Ks} channelopathies are also related to atrial fibrillation, and computer simulations have shed light on the understanding of the influence of I_{Ks} on this arrhythmia. The minK38G/S polymorphism has been linked to an increased prevalence of AF. A work that combines experimental and computational studies has showed that the minK38G/S polymorphism significantly reduces I_{Ks} membrane density.[199] Simulation of reduction of I_{Ks} maximal conductance predicted a prolongation of the APD and a reduction of the minimum frequency for alternans appearance.[199] Moreover, simulated minK38G/S polymorphism also favored the trigger of EADs under reduced repolarization reserve.[199]

In another study examining 231 individuals in the Vanderbilt AF Registry, two mutations linked with early onset lone AF in different kindreds were identified: a mutation in KCNQ1 and another mutation in natriuretic peptide precursor A (NPPA) gene (encoding atrial natriuretic peptide).[200] The functional effects of these variants on I_{Ks} were very similar. They elicited approximately threefold larger currents and with much faster activation than WT. To simulate the effects of WT and mutated I_{Ks}, Markov models were fitted to the current–voltage curves experimentally recorded for WT KCNQ1 cotransfected with KCNE1 for KCNQ1- and for IAP54–56 I_{Ks}, respectively. Computational simulations revealed that acceleration of the transitions into open states was the main alteration exerted by both mutations. Inclusion of this variation into an atrial AP model predicted a 37% shortening at 300 ms cycle length and a loss of the phase II dome.[200]

A multiscale modeling study (Fig. 12.4) comprising ion channel, 0D, 1D, 2D, and geometrically detailed 3D simulations investigated the mechanisms of AF in the presence of the S140G KCNQ1 mutation.[201] It is known that the major alterations on I_{Ks} produced by this variant are a large increase of current at the end of voltage clamp pulses, a marked instantaneous component through the whole time course, and a significant inward current at more hyperpolarized potentials. In this multiscale work, the mutant current was simulated by adding a nongated voltage-dependent instantaneous component of current to the WT formulation. I_{Ks} models were integrated into AP models to run single cell simulations to predict the APD and effective refractory period (ERP) restitution curves. 1D strand simulations were run to compute the conduction velocity (CV) restitution and the temporal vulnerability, and 2D tissue simulations were used to define critical substrates to initiate and sustain reentries (Fig. 12.5).

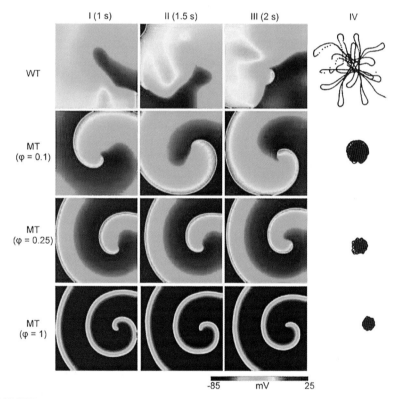

FIGURE 12.5 **Simulation of spiral waves in 2D model of human atrium.** Top panels show results from 2D reentrant wave simulation under WT condition; the second row of panels show results from $\phi = 0.1$ conditions; the third row of panels show results from $\phi = 0.25$ conditions; and the bottom row show results from $\phi = 1$. Frames from the 2D simulation at time $t = 1\,\mathrm{s}$ (column I), time $t = 1.5\,\mathrm{s}$ (column II), and time $t = 2\,\mathrm{s}$ (column III) are shown. Column IV shows the reentrant wave tip trajectories. *From Kharche S, et al. Pro-arrhythmogenic effects of the S140G KCNQ1 mutation in human atrial fibrillation - insights from modelling.* J Physiol 2012;**590**(Pt 18):4501–14.

The authors also constructed 3D models of the human atria to generate and maintain reentries and to compute the lifespan, tip meandering patterns, and dominant rotor frequency. Simulated S140G KCNQ1 mutation shortened atrial APD and ERP, flattened both APD and ERP restitution curves, reduced atrial CV at fast frequencies, and speeded it at slow frequencies, which eased the propagation at fast rates. Despite the fact that the variant slightly prolonged the vulnerable window for reentry, it significantly reduced the minimal size necessary for reentry generation and maintenance. At the atrial level, scroll waves degenerated into persistent and chaotic wavelets that produced fibrillation.[201]

The effects of the S140G KCNQ1 mutation were further investigated in a computational study.[202] This work employed the previously developed formulations of WT and S140G KCNQ1 mutated I_{Ks} for atrial cells and used a similar approach to include the effects of this mutation in ventricular cells. It also confirmed that the S140G KCNQ1 mutation shortened atrial AP duration and effective refractory period, which favors reentrant excitation. In addition, simulations predicted that selective K⁺ channel block could normalize the effects of the S140G KCNQ1 mutation on atrial cells when decreasing I_{Ks} but not I_{Kr} and, finally, the S140G KCNQ1 mutation could reduce ventricular AP duration, although it depended on the AP ventricular model.

Other computational work has investigated the genetic backgrounds of atrial fibrillation and their effects on ventricular electrophysiology in juvenile patients (less than 50 years).[203] In 30 juvenile-onset AF individuals, three mutant carriers were identified, and one mutation, the KCNQ1-G229D, which was novel, affected I_{Ks}. Functional analysis showed that its homozygous expression produced a large instantaneous activation component without deactivation after repolarization to −50 mV and the heterozygous expression induced both instantaneous and time-dependent activating I_{Ks}. The measured I_{Ks} steady state activation curve in the heterozygous expression was severely shifted to more negative potentials, and it was flatter. Moreover, the deactivation was slower. AP clamp experiments revealed a severe I_{Ks} gain-of-function. WT and WT/G229D I_{Ks} models were constructed using the Hodgkin–Huxley formulism and fitting the biophysical parameters to the results of the functional analysis. These I_{Ks} models were introduced into an atrial and a ventricular AP model. Computer simulations predicted a much larger WT/G229D-I_{Ks} in atrial than in ventricular cells. However, the shortening of the APD was marked in atrial simulations while it was not in ventricular simulations, due to the smaller contribution of the I_{Ks} to the ventricular AP. Simulated QT interval was not shortened by WT/G229D-I_{Ks} what was consistent with the phenotype of the patient.[203]

Computer simulations have also helped to explain how the V271F KCNQ1 variant, elicited two different phenotypes, atrial fibrillation and sinus bradycardia, in probands of the same family.[204] The V271F KCNQ1 mutation left shifts the activation curve to lower voltages and drastically slows deactivation. WT and V271F KCNQ1-mutated I_{Ks} were simulated by modifying the parameters of a formulation of I_{Ks} that follows the Hodgkin–Huxley formulism to reproduce the experimental recordings of I_{Ks} in HEK293 cells. Single cell simulations predicted that the V241F variant produced an excessive gain-of-function of I_{Ks}, which slowed the sinoatrial node pacemaking and resulted in a severe shortening of the atrial APD, which may cause AF.[204] Moreover, this study supports the role of sinus node dysfunction on AF.[204]

SUMMARY

Understanding how disruption in cardiac K$^+$ channels leads to arrhythmias is continually improving due to the development and implementation of computational modeling and simulation approaches as scales spanning the single atomistic ion channel scale to the high-resolution reconstruction of the whole heart. Connecting ion channelopathies to emergent behavior in the heart to cause disease is a critical area of research. A grand challenge of the current research era is to integrate data into meaningful models to reveal emergent mechanisms of disease and to facilitate prediction and development of novel therapeutic interventions. Mathematical modeling and simulation constitute some of the most promising, and necessary, methodologies to reveal fundamental biological principles and mechanisms, model effects of interactions between system components, and predict emergent disease processes and effects of new treatment modalities. There is no reasonable, efficient and cost-effective alternative experimental or clinical strategy that can achieve all of these goals. The ubiquity of disruption of K$^+$ channels in cardiac disorders of excitability emphasizes the importance of fundamental understanding of the associated mechanisms and disease processes to ultimately reveal new targets for human therapy.

Acknowledgments

S.Y. Noskov would like to acknowledge Dr. Laura Perissinotti and Mr. Yibo Wang for constructive suggestions and help with Figures 12.1 and 12.2. S.Y. Noskov acknowledges financial support from the Canadian Heart and Stroke Foundation and the Canadian Institutes of Health Research.

C.E. Clancy would like to acknowledge Pei-Chi Yang Ph.D., Steffen S. Docken, and Kevin R. DeMarco M.S. for constructive suggestions and editing. C.E. Clancy acknowledges financial support from the National Institutes of Health and the American Heart Association.

References

1. Noble D. The voltage dependence of the cardiac membrane conductance. *Biophys J* 1962;**2**:381–93.
2. Noble D. A modification of the Hodgkin–Huxley equations applicable to Purkinje fibre action and pace-maker potentials. *J Physiol* 1962;**160**:317–52.
3. Beeler GW, Reuter H. Reconstruction of the action potential of ventricular myocardial fibers. *J Physiol* 1977;**268**:177–210.
4. Luo CH, Rudy Y. A model of the ventricular cardiac action potential. Depolarization, repolarization, and their interaction. *Circ Res* 1991;**68**(6):1501–26.
5. Hodgkin AL, Huxley AF. A quantitative description of membrane current and its application to conduction and excitation in nerve. *J Physiol* 1952;**117**(4):500–44.
6. Pathmanathan P, et al. Uncertainty quantification of fast sodium current steady-state inactivation for multi-scale models of cardiac electrophysiology. *Prog Biophys Mol Biol* 2015;**117**(1):4–18.
7. Verkerk AO, Wilders R. Hyperpolarization-activated current, if, in mathematical models of rabbit sinoatrial node pacemaker cells. *Biomed Res Int* 2013;**2013**:872454.
8. Bueno-Orovio A, et al. Na/K pump regulation of cardiac repolarization: insights from a systems biology approach. *Pflugers Arch* 2014;**466**(2):183–93.
9. Greenstein JL, et al. Modeling CaMKII-mediated regulation of L-type Ca(2+) channels and ryanodine receptors in the heart. *Front Pharmacol* 2014;**5**:60.
10. Onal B, Unudurthi SD, Hund TJ. Modeling CaMKII in cardiac physiology: from molecule to tissue. *Front Pharmacol* 2014;**5**:9.
11. Qu Z, Nivala M, Weiss JN. Calcium alternans in cardiac myocytes: order from disorder. *J Mol Cell Cardiol* 2013;**58**:100–9.
12. Moreno JD, Clancy CE. Pathophysiology of the cardiac late Na current and its potential as a drug target. *J Mol Cell Cardiol* 2012;**52**(3):608–19.
13. Bett GC, Zhou Q, Rasmusson RL. Models of HERG gating. *Biophys J* 2011;**101**(3):631–42.
14. Ramirez E, et al. In silico ischaemia-induced reentry at the Purkinje-ventricle interface. *Europace* 2014;**16**(3):444–51.
15. Romero L, et al. Systematic characterization of the ionic basis of rabbit cellular electrophysiology using two ventricular models. *Prog Biophys Mol Biol* 2011;**107**(1):60–73.
16. Romero L, et al. Human and rabbit inter-species comparison of ionic mechanisms of arrhythmic risk: a simulation study. *Conf Proc IEEE Eng Med Biol Soc* 2010;**2010**:3253–6.
17. Glynn P, Unudurthi SD, Hund TJ. Mathematical modeling of physiological systems: an essential tool for discovery. *Life Sci* 2014;**111**(1–2):1–5.
18. Besse IM, et al. A computational investigation of cardiac caveolae as a source of persistent sodium current. *Front Physiol* 2011;**2**:87.
19. Tobon C, et al. A three-dimensional human atrial model with fiber orientation. Electrograms arrhythmic activation patterns relationship. *PLoS One* 2013;**8**(2):e50883.
20. Ferrero JM, Trenor B, Romero L. Multiscale computational analysis of the bioelectric consequences of myocardial ischaemia and infarction. *Europace* 2014;**16**(3):405–15.
21. Trayanova NA, et al. Computational cardiology: how computer simulations could be used to develop new therapies and advance existing ones. *Europace* 2012;**14**(Suppl. 5):v82–9.
22. Trayanova NA, Rantner LJ. New insights into defibrillation of the heart from realistic simulation studies. *Europace* 2014;**16**(5):705–13.
23. Henriquez CS. A brief history of tissue models for cardiac electrophysiology. *IEEE Trans Biomed Eng* 2014;**61**(5):1457–65.
24. Quail MA, Taylor AM. Computer modeling to tailor therapy for congenital heart disease. *Curr Cardiol Rep* 2013;**15**(9):395.
25. Duncker DJ, et al. Animal and in silico models for the study of sarcomeric cardiomyopathies. *Cardiovasc Res* 2015;**105**(4):439–48.

26. Trayanova NA, Boyle PM. Advances in modeling ventricular arrhythmias: from mechanisms to the clinic. *Wiley Interdiscip Rev Syst Biol Med* 2014;**6**(2):209–24.

27. Polakova E, Sobie EA. Alterations in T-tubule and dyad structure in heart disease: challenges and opportunities for computational analyses. *Cardiovasc Res* 2013;**98**(2):233–9.

28. Zhang P, Su J, Mende U. Cross talk between cardiac myocytes and fibroblasts: from multiscale investigative approaches to mechanisms and functional consequences. *Am J Physiol Heart Circ Physiol* 2012;**303**(12):H1385–96.

29. Roberts BN, et al. Computational approaches to understand cardiac electrophysiology and arrhythmias. *Am J Physiol Heart Circ Physiol* 2012;**303**(7):H766–83.

30. Sugiura S, et al. Multi-scale simulations of cardiac electrophysiology and mechanics using the University of Tokyo heart simulator. *Prog Biophys Mol Biol* 2012;**110**(2–3):380–9.

31. Bers DM, Grandi E. Human atrial fibrillation: insights from computational electrophysiological models. *Trends Cardiovasc Med* 2011;**21**(5):145–50.

32. Zhou L, O'Rourke B. Cardiac mitochondrial network excitability: insights from computational analysis. *Am J Physiol Heart Circ Physiol* 2012;**302**(11):H2178–89.

33. Niederer SA, Smith NP. At the heart of computational modelling. *J Physiol* 2012;**590**(Pt 6):1331–8.

34. Sato D, Clancy CE. Cardiac electrophysiological dynamics from the cellular level to the organ level. *Biomed Eng Comput Biol* 2013;**5**:69–75.

35. Gomez JF, et al. Electrophysiological and structural remodeling in heart failure modulate arrhythmogenesis. 1D simulation study. *PLoS One* 2014;**9**(9):e106602.

36. Romero L, et al. Non-uniform dispersion of the source-sink relationship alters wavefront curvature. *PLoS One* 2013;**8**(11):e78328.

37. Romero L, et al. The relative role of refractoriness and source-sink relationship in reentry generation during simulated acute ischemia. *Ann Biomed Eng* 2009;**37**(8):1560–71.

38. Trenor B, et al. Vulnerability to reentry in a regionally ischemic tissue: a simulation study. *Ann Biomed Eng* 2007;**35**(10):1756–70.

39. Gomez JF, et al. Electrophysiological and structural remodeling in heart failure modulate arrhythmogenesis. 2D simulation study. *PLoS One* 2014;**9**(7):e103273.

40. Abbott GW, et al. MiRP1 forms IKr potassium channels with HERG and is associated with cardiac arrhythmia. *Cell* 1999;**97**:175–87.

41. Zhou Z, et al. HERG channel dysfunction in human long QT syndrome. Intracellular transport and functional defects. *J Biol Chem* 1998;**273**(33):21061–6.

42. Shibasaki T. Conductance and kinetics of delayed rectifier potassium channels in nodal cells of the rabbit heart. *J Physiol* 1987;**387**:227–50.

43. McDonald TV, et al. A minK-HERG complex regulates the cardiac potassium current I(Kr). *Nature* 1997;**388**(6639):289–92.

44. Sanguinetti MC, et al. A mechanistic link between an inherited and an acquired cardiac arrhythmia: HERG encodes the IKr potassium channel. *Cell* 1995;**81**(2):299–307.

45. Sanguinetti MC, Jurkiewicz NK. Two components of cardiac delayed rectifier K$^+$ current. Differential sensitivity to block by class III antiarrhythmic agents. *J General Physiol* 1990;**96**(1):195–215.

46. Abbott GW. KCNE2 and the K(+) channel: the tail wagging the dog. *Channels (Austin)* 2012;**6**(1):1–10.

47. London B, et al. Two isoforms of the mouse ether-a-go-go-related gene coassemble to form channels with properties similar to the rapidly activating component of the cardiac delayed rectifier K$^+$ current. *Circ Res* 1997;**81**:870–8.

48. Lees-Miller JP, et al. Electrophysiological characterization of an alternatively processed ERG K$^+$ channel in mouse and human hearts. *Circ Res* 1997;**81**:719–26.

49. Pond AL, et al. Expression of distinct ERG proteins in rat, mouse, and human heart - relation to functional I-Kr channels. *J Biol Chem* 2000;**275**(8):5997–6006.

50. Pond AL, Nerbonne JM. ERG proteins and functional cardiac I-Kr channels in rat, mouse, and human heart. *Trends Cardiovasc Med* 2001;**11**(7):286–94.

51. Roden DM, Balser JR. A plethora of mechanisms in the HERG-related long QT syndrome - genetics meets electrophysiology. *Cardiovasc Res* 1999;**44**(2):242–6.
52. Splawski I, et al. Spectrum of mutations in long-QT syndrome genes : KVLQT1, HERG, SCN5A, KCNE1, and KCNE2. *Circulation* 2000;**102**(10):1178–85.
53. van Noord C, Eijgelsheim M, Stricker BH. Drug- and non-drug-associated QT interval prolongation. *Br J Clin Pharmacol* 2010;**70**(1):16–23.
54. Perrin MJ, et al. Human ether-a-go-go related gene (hERG) K+ channels: function and dysfunction. *Prog Biophys Mol Biol* 2008;**98**(2–3):137–48.
55. Petrecca K, et al. N-linked glycosylation sites determine HERG channel surface membrane expression. *J Physiol* 1999;**515**:41–8.
56. Sanguinetti MC, et al. Spectrum of HERG K+-channel dysfunction in an inherited cardiac arrhythmia. *Proc Natl Acad Sci USA* 1996;**93**(5):2208–12.
57. Ficker E, et al. Molecular determinants of dofetilide block of HERG K+ channels. *Circ Res* 1998;**82**:386–95.
58. Smith PL, Baukrowitz T, Yellen G. The inward rectification mechanism of the HERG cardiac potassium channel. *Nature* 1996;**379**(6568):833–6.
59. Eldstrom J, Fedida D. The voltage-gated channel accessory protein KCNE2: multiple ion channel partners, multiple ways to long QT syndrome. *Expert Rev Mol Med* 2011;**13**:e38.
60. Sesti F, et al. A common polymorphism associated with antibiotic-induced cardiac arrhythmia. *Proc Natl Acad Sci USA* 2000;**97**(19):10613–8.
61. Zhou M, et al. Potassium channel receptor site for the inactivation gate and quaternary amine inhibitors. *Nature* 2001;**411**(6838):657–61.
62. Jiang Y, et al. Crystal structure and mechanism of a calcium-gated potassium channel. *Nature* 2002;**417**(6888):515–22.
63. Jiang Y, et al. The open pore conformation of potassium channels. *Nature* 2002;**417**(6888):523–6.
64. Cabral JHM, et al. Crystal structure and functional analysis of the HERG potassium channel N terminus: a eukaryotic PAS domain. *Cell* 1998;**95**(5):649–56.
65. Doyle DA, et al. The structure of the potassium channel: molecular basis of K+ conduction and selectivity. *Science* 1998;**280**(5360):69–77.
66. Cordero-Morales JF, et al. Molecular driving forces determining potassium channel slow inactivation. *Nat Struct Mol Biol* 2007;**14**(11):1062–9.
67. Cuello LG, et al. Structural basis for the coupling between activation and inactivation gates in K(+) channels. *Nature* 2010;**466**(7303):272–5.
68. Cuello LG, et al. Structural mechanism of C-type inactivation in K(+) channels. *Nature* 2010;**466**(7303):203–8.
69. Pan AC, et al. Thermodynamic coupling between activation and inactivation gating in potassium channels revealed by free energy molecular dynamics simulations. *J Gen Physiol* 2011;**138**(6):571–80.
70. Pathak MM, et al. Closing in on the resting state of the shaker K+ channel. *Neuron* 2007;**56**(1):124–40.
71. Yarov-Yarovoy V, Baker D, Catterall WA. Voltage sensor conformations in the open and closed states in ROSETTA structural models of K+ channels. *Proc Natl Acad Sci USA* 2006;**103**(19):7292–7.
72. MacKinnon R. Potassium channels and the atomic basis of selective ion conduction (Nobel Lecture). *Angew Chem Int Ed Engl* 2004;**43**(33):4265–77.
73. MacKinnon R. Potassium channels. *FEBS Lett* 2003;**555**(1):62–5.
74. Yan GX, Antzelevitch C. Cellular basis for the Brugada syndrome and other mechanisms of arrhythmogenesis associated with ST-segment elevation. *Circulation* 1999;**100**(15):1660–6.
75. Splawski I, et al. Mutations in the hminK gene cause long QT syndrome and suppress I-Ks function. *Nat Genet* 1997;**17**(3):338–40.

76. Wang Q, et al. Positional cloning of a novel potassium channel gene: KVLQT1 mutations cause cardiac arrhythmias. *Nat Genet* 1996;**12**(1):17–23.

77. Stary A, et al. Toward a consensus model of the HERG potassium channel. *ChemMedChem* 2010;**5**(3):455–67.

78. Sanguinetti MC, Tristani-Firouzi M. hERG potassium channels and cardiac arrhythmia. *Nature* 2006;**440**(7083):463–9.

79. Clarke CE, et al. Effect of S5P alpha-helix charge mutants on inactivation of hERG K+ channels. *J Physiol* 2006;**573**(2):291–304.

80. Hill AP, et al. Mechanism of block of the hERG K+ channel by the scorpion toxin CnErg1. *Biophys J* 2007;**92**(11):3915–29.

81. Ju P, et al. The pore domain outer helix contributes to both activation and inactivation of the HERG K+ channel. *J Biol Chem* 2009;**284**(2):1000–8.

82. Mitcheson JS, et al. A structural basis for drug-induced long QT syndrome. *Proc Natl Acad Sci USA* 2000;**97**(22):12329–33.

83. Perry M, et al. Structural determinants of HERG channel block by clofilium and ibutilide. *Mol Pharmacol* 2004;**66**:240–9.

84. Stansfeld PJ, et al. Drug block of the hERG potassium channel: insight from modeling. *Proteins* 2007;**68**(2):568–80.

85. Durdagi S, et al. Insights into the molecular mechanism of hERG1 channel activation and blockade by drugs. *Curr Med Chem* 2010;**17**(30):3514–32.

86. Mitcheson JS, Chen J, Sanguinetti MC. Trapping of a methanesulfonanilide by closure of the HERG potassium channel activation gate. *J Gen Physiol* 2000;**115**(3):229–40.

87. Lees-Miller JP, et al. Molecular determinant of high-affinity dofetilide binding to HERG1 expressed in Xenopus oocytes: involvement of S6 sites. *Mol Pharmacol* 2000;**57**(2):367–74.

88. Durdagi S, et al. Modeling of open, closed, and open-inactivated states of the hERG1 channel: structural mechanisms of the state-dependent drug binding. *J Chem Inf Model* 2012;**52**(10):2760–74.

89. Durdagi S, Duff HJ, Noskov SY. Combined receptor and ligand-based approach to the universal pharmacophore model development for studies of drug blockade to the hERG1 pore domain. *J Chem Inf Model* 2011;**51**(2):463–74.

90. Anwar-Mohamed A, et al. A human ether-a-go-go-related (hERG) ion channel atomistic model generated by long supercomputer molecular dynamics simulations and its use in predicting drug cardiotoxicity. *Toxicol Lett* 2014;**230**(3):382–92.

91. Ferrer T, et al. Molecular coupling in the human ether-a-go-go-related gene-1 (hERG1) K+ channel inactivation pathway. *J Biol Chem* 2011;**286**(45):39091–9.

92. Fan JS, et al. Effects of outer mouth mutations on hERG channel function: a comparison with similar mutations in the Shaker channel. *Biophys J* 1999;**76**(6):3128–40.

93. Jiang M, et al. Dynamic conformational changes of extracellular S5-P linkers in the hERG channel. *J Physiol (London)* 2005;**569**(1):75–89.

94. Tseng GN, et al. Probing the outer mouth structure of the hERG channel with peptide toxin footprinting and molecular modeling. *Biophys J* 2007;**92**(10):3524–40.

95. Vandenberg JI, et al. hERG K(+) channels: structure, function, and clinical significance. *Physiol Rev* 2012;**92**(3):1393–478.

96. Lees-Miller JP, et al. Interactions of H562 in the S5 helix with T618 and S621 in the pore helix are important determinants of hERG1 potassium channel structure and function. *Biophys J* 2009;**96**(9):3600–10.

97. Perry MD, Ng CA, Vandenberg JI. Pore helices play a dynamic role as integrators of domain motion during Kv11.1 channel inactivation gating. *J Biol Chem* 2013;**288**(16):11482–91.

98. Harrell DT, et al. Genotype-dependent differences in age of manifestation and arrhythmia complications in short QT syndrome. *Int J Cardiol* 2015;**190**:393–402.

99. Itoh H, et al. Arrhythmogenesis in the short-QT syndrome associated with combined HERG channel gating defects: a simulation study. *Circ J* 2006;**70**(4):502–8.

100. Lees-Miller JP, et al. Ivabradine prolongs phase 3 of cardiac repolarization and blocks the hERG1 (KCNH2) current over a concentration-range overlapping with that required to block HCN4. *J Mol Cell Cardiol* 2015;**85**:71–8.

101. Stork D, et al. State dependent dissociation of HERG channel inhibitors. *Br J Pharmacol* 2007;**151**(8):1368–76.

102. Szabo G, et al. Enhanced repolarization capacity: new potential antiarrhythmic strategy based on HERG channel activation. *Curr Med Chem* 2011;**18**(24):3607–21.

103. Farkas AS, Nattel S. Minimizing repolarization-related proarrhythmic risk in drug development and clinical practice. *Drugs* 2010;**70**(5):573–603.

104. Dennis A, et al. hERG channel trafficking: novel targets in drug-induced long QT syndrome. *Biochem Soc Trans* 2007;**35**(Pt 5):1060–3.

105. Colenso CK, et al. Interactions between voltage sensor and pore domains in a hERG K⁺ channel model from molecular simulations and the effects of a voltage sensor mutation. *J Chem Inf Model* 2013;**53**(6):1358–70.

106. Colenso CK, et al. Voltage sensor gating charge transfer in a hERG potassium channel model. *Biophys J* 2014;**107**(10):L25–8.

107. Zhang M, et al. Interactions between charged residues in the transmembrane helices of Herg's voltage-sensing domain. *Biophys J* 2004;**86**(1):277a.

108. Zhang M, Liu J, Tseng GN. Gating charges in the activation and inactivation processes of the hERG channel. *J Gen Physiol* 2004;**124**(6):703–18.

109. Schoppa NE, et al. The size of gating charge in wild-type and mutant Shaker potassium channels. *Science* 1992;**255**(5052):1712–5.

110. Moss AJ, et al. Increased risk of arrhythmic events in long-QT syndrome with mutations in the pore region of the human ether-a-go-go-related gene potassium channel. *Circulation* 2002;**105**(7):794–9.

111. Tristani-Firouzi M, Chen J, Sanguinetti MC. Interactions between S4-S5 linker and S6 transmembrane domain modulate gating of HERG K⁺ channels. *J Biol Chem* 2002;**277**(21):18994–9000.

112. Teng GQ, et al. Homozygous missense N629D hERG (KCNH2) potassium channel mutation causes developmental defects in the right ventricle and its outflow tract and embryonic lethality. *Circ Res* 2008;**103**(12):1483. U270.

113. Teng GQ, et al. Role of mutation and pharmacologic block of human KCNH2 in vasculogenesis and fetal mortality partial rescue by transforming growth factor-beta. *Circ Arrhythm Electrophysiol* 2015;**8**(2):420–8.

114. Pillozzi S, et al. Differential expression of hERG1A and hERG1B genes in pediatric acute lymphoblastic leukemia identifies different prognostic subgroups. *Leukemia* 2014;**28**(6):1352–5.

115. Trudeau MC, et al. hERG1a N-terminal eag domain-containing polypeptides regulate homomeric hERG1b and heteromeric hERG1a/hERG1b channels: a possible mechanism for long QT syndrome. *J Gen Physiol* 2011;**138**(6):581–92.

116. McPate MJ, et al. hERG1a/1b heteromeric currents exhibit amplified attenuation of inactivation in variant 1 short QT syndrome. *Biochem Biophys Res Commun* 2009;**386**(1):111–7.

117. Larsen AP, et al. Characterization of hERG1a and hERG1b potassium channels-a possible role for hERG1b in the I(Kr) current. *Pflugers Arch* 2008;**456**(6):1137–48.

118. Robertson GA, Jones EM, Wang J. Gating and assembly of heteromeric hERG1a/1b channels underlying I(Kr) in the heart. *Novartis Found Symp* 2005;**266**:4–15. discussion 15-8, 44-5.

119. Gustina AS, Trudeau MC. A recombinant N-terminal domain fully restores deactivation gating in N-truncated and long QT syndrome mutant hERG potassium channels. *Proc Natl Acad Sci USA* 2009;**106**(31):13082–7.

120. Robertson GA, et al. hERG 1b as a potential target for inherited and acquired long QT syndrome. *Circulation* 2008;**118**(18):S525.

121. Larsen AP. Role of ERG1 isoforms in modulation of ERG1 channel trafficking and function. *Pflugers Arch* 2010;**460**(5):803–12.
122. Sale H, et al. Physiological properties of hERG 1a/1b heteromeric currents and a hERG 1b-specific mutation associated with long-QT syndrome. *Circ Res* 2008;**103**(7): e81–95.
123. Larsen AP, Bentzen BH, Grunnet M. Differential effects of Kv11.1 activators on Kv11.1a, Kv11.1b and Kv11.1a/Kv11.1b channels. *Br J Pharmacol* 2010;**161**(3):614–28.
124. Gustina AS, Trudeau MC. N-Terminal region regulation of hERG potassium channel gating: contributions of the eag/Pas domain and the proximal N-Terminus. *Biophys J* 2011;**102**(3):330a–1a.
125. Saenen J, et al. Modulation of HERG gating by a charge cluster in the N-terminal proximal domain. *Biophys J* 2006;**91**(2):4381–91.
126. Schönherr R, Heinemann SH. Molecular determinants for activation and inactivation of HERG, a human inward rectifier potassium channel. *J Physiol* 1996;**493**(3):635–42.
127. Wang X, et al. Kv11.1 channel subunit composition includes MinK and varies developmentally in mouse cardiac muscle. *Dev Dyn* 2008;**237**(9):2430–7.
128. Gustina AS, Trudeau MC. The eag domain regulates hERG channel inactivation gating via a direct interaction. *J Gen Physiol* 2013;**141**(2):229–41.
129. Haitin Y, Carlson AE, Zagotta WN. The structural mechanism of KCNH-channel regulation by the eag domain. *Nature* 2013;**501**(7467):444–8.
130. Perissinotti LL, et al. Kinetic model for NS1643 drug activation of WT and L529I variants of Kv11.1 (hERG1) potassium channel. *Biophys J* 2015;**108**(6):1414–24.
131. Clayton RH, et al. Re-entrant cardiac arrhythmias in computational models of long QT myocardium. *J Theor Biol* 2001;**208**(2):215–25.
132. Clancy CE, Rudy Y. Cellular consequences of HERG mutations in the long QT syndrome: precursors to sudden cardiac death. *Cardiovasc Res* 2001;**50**(2):301–13.
133. Luo CH, Rudy Y. A dynamic model of the cardiac ventricular action potential. I. Simulations of ionic currents and concentration changes. *Circ Res* 1994;**74**(6):1071–96.
134. Lees-Miller JP, et al. Novel gain-of-function mechanism in K$^+$ channel-related long-QT syndrome: altered gating and selectivity in the HERG1 N629D mutant. *Circ Res* 2000;**86**(5):507–13.
135. McBride CM, et al. Mechanistic basis for type 2 long QT syndrome caused by KCNH2 mutations that disrupt conserved arginine residues in the voltage sensor. *J Membr Biol* 2013;**246**(5):355–64.
136. Benson AP, Al Owais M, Holden AV. Quantitative prediction of the arrhythmogenic effects of de novo hERG mutations in computational models of human ventricular tissues. *Eur Biophys J* 2011;**40**(5):627–39.
137. Sun Z, et al. Role of a KCNH2 polymorphism (R1047 L) in dofetilide-induced Torsades de Pointes. *J Mol Cell Cardiol* 2004;**37**(5):1031–9.
138. Romero L, et al. In silico screening of the impact of hERG channel kinetic abnormalities on channel block and susceptibility to acquired long QT syndrome. *J Mol Cell Cardiol* 2014;**72**:126–37.
139. Perry M, Sachse FB, Sanguinetti MC. Structural basis of action for a human ether-a-go-go-related gene 1 potassium channel activator. *Proc Natl Acad Sci USA* 2007;**104**(34):13827–32.
140. Peitersen T, et al. Computational analysis of the effects of the hERG channel opener NS1643 in a human ventricular cell model. *Heart Rhythm* 2008;**5**(5):734–41.
141. Zhang H, et al. Modulation of hERG potassium channel gating normalizes action potential duration prolonged by dysfunctional KCNQ1 potassium channel. *Proc Natl Acad Sci USA* 2012;**109**(29):11866–71.
142. Zhang H, Hancox JC. In silico study of action potential and QT interval shortening due to loss of inactivation of the cardiac rapid delayed rectifier potassium current. *Biochem Biophys Res Commun* 2004;**322**(2):693–9.

143. Weiss DL, et al. Modelling of short QT syndrome in a heterogeneous model of the human ventricular wall. *Europace* 2005;**7**(Suppl. 2):105–17.

144. Brugada R, et al. Sudden death associated with short-QT syndrome linked to mutations in HERG. *Circulation* 2004;**109**(1):30–5.

145. Giustetto C, et al. Short QT syndrome: clinical findings and diagnostic-therapeutic implications. *Eur Heart J* 2006;**27**(20):2440–7.

146. Gollob MH, Redpath CJ, Roberts JD. The short QT syndrome: proposed diagnostic criteria. *J Am Coll Cardiol* 2011;**57**(7):802–12.

147. Watanabe H, et al. High prevalence of early repolarization in short QT syndrome. *Heart Rhythm* 2010;**7**(5):647–52.

148. Adeniran I, et al. Increased vulnerability of human ventricle to re-entrant excitation in hERG-linked variant 1 short QT syndrome. *PLoS Comput Biol* 2011;**7**(12):e1002313.

149. Cordeiro JM, et al. Modulation of I(Kr) inactivation by mutation N588K in KCNH2: a link to arrhythmogenesis in short QT syndrome. *Cardiovasc Res* 2005;**67**(3):498–509.

150. McPate MJ, et al. The N588K-HERG K⁺ channel mutation in the 'short QT syndrome': mechanism of gain-in-function determined at 37 degrees C. *Biochem Biophys Res Commun* 2005;**334**(2):441–9.

151. Adeniran I, Hancox JC, Zhang H. In silico investigation of the short QT syndrome, using human ventricle models incorporating electromechanical coupling. *Front Physiol* 2013;**4**:166.

152. Verkerk AO, et al. Role of sequence variations in the human ether-a-go-go-related gene (HERG, KCNH2) in the Brugada syndrome. *Cardiovasc Res* 2005;**68**(3):441–53.

153. Loewe A, et al. Arrhythmic potency of human ether-a-go-go-related gene mutations L532P and N588K in a computational model of human atrial myocytes. *Europace* 2014;**16**(3):435–43.

154. McPate MJ, et al. Comparative effects of the short QT N588K mutation at 37 degrees C on hERG K⁺ channel current during ventricular, Purkinje fibre and atrial action potentials: an action potential clamp study. *J Physiol Pharmacol* 2009;**60**(1):23–41.

155. Iyer V, Sampson KJ, Kass RS. Modeling tissue- and mutation-specific electrophysiological effects in the long QT syndrome: role of the Purkinje fiber. *PLoS One* 2014;**9**(6):e97720.

156. Nawathe PA, et al. An LQTS6 MiRP1 mutation suppresses pacemaker current and is associated with sinus bradycardia. *J Cardiovasc Electrophysiol* 2013;**24**(9):1021–7.

157. Hong K, et al. Short QT syndrome and atrial fibrillation caused by mutation in KCNH2. *J Cardiovasc Electrophysiol* 2005;**16**(4):394–6.

158. Barhanin J, et al. K(V)LQT1 and lsK (minK) proteins associate to form the I(Ks) cardiac potassium current. *Nature* 1996;**384**(6604):78–80.

159. Sanguinetti MC, et al. Coassembly of K(V)LQT1 and minK (IsK) proteins to form cardiac I(Ks) potassium channel. *Nature* 1996;**384**(6604):80–3.

160. Kass RS, Freeman LC. Potassium channels in the heart - cellular, molecular, and clinical implications. *Trends Cardiovasc Med* 1993;**3**(4):149–59.

161. Kurokawa J, Motoike HK, Kass RS. TEA(+)-sensitive KCNQ1 constructs reveal pore-independent access to KCNE1 in assembled I-Ks channels. *J Gen Physiol* 2001;**117**(1):43–52.

162. Walsh KB, Kass RS. Distinct voltage-dependent regulation of a heart-delayed IK by protein kinases A and C. *Am J Physiol* 1991;**261**(6 Pt 1):C1081–90.

163. Kurokawa J, Abriel H, Kass RS. Molecular basis of the delayed rectifier current I-Ks in heart. *J Mol Cell Cardiol* 2001;**33**(5):873–82.

164. Marx SO, et al. Requirement of a macromolecular signaling complex for beta adrenergic receptor modulation of the KCNQ1-KCNE1 potassium channel. *Science* 2002;**295**(5554):496–9.

165. Kurokawa J, Chen L, Kass RS. Requirement of subunit expression for cAMP-mediated regulation of a heart potassium channel. *Proc Natl Acad Sci USA* 2003;**100**(4):2122–7.

166. Zeng J, et al. Two components of the delayed rectifier K$^+$ current in ventricular myocytes of the guinea pig type. Theoretical formulation and their role in repolarization. *Circ Res* 1995;**77**(1):140–52.

167. Faber GM, Rudy Y. Action potential and contractility changes in [Na+](i) overloaded cardiac myocytes: a simulation study. *Biophys J* 2000;**78**(5):2392–404.

168. Volders P, et al. Probing the contribution of IKs to canine ventricular repolarization: key role for beta-adrenergic receptor stimulation. *Circulation* 2003;**107**(21):2753–60.

169. Stengl M, et al. Accumulation of slowly activating delayed rectifier potassium current (IKs) in canine ventricular myocytes. *J Physiol (Lond)* 2003;**551**(3):777–86.

170. Zaydman MA, Cui J. PIP2 regulation of KCNQ channels: biophysical and molecular mechanisms for lipid modulation of voltage-dependent gating. *Front Physiol* 2014;**5**:195.

171. Lang F, Shumilina E. Regulation of ion channels by the serum- and glucocorticoid-inducible kinase SGK1. *FASEB J* 2013;**27**(1):3–12.

172. Dvir M, et al. Recent molecular insights from mutated IKS channels in cardiac arrhythmia. *Curr Opin Pharmacol* 2014;**15**:74–82.

173. Shalaby FY, et al. Dominant-negative KvLQT1 mutations underlie the LQT1 form of long QT syndrome. *Circulation* 1997;**96**(6):1733–6.

174. Bianchi L, et al. Cellular dysfunction of LQT5-minK mutants: abnormalities of I-Ks, I-Kr and trafficking in long QT syndrome. *Hum Mol Genet* 1999;**8**(8):1499–507.

175. Bianchi L, et al. Mechanisms of I-Ks suppression in LQT1 mutants. *Am J Physiol Heart Circ Physiol* 2000;**279**(6):H3003–11.

176. Abbott GW, Goldstein SA. Disease-associated mutations in KCNE potassium channel subunits (MiRPs) reveal promiscuous disruption of multiple currents and conservation of mechanism. *FASEB J* 2002;**16**(3):390–400.

177. Chen QY, et al. Homozygous deletion in KVLQT1 associated with Jervell and Lange-Nielsen syndrome. *Circulation* 1999;**99**(10):1344–7.

178. Roden DM, et al. Multiple mechanisms in the long-QT syndrome. Current knowledge, gaps, and future directions. The SADS Foundation Task Force on LQTS. *Circulation* 1996;**94**(8):1996–2012.

179. Wollnik B, et al. Pathophysiological mechanisms of dominant and recessive KVLQT1 K$^+$ channel mutations found in inherited cardiac arrhythmias. *Hum Mol Genet* 1997;**6**(11):1943–9.

180. Neyroud N, et al. A novel mutation in the potassium channel gene KVLQT1 causes the Jervell and Lange-Nielsen cardioauditory syndrome. *Nat Genet* 1997;**15**(2):186–9.

181. Donger C, et al. KVLQT1 C-terminal missense mutation causes a forme fruste long-QT syndrome. *Circulation* 1997;**96**(9):2778–81.

182. Mohammad-Panah R, et al. Mutations in a dominant-negative isoform correlate with phenotype in inherited cardiac arrhythmias. *Am J Hum Genet* 1999;**64**(4):1015–23.

183. Chouabe C, et al. Functional expression of a KVLQT1 missense mutation that causes a forme fruste long QT syndrome. *Circulation* 1998;**98**(17):468–9.

184. Napolitano C, et al. Prevalence and clinical features of LQT5 and LQT6, two uncommon variants of the long QT-syndrome. *Eur Heart J* 2000;**21**:352.

185. Chouabe C, et al. Novel mutations in KvLQT1 that affect I-ks activation through interactions with Isk. *Cardiovasc Res* 2000;**45**(4):971–80.

186. Chouabe C, et al. Properties of KvLQT1 K$^+$ channel mutations in Romano-Ward and Jervell and Lange-Nielsen inherited cardiac arrhythmias. *EMBO J* 1997;**16**(17):5472–9.

187. Franqueza L, et al. Long QT syndrome-associated mutations in the S4-S5 linker of KvLQT1 potassium channels modify gating and interaction with minK subunits. *J Biol Chem* 1999;**274**(30):21063–70.

188. Roden DM, Spooner PM. Inherited Long QT syndromes: a paradigm for understanding arrhythmogenesis. *J Cardiovasc Electrophysiol* 1999;**10**:1664–83.

189. Gima K, Rudy Y. Ionic current basis of electrocardiographic waveforms - a model study. *Circ Res* 2002;**90**(8):889–96.

190. Jons C, et al. Use of mutant-specific ion channel characteristics for risk stratification of long QT syndrome patients. *Sci Transl Med* 2011;**3**(76):76ra28.

191. Hoefen R, et al. In silico cardiac risk assessment in patients with long QT syndrome: type 1: clinical predictability of cardiac models. *J Am Coll Cardiol* 2012;**60**(21):2182–91.

192. O.U J, et al. Impaired IKs channel activation by Ca(2+)-dependent PKC shows correlation with emotion/arousal-triggered events in LQT1. *J Mol Cell Cardiol* 2015;**79**:203–11.

193. Tsuji-Wakisaka K, et al. Identification and functional characterization of KCNQ1 mutations around the exon 7-intron 7 junction affecting the splicing process. *Biochim Biophys Acta* 2011;**1812**(11):1452–9.

194. Bartos DC, et al. A KCNQ1 mutation contributes to the concealed type 1 long QT phenotype by limiting the Kv7.1 channel conformational changes associated with protein kinase A phosphorylation. *Heart Rhythm* 2014;**11**(3):459–68.

195. O'Hara T, Rudy Y. Arrhythmia formation in subclinical ("silent") long QT syndrome requires multiple insults: quantitative mechanistic study using the KCNQ1 mutation Q357R as example. *Heart Rhythm* 2012;**9**(2):275–82.

196. Veerman CC, et al. Slow delayed rectifier potassium current blockade contributes importantly to drug-induced long QT syndrome. *Circ Arrhythm Electrophysiol* 2013;**6**(5):1002–9.

197. Bellocq C, et al. Mutation in the KCNQ1 gene leading to the short QT-interval syndrome. *Circulation* 2004;**109**(20):2394–7.

198. Zhang H, et al. Repolarisation and vulnerability to re-entry in the human heart with short QT syndrome arising from KCNQ1 mutation–a simulation study. *Prog Biophys Mol Biol* 2008;**96**(1–3):112–31.

199. Ehrlich JR, et al. Atrial fibrillation-associated minK38G/S polymorphism modulates delayed rectifier current and membrane localization. *Cardiovasc Res* 2005;**67**(3):520–8.

200. Abraham RL, et al. Augmented potassium current is a shared phenotype for two genetic defects associated with familial atrial fibrillation. *J Mol Cell Cardiol* 2010;**48**(1):181–90.

201. Kharche S, et al. Pro-arrhythmogenic effects of the S140G KCNQ1 mutation in human atrial fibrillation - insights from modelling. *J Physiol* 2012;**590**(Pt 18):4501–14.

202. Hancox JC, et al. In silico investigation of a KCNQ1 mutation associated with familial atrial fibrillation. *J Electrocardiol* 2014;**47**(2):158–65.

203. Hasegawa K, et al. A novel KCNQ1 missense mutation identified in a patient with juvenile-onset atrial fibrillation causes constitutively open IKs channels. *Heart Rhythm* 2014;**11**(1):67–75.

204. Ki CS, et al. A KCNQ1 mutation causes age-dependant bradycardia and persistent atrial fibrillation. *Pflugers Arch* 2014;**466**(3):529–40.

Connexins and Heritable Human Diseases

S.A. Bernstein, G.I. Fishman

New York University School of Medicine, New York, NY, United States

INTRODUCTION

Connexins are tetratransmembrane proteins that assemble into hexameric pore-forming structures known as connexons or hemichannels. Connexons typically dock with their counterparts in adjacent cells to form intercellular gap junction channels, but may remain unpaired as cell surface hemichannels. Gap junction channels and hemichannels allow for the passage of ions, secondary messengers, and other small molecules between cells or between intra- and extracellular compartments, respectively, playing key roles in embryonic development, tissue homeostasis, and response to pathologic stress.[15,18] The human genome contains 20 genes that encode distinct but structurally related connexin isoforms.[50] Most tissues express multiple connexin types and individual isoforms can assemble into homomeric and heteromeric hemichannels or homotypic and heterotypic gap junction channels. Given the large number of individual connexin genes, the potential combinatorial diversity of encoded channel proteins, and their complex developmental and tissue-restricted patterns of expression, it is not surprising that mutations in connexins are responsible for a significant burden of heritable human diseases.[23,50] In this chapter we will review the clinical disorders associated with genetic abnormalities of connexins and, where known, the molecular mechanisms responsible for the disease phenotype.

CONNEXINS, CONNEXONS, AND GAP JUNCTION CHANNELS

Gap junctions were first identified in electron micrographs of cardiomyocyte plasma membranes as five-layered membrane specializations between adjacent cells.[48] During the subsequent 50 years, multiple molecular, cellular, and imaging techniques have been utilized to more fully resolve the structure of gap junction channels.[17] Based upon molecular cloning and sequence analysis, all connexins are predicted to share a common topology, with four transmembrane domains, amino and carboxy-terminal cytoplasmic domains, a cytoplasmic loop and two extracellular loops (Fig. 13.1). Although the overall homology between connexins is high, important differences between these proteins are found in the intracellular loop and particularly the length and sequence of the carboxyl terminus, where many regulatory elements act, including sites of posttranslational modification such as phosphorylation and the binding of various connexin interacting proteins.[33,51] 3D electron crystallography, Cryo-EM, and molecular dynamics simulations have provided structural information at 3.5 Å resolution,[24,28,37,56,62] as illustrated in Fig. 13.1. These studies indicate that the diameter of the pore is about 40 Å at the cytoplasmic side of the channel and narrows to 14 Å near the extracellular membrane surface. The first transmembrane domain (TM1) appears to be the primary pore-lining domain and the N-terminal regions of the six subunits line the entrance to the cytoplasmic pore and restrict its size to 14 Å. Moreover, the six amino termini appear positioned to sense transjunctional voltage and form the plug to close the pore.[28] These data are consistent with experimental data indicating permeation of molecules up to approximately 1.8 kDa in size through hemichannels and gap junction channels.[18]

CONNEXINS AND HERITABLE HUMAN DISEASES

Remarkably, of the 20 human connexin genes, mutations associated with heritable disease states have been identified in as many as 11 family members (Table 13.1). Broadly speaking, these diseases fall into six major categories, including myelin disorders; nonsyndromic and syndromic deafness; skin diseases; cataracts; oculodentodigital dysplasia (ODDD); and atrial fibrillation (AF).[12,21] As with other disease phenotypes arising from mutations in ion channel subunits, a range of mechanisms have been implicated in the underlying molecular pathogenesis. These include truncation defects that lead to simple haploinsufficiency; truncations or missense mutations that produce trafficking defects; missense mutations that interfere with the assembly of individual connexins into hemichannels or the docking of hemichannels into full channels; others that produce gap junction channels that traffic and assemble but with defects in permeability and/or gating; and finally, mutations that result in hemichannels with abnormal permeability, most notably those that are "leaky" and potentially induce premature cell death.[41,52] Later we review the major clinical

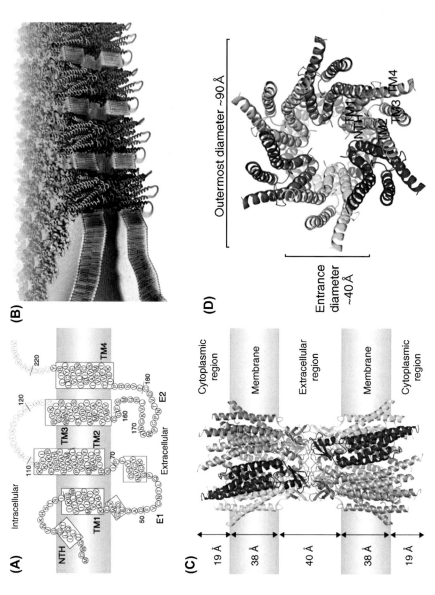

FIGURE 13.1 **Structure of connexins and gap junction channels.** (A) Topological diagram of a Cx26 monomer within the cell membrane. (B) Schematic of the gap junction. The gap junction channels are drawn spanning two adjoining membranes with an intermembrane space of 40 Å. (C) Overall structure of Cx26 gap junction channel viewed parallel to the membrane plane. (D) View of Cx26 from the cytoplasmic side along the pore axis. *Modified from Curr Opin Struct Biol 2010;20:423–30.*

TABLE 13.1   Human Connexin Genes and Diseases

Gene	Connexin	Human Disease	Initial Description
GJA1	Cx43	Oculodentodigital dysplasia	38
		Atrial fibrillation	53
GJA3	Cx46	Zonular pulverulent cataract-3 (CZP3)	27
GJA4	Cx37	None to date	
GJA5	Cx40	Atrial fibrillation	14
GJA8	Cx50	Zonular pulverulent cataract-1 (CZP1)	47
GJA9	Cx59	None to date	
GJA10	Cx62	None to date	
GJB1	Cx32	X-linked Charcot-Marie-Tooth disease (CMTX)	6
GJB2	Cx26	Nonsyndromic and syndromic deafness Deafness and palmoplantar keratoderma Keratitis-icthyosis-deafness syndrome Bart–Pumphrey syndrome Vohwinkel syndrome	21a 43 51a 1a 29 45a
GJB3	Cx31	Erythrokeratodermia variabilis	42,43
		Nonsyndromic hearing impairment	33a
GJB4	Cx30.3	Erythrokeratodermia variabilis	26
GJB5	Cx31.1	None to date	
GJB6	Cx30	Nonsyndromic deafness	16a
		Clouston syndrome	49
		Hidrotic ectodermal dysplasia	61a
GJB7	Cx25	None to date	
GJC1	Cx45	None to date	
GJC2	Cx47	Pelizaeus-Merzbacher-like disease (PMLD1)	55
GJC3	Cx29	Nonsyndromic deafness	59a
GJD2	Cx36	None to date	
GJD3	Cx31.9	None to date	
GJD4	Cx40.1	None to date	

disorders associated with genetic abnormalities of connexins and, where known, the molecular mechanisms responsible for the disease phenotype.

Myelin Disorders and Neuropathies

X-linked Charcot-Marie-Tooth disease (CMTX1) is one of the more common inherited neuropathies and the first disease associated with mutations in a connexin gene.[6] CMTX1 is caused by mutations in the GJB1 gene, which encodes connexin 32 (Cx32). CMTX1 produces limb weakness and is associated with progressive demyelination of peripheral axons.[2,22] Since its original description in 1993, over 400 individual GJB1 mutations have been described in CMT1X patients.[22] The clinical hallmarks of CMT1X include slowly progressive weakness and muscle atrophy beginning in the distal lower extremities, often presenting as early as the age of 10. As an X-linked syndrome, in comparison to affected males, heterozygous affected females may be asymptomatic, have very mild disease, or have only subclinical slowing of nerve conduction as determined by nerve conduction studies.

Cx32 is robustly expressed by Schwann cells in the peripheral nervous system, which form the protective myelin sheath around peripheral nerves.[32] Schwann cells generate a flattened membrane extension that spirals around the long axis of the neuronal axon. Cytoplasm remains in the inner and outermost layers, with gap junctions forming a direct radial route for ions and metabolites. This localization has been experimentally confirmed by freeze-fracture immunogold labeling and is approximately a thousand times shorter than the circumferential pathway following the Schwann cell cytoplasm.[32] Nerve biopsies in CMT patients reveal a variety of pathological findings including demyelination and axonal changes in peripheral nerves. Mice with a targeted deletion of the *Gjb1* gene show a progressive demyelinating peripheral neuropathy by age 3 months,[3] similar to the human condition. There are also more subtle findings in the central nervous system in tissues that express Cx32, characterized by demyelination and thinning of nerve fibers, attributed to one of several mechanisms including alterations in axonal cytoskeleton signaling, abnormal calcium handling, lack of second messenger transport, and impaired energy supply. Although Cx32 is expressed in many tissue types, clinical disease associated with GJB1 mutations appear largely restricted to neurological disorders. Compensatory expression of alternative connexin isoforms in other tissues may provide some degree of protection against pathologic sequelae.

Studies of cell lines expressing Cx32 mutations demonstrate a range of molecular phenotypes, including absence of detectable protein, retention of Cx32 in the endoplasmic reticulum or Golgi, or normal trafficking of Cx32 to the junctional membrane but abnormal biophysical properties.

Mutant channels with altered properties fall into two distinct categories: those displaying altered voltage–conductance relationship and those with diminished effective pore size. For example, a single serine to leucine substitution at residue 26 can reduce the effective pore size from 7 to 3 Å, potentially rendering the channel impermeable to certain cell-signaling molecules. How GJB1 mutations might specifically impact cell signaling and contribute to CMTX pathology remains uncertain. In addition, a subset of mutations are thought to increase hemichannel permeability, leading to premature cell death.[1]

Pelizaeus-Merzbacher-like disease (PMLD1) is a second myelin disorder resulting from recessively inherited mutations in GJC2, which encodes the Cx47 protein.[55] Patients with PMLD1 develop severe demyelination in the central nervous system resulting in nystagmus, developmental delay, ataxia, and spasticity. Heterologous expression studies have demonstrated a subset of mutations produce trafficking defects with accumulation of mutant protein in the endoplasmic reticulum and failure to form normal channels.[36] A Glu260Lys mutation was hypothesized by molecular modeling to disrupt connexin–connexin interactions and connexon formation.[8]

Deafness

Deafness is the most common heritable connexin-associated disease and associated with mutations in as many as six individual genes. Mutations in GJB2 (Cx26) have been the most frequently reported etiology, with more than 220 different mutations accounting for more than 50% of cases of autosomal recessive nonsyndromic (isolated) hearing loss.[40] However, both nonsyndromic and syndromic (accompanied by skin disease; see later discussion) forms of hearing loss are also associated with mutations in GJB1, GJB3, GJB6, GJA1, and GJC3.[31] Cx26 and Cx30 proteins are richly expressed in the inner ear and are postulated to play essential roles in maintenance of the high endolymphatic K^+ concentration critical for preserving the high sensitivity of hair cells to sonic stimuli.[59] Recently, however, this hypothesis has been challenged. Instead, it has been proposed that Cx26 is essential for normal cochlear development and that Cx30 mutations indirectly produce hearing loss by decreasing the abundance of Cx26.[59] Heterologous expression studies of GJB2 mutants demonstrate that the majority of mutant Cx26 proteins show trafficking defects and fail to form functional gap junction channels. A subset of mutations appear to dominantly inhibit the function of wild-type Cx26. Interestingly, a V84L mutation is reportedly permeable to ions but not to the second messenger inositol triphosphate (IP_3).[5] This mutation, as well as a V95M mutant, reportedly forms homotypic channels but not heterotypic channels.[57] Finally, there are a class of GJB2 mutations, such as the missense mutation G45E, that affect hemichannel function, opening at normal extracellular

Ca^{+2} concentrations and resulting in cell death.[52] GJB3 mutations can cause both recessive and dominant nonsyndromic hearing loss. Expression studies have identified one class of mutants that target normally to the junctional membrane but are nonpermeant to dyes and ions,[35] whereas a second group appears to accumulate in lysosomes.[25]

Skin Diseases

Skin comprises the largest organ in the body, and the keratinocytes are rich in gap junctions formed from as many as eight different connexins with highly regulated patterns of expression.[10,30] Of these different isoforms, mutations in four have been identified in various skin disorders including GJB2, GJB3, GJB4, and GJB6. As some of these same connexins are also expressed in the inner ear, a number of skin diseases are syndromic and also associated with deafness. Vohwinkel syndrome, an autosomal dominant condition characterized by mutilating keratoderma associated with sensorineural hearing loss, was the first such syndrome attributed to a connexin mutation, a D66H missense mutation in the GJB2 gene. Bart–Pumphrey syndrome is a rare autosomal dominant disorder characterized by congenital deafness, palmoplantar hyperkeratosis, knuckle pads and leukonychia, ascribed to a G59S missense mutation in GJB2. Finally, keratitis-ichthyosis-deafness (KID) syndrome and the related hystrix-like ichthyosis-deafness (HID) syndrome have both been ascribed to a D50N mutation in GJB2. Additional GJB2 mutations associated with skin disease have also been identified, and as with the syndromic conditions, many of the mutations cluster in the first extracellular domain of the Cx26 protein. Mutations in other β-family connexins also cause skin disorders, including erythrokeratoderma variabilis (EKV), an autosomal dominant disorder with palmoplantar keratoderma, most notably in GJB3 but also reported to arise in individuals with mutations in the nearby GJB4 gene.[26,42] Finally, mutations in GJB6, encoding Cx30, have been described in hidrotic ectodermal dysplasia, or Clouston syndrome, another autosomal dominant disorder with diffuse palmoplantar keratoderma, nail dystrophy and sparse scalp, and body hair.[49]

A variety of molecular mechanisms have been proposed to account for the phenotypic manifestations associated with mutations in the four β-family connexins. These include dominant-negative GJB2 missense mutations (R75W and W77R) that fail to produce functional channels and suppress the activity of coexpressed wild-type Cx26[43]; a missense mutation (D66H), identified in 10 patients with Vohwinkel syndrome that interferes with gap junction assembly[29]; as well as a missense mutation (D50N), identified in patients with KID syndrome that reportedly shifts the voltage dependence of hemichannel activation and inhibits full gap junction channel function.[45] Mutations in GJB6 have been identified

(G11R, V37E, and A88V) that produce Cx30 proteins that are trapped in the cytoplasm and fail to normally traffic to the junctional membrane.[9] Heterologous expression studies have also been reported for GJB3 mutations identified in patients with erythrokeratoderma variabilis. In contrast to the normal membrane localization of a wild-type Cx31-GFP fusion protein, constructs with the disease-causing mutations R42P, C86S, G12R, or G12D all showed cytoplasmic localization and produced cell death. A similar result was reported with an L34P mutation residing in the first transmembrane domain.[16] In contrast, cells transfected with R32W-Cx31-EGFP and 66delD-Cx31-EGFP were viable, but showed reduced dye transfer, indicative of failure to form functional gap junction channels.[11]

Cataracts

Mutations in GJA8 encoding Cx50 and GJA3 encoding Cx46 result in zonular pulverulent cataracts type 1 (CZP1) and type 3 (CZP3) respectively.[27,47] The lens itself is avascular and the fiber cells are devoid of organelles; therefore these cells are uniquely dependent upon diffusion of nutrients through gap junctional coupling for their survival.[4,7] Both Cx50 and Cx46 are expressed in the mature fiber cells at the core of the lens.[34,58] Moreover, the two connexins appear capable of interacting and forming functional heteromeric and heterotypic channels, at least as assayed in *Xenopus* expression systems.[19] The most common molecular perturbation of connexins associated with cataracts is a trafficking defect and the failure to form gap junction plaques. This abnormality has been described in Cx50 R23T, D47N, P88S, and P88Q mutants, as well as a Cx46 fs380 mutant. A second class of mutants forms plaques, but nonfunctional channels, ie, Cx50W45S, Cx46D3Y, and Cx46L11S. Finally, one especially interesting class of mutations results in hemichannels with altered gating properties, including Cx46N63S which showed increased sensitivity to extracellular Mg^{+2} concentration and Cx46D3Y with altered charge selectivity and voltage gating.[13,54]

Oculodentodigital Dysplasia

ODDD is an uncommon autosomal dominant multisystem disorder characterized by variable expression of craniofacial and limb dysmorphisms including abnormalities of teeth and eyes, neurodegeneration, and webbing of the hands and feet. ODDD results from mutations in the GJA1 gene, which encodes the widely expressed Cx43 gap junctional protein.[38] Upwards of more than 75 distinct mutations in GJA1 have been reported in patients with ODDD, the great majority of which are missense mutations, and almost all are confined to regions other than the less well-conserved carboxyl terminal domain of the Cx43 protein. Many individual mutations have been examined in heterologous expression systems and not surprisingly, these studies demonstrate a range of molecular phenotypes, with examples of trafficking defects, mutants that form gap junction plaques but couple poorly or not at all, as well as those that form channels with altered permeability or gating properties.[39,46] Interestingly, at least one mutation responsible for ODDD, I130T, appears to interfere with posttranslational phosphorylation of Cx43 and inhibit its normal delivery to the junctional membrane.[20]

Atrial Fibrillation

Atrial myocardium is richly coupled by gap junctions comprised primarily of Cx40 and Cx43. The first report linking mutations in connexin genes to cardiac arrhythmias described four patients with lone AF and mutations in GJA5. However, surprisingly, three of these patients apparently had somatic, rather than germline mutations, including one with two nonallelic mutations. Heterologous expression of the mutant proteins demonstrated defects in trafficking and/or ability to form functional gap junction channels. Moreover, the mutants all dominantly inhibited coupling of coexpressed wild-type Cx40 or Cx43.[14] However, as a cautionary note, a follow-up deep sequencing study of DNA prepared from left atrial appendages and lymphocytes of 25 lone AF patients did not confirm these findings, suggesting that somatic mosaicism is unlikely to play a prominent role in AF pathogenesis.[44] Subsequent studies have identified very rare germline connexin mutations in patients with lone AF, including one family with a frameshift mutation resulting in a premature truncation (Q49X) and three other families with missense mutations (V85I, L221I, and L229M) in GJA5.[60,61] In addition, a single patient with a frameshift mutation in GJA1 has been reported.[53]

SUMMARY

Connexins encode a family of structurally related proteins that participate in a dizzying array of functions, including developmental programming, tissue homeostasis, electrotonic coupling in excitable tissues, and response to pathologic stressors. At least one and often multiple connexin isoforms are expressed in almost every cell type during the life cycle of an organism, and therefore it is no surprise that mutations affecting connexin expression or function are associated with a remarkably diverse group of diseases. As detailed earlier, to date the most prominent manifestations of connexin mutations have been identified in patients with myelin disorders, deafness, skin diseases, cataracts, ODDD, and to a lesser extent, atrial arrhythmias. These syndromes arise from mutations in 11 of the 20 known connexins in the human genome. Given the remarkable pace of discovery with next generation sequencing and human genetics, it will not be surprising if additional disease states are linked to the remaining nine connexin isoforms in the coming days.

The mechanisms through which mutations in connexin genes produce disease are quite broad and include truncations that are rapidly degraded and produce haploinsufficiency; trafficking mutants that are retained in the endoplasmic reticulum or Golgi apparatus; mutants that form connexons but are incompetent to dock with their counterparts in adjacent cells; and mutants that form gap junction channels with diminished or absent permeability or altered gating properties. Indeed, examples of each of these molecular pathogenesis mechanisms are seen with connexin mutations. Perhaps somewhat more surprising is the relatively recent realization that hemichannels may contribute to human disease states, with gain-of-function mutations producing leaky channels and premature cell death. Looking to the future, clearly the hope is that a more complete understanding of molecular pathogenesis will lead to novel strategies targeting connexin channelopathies.

Acknowledgment

Research in the Fishman laboratory is supported by National Institutes of Health grants R01HL105983 and R01HL82727.

References

1. Abrams CK, Oh S, Ri Y, Bargiello TA. Mutations in connexin 32: the molecular and biophysical bases for the X-linked form of Charcot-Marie-Tooth disease. *Brain Res Brain Res Rev* 2000;**32**:203–14.

1a. Alexandrino F, Sartorato EL, Marques-de-Faria AP, Steiner CE. G59s mutation in the gjb2 (connexin 26) gene in a patient with bart-pumphrey syndrome. *Am J Med Genet A* 2005;**136**:282–4.

2. Altevogt BM, Kleopa KA, Postma FR, Scherer SS, Paul DL. Connexin29 is uniquely distributed within myelinating glial cells of the central and peripheral nervous systems. *J Neurosci* 2002;**22**:6458–70.

3. Anzini P, Neuberg DH, Schachner M, Nelles E, Willecke K, Zielasek J, et al. Structural abnormalities and deficient maintenance of peripheral nerve myelin in mice lacking the gap junction protein connexin 32. *J Neurosci* 1997;**17**:4545–51.

4. Beebe DC. Maintaining transparency: a review of the developmental physiology and pathophysiology of two avascular tissues. *Semin Cell Dev Biol* 2008;**19**:125–33.

5. Beltramello M, Piazza V, Bukauskas FF, Pozzan T, Mammano F. Impaired permeability to Ins(1,4,5)P3 in a mutant connexin underlies recessive hereditary deafness. *Nat Cell Biol* 2005;**7**:63–9.

6. Bergoffen J, Scherer SS, Wang S, Scott MO, Bone LJ, Paul DL, et al. Connexin mutations in X-linked Charcot-Marie-Tooth disease. *Science* 1993;**262**:2039–42.

7. Beyer EC, Ebihara L, Berthoud VM. Connexin mutants and cataracts. *Front Pharmacol* 2013;**4**:43.

8. Biancheri R, Rosano C, Denegri L, Lamantea E, Pinto F, Lanza F, et al. Expanded spectrum of Pelizaeus-Merzbacher-like disease: literature revision and description of a novel GJC2 mutation in an unusually severe form. *Eur J Hum Genet* 2013;**21**:34–9.

9. Common JE, Becker D, Di WL, Leigh IM, O'Toole EA, Kelsell DP. Functional studies of human skin disease- and deafness-associated connexin 30 mutations. *Biochem Biophys Res Commun* 2002;**298**:651–6.

10. Di WL, Common JE, Kelsell DP. Connexin 26 expression and mutation analysis in epidermal disease. *Cell Commun Adhes* 2001;**8**:415–8.

11. Di WL, Monypenny J, Common JE, Kennedy CT, Holland KA, Leigh IM, et al. Defective trafficking and cell death is characteristic of skin disease-associated connexin 31 mutations. *Hum Mol Genet* 2002;**11**:2005–14.

12. Dobrowolski R, Willecke K. Connexin-caused genetic diseases and corresponding mouse models. *Antioxid Redox Signal* 2009;**11**:283–95.

13. Ebihara L, Liu X, Pal JD. Effect of external magnesium and calcium on human connexin46 hemichannels. *Biophys J* 2003;**84**:277–86.

14. Gollob MH, Jones DL, Krahn AD, Danis L, Gong XQ, Shao Q, et al. Somatic mutations in the connexin 40 gene (GJA5) in atrial fibrillation. *N Engl J Med* 2006;**354**:2677–88.

15. Goodenough DA, Paul DL. Gap junctions. *Cold Spring Harb Perspect Biol* 2009;**1**:a002576.

16. Gottfried I, Landau M, Glaser F, Di WL, Ophir J, Mevorah B, et al. A mutation in GJB3 is associated with recessive erythrokeratodermia variabilis (EKV) and leads to defective trafficking of the connexin 31 protein. *Hum Mol Genet* 2002;**11**:1311–16.

16a. Grifa A, Wagner CA, D'Ambrosio L, Melchionda S, Bernardi F, Lopez-Bigas N, et al. Mutations in gjb6 cause nonsyndromic autosomal dominant deafness at dfna3 locus. *Nat Genet.* 1999;**23**:16–8.

17. Grosely R, Sorgen PL. A history of gap junction structure: hexagonal arrays to atomic resolution. *Cell Commun Adhes* 2013;**20**:11–20.

18. Harris AL. Emerging issues of connexin channels: biophysics fills the gap. *Q Rev Biophys* 2001;**34**:325–472.

19. Hopperstad MG, Srinivas M, Spray DC. Properties of gap junction channels formed by Cx46 alone and in combination with Cx50. *Biophys J* 2000;**79**:1954–66.

20. Kalcheva N, Qu J, Sandeep N, Garcia L, Zhang J, Wang Z, et al. Gap junction remodeling and cardiac arrhythmogenesis in a murine model of oculodentodigital dysplasia. *Proc Natl Acad Sci USA* 2007;**104**:20512–6.

21. Kelly JJ, Simek J, Laird DW. Mechanisms linking connexin mutations to human diseases. *Cell Tissue Res* 2015;**360**:701–21.

21a. Kelsell DP, Dunlop J, Stevens HP, Lench NJ, Liang JN, Parry G, Mueller RF, Leigh IM. Connexin 26 mutations in hereditary non-syndromic sensorineural deafness. *Nature* 1997;**387**:80–3.

22. Kleopa KA. The role of gap junctions in Charcot-Marie-Tooth disease. *J Neurosci* 2011;**31**:17753–60.

23. Koval M, Molina SA, Burt JM. Mix and match: investigating heteromeric and heterotypic gap junction channels in model systems and native tissues. *FEBS Lett* 2014;**588**:1193–204.

24. Kumar NM, Gilula NB. Cloning and characterization of human and rat liver cDNAs coding for a gap junction protein. *J Cell Biol* 1986;**103**:767–76.

25. Li TC, Kuan YH, Ko TY, Li C, Yang JJ. Mechanism of a novel missense mutation, p.V174M, of the human connexin31 (GJB3) in causing nonsyndromic hearing loss. *Biochem Cell Biol* 2014;**92**:251–7.

26. Macari F, Landau M, Cousin P, Mevorah B, Brenner S, Panizzon R, et al. Mutation in the gene for connexin 30.3 in a family with erythrokeratodermia variabilis. *Am J Hum Genet* 2000;**67**:1296–301.

27. Mackay D, Ionides A, Kibar Z, Rouleau G, Berry V, Moore A, et al. Connexin46 mutations in autosomal dominant congenital cataract. *Am J Hum Genet* 1999;**64**:1357–64.

28. Maeda S, Nakagawa S, Suga M, Yamashita E, Oshima A, Fujiyoshi Y, et al. Structure of the connexin 26 gap junction channel at 3.5 Å resolution. *Nature* 2009;**458**:597–602.

29. Maestrini E, Korge BP, Ocana-Sierra J, Calzolari E, Cambiaghi S, Scudder PM, et al. A missense mutation in connexin26, D66H, causes mutilating keratoderma with sensorineural deafness (Vohwinkel's syndrome) in three unrelated families. *Hum Mol Genet* 1999;**8**:1237–43.

30. Martin PE, Easton JA, Hodgins MB, Wright CS. Connexins: sensors of epidermal integrity that are therapeutic targets. *FEBS Lett* 2014;**588**:1304–14.

31. Martinez AD, Acuna R, Figueroa V, Maripillan J, Nicholson B. Gap-junction channels dysfunction in deafness and hearing loss. *Antioxid Redox Signal* 2009;**11**:309–22.

32. Meier C, Dermietzel R, Davidson KG, Yasumura T, Rash JE. Connexin32-containing gap junctions in Schwann cells at the internodal zone of partial myelin compaction and in Schmidt-Lanterman incisures. *J Neurosci* 2004;**24**:3186–98.

33. Mese G, Richard G, White TW. Gap junctions: basic structure and function. *J Invest Dermatol* 2007;**127**:2516–24.

33a. Mhatre AN, Weld E, Lalwani AK. Mutation analysis of connexin 31 (gjb3) in sporadic non-syndromic hearing impairment. *Clin Genet*. 2003;**63**:154–9.

34. Musil LS, Beyer EC, Goodenough DA. Expression of the gap junction protein connexin43 in embryonic chick lens: molecular cloning, ultrastructural localization, and post-translational phosphorylation. *J Membr Biol* 1990;**116**:163–75.

35. Oh SK, Choi SY, Yu SH, Lee KY, Hong JH, Hur SW, et al. Evaluation of the pathogenicity of GJB3 and GJB6 variants associated with nonsyndromic hearing loss. *Biochim Biophys Acta* 2013;**1832**:285–91.

36. Orthmann-Murphy JL, Enriquez AD, Abrams CK, Scherer SS. Loss-of-function GJA12/Connexin47 mutations cause Pelizaeus-Merzbacher-like disease. *Mol Cell Neurosci* 2007;**34**:629–41.

37. Paul DL. Molecular cloning of cDNA for rat liver gap junction protein. *J Cell Biol* 1986;**103**:123–34.

38. Paznekas WA, Boyadjiev SA, Shapiro RE, Daniels O, Wollnik B, Keegan CE, et al. Connexin 43 (GJA1) mutations cause the pleiotropic phenotype of oculodentodigital dysplasia. *Am J Hum Genet* 2003;**72**:408–18.

39. Paznekas WA, Karczeski B, Vermeer S, Lowry RB, Delatycki M, Laurence F, et al. GJA1 mutations, variants, and connexin 43 dysfunction as it relates to the oculodentodigital dysplasia phenotype. *Hum Mutat* 2009;**30**:724–33.

40. Rabionet R, Zelante L, Lopez-Bigas N, D'Agruma L, Melchionda S, Restagno G, et al. Molecular basis of childhood deafness resulting from mutations in the GJB2 (connexin 26) gene. *Hum Genet* 2000;**106**:40–4.

41. Retamal MA, Reyes EP, Garcia IE, Pinto B, Martinez AD, Gonzalez C. Diseases associated with leaky hemichannels. *Front Cell Neurosci* 2015;**9**:267.

42. Richard G, Smith LE, Bailey RA, Itin P, Hohl D, Epstein Jr EH, et al. Mutations in the human connexin gene GJB3 cause erythrokeratodermia variabilis. *Nat Genet* 1998a;**20**:366–9.

43. Richard G, White TW, Smith LE, Bailey RA, Compton JG, Paul DL, et al. Functional defects of Cx26 resulting from a heterozygous missense mutation in a family with dominant deaf-mutism and palmoplantar keratoderma. *Hum Genet* 1998b;**103**:393–9.

44. Roberts JD, Longoria J, Poon A, Gollob MH, Dewland TA, Kwok PY, et al. Targeted deep sequencing reveals no definitive evidence for somatic mosaicism in atrial fibrillation. *Circ Cardiovasc Genet* 2015;**8**:50–7.

45. Sanchez HA, Villone K, Srinivas M, Verselis VK. The D50N mutation and syndromic deafness: altered Cx26 hemichannel properties caused by effects on the pore and intersubunit interactions. *J Gen Physiol* 2013;**142**:3–22.

45a. Serrano Castro PJ, Naranjo Fernandez C, Quiroga Subirana P, Payan Ortiz M. Vohwinkel syndrome secondary to missense mutation d66h in gjb2 gene (connexin 26) can include epileptic manifestations. *Seizure* 2010;**19**:129–31.

46. Shibayama J, Paznekas W, Seki A, Taffet S, Jabs EW, Delmar M, et al. Functional characterization of connexin43 mutations found in patients with oculodentodigital dysplasia. *Circ Res* 2005;**96**:e83–91.

47. Shiels A, Mackay D, Ionides A, Berry V, Moore A, Bhattacharya S. A missense mutation in the human connexin50 gene (GJA8) underlies autosomal dominant "zonular pulverulent" cataract, on chromosome 1q. *Am J Hum Genet* 1998;**62**:526–32.

48. Sjostrand FS, Andersson-Cedergren E, Dewey MM. The ultrastructure of the intercalated discs of frog, mouse and guinea pig cardiac muscle. *J Ultrastruct Res* 1958;**1**:271–87.

49. Smith FJ, Morley SM, McLean WH. A novel connexin 30 mutation in Clouston syndrome. *J Invest Dermatol* 2002;**118**:530–2.

50. Sohl G, Willecke K. Gap junctions and the connexin protein family. *Cardiovasc Res* 2004;**62**:228–32.

51. Solan JL, Lampe PD. Connexin43 phosphorylation: structural changes and biological effects. *Biochem J* 2009;**419**:261–72.

51a. van Steensel MA, van Geel M, Nahuys M, Smitt JH, Steijlen PM. A novel connexin 26 mutation in a patient diagnosed with keratitis-ichthyosis-deafness syndrome. *J Invest Dermatol* 2002;**118**:724–7.

52. Stong BC, Chang Q, Ahmad S, Lin X. A novel mechanism for connexin 26 mutation linked deafness: cell death caused by leaky gap junction hemichannels. *Laryngoscope* 2006;**116**:2205–10.

53. Thibodeau IL, Xu J, Li Q, Liu G, Lam K, Veinot JP, et al. Paradigm of genetic mosaicism and lone atrial fibrillation: physiological characterization of a connexin 43-deletion mutant identified from atrial tissue. *Circulation* 2010;**122**:236–44.

54. Tong JJ, Sohn BC, Lam A, Walters DE, Vertel BM, Ebihara L. Properties of two cataract-associated mutations located in the NH2 terminus of connexin 46. *Am J Physiol Cell Physiol* 2013;**304**:C823–32.

55. Uhlenberg B, Schuelke M, Ruschendorf F, Ruf N, Kaindl AM, Henneke M, et al. Mutations in the gene encoding gap junction protein alpha 12 (connexin 46.6) cause Pelizaeus-Merzbacher-like disease. *Am J Hum Genet* 2004;**75**:251–60.

56. Unger VM, Kumar NM, Gilula NB, Yeager M. Three-dimensional structure of a recombinant gap junction membrane channel. *Science* 1999;**283**:1176–80.

57. Wang HL, Chang WT, Li AH, Yeh TH, Wu CY, Chen MS, et al. Functional analysis of connexin-26 mutants associated with hereditary recessive deafness. *J Neurochem* 2003;**84**:735–42.

58. White TW, Bruzzone R, Goodenough DA, Paul DL. Mouse Cx50, a functional member of the connexin family of gap junction proteins, is the lens fiber protein MP70. *Mol Biol Cell* 1992;**3**:711–20.

59. Wingard JC, Zhao HB. Cellular and deafness mechanisms underlying connexin mutation-induced hearing loss - a common hereditary deafness. *Front Cell Neurosci* 2015;**9**:202.

59a. Yang JJ, Huang SH, Chou KH, Liao PJ, Su CC, Li SY. Identification of mutations in members of the connexin gene family as a cause of nonsyndromic deafness in taiwan. *Audiol Neurootol* 2007;**12**:198–208.

60. Yang YQ, Liu X, Zhang XL, Wang XH, Tan HW, Shi HF, et al. Novel connexin40 missense mutations in patients with familial atrial fibrillation. *Europace* 2010a;**12**:1421–7.

61. Yang YQ, Zhang XL, Wang XH, Tan HW, Shi HF, Jiang WF, et al. Connexin40 nonsense mutation in familial atrial fibrillation. *Int J Mol Med* 2010b;**26**:605–10.

61a. Zhang XJ, Chen JJ, Yang S, Cui Y, Xiong XY, He PP, et al. A mutation in the connexin 30 gene in chinese han patients with hidrotic ectodermal dysplasia. *J Dermatol Sci.* 2003;**32**:11–7.

62. Zonta F, Polles G, Sanasi MF, Bortolozzi M, Mammano F. The 3.5 angstrom X-ray structure of the human connexin26 gap junction channel is unlikely that of a fully open channel. *Cell Commun Signal* 2013;**11**:15.

Complexity of Molecular Genetics in the Inherited Cardiac Arrhythmias

N. Lahrouchi[1], A.A.M. Wilde[1,2]

[1]Amsterdam Medical Center, University of Amsterdam, Amsterdam, The Netherlands; [2]Princess Al-Jawhara Al-Brahim Centre of Excellence in Research of Hereditary Disorders, Jeddah, Kingdom of Saudi Arabia

INTRODUCTION

The cardiac channelopathies encompass a group of heritable disorders that are associated with arrhythmia and sudden cardiac death (SCD) in the presence of a structurally normal heart. These disorders are caused by mutations in genes that encode cardiac ion channel subunits or proteins that interact with, and regulate, ion channels (Fig. 14.1). These genetic variants thereby disrupt the delicate balance of ionic currents across the plasma membrane or the sarcoplasmic reticulum (SR) and predispose to the occurrence of life-threatening arrhythmias (Fig. 14.2). The identification of the causal mutation in the proband allows for genetic testing in family members that carry the disease-causing mutation and are at an increased risk for arrhythmia and SCD.[1] The presently known cardiac channelopathies include long QT syndrome (LQTS), short QT syndrome (SQTS), sinus node disease (SND), cardiac conduction disease (CCD), early repolarization syndrome (ERS), Brugada syndrome (BrS), and catecholaminergic polymorphic ventricular tachycardia (CPVT).[2] In 1995 and 1996, the genetic and molecular underpinnings of the LQTS started to be identified.[3–5] Since these milestone discoveries the genetic advances in recent years have provided much insight into the genetics and mechanisms underlying these disorders. In this chapter the genetic substrate and the increasing insight into

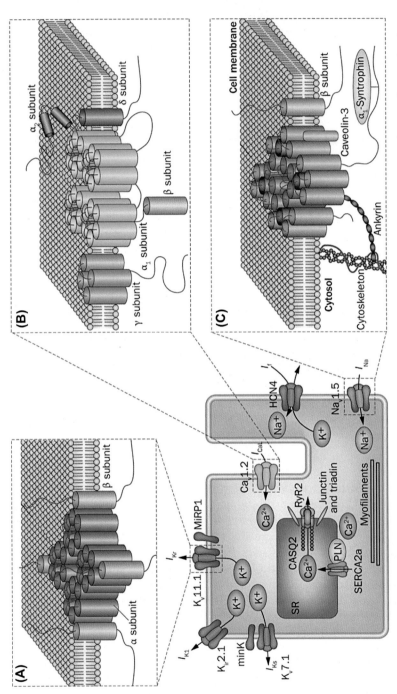

FIGURE 14.1 **The main proteins involved in the channelopathies.** Schematic representation of a cardiomyocyte showing the proteins that are involved in the pathogenesis of the channelopathies (A) potassium (I_{Kr}), (B) calcium (I_{CaL}), and (C) sodium (I_{Na}) channel subunits and structures are shown. *SR*, sarcoplasmic reticulum; $_{Na}$SERCA2a, sarcoplasmic/endoplasmic reticulum calcium ATPase 2a; $_{Ca}$RyR, ryanodine receptor 2; *CASQ2*, calsequentrin-2; *PLN*, cardiac phospholamban.*CASQ2PLN*

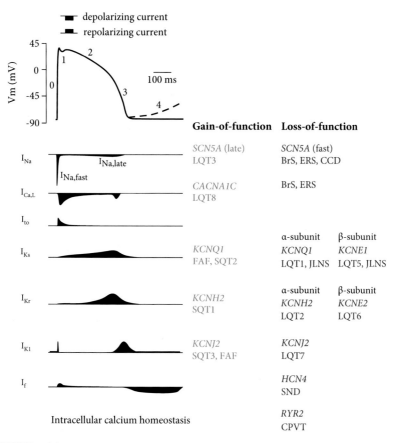

FIGURE 14.2 **The ventricular action potential.** Schematic representation of the ionic currents that contribute to the ventricular action potential. Disorders associated to mutations in genes encoding for the respective ion channels are listed according to their underlying mutational effect (left: gain-of-function mutations, right: loss-of-function mutations). *BrS,* Brugada syndrome; *ERS,* early repolarization syndrome; *CCD,* cardiac conduction disease; *LQT,* long QT syndrome; *SQT,* short QT syndrome; *JLNS,* Jervell and Lange-Nielsen syndrome; *FAF,* familial atrial fibrillation; *SND,* sinus node disease; *CPVT,* catecholaminergic polymorphic ventricular tachycardia.

the genetic complexity of the cardiac channelopathies are discussed. We will furthermore discuss the role of genetic testing, including next-generation sequencing (NGS) and genotype–phenotype correlations. We will not address clinical recommendations for diagnosis and genetic testing in detail as these have been gathered in consensus guidelines elsewhere.[2,6]

GENETIC ARCHITECTURE OF THE CARDIAC CHANNELOPATHIES

The cardiac channelopathies are considered Mendelian disorders that present with familial inheritance and wherein a rare disease-causing variant mediates the risk of arrhythmia and SCD. Using classical linkage analysis in these pedigrees, multiple genes harboring rare variants with a large effect size have been identified for a number of these disorders.[3-5] The genetic basis of the cardiac channelopathies is heterogeneous as mutations in different genes predispose to the same cardiac phenotype. In addition, these disorders are characterized by large allelic heterogeneity in that a large number of different mutations within one gene may lead to the same clinical manifestations. The observed mutations are mostly unique (ie, "private") to one or a limited number of families.[7] Evidence for causality of an identified mutation is obtained by virtue of the novelty of the genetic defect (not previously identified in large samples of the general population), by segregation analysis within the family (ie, mutation carriers express the disease whereas nonmutation carriers are unaffected) or by functional studies (usually in the research setting at tertiary clinical centers). The mutation detection rate varies across the disorders but is commonly higher when the disorder is familial in comparison to isolated cases.[8] Even though negative selection has kept the allele frequency of most of these mutations low, founder mutations have been described in extended families in South Africa,[9] Finland,[10] Sweden,[11] and the Netherlands.[12] The genetic breakthroughs in the cardiac channelopathies have advanced from basic scientific discovery to an established role in patient care contributing to diagnosis, prognosis, and therapeutic decision making.[6] Yet, despite these achievements, the clinical management of patients with cardiac channelopathies is complicated by the large variability in disease expression (eg, the extent of QTc prolongation and occurrence of arrhythmia in LQTS) among mutation carriers. In some cases, family members who carry the familial disease-causing mutation do not display clinical signs of the disease (this observation is referred to as reduced penetrance). These observations indicate that alongside the primary genetic defect, other genetic and nongenetic factors modulate the expression of the disease.[7] Common nongenetic factors are gender,[13] age,[14] and the use of drugs.[15] Genetic factors include compound mutations (ie, the inheritance of more than one mutation) or the specific type and location of the mutation, all of which have been shown to explain severity of disease in some patients.[16] In addition to the coinheritance of a rare variant with a large effect size, the role of common genetics variants such as single nucleotide polymorphisms (SNPs) has been explored in a number of studies.[17-21] It is hoped that as knowledge concerning the identity of such modulatory genetic factors continues to increase, this may ultimately be employed

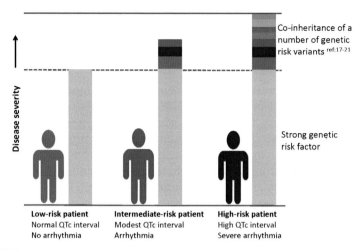

FIGURE 14.3   **Genetic risk variants in long QT syndrome.** Schematic representation of the coinheritance of a number of genetic risk variants in addition to a strong genetic risk factor in LQTS (ie, LQTS-causing mutations). The identification of these modulatory genetic variants may ultimately be employed for improved risk stratification of LQTS patient, as illustrated in the figure.

for improved risk stratification (Fig. 14.3). The application of the genome-wide association study (GWAS) approach in the cardiac channelopathies may play an important role in the identification of common genetic variants that affect disease susceptibility and severity.[22]

THE CARDIAC ACTION POTENTIAL

Ion channels are transmembrane proteins that selectively allow for ionic currents to cross the membrane and orchestrate the cardiac action potential (AP, Fig. 14.2). The ventricular AP starts with phase 0 (upstroke) brought about by the activation of the voltage-gated sodium channels which leads to membrane permeability for sodium ions (I_{Na}). The rapid influx of positively charged sodium ions (Na^+) into the negatively charged cardiomyocyte depolarizes the membrane potential to positive voltages. This is followed by phase 1 repolarization, brought about by the efflux of potassium ions (K^+) by means of the transient outward K^+ current (I_{to}). In the next phase (phase 2) voltage-gated L-type calcium ion (Ca^{2+}) channels are activated and lead to an inward movement of calcium ions ($I_{Ca,L}$) establishing a delicate balance of outward K^+ and inward Ca^{2+} currents. This phase is generally referred to as the plateau phase. Calcium ions have a crucial role in excitation–contraction coupling, a process whereby an electrical stimulus is converted into mechanical activation.[23] The activation of two outward K^+ currents (I_{Kr} and I_{Ks}) shifts the balance to an outward movement of positively charged K$^+$ that is the main effector of phase 3 repolarization. In addition and importantly, the inward rectifying K$^+$ current (I_{K1}) contributes to the terminal portion of phase 3. This repolarizes the membrane potential toward the negative resting membrane potential of approximately −85 mV referred to as phase 4. In this phase I_{K1} stabilizes the resting membrane potential. This completes the AP cycle until the subsequent activation of the sodium channels by the next electrical impulses from neighboring cells. Mutations in genes that encode the ion channels underlying the AP, or their interacting proteins, can have a profound effect on channel function and thereby lead to arrhythmia and the characteristic features associated to the corresponding channelopathy (eg, ECG morphology and triggers for arrhythmia, Fig. 14.2).

DISORDERS RELATED TO POTASSIUM CHANNELS AND THEIR ASSOCIATED PROTEINS

Mutations in genes encoding potassium channels have been associated to three distinct cardiac disorders namely the LQTS, SQTS, and familial atrial fibrillation (AF). These mutations can induce a gain or loss of channel function. Loss-of-function mutations lead to LQTS, whereas gain-of-function mutations have been associated to SQTS and AF.

Long QT Syndrome

The congenital LQTS is a cardiac channelopathy that is characterized by QT-interval prolongation on the ECG in association with syncope and SCD due to *torsades des pointes* ventricular tachycardia (TdP).[24] The disease is generally inherited in an autosomal dominant trait (previously described as the Romano-Ward syndrome) and has been associated to mutations in 15 different genes.[25] In the mid-1990s the first disease-causing genes, namely *KCNQ1*, *KCNH2*, and *SCN5A*, were discovered in families affected by the LQTS. Together, these three genes account for ~90% of genotype-positive LQTS patients.[6,26] LQTS is a cardiac repolarization disease which is most commonly caused by repolarization abnormalities due to mutations in potassium channels (*KCNQ1*, *KCNH2*, *KCNE1*, *KCNE2*, *KCNJ2*, and *KCNJ5*). Mutations in *KCNQ1* are the most frequent cause of LQTS and accounts for 40–55% of cases. It is referred to as long QT syndrome type 1 (LQT1) and arises from loss-of-function mutations in *KCNQ1*, encoding the α-subunit of the slowly activating delayed rectifier current (I_{Ks}).[5] *KCNE1* encodes for the β-subunit of the potassium channel I_{Ks} and mutations in this gene are associated to LQT5. This channel has an important role in the adaptation of the QT-interval during increased heart rates. In addition, homozygous or compound heterozygous mutations in $KCNQ1^{27}$ and $KCNE1^{28}$ cause a clinical autosomal recessive variant of LQTS named the Jervell and Lange-Nielsen syndrome that, besides severely prolonged QTc-interval, additionally manifests with deafness. However, homozygous or compound heterozygous mutations in the absence of deafness have also been reported, and these cases are referred to as autosomal recessive LQTS.[16,29] Loss-of-function mutations in $KCNH2$ (also known as HERG), encoding the α-subunit of the rapidly activating delayed rectifier current (I_{Kr}), account for 30–45% of cases (LQT2).[3] $KCNE2$ encodes for the β-subunit of the I_{Kr} potassium channel, and mutations in this gene are associated with LQT6.[30] I_{Kr} and I_{Ks} are both part of the delayed rectifier current I_K which has an important role in the plateau phase of cardiac repolarization. The Anderson-Tawil syndrome (LQT7) is a distinct clinical variant of the LQTS presenting with U-wave accentuation (not always easy to distinguish from QT-interval prolongation), dysmorphic features, and hypokalemic periodic paralysis. This disorder is caused by loss-of-function mutations in the potassium channel gene $KCNJ2$.[31] This gene encodes a pore-forming subunit of inwardly rectifying potassium channels (I_{K1}).[32,33] LQTS has been associated to several genes that interact with and/or regulate channel function. Patients with LQT4 present with a prolonged QT-interval and in some cases paroxysmal AF, sinus bradycardia, and a CPVT-like phenotype. Genetic defects in $ANK2$, an adapter protein, have been linked to this type of LQTS.[34,35] In 2007, a mutation (S1570L) in the $AKAP9$ gene has been reported in a family affected by the LQTS (LQT11).[36] This mutation was found to influence the assembly of $AKAP9$ (Yotiao) with the I_{Ks} potassium channel α subunit and thereby affects the sympathetic regulation of the cardiac AP.

Since the earliest descriptions of the LQTS, it has been recognized that variable disease expressivity and reduced penetrance are present among mutation carriers.[37] Even though family members who carry the causal familial LQTS mutation may not display characteristics of the disease (eg, QTc < 440 m), they still have a tenfold increased risk of cardiac events in comparison to noncarriers.[13,38] Therefore, even when clinical risk factors are absent, mutation analysis is essential to identify family members that are at risk. Established risk predictors in LQTS are previous cardiac events including syncope, *torsades des pointes* or aborted SCD, a QTc interval >500 ms, and age (prepubertal males and adult females being most at risk).[13,39–41] Genotype–phenotype studies have established specific characteristics for some of the LQTS subtypes.[42,43] Patients with LQT1 have broad-based T-waves on their ECG and cardiac events typically occur during exercise, especially during swimming and diving (Fig. 14.4, panel A). Patients with LQT2 display low-amplitude and/or notched T-waves, and arrhythmia may be triggered by sudden auditory stimuli (Fig. 14.4, panel B).[44] In some cases, disease severity can be explained by the inheritance of more than one mutation in one or more LQTS genes, which has

FIGURE 14.4 **Characteristic ECG recordings of LQT1, 2, and 3.** Electrocardiograms displaying the characteristics for long QT syndrome subtypes 1–3 at baseline. LQTS type 1 typically presents with broad-based T-waves (panel A), LQTS type 2 with low-amplitude T-waves (panel B), LQTS type 3 is characterized by a long and flat ST segment followed by a peaked T-wave (panel C).

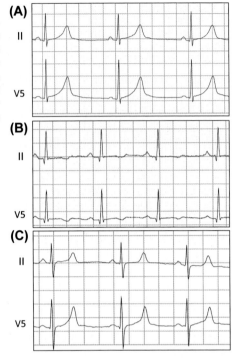

been established to occur in 5–10% of LQTS probands.[16] These patients do not only have longer QT-intervals but are also more susceptible to arrhythmias. Features of the primary mutation such as mutation type and location may impact on the severity of the biophysical defect and modulate disease severity. These data may lead to *intragenic risk stratification* (ie, the characteristics of the primary mutation are used to predict risk). This notion has been supported by studies conducted in patients with *KCNQ1* mutations; both mutations at transmembrane regions (versus C-terminus mutations) and mutations with a dominant-negative effect (versus those resulting in haploinsufficiency) were found to be associated with a higher risk of cardiac events.[45] Furthermore, in patients with LQTS due to a *KCNH2* mutation, those with missense mutations in the pore region of the channel appear to display the highest risk of life-threatening arrhythmias.[46] Despite these developments, their use in clinical practice is not fully explored.[7]

The modulating effects of SNPs in candidate genes (including *KCNE1*, *KCNQ1, and AKAP9)* on QTc interval and the occurrence of symptoms have been studied in LQTS patients.[20,21,47,48] In line with our expectations, common genetic variants located at genetic loci associated to the LQTS have been identified to modulate the QT-interval in the general population.[49] This observation highlights the overlap of common and rare genetic variants in the determination of QT-interval duration. Common variants modulating ECG indices (heart rate, PR, QRS, and QT-interval) identified through GWAS in the general population could theoretically have an effect on disease severity in addition to the primary mutation in the cardiac channelopathies.[49–52] This hypothesis is supported by studies were common variants at *KCNQ1*[19] and *NOS1AP*[17,18] identified through GWAS of the QT-interval in the general population have been shown to affect the extent of QT-interval prolongation and the risk of cardiac events in patients with LQTS.[20] In addition to these studies, common genetic variants in the 3′ untranslated (3′UTR) region of the *KCNQ1* have been shown to affect disease severity in an allele specific manner. Patients who carried the *KCNQ1* variants on the allele harboring the LQTS causing mutation (*in cis*) had a less severe phenotype in comparison to patients who carried these variants on the other allele (*in trans*).[21]

Short QT Syndrome

The SQTS is characterized by a short QT-interval (QTc < 350 ms) on the ECG, a predisposition to supraventricular and ventricular arrhythmia, and a high incidence of SCD.[53–55] It is a genetically heterogeneous disorder which is inherited as an autosomal dominant trait. Three potassium channel genes, also implicated in the LQTS, have been associated with SQTS: *KCNH2 (SQT1)*, *KCNQ1 (SQT2)*, and *KCNJ2 (SQT3)*.[56–58] These mutations lead to *gain-of-function* defects in the encoded channel and thereby lead to a hastened cardiac repolarization which is in contrast to the *loss-of-function* mutations in these genes that prolong repolarization and lead to LQTS.[56–60] The yield of genetic testing is particularly low as an underlying mutation can only be identified in around 14% of SQTS patients.[55]

Atrial Fibrillation

Despite the frequent manifestation of AF in the general population, the occurrence of AF at young age in the absence of structural heart disease and with clear inheritance across the generations is rare. Due to the scarcity of these AF pedigrees, classical approaches, such as linkage analysis, have not been as successful as for the LQTS and other inherited disorders. In 2003, a causative gain-of-function mutation in *KCNQ1* (S140G) has been identified in a large pedigree of Chinese descent with persistent AF.[61] The discovery of a mutation in *KCNQ1*, a gene previously associated to the LQTS, boosted the use of candidate-based approaches in small AF families, and through these studies several variants have been identified in genes encoding potassium channels (*KCNA5*, *KCND3*, and *KCNJ2*) and their subunits (*KCNE1*, *KCNE2*, *KCNE3*, and *KCNE5*).[62] The interpretation of the pathogenicity for some of these variants, however, is challenging, and evidence for causality is not always robust. In 1997, a genetic locus (10q22–q24) for AF was identified with the use of several families that presented with early onset AF inherited as an autosomal dominant trait.[63] Even though the focus of this section is on familial AF, it has been well demonstrated that a heritable component underlies AF in the general population.[62,64] To identify the underlying heritability, beyond the use of AF pedigrees with a clear heritable component, GWAS has been used in large AF cohorts. This approach has led to the discovery of several susceptibility loci which among others includes a region intronic to the *KCNN3* gene that encodes for a potassium channel.[62,65,66]

DISORDERS RELATED TO SODIUM CHANNELS AND THEIR ASSOCIATED PROTEINS

Mutations in *SCN5A*, encoding the pore-forming subunit of the cardiac sodium channel ($Na_v1.5$) have been associated to several hereditary cardiac disorders, including LQTS, BrS, SND, CCD, familial AF, and dilated cardiomyopathy. Mutations in *SCN5A* do not only lead to a variety of cardiac disorders but may also give rise to so-called overlap syndromes, wherein single families or patients are affected by multiple cardiac arrhythmia syndromes (eg, LQT3, BrS, and CCD) despite the identical genetic mutation in these patients.[67–69]

Brugada Syndrome

The BrS is associated with syncope and cardiac arrest due to polymorphic ventricular tachycardia (VT) that degenerates into ventricular fibrillation (VF).[70] Most patients develop cardiac events at rest or during sleep.[2] Symptoms mostly occur after the fourth decade of life and predominately affect males.[71] According to the current guidelines the disorder is diagnosed when a type I pattern is present on the ECG: a coved ST-segment and J-point elevation ≥0.2 mV, in ≥1 precordial lead positioned in the 2nd, 3rd, or 4th intercostal space (Fig. 14.5, panel A). For diagnostic purpose, this ECG pattern may either occur spontaneously or after a provocative drug test with intravenous administration of a sodium channel blocking agent. The typical ECG pattern can be present intermittently or in all recordings in some, whereas in others the pattern may only be evoked by sodium channel blockers or fever.[71] Loss-of-function mutations in the *SCN5A* gene that encodes the α-subunit of the cardiac voltage-gated sodium channel ($Na_v1.5$) were the first described mutations linked to BrS and lead to a reduction of inward sodium (I_{Na}) current.[72] According to a

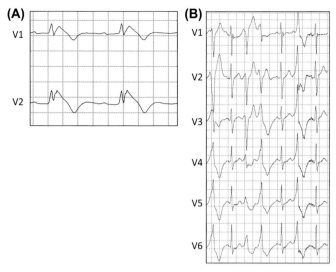

FIGURE 14.5 **Characteristic ECG recordings of BrS and CPVT.** Typical type I Brugada syndrome electrocardiogram with right precordial ST-segment elevation followed by negative T-waves in lead V1 and V2 (panel A), onset of polymorphic ventricular ectopy during exercise in catecholaminergic polymorphic ventricular tachycardia (panel B).

mutational analysis in a large cohort of unrelated BrS patients, *SCN5A* mutations are identified in approximately 16% of patients.[73] In a study conducted in 150 BrS cases, mutations in *SCN10A*, a neuronal sodium channel gene encoding $Na_v1.8$, have been reported to account for 16% of patients with BrS.[74] However, two subsequent studies were not able to reproduce this observation and concluded that rare variants in *SCN10A* are not implicated in the BrS.[75,76] Mutations in β-subunits of $Na_v1.5$, encoded by *SCN1B* and *SCN3B*, and the channel interacting protein *GPD1L* have been associated to BrS in a minority of patients. These mutations lead to a reduction of the inward sodium (I_{Na}) current.[77,78] Despite the low yield of genetic testing, mutation analysis can contribute to disease confirmation, especially in clinically uncertain cases.[2,6] Cascade testing of family members can identify those who need preventive measures (eg, sodium channel blockers or antipyretic treatment during fever) and clinical follow-up (for development of cardiac conduction disease or cardiac events). In the search for genotype–phenotype relations, one study identified that mutations that lead to a premature truncated protein (frame shift, stop codon) were associated with a higher incidence of syncope and prolongation of PR and QRS interval, compared with patients carrying missense mutations.[79] Despite these developments the majority of BrS remains genetically elusive and in addition, in many families the BrS phenotype often does not segregate with mutation carriership (eg, some affected family members do not carry the familial mutation) suggesting a more complex form of inheritance.[80] GWAS of electrocardiographic traits conducted in the general population has identified multiple common genetic variants at the *SCN5A* and *SCN10A* locus that modulate PR interval and QRS duration.[81–84] GWAS conducted in BrS patients uncovered three independent loci associated to disease probability.[22] Two of these loci (rs10428132 and rs11708996) reside at the *SCN5A/SCN10A* locus and have previously been associated to PR and QRS duration in the general population.[81–84] The third locus (rs9388451) resides near *HEY2*, which encodes the transcriptional repressor hairy/enhancer-of-split related with YRPW motif protein. Functional studies in mice indicated a role for this gene in regulation of *SCN5A* expression and electrical activity.[22] This study contributed to the hypothesis that a more complex genetic architecture underlies BrS and emphasized that common genetic variants that modulate ECG measures in the general population may impact on disease susceptibility to the inherited cardiac channelopathies.

Long QT Syndrome

Mutations in *SCN5A* cause LQT3 by gain-of-function defects that disrupt the fast inactivation of the sodium channel which leads to a persistent inward sodium current (I_{Na}) during the plateau phase of the AP and thereby prolong repolarization.[85] Other described mechanisms for LQT3 include an increase in "window current", faster recovery from inactivation, slowed inactivation, and increased peak I_{Na} density.[86] Most LQTS causing mutations in *SCN5A* are missense mutations and are found in ~10% of LQTS probands.[87,88] Some mutations are inframe deletions, among which the "classical" *SCN5A* ΔKPQ mutation.[85] LQT3 patients often display bradycardia with long isoelectric ST-segments and experience most of their cardiac events (~40%) during rest or sleep (Fig. 14.4, panel C).[42] Gain-of-function mutations in *CAV3*,[89] *SCN4B*,[90] and *SNTN1*[91] have been associated to respectively LQT9, LQT10, and LQT12. The mutations in these genes lead to an increase in late I_{Na} resembling the mechanism underlying LQT3.

Cardiac Conduction Disease

Cardiac conduction disorders are characterized by an impairment in cardiac conduction at the level of the atrium, atrioventricular (AV) node, and/or the ventricles. The ECG of affected patients can display prolongation of the P-wave or PR interval, QRS widening, bundle branch blocks, and AV-block. In 1999, through linkage analysis, a splice donor site mutation in exon 22 of the *SCN5A* gene was identified in a large pedigree with several members affected by multiple cardiac conduction defects.[92] In 2001, functional testing of a mutation that caused familial isolated CCD showed that the familial mutation affected biophysical properties of the ion channel.[93] GWAS of cardiac conduction parameters (eg, PR and QRS interval) in the general population has identified several common genetic variants at the *SCN5A* locus that have an effect on parameters of cardiac conduction.[81–84]

Atrial Fibrillation

AF has been associated to both gain-of-function and loss-of function mutations in *SCN5A*.[94–96] Loss-of-function mutations in the β-subunits (encoded by *SCN1B-4B*) have also been described for AF.[62] An overlap syndrome consisting of AF and LQT3 linked to an *SCN5A* mutation has been found in a single family.[97]

DISORDERS RELATED TO CALCIUM CHANNELS AND THEIR ASSOCIATED PROTEINS

Mutations in calcium channels and their interacting proteins have been associated with three disorders, namely, CPVT, Timothy syndrome (TS or LQT8), and a disorder presenting with both BrS and a shorter than normal QT-interval. Disease-causing mutations in these genes either affect the intracellular (*RYR2* and *CASQ2* for CPVT) or the transmembrane part of calcium handling (eg, *CACNA1C*, *CACNB2*, and *CACANA2D1* for LQT8 and BrS with short QT-interval).[98]

Catecholaminergic Polymorphic Ventricular Tachycardia

CPVT is an inherited arrhythmia syndrome characterized by the onset of life-threatening arrhythmias triggered by physical or emotional stress (ie, sudden increase in sympathetic tone).[99] Patients present with an unre-markable resting ECG but exercise-stress testing or Holter recording can reveal distinctive bidirectional or polymorphic VT (Fig. 14.5, panel B).[2] CPVT is a highly lethal disorder with a high incidence of cardiac events (79%) and SCD (30%) before the age of 40 among untreated patients.[100,101] Mutations in *RYR2*[102] encoding the ryanodine receptor (or SR Ca^{2+} chan-nel) and *CASQ2*[103] encoding for the calsequestrin-2 protein have been asso-ciated to respectively an autosomal dominant and recessive form of CPVT. These genes have a crucial role in the control of calcium release from the SR.[98] Approximately 60% of CPVT patients with a typical diagnosis carry a mutation in the *RYR2* gene and the majority of these mutations are clus-tered around specific regions within the channel.[102,104,105] Compound het-erozygous or homozygous mutations in *CASQ2* underlie the autosomal recessive form of CPVT.[106–108] *RYR2* and *CASQ2* mutations cause diastolic calcium leakage from the SR resulting in an intracellular calcium over-load and the development of delayed afterdepolarizations and triggered arrhythmias.[98] Two genes that encode for calcium-handling proteins, namely *TRDN* and *CALM1*, have been associated to CPVT respectively in a candidate gene and linkage approach.[109,110] *KCNJ2* and *ANKB* mutations have been identified in a small proportion of patients.[35,111] The yield of genetic testing for CPVT is high and mutation analysis has an important role in disease confirmation in probands. First-degree relatives should be offered genetic testing when a pathogenic mutation is identified in the proband.[6] Lethal arrhythmias can be the first expression of CPVT and in approximately 15% of autopsy negative unexplained SCD cases, CPVT is found as the underlying genetic substrate.[101,112] Therefore, genetic evalu-ation of family members can facilitate presymptomatic diagnosis and ini-tiation of preventive measures, such as beta blocker therapy.[6] In the search for genotype–phenotype correlations, preliminary data suggest that *RYR2* mutations in the C-terminal portion of the protein may be associated to a higher risk of arrhythmias in comparison to mutations in the N-terminal domain.[113]

Long QT Syndrome

TS or LQT8 is a rare multisystem disorder presenting with both cardiac as noncardiac abnormalities (Fig. 14.6). The syndrome typically presents with fetal bradycardia, 2:1 AV-block, marked prolongation of the QT-inter-val, and alternating T-waves.[114,115] Besides these ECG abnormalities, LQT8 may present with congenital heart defects, cardiomyopathy, and most patients develop severe ventricular arrhythmia (eg, polymorphic VT and VF) in early infancy. Extra cardiac abnormalities include syndactyly, autism, malignant hypoglycemia, and an abnormal immune system. The majority of LQT8 cases are caused by the same gain-of-function mutation (G406R) in *CACNA1C* which encodes the L-type Ca^{2+} channel $Ca_v1.2$.[114] Due to the high lethality of this disorder, the majority of patients do not reach repro-ductive age. Most cases are the result of de novo mutations but parental mosaicism has been described in single cases and in one family with two affected siblings.[114,116] Whole exome sequencing using parent–child trios has been successful in the discovery of the genetic underpinnings in two unrelated patients who displayed severe prolongation of the QT-interval and arrhythmia.[117] Through this analysis the authors found mutations in *CALM1* and *CALM2*, encoding the ubiquitously expressed Ca^{2+}-signaling protein calmodulin (CaM), that were present in both patients but absent in their unaffected parents (ie, de novo mutation). Screening of *CALM1*, *CALM2*, and *CALM3* in a cohort of genotype-negative LQTS cases identi-fied two additional mutations in *CALM1*. An independent study of geno-type-negative LQTS probands, of whom two presented with overlapping

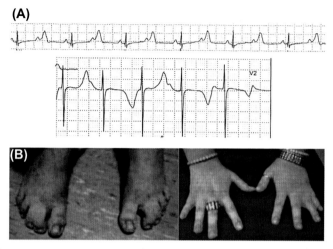

FIGURE 14.6   **Timothy syndrome (LQT8).** 2:1 atrioventricular block and evident T-wave alternans in an electrocardiogram from a patient with Timothy syndrome (panel A), two Timothy syndrome patients with syndactyly (panel B).

features of both CPVT as LQTS, described 5 de novo mutations in *CALM2*.[118] In addition, *CALM1* has previously been associated to a CPVT-like phenotype in a large Swedish pedigree through the use of linkage analysis.[109] These observations suggest that CaM mutations can underlie LQTS, CPVT, and an overlap syndrome that includes characteristics of both CPVT and LQTS. Most of the CaM mutations are found in conserved residues that are known to have a role in Ca^{2+} binding, and biochemical studies have indeed revealed decreased Ca^{2+} binding affinity.[109,117] CaM binds to RyR2 channels in the cardiomyocyte and impairs channel activity by altering ion channel kinetics.[119] A study found that CPVT-CaM mutations lead to Ca^{2+} handling abnormalities, which were not seen for LQTS-CaM mutations, that are similar to those found in CPVT associated to *CASQ2* and *RYR2*.[120] Functional studies showed that LQTS-CaM mutations affect the inactivation of the L-type calcium currents, whereas CPVT-CaM mutations had no or a smaller effect on this current.[121,122] Interestingly, a subsequent *CALM1* mutation has been found in a family affected by idiopathic VF through an exome sequencing approach.[123]

DISORDERS RELATED TO HYPERPOLARIZATION-ACTIVATED CYCLIC NUCLEOTIDE–GATED CHANNELS

Mutations in *HCN4* encoding for the hyperpolarization-activated cyclic nucleotide-gated channel (HCN) have been identified for SND by a candidate-based approach.[124] This channel mediates the cardiac hyperpolarization-activated "pacemaker current" also referred to as "funny current" (If).[125] *HCN4* mutations have been linked to sinus node dysfunction in additional cases/families, and these studies have been reviewed extensively elsewhere.[126] Two independent studies published concurrently described loss-of-function mutations in *HCN4* in families displaying left ventricular non compaction and sinus bradycardia.[127,128] Interestingly, through GWAS in the general population, common genetic variants at the *HCN4* locus have been shown to modulate heart rate.[129]

OTHER DISORDERS

Other mostly rare disorders have also been associated to mutations in genes encoding for ion channels. Loss-of-function mutations in the $Ca_v1.2$ L-type calcium channel subunits encoded by *CACNA1C* or *CACNB2* have been associated to an overlap clinical presentation consisting of familial SCD, the BrS, and a shorter than normal QT-interval.[130] The ERS is characterized by an early repolarization (ER) pattern on the ECG in association with ventricular arrhythmia.[131] Candidate-based approaches in patients with idiopathic VF and the ER pattern have identified variants in the potassium channel *KCNJ8*, L-type calcium channels (*CACNA1C*, *CACNB2B*, and *CACNA2D1*), and the *SCN5A* gene.[131] A study screened relatives of four families affected by ERS and observed that the phenotype (ER-pattern associated to SCD) was inherited in autosomal dominant fashion.[132] Due to the high prevalence of the ER pattern in the general population a conservative approach should be applied in the diagnosis of this disorder.[2] *TRPM4* mutations have been implicated in multiple families with CCD.[133–135] A haplotype in the *DPP6* gene has been associated to idiopathic ventricular fibrillation in several Dutch families.[136] This finding has made it possible to identify family members at risk before the onset of symptoms. Functional studies indicated an important role for *DPP6* in the transient outward current (I_{to}).[137]

NEXT-GENERATION SEQUENCING FOR THE DIAGNOSIS OF CARDIAC CHANNELOPATHIES

The molecular and genetic basis of the cardiac channelopathies has been discussed in the previous sections. Consensus guidelines concerning recommendations on genetic testing and clinical diagnosis of the channelopathies are available for clinical use.[2,6] The technical developments in DNA sequencing, which are referred to as NGS, have brought with them the possibility to sequence large sets of channelopathy-associated genes, the whole exome or even the whole genome in clinically affected patients.[138] Despite the major advantages, these technical developments have complicated the interpretation of identified variants and introduced a large number of so-called "variants of unknown significance" (VUS) into clinical practice. As an illustration, sequencing of an exome in one patient will generally lead to approximately 200 novel protein-altering single nucleotide variants.[139] This large amount of variants makes it difficult to relate a specific genetic defect to the patient's phenotype especially when no affected family members are available for segregation analysis, which is often the case. One study found that approximately 5% of healthy individuals carry a rare protein-altering variant in one of the three main LQTS-causing genes (*KCNQ1*, *KCNH2*, and *SCN5A*).[140] Studies identified an overexpression of previous LQTS- or CPVT-associated variants in large exome databases consisting of healthy subjects.[141,142] These studies emphasized that variants previously associated to disease might actually be uncommon or rare benign variants. Although the role of NGS in diagnosis is complicated, these techniques have made it possible to uncover the genetic underpinnings in multiple patients and families with rare cardiac disease that remained genetically unsolved.

References

1. Wilde AAM, Behr ER. Genetic testing for inherited cardiac disease. *Nat Rev Cardiol* 2013;**10**:571–83.
2. Priori SG, Wilde AA, Horie M, et al. HRS/EHRA/APHRS expert consensus statement on the diagnosis and management of patients with inherited primary arrhythmia syndromes: document endorsed by HRS, EHRA, and APHRS in May 2013 and by ACCF, AHA, PACES, and AEPC in June 2013. *Heart Rhythm* 2013;**10**:1932–63.
3. Curran ME, Splawski I, Timothy KW, Vincent GM, Green ED, Keating MT. A molecular basis for cardiac arrhythmia: HERG mutations cause long QT syndrome. *Cell* 1995;**80**:795–803.
4. Wang Q, Shen J, Splawski I, et al. SCN5A mutations associated with an inherited cardiac arrhythmia, long QT syndrome. *Cell* 1995;**80**:805–11.
5. Wang Q, Curran ME, Splawski I, et al. Positional cloning of a novel potassium channel gene: KVLQT1 mutations cause cardiac arrhythmias. *Nat Genet* 1996;**12**:17–23.
6. Ackerman MJ, Priori SG, Willems S, et al. HRS/EHRA expert consensus statement on the state of genetic testing for the channelopathies and cardiomyopathies this document was developed as a partnership between the Heart Rhythm Society (HRS) and the European Heart Rhythm Association (EHRA). *Heart Rhythm* 2011;**8**:1308–39.
7. Schwartz PJ, Ackerman MJ, George AL, Wilde AAM. Impact of genetics on the clinical management of channelopathies. *J Am Coll Cardiol* 2013;**62**:169–80.
8. Hofman N, Tan HL, Alders M, et al. Yield of molecular and clinical testing for arrhythmia syndromes: report of 15 years' experience. *Circulation* 2013;**128**:1513–21.
9. Brink PA, Crotti L, Corfield V, et al. Phenotypic variability and unusual clinical severity of congenital long-QT syndrome in a founder population. *Circulation* 2005;**112**:2602–10.
10. Marjamaa A, Salomaa V, Newton-Cheh C, et al. High prevalence of four long QT syndrome founder mutations in the Finnish population. *Ann Med* 2009;**41**:234–40.

11. Winbo A, Diamant U-B, Rydberg A, Persson J, Jensen SM, Stattin E-L. Origin of the Swedish long QT syndrome Y111C/KCNQ1 founder mutation. *Heart Rhythm* 2011;**8**:541–7.
12. Bezzina C, Veldkamp MW, van Den Berg MP, et al. A single Na(+) channel mutation causing both long-QT and Brugada syndromes. *Circ Res* 1999;**85**:1206–13.
13. Priori SG, Schwartz PJ, Napolitano C, et al. Risk stratification in the long-QT syndrome. *N Engl J Med* 2003;**348**:1866–74.
14. Makita N, Sumitomo N, Watanabe I, Tsutsui H. Novel SCN5A mutation (Q55X) associated with age-dependent expression of Brugada syndrome presenting as neurally mediated syncope. *Heart Rhythm* 2007;**4**:516–9.
15. Roden DM. Drug-induced prolongation of the QT interval. *N Engl J Med* 2004;**350**:1013–22.
16. Westenskow P, Splawski I, Timothy KW, Keating MT, Sanguinetti MC. Compound mutations: a common cause of severe long-QT syndrome. *Circulation* 2004;**109**:1834–41.
17. Crotti L, Monti MC, Insolia R, et al. NOS1AP is a genetic modifier of the long-QT syndrome. *Circulation* 2009;**120**:1657–63.
18. Tomás M, Napolitano C, De Giuli L, et al. Polymorphisms in the NOS1AP gene modulate QT interval duration and risk of arrhythmias in the long QT syndrome. *J Am Coll Cardiol* 2010;**55**:2745–52.
19. Duchatelet S, Crotti L, Peat RA, et al. Identification of a KCNQ1 polymorphism acting as a protective modifier against arrhythmic risk in long-QT syndrome. *Circ Cardiovasc Genet* 2013;**6**:354–61.
20. Kolder ICRM, Tanck MWT, Postema PG, et al. Analysis for genetic modifiers of disease severity in patients with long QT syndrome type 2. *Circ Cardiovasc Genet* 2015;**8**. Published online March 3.
21. Amin AS, Giudicessi JR, Tijsen AJ, et al. Variants in the 3′ untranslated region of the KCNQ1-encoded Kv7.1 potassium channel modify disease severity in patients with type 1 long QT syndrome in an allele-specific manner. *Eur Heart J* 2012;**33**:714–23.
22. Bezzina CR, Barc J, Mizusawa Y, et al. Common variants at SCN5A-SCN10A and HEY2 are associated with Brugada syndrome, a rare disease with high risk of sudden cardiac death. *Nat Genet* 2013;**45**:1044–9.
23. Bers DM. Cardiac excitation-contraction coupling. *Nature* 2002;**415**:198–205.
24. Schwartz PJ, Moss AJ, Vincent GM, Crampton RS. Diagnostic criteria for the long QT syndrome. An update. *Circulation* 1993;**88**:782–4.
25. Mizusawa Y, Horie M, Wilde AA. Genetic and clinical advances in congenital long QT syndrome. *Circ J* 2014;**78**. Published online October 1.
26. Hedley PL, Jørgensen P, Schlamowitz S, et al. The genetic basis of long QT and short QT syndromes: a mutation update. *Hum Mutat* 2009;**30**:1486–511.
27. Neyroud N, Tesson F, Denjoy I, et al. A novel mutation in the potassium channel gene KVLQT1 causes the Jervell and Lange-Nielsen cardioauditory syndrome. *Nat Genet* 1997;**15**:186–9.
28. Schulze-Bahr E, Wang Q, Wedekind H, et al. KCNE1 mutations cause Jervell and Lange-Nielsen syndrome. *Nat Genet* 1997;**17**:267–8.
29. Priori SG, Schwartz PJ, Napolitano C, et al. A recessive variant of the Romano-Ward long-QT syndrome? *Circulation* 1998;**97**:2420–5.
30. Abbott GW, Sesti F, Splawski I, et al. MiRP1 forms I_{Kr} potassium channels with HERG and is associated with cardiac arrhythmia. *Cell* 1999;**97**:175–87.
31. Plaster NM, Tawil R, Tristani-Firouzi M, et al. Mutations in Kir2.1 cause the developmental and episodic electrical phenotypes of Andersen's syndrome. *Cell* 2001;**105**:511–9.
32. Kubo Y, Baldwin TJ, Jan YN, Jan LY. Primary structure and functional expression of a mouse inward rectifier potassium channel. *Nature* 1993;**362**:127–33.
33. Raab-Graham KF, Radeke CM, Vandenberg CA. Molecular cloning and expression of a human heart inward rectifier potassium channel. *Neuroreport* 1994;**5**:2501–5.
34. Schott JJ, Charpentier F, Peltier S, et al. Mapping of a gene for long QT syndrome to chromosome 4q25-27. *Am J Hum Genet* 1995;**57**:1114–22.

35. Mohler PJ, Splawski I, Napolitano C, et al. A cardiac arrhythmia syndrome caused by loss of ankyrin-B function. *Proc Natl Acad Sci USA* 2004;**101**:9137–42.

36. Chen L, Marquardt ML, Tester DJ, Sampson KJ, Ackerman MJ, Kass RS. Mutation of an A-kinase-anchoring protein causes long-QT syndrome. *Proc Natl Acad Sci USA* 2007;**104**:20990–5.

37. Priori SG, Napolitano C, Schwartz PJ. Low penetrance in the long-QT syndrome: clinical impact. *Circulation* 1999;**99**:529–33.

38. Goldenberg I, Horr S, Moss AJ, et al. Risk for life-threatening cardiac events in patients with genotype-confirmed long-QT syndrome and normal-range corrected QT intervals. *J Am Coll Cardiol* 2011;**57**:51–9.

39. Kimbrough J, Moss AJ, Zareba W, et al. Clinical implications for affected parents and siblings of probands with long-QT syndrome. *Circulation* 2001;**104**:557–62.

40. Hobbs JB, Peterson DR, Moss AJ, et al. Risk of aborted cardiac arrest or sudden cardiac death during adolescence in the long-QT syndrome. *JAMA* 2006;**296**:1249–54.

41. Sauer AJ, Moss AJ, McNitt S, et al. Long QT syndrome in adults. *J Am Coll Cardiol* 2007;**49**:329–37.

42. Schwartz PJ, Priori SG, Spazzolini C, et al. Genotype-phenotype correlation in the long-QT syndrome: gene-specific triggers for life-threatening arrhythmias. *Circulation* 2001;**103**:89–95.

43. Zhang L, Timothy KW, Vincent GM, et al. Spectrum of ST-T-wave patterns and repolarization parameters in congenital long-QT syndrome: ECG findings identify genotypes. *Circulation* 2000;**102**:2849–55.

44. Wilde AA, Jongbloed RJ, Doevendans PA, et al. Auditory stimuli as a trigger for arrhythmic events differentiate HERG-related (LQTS2) patients from KVLQT1-related patients (LQTS1). *J Am Coll Cardiol* 1999;**33**:327–32.

45. Moss AJ, Shimizu W, Wilde AAM, et al. Clinical aspects of type-1 long-QT syndrome by location, coding type, and biophysical function of mutations involving the KCNQ1 gene. *Circulation* 2007;**115**:2481–9.

46. Shimizu W, Moss AJ, Wilde AAM, et al. Genotype-phenotype aspects of type 2 long QT syndrome. *J Am Coll Cardiol* 2009;**54**:2052–62.

47. Lahtinen AM, Marjamaa A, Swan H, Kontula K. KCNE1 D85N polymorphism–a sex-specific modifier in type 1 long QT syndrome? *BMC Med Genet* 2011;**12**:11.

48. De Villiers CP, van der Merwe L, Crotti L, et al. AKAP9 is a genetic modifier of congenital long-QT syndrome type 1. *Circ Cardiovasc Genet* 2014;**7**:599–606.

49. Arking DE, Pulit SL, Crotti L, et al. Genetic association study of QT interval highlights role for calcium signaling pathways in myocardial repolarization. *Nat Genet* 2014;**46**:826–36.

50. Arking DE, Pfeufer A, Post W, et al. A common genetic variant in the NOS1 regulator NOS1AP modulates cardiac repolarization. *Nat Genet* 2006;**38**:644–51.

51. Pfeufer A, Sanna S, Arking DE, et al. Common variants at ten loci modulate the QT interval duration in the QTSCD Study. *Nat Genet* 2009;**41**:407–14.

52. Newton-Cheh C, Eijgelsheim M, Rice KM, et al. Common variants at ten loci influence QT interval duration in the QTGEN Study. *Nat Genet* 2009;**41**:399–406.

53. Gussak I, Brugada P, Brugada J, et al. Idiopathic short QT interval: a new clinical syndrome? *Cardiology* 2000;**94**:99–102.

54. Gaita F, Giustetto C, Bianchi F, et al. Short QT Syndrome: a familial cause of sudden death. *Circulation* 2003;**108**:965–70.

55. Mazzanti A, Kanthan A, Monteforte N, et al. Novel insight into the natural history of short QT syndrome. *J Am Coll Cardiol* 2014;**63**:1300–8.

56. Brugada R, Hong K, Dumaine R, et al. Sudden death associated with short-QT syndrome linked to mutations in HERG. *Circulation* 2004;**109**:30–5.

57. Bellocq C, van Ginneken ACG, Bezzina CR, et al. Mutation in the KCNQ1 gene leading to the short QT-interval syndrome. *Circulation* 2004;**109**:2394–7.

58. Priori SG, Pandit SV, Rivolta I, et al. A novel form of short QT syndrome (SQT3) is caused by a mutation in the KCNJ2 gene. *Circ Res* 2005;**96**:800–7.

59. Deo M, Ruan Y, Pandit SV, et al. KCNJ2 mutation in short QT syndrome 3 results in atrial fibrillation and ventricular proarrhythmia. *Proc Natl Acad Sci USA* 2013;**110**: 4291–6.

60. Cordeiro JM, Brugada R, Wu YS, Hong K, Dumaine R. Modulation of I(Kr) inactivation by mutation N588K in KCNH2: a link to arrhythmogenesis in short QT syndrome. *Cardiovasc Res* 2005;**67**:498–509.

61. Chen Y-H, Xu S-J, Bendahhou S, et al. KCNQ1 gain-of-function mutation in familial atrial fibrillation. *Science* 2003;**299**:251–4.

62. Tucker NR, Ellinor PT. Emerging directions in the genetics of atrial fibrillation. *Circ Res* 2014;**114**:1469–82.

63. Brugada R, Tapscott T, Czernuszewicz GZ, et al. Identification of a genetic locus for familial atrial fibrillation. *N Engl J Med* 1997;**336**:905–11.

64. Lubitz SA, Yin X, Fontes JD, et al. Association between familial atrial fibrillation and risk of new-onset atrial fibrillation. *JAMA* 2010;**304**:2263–9.

65. Lubitz SA, Lunetta KL, Lin H, et al. Novel genetic markers associate with atrial fibrillation risk in Europeans and Japanese. *J Am Coll Cardiol* 2014;**63**:1200–10.

66. Sinner MF, Tucker NR, Lunetta KL, et al. Integrating genetic, transcriptional, and functional analyses to identify 5 novel genes for atrial fibrillation. *Circulation* 2014;**130**:1225–35.

67. Veldkamp MW, Viswanathan PC, Bezzina C, Baartscheer A, Wilde AAM, Balser JR. Two distinct congenital arrhythmias evoked by a multidysfunctional Na⁺ channel. *Circ Res* 2000;**86**:e91–7.

68. Grant AO, Carboni MP, Neplioueva V, et al. Long QT syndrome, Brugada syndrome, and conduction system disease are linked to a single sodium channel mutation. *J Clin Invest* 2002;**110**:1201–9.

69. Kyndt F, Probst V, Potet F, et al. Novel SCN5A mutation leading either to isolated cardiac conduction defect or Brugada syndrome in a large French family. *Circulation* 2001;**104**:3081–6.

70. Brugada P, Brugada J. Right bundle branch block, persistent ST segment elevation and sudden cardiac death: a distinct clinical and electrocardiographic syndrome. *J Am Coll Cardiol* 1992;**20**:1391–6.

71. Antzelevitch C, Brugada P, Borggrefe M, et al. Brugada syndrome: report of the second consensus conference. *Heart Rhythm* 2005;**2**:429–40.

72. Chen Q, Kirsch GE, Zhang D, et al. Genetic basis and molecular mechanism for idiopathic ventricular fibrillation. *Nature* 1998;**392**:293–6.

73. Crotti L, Marcou CA, Tester DJ, et al. Spectrum and prevalence of mutations involving BrS1- through BrS12-susceptibility genes in a cohort of unrelated patients referred for Brugada syndrome genetic testing: implications for genetic testing. *J Am Coll Cardiol* 2012;**60**:1410–8.

74. Hu D, Barajas-Martínez H, Pfeiffer R, et al. Mutations in SCN10A are responsible for a large fraction of cases of Brugada syndrome. *J Am Coll Cardiol* 2014;**64**:66–79.

75. Behr ER, Savio-Galimberti E, Barc J, et al. Role of common and rare variants in SCN10A: results from the Brugada syndrome QRS locus gene discovery collaborative study. *Cardiovasc Res* 2015;**106**. Published online February 17.

76. Le Scouarnec S, Karakachoff M, Gourraud J-B, et al. Testing the burden of rare variation in arrhythmia-susceptibility genes provides new insights into molecular diagnosis for Brugada syndrome. *Hum Mol Genet* 2015;**24**. http://dx.doi.org/10.1093/hmg/ddv036. Published online February 03.

77. Watanabe H, Koopmann TT, Le Scouarnec S, et al. Sodium channel β1 subunit mutations associated with Brugada syndrome and cardiac conduction disease in humans. *J Clin Invest* 2008;**118**:2260–8.

78. Van Norstrand DW, Valdivia CR, Tester DJ, et al. Molecular and functional characterization of novel glycerol-3-phosphate dehydrogenase 1 like gene (GPD1-L) mutations in sudden infant death syndrome. *Circulation* 2007;**116**:2253–9.

79. Meregalli PG, Tan HL, Probst V, et al. Type of SCN5A mutation determines clinical severity and degree of conduction slowing in loss-of-function sodium channelopathies. *Heart Rhythm* 2009;**6**:341–8.

80. Probst V, Wilde AAM, Barc J, et al. SCN5A mutations and the role of genetic background in the pathophysiology of Brugada syndrome. *Circ Cardiovasc Genet* 2009;**2**:552–7.

81. Chambers JC, Zhao J, Terracciano CMN, et al. Genetic variation in SCN10A influences cardiac conduction. *Nat Genet* 2010;**42**:149–52.

82. Holm H, Gudbjartsson DF, Arnar DO, et al. Several common variants modulate heart rate, PR interval and QRS duration. *Nat Genet* 2010;**42**:117–22.

83. Pfeufer A, van Noord C, Marciante KD, et al. Genome-wide association study of PR interval. *Nat Genet* 2010;**42**:153–9.

84. Sotoodehnia N, Isaacs A, de Bakker PIW, et al. Common variants in 22 loci are associated with QRS duration and cardiac ventricular conduction. *Nat Genet* 2010;**42**:1068–76.

85. Bennett PB, Yazawa K, Makita N, George AL. Molecular mechanism for an inherited cardiac arrhythmia. *Nature* 1995;**376**:683–5.

86. Remme CA. Cardiac sodium channelopathy associated with SCN5A mutations: electrophysiological, molecular and genetic aspects. *J Physiol* 2013;**591**:4099–116.

87. Wang Q, Shen J, Li Z, et al. Cardiac sodium channel mutations in patients with long QT syndrome, an inherited cardiac arrhythmia. *Hum Mol Genet* 1995;**4**:1603–7.

88. Tester DJ, Will ML, Haglund CM, Ackerman MJ. Compendium of cardiac channel mutations in 541 consecutive unrelated patients referred for long QT syndrome genetic testing. *Heart Rhythm* 2005;**2**:507–17.

89. Vatta M, Ackerman MJ, Ye B, et al. Mutant caveolin-3 induces persistent late sodium current and is associated with long-QT syndrome. *Circulation* 2006;**114**:2104–12.

90. Medeiros-Domingo A, Kaku T, Tester DJ, et al. SCN4B-encoded sodium channel beta4 subunit in congenital long-QT syndrome. *Circulation* 2007;**116**:134–42.

91. Wu G, Ai T, Kim JJ, et al. Alpha-1-syntrophin mutation and the long-QT syndrome: a disease of sodium channel disruption. *Circ Arrhythm Electrophysiol* 2008;**1**:193–201.

92. Schott JJ, Alshinawi C, Kyndt F, et al. Cardiac conduction defects associate with mutations in SCN5A. *Nat Genet* 1999;**23**:20–1.

93. Tan HL, Bink-Boelkens MT, Bezzina CR, et al. A sodium-channel mutation causes isolated cardiac conduction disease. *Nature* 2001;**409**:1043–7.

94. Darbar D, Kannankeril PJ, Donahue BS, et al. Cardiac sodium channel (SCN5A) variants associated with atrial fibrillation. *Circulation* 2008;**117**:1927–35.

95. Ellinor PT, Nam EG, Shea MA, Milan DJ, Ruskin JN, MacRae CA. Cardiac sodium channel mutation in atrial fibrillation. *Heart Rhythm* 2008;**5**:99–105.

96. Makiyama T, Akao M, Shizuta S, et al. A novel SCN5A gain-of-function mutation M1875T associated with familial atrial fibrillation. *J Am Coll Cardiol* 2008;**52**:1326–34.

97. Benito B, Brugada R, Perich RM, et al. A mutation in the sodium channel is responsible for the association of long QT syndrome and familial atrial fibrillation. *Heart Rhythm* 2008;**5**:1434–40.

98. Venetucci L, Denegri M, Napolitano C, Priori SG. Inherited calcium channelopathies in the pathophysiology of arrhythmias. *Nat Rev Cardiol* 2012;**9**:561–75.

99. Leenhardt A, Lucet V, Denjoy I, Grau F, Ngoc DD, Coumel P. Catecholaminergic polymorphic ventricular tachycardia in children. A 7-year follow-up of 21 patients. *Circulation* 1995;**91**:1512–9.

100. Priori SG, Napolitano C, Memmi M, et al. Clinical and molecular characterization of patients with catecholaminergic polymorphic ventricular tachycardia. *Circulation* 2002;**106**:69–74.

101. Hayashi M, Denjoy I, Extramiana F, et al. Incidence and risk factors of arrhythmic events in catecholaminergic polymorphic ventricular tachycardia. *Circulation* 2009;**119**:2426–34.

102. Priori SG, Napolitano C, Tiso N, et al. Mutations in the cardiac ryanodine receptor gene (hRyR2) underlie catecholaminergic polymorphic ventricular tachycardia. *Circulation* 2001;**103**:196–200.

103. Lahat H, Pras E, Olender T, et al. A missense mutation in a highly conserved region of CASQ2 is associated with autosomal recessive catecholamine-induced polymorphic ventricular tachycardia in Bedouin families from Israel. *Am J Hum Genet* 2001;**69**:1378–84.

104. Laitinen PJ, Brown KM, Piippo K, et al. Mutations of the cardiac ryanodine receptor (RyR2) gene in familial polymorphic ventricular tachycardia. *Circulation* 2001;**103**:485–90.

105. Medeiros-Domingo A, Bhuiyan ZA, Tester DJ, et al. The RYR2-encoded ryanodine receptor/calcium release channel in patients diagnosed previously with either catecholaminergic polymorphic ventricular tachycardia or genotype negative, exercise-induced long QT syndrome: a comprehensive open reading frame muta. *J Am Coll Cardiol* 2009;**54**:2065–74.

106. Di Barletta MR, Viatchenko-Karpinski S, Nori A, et al. Clinical phenotype and functional characterization of CASQ2 mutations associated with catecholaminergic polymorphic ventricular tachycardia. *Circulation* 2006;**114**:1012–9.

107. Postma AV, Denjoy I, Hoorntje TM, et al. Absence of calsequestrin 2 causes severe forms of catecholaminergic polymorphic ventricular tachycardia. *Circ Res* 2002;**91**:e21–6.

108. De la Fuente S, Van Langen IM, Postma AV, Bikker H, Meijer AA. Case of catecholaminergic polymorphic ventricular tachycardia caused by two calsequestrin 2 mutations. *Pacing Clin Electrophysiol* 2008;**31**:916–9.

109. Nyegaard M, Overgaard MT, Søndergaard MT, et al. Mutations in calmodulin cause ventricular tachycardia and sudden cardiac death. *Am J Hum Genet* 2012;**91**:703–12.

110. Roux-Buisson N, Cacheux M, Fourest-Lieuvin A, et al. Absence of triadin, a protein of the calcium release complex, is responsible for cardiac arrhythmia with sudden death in human. *Hum Mol Genet* 2012;**21**:2759–67.

111. Tester DJ, Arya P, Will M, et al. Genotypic heterogeneity and phenotypic mimicry among unrelated patients referred for catecholaminergic polymorphic ventricular tachycardia genetic testing. *Heart Rhythm* 2006;**3**:800–5.

112. Tester DJ, Medeiros-Domingo A, Will ML, Haglund CM, Ackerman MJ. Cardiac channel molecular autopsy: insights from 173 consecutive cases of autopsy-negative sudden unexplained death referred for postmortem genetic testing. *Mayo Clin Proc* 2012;**87**:524–39.

113. Van der Werf C, Nederend I, Hofman N, et al. Familial evaluation in catecholaminergic polymorphic ventricular tachycardia: disease penetrance and expression in cardiac ryanodine receptor mutation-carrying relatives. *Circ Arrhythm Electrophysiol* 2012;**5**:748–56.

114. Splawski I, Timothy KW, Sharpe LM, et al. Ca(V)1.2 calcium channel dysfunction causes a multisystem disorder including arrhythmia and autism. *Cell* 2004;**119**:19–31.

115. Splawski I, Timothy KW, Decher N, et al. Severe arrhythmia disorder caused by cardiac L-type calcium channel mutations. *Proc Natl Acad Sci USA* 2005;**102**:8089–96. discussion 8086–8.

116. Etheridge SP, Bowles NE, Arrington CB, et al. Somatic mosaicism contributes to phenotypic variation in Timothy syndrome. *Am J Med Genet A* 2011;**155A**:2578–83.

117. Crotti L, Johnson CN, Graf E, et al. Calmodulin mutations associated with recurrent cardiac arrest in infants. *Circulation* 2013;**127**:1009–17.

118. Makita N, Yagihara N, Crotti L, et al. Novel calmodulin mutations associated with congenital arrhythmia susceptibility. *Circ Cardiovasc Genet* 2014;**7**:466–74.

119. Xu L, Meissner G. Mechanism of calmodulin inhibition of cardiac sarcoplasmic reticulum Ca^{2+} release channel (ryanodine receptor). *Biophys J* 2004;**86**:797–804.

120. Hwang HS, Nitu FR, Yang Y, et al. Divergent regulation of ryanodine receptor 2 calcium release channels by arrhythmogenic human calmodulin missense mutants. *Circ Res* 2014;**114**:1114–24.

121. Yin G, Hassan F, Haroun AR, et al. Arrhythmogenic calmodulin mutations disrupt intracellular cardiomyocyte Ca^{2+} regulation by distinct mechanisms. *J Am Heart Assoc* 2014;**3**:e000996.

122. Limpitikul WB, Dick IE, Joshi-Mukherjee R, Overgaard MT, George AL, Yue DT. Calmodulin mutations associated with long QT syndrome prevent inactivation of cardiac L-type Ca(2+) currents and promote proarrhythmic behavior in ventricular myocytes. *J Mol Cell Cardiol* 2014;**74**:115–24.

123. Marsman RF, Barc J, Beekman L, et al. A mutation in CALM1 encoding calmodulin in familial idiopathic ventricular fibrillation in childhood and adolescence. *J Am Coll Cardiol* 2014;**63**:259–66.

124. Schulze-Bahr E, Neu A, Friederich P, et al. Pacemaker channel dysfunction in a patient with sinus node disease. *J Clin Invest* 2003;**111**:1537–45.

125. DiFrancesco D. Funny channel gene mutations associated with arrhythmias. *J Physiol* 2013;**591**:4117–24.

126. Verkerk A, Wilders R. Pacemaker activity of the human Sinoatrial node: an update on the effects of mutations in HCN4 on the hyperpolarization-activated current. *Int J Mol Sci* 2015;**16**:3071–94.

127. Milano A, Vermeer AMC, Lodder EM, et al. HCN4 mutations in multiple families with bradycardia and left ventricular noncompaction cardiomyopathy. *J Am Coll Cardiol* 2014;**64**:745–56.

128. Schweizer PA, Schröter J, Greiner S, et al. The symptom complex of familial sinus node dysfunction and myocardial noncompaction is associated with mutations in the HCN4 channel. *J Am Coll Cardiol* 2014;**64**:757–67.

129. Den Hoed M, Eijgelsheim M, Esko T, et al. Identification of heart rate-associated loci and their effects on cardiac conduction and rhythm disorders. *Nat Genet* 2013;**45**:621–31.

130. Antzelevitch C, Pollevick GD, Cordeiro JM, et al. Loss-of-function mutations in the cardiac calcium channel underlie a new clinical entity characterized by ST-segment elevation, short QT intervals, and sudden cardiac death. *Circulation* 2007;**115**:442–9.

131. Mizusawa Y, Bezzina CR. Early repolarization pattern: its ECG characteristics, arrhythmogeneity and heritability. *J Interv Card Electrophysiol* 2014;**39**:185–92.

132. Gourraud J-B, Le Scouarnec S, Sacher F, et al. Identification of large families in early repolarization syndrome. *J Am Coll Cardiol* 2013;**61**:164–72.

133. Kruse M, Schulze-Bahr E, Corfield V, et al. Impaired endocytosis of the ion channel TRPM4 is associated with human progressive familial heart block type I. *J Clin Invest* 2009;**119**:2737–44.

134. Stallmeyer B, Zumhagen S, Denjoy I, et al. Mutational spectrum in the Ca(2+)–activated cation channel gene TRPM4 in patients with cardiac conductance disturbances. *Hum Mutat* 2012;**33**:109–17.

135. Baruteau A-E, Probst V, Abriel H. Inherited progressive cardiac conduction disorders. *Curr Opin Cardiol* 2015;**30**:33–9.

136. Alders M, Koopmann TT, Christiaans I, et al. Haplotype-sharing analysis implicates chromosome 7q36 harboring DPP6 in familial idiopathic ventricular fibrillation. *Am J Hum Genet* 2009;**84**:468–76.

137. Xiao L, Koopmann TT, Ördög B, et al. Unique cardiac Purkinje fiber transient outward current β-subunit composition: a potential molecular link to idiopathic ventricular fibrillation. *Circ Res* 2013;**112**:1310–22.

138. Lubitz SA, Ellinor PT. Next generation sequencing for the diagnosis of cardiac arrhythmia syndromes. *Heart Rhythm* 2015;**12**. Published online January 24.

139. Abecasis GR, Altshuler D, Auton A, et al. A map of human genome variation from population-scale sequencing. *Nature* 2010;**467**:1061–73.
140. Kapa S, Tester DJ, Salisbury BA, et al. Genetic testing for long-QT syndrome: distinguishing pathogenic mutations from benign variants. *Circulation* 2009;**120**:1752–60.
141. Risgaard B, Jabbari R, Refsgaard L, et al. High prevalence of genetic variants previously associated with Brugada syndrome in new exome data. *Clin Genet* 2013;**84**:489–95.
142. Jabbari J, Jabbari R, Nielsen MW, et al. New exome data question the pathogenicity of genetic variants previously associated with catecholaminergic polymorphic ventricular tachycardia. *Circ Cardiovasc Genet* 2013;**6**:481–9.

Other Volumes in the Perspectives on Translational Cell Biology Series

The Physiology of Developing Fish: Viviparity and Posthatching Juveniles
Edited by W. S. Hoar and D. J. Randall

The Cardiovascular System
Edited by W. S. Hoar, D. J. Randall, and A. P. Farrell

The Cardiovascular System
Edited by W. S. Hoar, D. J. Randall, and A. P. Farrell

Molecular Endocrinology of Fish
Edited by N. M. Sherwood and C. L. Hew

Cellular and Molecular Approaches to Fish Ionic Regulation
Edited by Chris M. Wood and Trevor J. Shuttleworth

The Fish Immune System: Organism, Pathogen, and Environment
Edited by George Iwama and Teruyuki Nakanishi

Deep Sea Fishes
Edited by D. J. Randall and A. P. Farrell

Fish Respiration
Edited by Steve F. Perry and Bruce L. Tufts

Muscle Development and Growth
Edited by Ian A. Johnston

Tuna: Physiology, Ecology, and Evolution
Edited by Barbara A. Block and E. Donald Stevens

Nitrogen Excretion
Edited by Patricia A. Wright and Paul M. Anderson

The Physiology of Tropical Fishes
Edited by Adalberto L. Val, Vera Maria F. De Almeida-Val, and David J. Randall

The Physiology of Polar Fishes
Edited by Anthony P. Farrell and John F. Steffensen

Fish Biomechanics
Edited by Robert E. Shadwick and George V. Lauder

Behavior and Physiology of Fish
Edited by Katharine A. Sloman, Rod W. Wilson, and Sigal Balshine

Sensory Systems Neuroscience
Edited by Toshiaki J. Hara and Barbara S. Zielinski

Primitive Fishes
Edited by David J. McKenzie, Anthony P. Farrell, and Collin J Brauner

Hypoxia
Edited by Jeffrey G. Richards, Anthony P. Farrell, and Colin J. Brauner

Fish Neuroendocrinology
Edited by Nicholas J. Bernier, Glen Van Der Kraak, Anthony P. Farrell, and Colin J. Brauner

Zebrafish
Edited by Steve F. Perry, Marc Ekker, Anthony P. Farrell, and Colin J Brauner

The Multifunctional Gut of Fish
Edited by Martin Grosell, Anthony P. Farrell, and Colin J. Brauner

Homeostasis and Toxicology of Essential Metals
Edited by Chris M. Wood, Anthony P. Farrell, and Colin J. Brauner

Homeostasis and Toxicology of Non-Essential Metals
Edited by Chris M. Wood, Anthony P. Farrell, and Colin J. Brauner

Euryhaline Fishes
Edited by Stephen C McCormick, Anthony P. Farrell, and Colin J. Brauner

Organic Chemical Toxicology of Fishes
Edited by Keith B. Tierney, Anthony P. Farrell, and Colin J. Brauner

Index

Printed in the United States
By Bookmasters